T0135084

Communications in Computer and Information Science 1337

Editorial Board Members

Joaquim Filipe
Polytechnic Institute of Setúbal, Setúbal, Portugal

Ashish Ghosh
Indian Statistical Institute, Kolkata, India

Raquel Oliveira Prates
Federal University of Minas Gerais (UFMG), Belo Horizonte, Brazil

Lizhu Zhou
Tsinghua University, Beijing, China

More information about this series at http://www.springer.com/series/7899

Vladimir M. Vishnevskiy ·
Konstantin E. Samouylov ·
Dmitry V. Kozyrev (Eds.)

Distributed Computer and Communication Networks: Control, Computation, Communications

23rd International Conference, DCCN 2020
Moscow, Russia, September 14–18, 2020
Revised Selected Papers

Editors
Vladimir M. Vishnevskiy
V.A.Trapeznikov Institute of Control Sciences of Russian Academy of Sciences
Moscow, Russia

Konstantin E. Samouylov
Peoples' Friendship University of Russia (RUDN University)
Moscow, Russia

Dmitry V. Kozyrev
V.A.Trapeznikov Institute of Control Sciences of Russian Academy of Sciences
Moscow, Russia

Peoples' Friendship University of Russia (RUDN University)
Moscow, Russia

ISSN 1865-0929 ISSN 1865-0937 (electronic)
Communications in Computer and Information Science
ISBN 978-3-030-66241-7 ISBN 978-3-030-66242-4 (eBook)
https://doi.org/10.1007/978-3-030-66242-4

© Springer Nature Switzerland AG 2020
This work is subject to copyright. All rights are reserved by the Publisher, whether the whole or part of the material is concerned, specifically the rights of translation, reprinting, reuse of illustrations, recitation, broadcasting, reproduction on microfilms or in any other physical way, and transmission or information storage and retrieval, electronic adaptation, computer software, or by similar or dissimilar methodology now known or hereafter developed.
The use of general descriptive names, registered names, trademarks, service marks, etc. in this publication does not imply, even in the absence of a specific statement, that such names are exempt from the relevant protective laws and regulations and therefore free for general use.
The publisher, the authors and the editors are safe to assume that the advice and information in this book are believed to be true and accurate at the date of publication. Neither the publisher nor the authors or the editors give a warranty, expressed or implied, with respect to the material contained herein or for any errors or omissions that may have been made. The publisher remains neutral with regard to jurisdictional claims in published maps and institutional affiliations.

This Springer imprint is published by the registered company Springer Nature Switzerland AG
The registered company address is: Gewerbestrasse 11, 6330 Cham, Switzerland

Preface

This volume contains a collection of revised selected full-text papers presented at the 23rd International Conference on Distributed Computer and Communication Networks (DCCN 2020), held in Moscow, Russia, September 14–18, 2020.

The conference is a continuation of traditional international conferences of the DCCN series, which took place in Sofia, Bulgaria (1995, 2005, 2006, 2008, 2009, 2014), Tel Aviv, Israel (1996, 1997, 1999, 2001), and Moscow, Russia (1998, 2000, 2003, 2007, 2010, 2011, 2013, 2015, 2016, 2017, 2018, 2019) in the last 25 years. The main idea of the conference is to provide a platform and forum for researchers and developers from academia and industry from various countries working in the area of theory and applications of distributed computer and communication networks, mathematical modeling, methods of control and optimization of distributed systems, by offering them a unique opportunity to share their views as well as discuss the prospective developments and pursue collaboration in this area. The content of this volume is related to the following subjects:

1. Communication networks algorithms and protocols;
2. Wireless and mobile networks;
3. Computer and telecommunication networks control and management;
4. Performance analysis, QoS/QoE evaluation and network efficiency;
5. Analytical modeling and simulation of communication systems;
6. Evolution of wireless networks toward 5G;
7. Internet of Things and Fog Computing;
8. Machine learning, big data, artificial intelligence;
9. Probabilistic and statistical models in information systems;
10. Queuing theory and reliability theory applications;
11. Quantum Information and quantum communication;
12. High-altitude telecommunications platforms;
13. Security in infocommunication systems.

The DCCN 2020 conference gathered 167 submissions from authors from 25 different countries. From these, 140 high-quality papers in English were accepted and presented during the conference. The current volume contains 43 extended mostly application-oriented papers which were recommended by session chairs and selected by the Program Committee for the Springer post-proceedings.

All the papers selected for the post-proceedings volume are given in the form presented by the authors. These papers are of interest to everyone working in the field of computer and communication networks.

We thank all the authors for their interest in DCCN, the members of the Program Committee for their contributions, and the reviewers for their peer-reviewing efforts.

September 2020

Vladimir Vishnevskiy
Konstantin Samouylov

Organization

DCCN 2020 was jointly organized by the Russian Academy of Sciences (RAS), the V.A. Trapeznikov Institute of Control Sciences of RAS (ICS RAS), the Peoples' Friendship University of Russia (RUDN University), the National Research Tomsk State University, and the Institute of Information and Communication Technologies of the Bulgarian Academy of Sciences (IICT BAS).

International Program Committee

V. M. Vishnevskiy (Chair)	ICS RAS, Russia
K. E. Samouylov (Co-chair)	RUDN University, Russia
S. M. Abramov	Program Systems Institute of RAS, Russia
S. D. Andreev	Tampere University, Finland
A. M. Andronov	Riga Technical University, Latvia
N. Balakrishnan	McMaster University, Canada
A. S. Bugaev	Moscow Institute of Physics and Technology, Russia
S. R. Chakravarthy	Kettering University, USA
T. Czachorski	Institute of Computer Science of the Polish Academy of Sciences, Poland
D. Deng	National Changhua University of Education, Taiwan, ROC
A. N. Dudin	Belarusian State University, Belarus
A. V. Dvorkovich	Moscow Institute of Physics and Technology, Russia
Yu. V. Gaidamaka	RUDN University, Russia
P. Gaj	Silesian University of Technology, Poland
D. Grace	University of York, UK
Yu. V. Gulyaev	Kotelnikov Institute of Radioengineering and Electronics of RAS, Russia
J. Hosek	Brno University of Technology, Czech Republic
V. C. Joshua	CMS College, India
H. Karatza	Aristotle University of Thessaloniki, Greece
N. Kolev	University of São Paulo, Brazil
J. Kolodziej	NASK, Poland
G. Kotsis	Johannes Kepler University Linz, Austria
A. E. Koucheryavy	Bonch-Bruevich Saint-Petersburg State University of Telecommunications, Russia
Ye. A. Koucheryavy	Tampere University, Finland
T. Kozlova Madsen	Aalborg University, Denmark
U. Krieger	University of Bamberg, Germany
A. Krishnamoorthy	Cochin University of Science and Technology, India
N. A. Kuznetsov	Moscow Institute of Physics and Technology, Russia

L. Lakatos	Budapest University of Technology and Economics, Hungary
E. Levner	Holon Institute of Technology, Israel
S. D. Margenov	Institute of Information and Communication Technologies of Bulgarian Academy of Sciences, Bulgaria
N. Markovich	ICS RAS, Russia
A. Melikov	Institute of Cybernetics of the Azerbaijan National Academy of Sciences, Azerbaijan
G. K. Miscoi	Academy of Sciences of Moldova, Moldova
E. V. Morozov	Institute of Applied Mathematical Research of the Karelian Research Centre RAS, Russia
V. A. Naumov	Service Innovation Research Institute (PIKE), Finland
A. A. Nazarov	Tomsk State University, Russia
I. V. Nikiforov	Université de Technologie de Troyes, France
P. Nikitin	University of Washington, USA
S. A. Nikitov	Institute of Radioengineering and Electronics of RAS, Russia
D. A. Novikov	ICS RAS, Russia
M. Pagano	University of Pisa, Italy
E. Petersons	Riga Technical University, Latvia
V. V. Rykov	Gubkin Russian State University of Oil and Gas, Russia
L. A. Sevastianov	RUDN University, Russia
M. A. Sneps-Sneppe	Ventspils University College, Latvia
P. Stanchev	Kettering University, USA
S. N. Stepanov	Moscow Technical University of Communication and Informatics, Russia
S. P. Suschenko	Tomsk State University, Russia
J. Sztrik	University of Debrecen, Hungary
H. Tijms	Vrije Universiteit Amsterdam, The Netherlands
S. N. Vasiliev	ICS RAS, Russia
M. Xie	City University of Hong Kong, Hong Kong, SAR, China
A. Zaslavsky	Deakin University, Australia
Yu. P. Zaychenko	Kyiv Polytechnic Institute, Ukraine

Organizing Committee

V. M. Vishnevskiy (Chair)	ICS RAS, Russia
K. E. Samouylov (Vice Chair)	RUDN University, Russia
D. V. Kozyrev	ICS RAS and RUDN University, Russia
A. A. Larionov	ICS RAS, Russia
S. N. Kupriyakhina	ICS RAS, Russia
S. P. Moiseeva	Tomsk State University, Russia

T. Atanasova	IICT BAS, Bulgaria
I. A. Kochetkova	RUDN University, Russia

Organizers and Partners

Organizers

Russian Academy of Sciences
RUDN University
V.A. Trapeznikov Institute of Control Sciences of RAS
National Research Tomsk State University
Institute of Information and Communication Technologies of the Bulgarian Academy of Sciences
Research and Development Company "Information and Networking Technologies"

Support

Information support was provided by the Russian Academy of Sciences. The conference was organized with the support of the "RUDN University Program 5-100."

Contents

Computer and Communication Networks

Computer and Communication Networks

Challenges and Performance Evaluation of Multicast Transmission in 60 GHz mmWave

Nadezhda Chukhno[1,2(✉)], Olga Chukhno[1,3], Giuseppe Araniti[1], Antonio Iera[4], Antonella Molinaro[1,5], and Sara Pizzi[1]

[1] University Mediterranea of Reggio Calabria, Reggio Calabria, Italy
{nadezda.chukhno,olga.chukhno,araniti,antonella.molinaro, sara.pizzi}@unirc.it

[2] Universitat Jaume I, Castelló de la Plana, Spain

[3] Tampere University, Tampere, Finland

[4] University of Calabria, Rende, Italy
antonio.iera@dimes.unical.it

[5] Université Paris-Saclay, Gif-sur-Yvette, France

Abstract. Recently, millimeter-wave (mmWave) technology has attracted significant attention due to its ambitious promise to deal with the rapid growth in wireless data traffic. Moreover, mmWave is expected to constitute a foundation for the fifth-generation (5G) communication systems' services, claimed to efficiently and effectively support both unicast and multicast transmission modes. However, the use of highly directional antennas at both user and access point sides is required to compensate for the severe path loss, high attenuation, and atmospheric absorption at extremely high-frequency bands, e.g., mmWave. Hence, multicast transmission needs special attention in directional systems due to the nature of group-oriented services, wherein a single beam simultaneously feeds receivers located at different positions. Since the widest possible beams at 60 GHz band are limited in terms of range and data rate and cannot serve all users, and, inversely, the use of only fine beams steered toward each user in unicast fashion requires long data transmission duration, the design of efficient directional multicast schemes is of utmost importance. Further, a slight beam misalignment due to mobility can generate a significant signal drop even between devices communicating in unicast fashions. The mission of this paper is to discuss the main challenges that must be faced to take advantage of mmWave communication for multicast data delivery. To this end, we investigate the performance of such systems in terms of data rate and data transmission duration via simulations considering both static and dynamic scenarios.

Keywords: 5G · mmWave · 802.11ad/ay · Unicast · Multicast

© Springer Nature Switzerland AG 2020
V. M. Vishnevskiy et al. (Eds.): DCCN 2020, CCIS 1337, pp. 3–17, 2020.
https://doi.org/10.1007/978-3-030-66242-4_1

1 Introduction

Millimeter wave (mmWave) communications are considered promising candidates for future wireless networks to meet the ever-increasing demands for high-data-rate multimedia access and spectrum requirement of fifth-generation (5G) mobile communication systems [1]. Furthermore, according to both academic and industrial communities, mmWave technology is expected to play a fundamental role even in beyond-5G (B5G) networks to ensure efficient massive data transmissions [2]. For example, the 3GPP New Radio (NR) technology will exploit the mmWave spectrum to achieve increased bandwidths and higher data rates [3]. IEEE 802.11ad/ay specifications use a similar approach and claim to achieve up to 100 Gbps rate communication [4]. Therefore, the envisioned performance of mmWave systems is ideal for satisfying the typical demands of the bandwidth-hungry 5G and B5G services and emerging applications, which mainly require disseminating a large amount of data traffic with low latency, e.g., autonomous driving, mobile video streaming, virtual/augmented/mixed reality (VR/AR/XR) applications, public/road safety, road infotainment, among others. Moreover, the throughput gain enhancement is an essential target in mmWave communication development.

Meanwhile, a beneficial technique for the system bandwidth efficiency improvement is multicast communication, wherein the same packet is delivered from a transmitter to an arbitrary number of receivers simultaneously by utilizing the same frequency and modulation and coding scheme (MCS). However, to compensate for the high path loss at extremely high frequency (EHF) bands and guarantee the gigabit capabilities, highly directional antennas are required for mmWave transmissions where the signal is sensitive to rapid channel variations, atmospheric absorption, and severe attenuation. This requirement makes multicasting more complex to implement compared to microwave networks where omnidirectional antennas are typically used. The beam steering, the size of multicast subgroups, and the beamwidth to cover all users have to be properly selected. Further, the presence of mobile users poses an additional challenge to mmWave wireless systems with directional group-oriented transmissions. To this end, the paper dedicates to discuss and analyze the challenging issues and advantages of mmWave multicast communication with a particular focus on WiGig/IEEE 802.11 specifications.

The paper is organized as follows. The description of IEEE 802.11ad/ay standards is given in Sect. 2. The design challenges of multicast and unicast modes with directional mmWave transmissions are discussed in Sect. 3. Section 4 describes the system model under analysis. Simulation results are discussed in Sect. 5. The conclusions of our study are given in the last section.

2 IEEE 802.11ad/ay Specifications

The WiFi/WiGig standards (i.e., IEEE 802.11ad/ay) support wireless networking at 60 GHz. The IEEE 802.11ad standard, ratified in 2012, offers real multi-gigabit data rates. Its successor, IEEE 802.11ay, excels the capabilities of 802.11.ad by exploiting the same band and provides ultra-high-speed and super-low-latency services by introducing advanced physical layer (PHY) features. The second WiGig standard quadruples the bandwidth, adds MIMO up to 8 spatial streams, channel bonding, channel aggregation, and ensures non-uniform modulation constellation [5]. Besides, 802.11ay advanced power-saving feature makes it ideal for wearable devices. For instance, these kinds of requirements may be suitable for AR/VR applications [6].

2.1 IEEE 802.11ad

In this section, a more in-depth investigation and analysis of the beacon interval structure in IEEE 802.11ad (see Fig. 1) is conducted. The medium access control (MAC) design for 802.11ad may utilize both carrier sensing multiple access with collision avoidance (CSMA/CA) and scheduled service periods (SPs) channel access schemes depending on the type of application. In the case of SPs, 802.11ad uses time division multiple access (TDMA), where a personal basic service set (PBSS) central point (PCP) or access point (AP) utilizes the polling mechanism by asking devices and receiving their feedback. Alternatively, CSMA/CA is used for contention-based periods (CBP), where devices are allowed to use the same radio channel without pre-coordination with the help of *listen-before-talk* operating procedure. In this work, we consider SPs only.

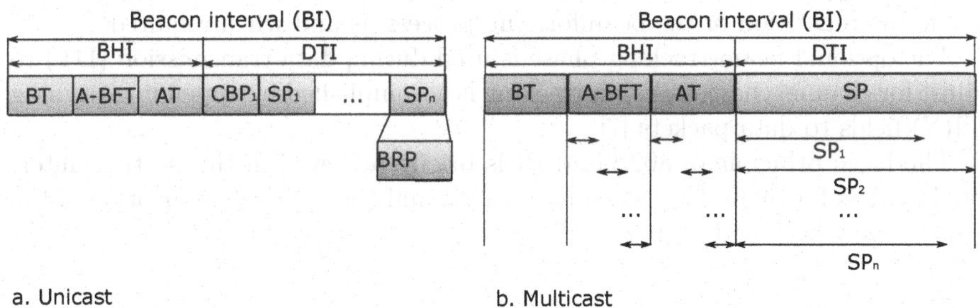

Fig. 1. IEEE 802.11.ad beacon structure.

The time is divided into beacon intervals (BIs) of total length T; each BI incorporates: (i) a beacon header interval (BHI), where devices perform initial beamforming, which generally involves sectored antennas (aka sector-level sweep, SLS), and adjust their wider transmit beams, and (ii) a data transmission interval (DTI), including SPs of different connected clients, while containing a beam refinement protocol (BRP) to improve the resulting instantaneous data rate.

More specifically, on the MAC layer each BI starts with a beacon time (BT) interval during which the *initiator* transmits sector sweep (I-TXSS) beacons across all M_{SLS} sectors with half-power beamwidth (HPBW) of $\theta_{\text{SLS,Tx}} =$

$2\pi/M_{\text{SLS}}$. The receive sector sweep (RSS) process and feedback during the association beamforming training (A-BFT) announced by the initiator are performed after I-TXSS by the *receiver.* Practically, in A-BFT interval, if more than one client selects the same transmission opportunity (up to eight slots for 802.11ad), the signals collide, and devices cannot establish a connection in the current BI.

The receive antenna operates in an omnidirectional mode during SLS and, after measuring the receive signal strength (RSS) across all N_{SLS} sectors, with $\theta_{\text{SLS,Rx}} = 2\pi/N_{\text{SLS}}$, it provides the SLS feedback to the transmitter identifying the sector with maximum RSS value. Based on the RSS indicator as well as by using an angle of arrival (AoA) or time difference of arrival (TDoA), the AP can determine the user location information. The training packets are transmitted with the low-power low-rate MCS 0, which provides the reliable communication required to establish the initial beamformed link.

In the announcement time interval (ATI), management information is exchanged between the PCP/AP and the receivers. Once the best sector pair is identified, the beam refinement phase (BRP) iteratively trains the transmit and receive antenna beams found during the SLS to select a beam pattern pair with finer beamwidths determined by the beam refinement factor b, $b > 1$. Therefore, for the transmit antenna training, both devices sweep through exactly b narrower beams (within the initial transmit sector), while during the receive training, all $M = bM_{\text{SLS}}$ or $N = bN_{\text{SLS}}$ directions should be covered.

We remark that the SLS and BRP phases of beamforming usually precede data transmission. They are always executed at the beginning of the beamforming process. However, the DTI can be used for all the beamforming phases to enable the repetition of the beamforming process as and when required.

The optional beam tracking phase is used during data transmission (DT) to adjust for channel changes. Beam tracking is accomplished by appending training (TRN) fields to data packets [7].

The basic principle of 802.11.ad BI is briefly reviewed in this section, interested readers for more details can refer to [8], and the detailed description of the protocol may be found in [9].

2.2 IEEE 802.11ay

IEEE 802.11ay is an amendment of great interest for applications ranging from high-speed short-range links to wireless backhaul that enable 100 Gbps communications in the 60 GHz mmWave band. IEEE 802.11ay incorporates a variety of technical advancements at the PHY over IEEE 802.11ad standard, such as channel bonding and aggregation, single-user (SU) and downlink (DL) multi-user (MU) Multiple-Input Multiple-Output (MIMO) transmissions, and nonuniform modulation constellation, as well as improved channel access and enhanced beamforming training.

In 802.11ay, the enhanced directional multi-gigabit (EDMG) PCP/AP can allocate multiple clients on different channels to communicate with the PCP/AP simultaneously. Moreover, two EDMG clients can communicate with each other on a bonded channel or an aggregated channel to achieve higher throughput and improve channel utilization.

A new packet structure is defined in IEEE 802.11ay to support MIMO wireless links and channel bonding. The EDMG packet contains new fields necessary to support the additional capabilities defined for EDMG stations and a redefined TRN field that is more flexible and efficient than the one specified in 802.11ad.

In this section, we summarized the main improvements of the IEEE 802.11ay standard. The detailed overview of the IEEE 802.11ay can be found in [4,5].

3 Design Challenges of mmWave Multicast and Unicast Modes

3.1 Unicast in Directional Networks

A considerable amount of research recently investigated mmWave communications by focusing on unicast data transmission optimization [10,11], where the AP performs serial TDMA transmissions. In mmWave wireless personal area networks (WPAN), each user is served independently of the others with minimum beamwidth (on the order of 10–20° or less) to provide high data rates. The reliability of such transmissions is very high since the beamwidth is equal to the resolution and provides the maximum available signal-to-noise ratio (SNR). However, the AP requires a long time to serve all users, as it generates a separate beam for each user to transmit the data sequentially.

3.2 Multicast in Traditional Networks

Multicast is a bandwidth-conserving technology, the basic concept of thereof consists in the traffic reduction by simultaneous delivering a single stream of information to a group of users, i.e., data packets are transmitted only once. Consequently, it considerably improves the bandwidth efficiency compared to unicast mode since all users are served simultaneously by using a single wide transmission beam (omnidirectional), which generally provides very short transmission duration. In the past literature, several efficient solutions improving traditional multicast schemes in omnidirectional networks were proposed [12–14]. However, pure multicast schemes are almost infeasible in mmWave directional systems due to the propagation specifics at EHF bands. The adaptation of the directional nature of mmWave communication for existing methods, as well as the development of novel mmWave-specific schemes [15,16], is of particular interest for the research community.

3.3 Multicast in Directional Networks

The use of highly directional transmission beams in mmWave systems represents an outstanding feature with respect to traditional networks and allows coping with high attenuation at EHF bands. Most phased array antennas that have been designed in past years have used analog beamforming where the phase adjustment is performed at RF, and there is one set of data converters for the entire antenna (one RF chain). Note that analog mmWave systems utilize sequential multicast. The beam orientation and the beam resolution (beamwidth), which need to be adjusted in addition to the beam radius, make these systems different from conventional omnidirectional networks. By contrast, in the case of hybrid or digital beamforming techniques, which allow transmitting to more than one user at a time, the power budget constraint at the transmitter side has to be taken into consideration [17]. Therefore, when multiple RF chains are available, new opportunities and challenges (e.g., determining the shape of numerous beams to be swept simultaneously under the total transmission power constraint) appear.

In alignment with the Friis transmission equation, the received power is directly proportional to the transmitter channel gain, which, in its turn, strongly depends on the beam orientation and resolution.

Then, multicast traffic delivery in mmWave systems presents the following problems, which have to be considered [17]:

1. Wide beams are more likely to reach all multicast receivers since they can cover a larger angle range and, thus, serve more users simultaneously. However, due to the lower antenna gain that wide beams provide, the supported transmission rate is limited.
2. Narrow beams provide higher antenna gain and thus can support higher transmission rates. However, they are limited in coverage in terms of the aperture angle and may not simultaneously serve a number of users. As a consequence, multiple unicast transmissions are required to reach all multicast users.
3. The presence of moving users is more challenging for multicast transmission. In fact, in the case of unicast, the AP is beamformed toward the only receiver, and small movements of the receiver still allow to guarantee a good reception. Differently, in the case of multicast, beams are steered in between users. Hence, some receivers may be close to the edge of a beam's coverage area and, due to the even small mobility, can be out of the beam coverage.

The integrated problem of directional beamforming and multicast communications is under investigation in this paper.

4 System Model

In this paper, we consider a general public scenario where owners of *wearable devices* are interested in receiving the same content, i.e., the AP transmits data to multiple users thought multicast mmWave links. We assume analog beamforming only to analyze the performance of sequential multicast in TDMA fashion. This means that the AP can transmit through a single beam at a time to serve the users.

4.1 Antenna and Channel Models

In what follows, we assume that devices transmit directionally with the same antenna beam pattern, which is symmetrical w.r.t. the boresight [18]. By this symmetrical assumption, we mean that antennas have a unique beam shape in both elevation and azimuth planes, i.e., their antenna pattern is akin to a conical shape.

In terms of the channel model, when HPBW θ is used, the received signal power at receiver i is calculated by the Friis equation:

$$P_{\text{rx},i} = \frac{P_{\text{tx}} D_0 \rho(\alpha_i) \lambda^2}{(4\pi)^2 r_i^{\kappa}}, \tag{1}$$

where P_{tx} is the transmit power, α_i is the current angular deviation of the transmit/receive direction from the antenna boresight for receiver i, $\rho(\alpha_i) \in [0;1]$ is a piece-wise linear function that scales the antenna directivity D_0 [18][1], λ is the wavelength, r_i is the separation distance between the transmitter (Tx) and receiver (Rx) i, and κ is the path loss exponent.

We assume the line of sight (LoS) path only. Hence, the maximum achievable rate D_i of the Tx-Rx$_i$ link could be estimated according to Shannon's channel capacity as:

$$D_i = W \log_2 \left(1 + \frac{P_{\text{rx},i}}{P_{\text{noise}}}\right), \tag{2}$$

where P_{rx} incorporates both transmit and receive antenna gains after the BRP phase, W is the bandwidth, P_{noise} is noise power in the channel, which corresponds to

$$P_{\text{noise}} = W N_0 \text{NF}, \tag{3}$$

where N_0 is the power spectral density of noise per 1 Hz, and NF is the noise figure.

For a multicast group containing n receivers, the overall performance of multicast transmission depends on the user with the worst channel. Thus, the achievable rate of a multicast transmission is given by

$$D = W \log_2 \left(1 + \min_i \left(\frac{P_{\text{rx},i}}{P_{\text{noise}}}, 0 | P_{\text{rx},i} < P_{\text{thr}}\right)\right), \tag{4}$$

where P_{thr} guarantees the minimum required received power for data transmission.

The data transmission duration for multicast transmission is given by

$$T_{DT} = \frac{B}{D}, \tag{5}$$

[1] $\rho(\alpha_i) = 1 - \frac{\alpha_i}{\theta}$, if $\alpha_i \leq \theta$, otherwise $\rho(\alpha_i) = 0$; $\rho(\alpha_i) = 1$ corresponds to the antenna boresight in the case of perfect alignment (e.g., unicast transmission after the beamforming procedure). In the case of multicast transmission, each user deviates on angle α from the boresight of the transmitter.

where B is the packet size.

Then, the total duration of data transmission can be calculated as

$$T = T_{\mathrm{SLS}} + U(T_{\mathrm{BRP}} + T_{\mathrm{DT}}) + T_0, \tag{6}$$

where $T_{\mathrm{SLS}} + U(T_{\mathrm{BRP}})$ is the overhead on the beam training, U is the average number of clients per AP, and T_0 is the total signaling overhead, which is independent of the number of beams.

5 Performance Analysis

In this section, we evaluate the multicast transmission performance in directional mmWave networks via simulations. To this end, we focus on a scenario composed of a group of people in a museum interested in receiving the same multimedia content by means of high-end wearable devices equipped with IEEE 802.11ad/ay chipsets operating at 60 GHz. We consider an analog beamforming technique (one RF chain), which means that the AP can transmit over a single beam at a time, to assess the performance of sequential multicast.

For the evaluation purpose, we consider resource-hungry applications in scenarios with and without mobility. We also analyze two patterns of users' distribution: (i) within a sector (see Sect. 5.1) to investigate sequential multicast performance characteristics, and (ii) in a line (see Sects. 5.2 and 5.3) to explore the angle coverage of the antenna. The transmit power is fixed at the level of $P_{\mathrm{tx}} = 23$ dBm, whereas $P_{\mathrm{thr}} = -68$ dBm (MCS 1) [9]. Main simulation parameters are summarized in Table 1.

Table 1. Main simulation parameters

Notation	Parameter	Value
f	Carrier frequency	60 GHz
λ	Wavelength	0.005 m
k	Propagation exponent	2
C	Propagation constant	$6.3165 \cdot 10^6$
P_{thr}	Sensitivity	−68 dBm [9]
P_{tx}	Transmit power	23 dBm
θ	Beamwidth	var
R_d	Radius of the area of interest	40 m
$\mathrm{SNR}_{\mathrm{max}}$	SNR corresponding to choosing MCS19 (rate 13/16)	20 dB
NF	Noise figure	6 dB
B	Packet size	1 Gb

5.1 Static Scenario: Unicast vs. Multicast Performance

To simulate sequential multicasting, we assume that the AP sweeps beams of equal resolution (i.e., fixed predefined beams) to cover the sector under analysis, as illustrated in Fig. 2.

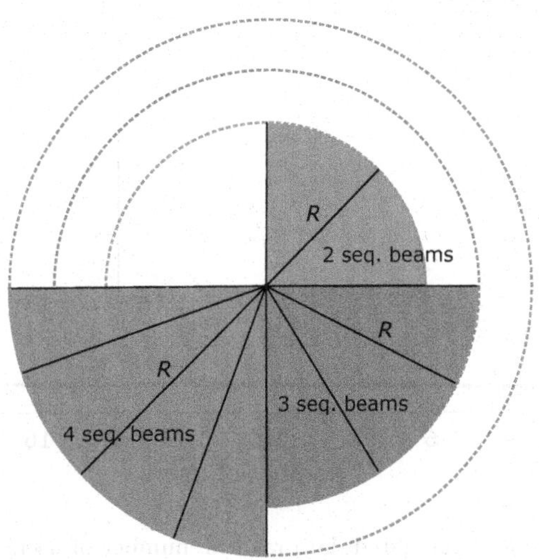

Fig. 2. Illustration of the multicast scheme with fixed beams.

We first analyze the system performance in terms of aggregated data rate (ADR) for sequential multicast and unicast schemes (see Fig. 3). For this purpose, we uniformly distribute users within a sector of 90° of radius $Rd = 40$ m. The choice of the service area radius can be explained by the fact that sequential multicast with the widest width (i.e., $\theta = 58°$) requires the lowest number of sequential beams, whereas it can cover the smallest Rd in comparison with narrower beams. More precisely, for given system parameters, the beam of 58° can cover only a distance of 45 m. However, due to the beam misalignment in multicast, we set Rd to 40 m to guarantee reliable communications for every considered transmission scheme. We recall that the wider HPBW of the predefined beams, the less number of beams needs to be swept by the AP to cover the considered sector (in terms of the angle coverage). One may deduce that the wides beams, e.g., HPBW = 58°, can improve the system throughput.

The effects of sequential transmissions on the total delay in mmWave systems are further highlighted in Fig. 4. As one may observe, sequential multicast with $\theta = 58°$ guarantees the shortest data transmission duration; hence, the lower total transmission delay. However, it provides lower SNR value as well as higher outage probability due to the lower antenna directionality. Using the widest possible beam at EHF bands severely limits the data rate and transmission range. In contrast, narrow beams require a longer data transmission duration. Therefore, we may conclude that the resource management algorithms are of crucial

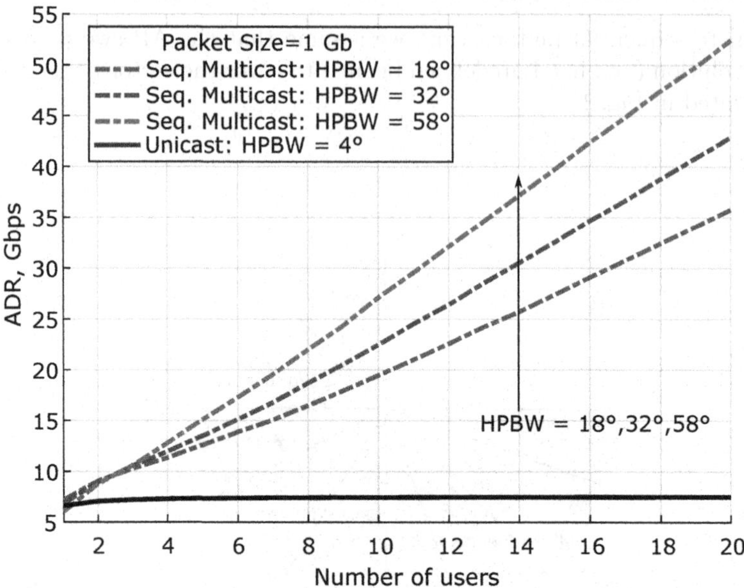

Fig. 3. Aggregated data rate vs. number of users.

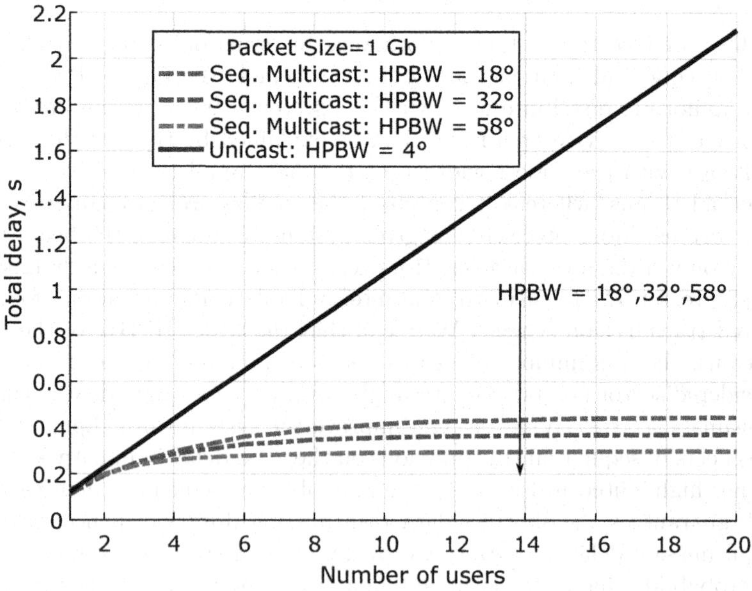

Fig. 4. Total transmission duration vs. number of users.

importance since they allow to dynamically make decisions based on the number and resolution of beams in directional multicast by taking into consideration (i) the multicast group size, (ii) the shape and size of the service area, (iii) users' locations and density, as well as (iv) QoS requirements. In addition, in the case of hybrid beamforming, power models have to be applied to split the power among composite multi beam patterns.

5.2 Static Scenario: Coverage Area Estimation

In this section, we show results in terms of achievable data rate and total transmission delay for multicast, unicast, and sequential multicast schemes when a group of *ten users* is located within a line of length d. Here, in the case of sequential multicast, we assume that the group of users is always served by a fixed number of beams $N = 3$, independently from the beam resolution. We investigate the maximum possible distance d at which the connection can be established, for the three considered schemes.

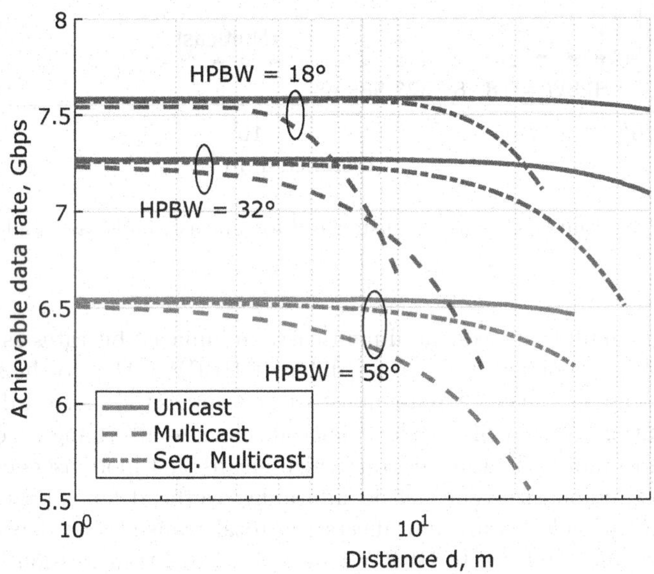

Fig. 5. Data rate vs. distance d for unicast, multicast, and sequential multicast.

By analyzing the results presented in Fig. 5, we learn that the distance d does not affect unicast transmission as a separate aligned beam is swept to serve each user. Regarding the pure multicast scheme, we can see that there is a threshold that determines the maximum coverage angle of the beam. For example, HPBW = 58° provides the larger distance d, whereas the narrowest beam (i.e., HPBW = 18°) covers the lowest distance d. We also highlight that sequential multicast achieves higher data rates for the same distance d compared to the multicast scheme. It can be explained by the fact that multicast (with a single beam only) experiences difficulties or even fails to provide sufficient data rates to users located far apart in terms of the angle coverage.

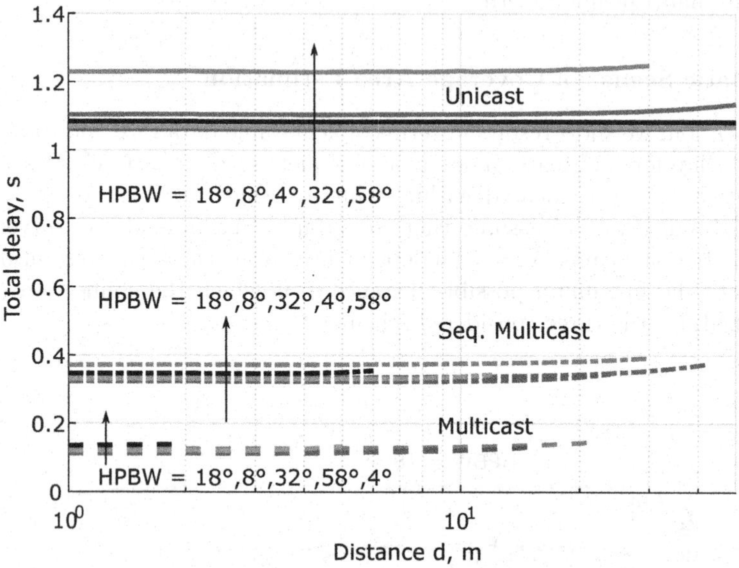

Fig. 6. Total transmission delay vs. distance d for unicast, multicast, and sequential multicast.

Achievable multicast, sequential multicast, and unicast bit rates as functions of distance d are depicted in Fig. 6 for different HPBWs. Due to the sequential nature of unicast in mmWave, pure multicast guarantees the shortest data transmission duration. Meanwhile, both sequential schemes ensure higher distance d with respect to multicast transmission, which means that more users can receive data. We also emphasize that the beamforming overhead for narrow beams is greater than for wide beams and affects the total transmission delay (see, for instance, the curves related to HPBW $= 4, 8°$ for three transmission fashions).

5.3 Dynamic Scenario: Mobility Impact on Multicast Transmission

To facilitate a more detailed performance evaluation, we proceed by considering the scenario with mobile users.

The mobility influence on the three transmission modes is evaluated in Fig. 7. We assume a simple case with two devices located in a line of size $d = 16$ m for $\theta = 8°$. We analyze the behavior of unicast and multicast transmissions in the presence of mobile users by considering a rectilinear mobility pattern with user speeds of $v = 0.5$ m/s (as per walking or rolling stairs), $v = 11$ m/s (as per

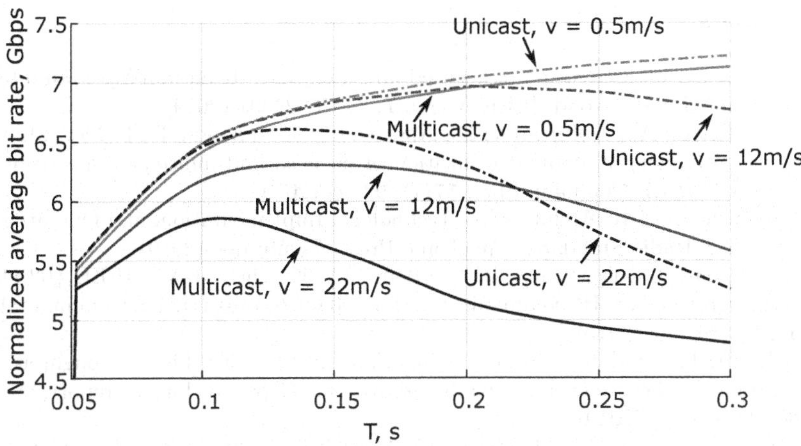

Fig. 7. Normalized average achievable data rate over time T for different speeds, $\theta = 8^{\circ}$.

segway), and $v = 22\,\text{m/s}$ (as per bicycle). We also consider the overhead on beam training and recalculate the data rate over the total BI duration, i.e., $D_{\text{norm}} = D \cdot T_{\text{DT}}/T$. As one may notice from Fig. 7, the multicast transmission is more vulnerable to dynamic users' behaviors due to the fact that the beam is steered in between users, and some of them may be located at the edge of the beam. Hence, mobility management in directional multicast systems is even more challenging compared to unicast transmissions.

6 Conclusion

This paper presented an initial investigation of multicast transmissions using mmWave links with particular emphasis on the channel access of 802.11ad/ay standards operating at the unlicensed band. We experimentally evaluate multicast and unicast performance in directional mmWave networks. From the numerical evaluation of the model, it is clear that multicast in directional networks poses additional challenges compared to conventional systems. Since using the widest possible omnidirectional beam at EHF bands severely limits the data rate and range, the AP needs to transmit data sequentially by utilizing a resource management algorithm, which is our future research task. In addition, scenarios with the presence of mobile users receiving group-oriented services in directional mmWave networks is an essential issue to be faced by the research community.

Acknowledgement. The authors gratefully acknowledge funding from European Union's Horizon 2020 Research and Innovation programme under the Marie Skłodowska Curie grant agreement No. 813278 (A-WEAR: A network for dynamic wearable applications with privacy constraints, http://www.a-wear.eu/).

References

1. Lu, X., et al.: Integrated use of licensed-and unlicensed-band mmWave radio technology in 5G and beyond. IEEE Access **7**, 24376–24391 (2019)
2. Lu, X., Petrov, V., Moltchanov, D., Andreev, S., Mahmoodi, T., Dohler, M.: 5G-U: conceptualizing integrated utilization of licensed and unlicensed spectrum for future IoT. IEEE Commun. Mag. **57**(7), 92–98 (2019)
3. Ahmadi, S.: 5G NR: Architecture, Technology, Implementation, and Operation of 3GPP New Radio Standards. Academic Press, Cambridge (2019)
4. Ghasempour, Y., da Silva, C.R., Cordeiro, C., Knightly, E.W.: IEEE 802.11ay: next-generation 60 GHz communication for 100 Gb/s Wi-Fi. IEEE Commun. Mag. **55**(12), 186–192 (2017)
5. IEEE 802.11 Working Group: Enhancements for very high throughput for operation in license-exempt bands above 45 GHz. Technical report, IEEE P802.11ay/D3.0 (2019)
6. da Silva, C.R., Lomayev, A., Chen, C., Cordeiro, C.: Analysis and simulation of the IEEE 802.11ay single-carrier PHY. In: 2018 IEEE International Conference on Communications (ICC), pp. 1–6. IEEE (2018)
7. Kutty, S., Sen, D.: Beamforming for millimeter wave communications: an inclusive survey. IEEE Commun. Surv. Tutor. **18**(2), 949–973 (2015)
8. Chukhno, N., Chukhno, O., Shorgin, S., Samouylov, K., Galinina, O., Gaidamaka, Y.: Maximizing achievable data rate in unlicensed mmWave networks with mobile clients. In: Galinina, O., Andreev, S., Balandin, S., Koucheryavy, Y. (eds.) NEW2AN/ruSMART 2019. LNCS, vol. 11660, pp. 282–294. Springer, Cham (2019). https://doi.org/10.1007/978-3-030-30859-9_24
9. IEEE 802.11 Working Group: Wireless LAN Medium Access Control (MAC) and Physical Layer (PHY) Specifications. Amendment 3: Enhancements for Very High Throughput in the 60 GHz Band (2012)
10. Al-samman, A.M., Azmi, M.H., Rahman, T.A.: A survey of millimeter wave (mm-Wave) communications for 5G: channel measurement below and above 6 GHz. In: Saeed, F., Gazem, N., Mohammed, F., Busalim, A. (eds.) IRICT 2018. AISC, vol. 843, pp. 451–463. Springer, Cham (2018)
11. Sanfilippo, G., Galinina, O., Andreev, S., Pizzi, S., Araniti, G.: A concise review of 5G new radio capabilities for directional access at mmWave frequencies. In: Galinina, O., Andreev, S., Balandin, S., Koucheryavy, Y. (eds.) NEW2AN/ruSMART 2018. LNCS, vol. 11118, pp. 340–354. Springer, Cham (2018). https://doi.org/10.1007/978-3-030-01168-0_32
12. Rinaldi, F., Pizzi, S., Orsino, A., Iera, A., Molinaro, A., Araniti, G.: A novel approach for MBSFN area formation aided by D2D communications for eMBB service delivery in 5G NR systems. IEEE Trans. Veh. Technol. **69**(2), 2058–2070 (2019)
13. Araniti, G., Rinaldi, F., Scopelliti, P., Molinaro, A., Iera, A.: A dynamic MBSFN area formation algorithm for multicast service delivery in 5G NR networks. IEEE Trans. Wirel. Commun. **19**(2), 808–821 (2019)
14. Pizzi, S., Suraci, C., Iera, A., Molinaro, A., Araniti, G.: A sidelink-aided approach for secure multicast service delivery: from human-oriented multimedia traffic to machine type communications. IEEE Trans. Broadcast. **PP**, 1–11 (2020)
15. Naribole, S., Knightly, E.: Scalable multicast in highly-directional 60-GHz WLANs. IEEE/ACM Trans. Network. **25**(5), 2844–2857 (2017)
16. Niu, Y., Yu, L., Li, Y., Zhong, Z., Ai, B.: Device-to-device communications enabled multicast scheduling for mmWave small cells using multi-level codebooks. IEEE Trans. Veh. Technol. **68**(3), 2724–2738 (2018)

17. Biason, A., Zorzi, M.: Multicast via point to multipoint transmissions in directional 5G mmWave communications. IEEE Commun. Mag. **57**(2), 88–94 (2019)
18. Chukhno, O., Chukhno, N., Galinina, O., Gaidamaka, Y., Andreev, S., Samouylov, K.: Analysis of 3D deafness effects in highly directional mmWave communications. In: 2019 IEEE Global Communications Conference (GLOBECOM), pp. 1–6. IEEE (2019)

Deep Learning for IoT Traffic Prediction Based on Edge Computing

Ali R. Abdellah[1,2], Volkov Artem[2], Ammar Muthanna[2(✉)], Denis Gallyamov[2], and Andrey Koucheryavy[2]

[1] Electronics and Communications Engineering, Electrical Engineering Department, Al-Azhar University, Qena 83513, Egypt
alirefaee@azhar.edu.eg

[2] The Bonch-Bruevich Saint-Petersburg State University of Telecommunications, Pr. Bolshevikov, 22, St. Petersburg 193232, Russia
artemanv.work@gmail.com, ammarexpress@gmail.com, gallyamovda@yandex.ru, akouch@mail.ru

Abstract. The 5G network is the latest wireless mobile communication technology. Nowadays, the emerging of many network applications has led to a massive amount of network traffic. Many researchers have devoted their studies to the accurate prediction of network traffic applications. Network management requires technology for the prediction of network traffic without network operator intervention. In practice, 5G uses the Internet of Things (IoT) for working in high-traffic networks with multiple sensors to send their packets to a destination simultaneously, which is a feature of IoT applications. Therefore, 5G offers wide bandwidth, low delay, and extremely high data throughput. Predicting network traffic is more important for IoT networks to provide reliable communication. The efficient 5G network cannot be complete without including artificial intelligence (AI) procedures. Machine learning (ML) has been successfully applied to traffic prediction. In this paper, we implement the prediction of IoT traffic in time series using deep learning. The prediction accuracy has been evaluated using the (RMSE) as a merit function and mean absolute percentage of error (MAPE).

Keywords: Internet of Things · Artificial intelligence · Edge computing · Deep learning · 5G · IMT-2030

1 Introduction

In 2020, the beginning of a new era in telecommunications is expected - fifth-generation 5G communication networks. 5G networks are expected to enable the effective and cost-effective implementation and support of many new services [1–3]. 5G networks should create an ecosystem for technical and business innovations, give a breakthrough to the further development of technologies in the field of ICT. It is worth noting that the designation "5G", usually representing the technological stage of mobile networks, at the moment at the international level de facto reflects a new era of communication networks and services

© Springer Nature Switzerland AG 2020
V. M. Vishnevskiy et al. (Eds.): DCCN 2020, CCIS 1337, pp. 18–29, 2020.
https://doi.org/10.1007/978-3-030-66242-4_2

in general [4]. The abbreviation 5G is usually given by the 3GPP consortium (3rd Generation Partnership Project), however, in this case, taking into account many factors (the trend of technology development, the emerging need for new services, etc.), the world community "attached" the abbreviation 5G/IMT-2020. The fifth generation communications network abbreviation IMT-2020 was given by the International telecommunication Union.

Based on the set of international recommendations and standards at this time, we can conclude that the concept of IMT-2020 communication networks or otherwise fifth-generation communication networks includes a whole range of concepts and technologies, and not only describes the principles and technologies for organizing a mobile access network. At the time of 2020, it is planned to complete the standardization of communication networks and fifth-generation systems 5G/IMT-2020, as well as their widespread adoption on communication networks. 5G/IMT-2020 networks are designed to integrate all the achievements of mobile and fixed communication networks, provide data transfer speeds of 10 Gb/s and higher, as well as bring the capabilities of new cloud computing structures (Edge computing and FoG Computing) directly to the user [5].

The main feature of the fifth generation networks is due to the emergence of the concept of the Internet of Things. This feature is associated with a fundamentally different number of potential things that can be connected to the communication network and be uniquely identifiable in it, compared to traditional ideas about the volume of device databases in the network. According to the well-known forecast, the limit of the Internet of Things is considered at the level of 50 trillion. Such a number of connected devices implies a high density per unit of space. Therefore, 5G networks are called high-density networks - Super High Dense. This direction of fifth-generation communication networks has been given the name - "Communication Networks with Ultra-low latency". However, on the basis of many scientific studies, the community came to the conclusion that at the moment of development of the elemental and technological base, such indicators are difficult to achieve to provide this type of service on an equal basis with others. However, this direction of research continues its development and smoothly proceeds to the next generation of networks, a description of which the International Telecommunication Union has already given in its document in the summer of 2018, namely, the IMT-2030 communication network. To meet the requirements for a number of new services based on the Internet of Things, to ensure the stability of communication networks as part of the development of new services, as well as the ability to automate business and system processes on networks, the International Telecommunication Union has proposed the use of the following communication network building technologies, such as SDN and NFV.

Thus, when integrating new technologies for building communication networks, as well as technologies for organizing cloud computing, in particular Edge Computing, it is possible to achieve a greater synergistic effect by using artificial intelligence algorithms. It is worth noting that the approaches that are implemented in these technologies provide such a theoretical and technical opportunity, thanks to the abstraction of the level of transmission/processing and storage of data from the control level.

To solve the problems of detection and forecasting, the approaches that are offered are mostly based on periodic traffic capture and analysis of the headers or a deeper analysis. However, in the era of new technologies for organizing network and computing (cloud) infrastructure, such as Software-defined networking, Network Function Virtualization (network technologies of 5G & Beyond communication networks), as well as Edge Computing & FoG Computing (Cloud computing technologies), there are new opportunities for developing methods for detecting the type of traffic, as well as its forecasting.

At the same time, the predictive models, due to the technical capabilities of the modern hardware part, can take into account many characteristics of the studied traffic/system model in its general sense: for example, Tactile Internet services of one of the providers. These technical capabilities allow you to implement algorithms that require much more computing power, while these algorithms can be adapted to the particular geographic location of the forecasting system (for example, different cities/countries where the services of different providers are used, etc.), as well as time (night/morning/working day/evening).

This type of algorithms, at present, is referred to as machine learning algorithms, as well as to a higher level of classification - Artificial Intelligence algorithms. Yes, of course, these algorithms, at the moment, will not be able to fully realize the AI functions that are laid down in the understanding of the world scientific community. However, already at this stage, a class of algorithms capable of processing a large amount of data, capable of adapting to a specific system, self-learning and improving, allows you to solve labor-intensive tasks, the solution of which previously the scientific and technical community could only dream of.

The introduction of AI algorithms in automation tasks and a certain level of telecommunication systems intellectualization will allow reaching a new level in ICT and solving those tasks that were set for fifth-generation and subsequent communication networks (IMT-2030). For example, as part of meeting the delay requirement for a number of Internet of Things applications.

2 Related Works

Currently, advances in computing power and the development of deep learning algorithms [6–8], artificial intelligence (AI) has shown promising results in various [17,18], including in the areas of security [6,9,10], networks [11], information retrieval [12], etc. And there is no doubt that AI reduces human intervention and ensures reliable results, regardless of functioning areas. Subsequently, the development of AI can also contribute to the creation of new solutions for the efficient organization and management of data flows. With the advent of new technologies and the abundance of modern wireless devices, today there is high growth in the development of data flow.

In addition, the production of devices with changing requirements is actively developing, that is, for the transition from 3G to 4G, and from 4G to 5G, the market is torn by smartphones, portable devices, and another device, which poses many tasks for network design. It is also expected that in the near future there will be 50 billion connected devices [13]. Therefore, this huge number of devices will impose restrictions on existing communication systems. To handle such large numbers of devices that are likely to increase the amount of transfer and connectivity, developing 5G networks will be the main and much-needed goal. It is also expected that in the coming years the data flow will be 200 times higher than in 2010, and in 2030 this will increase 20,000 times, compared with 2010 [13].

To ensure the operation of such a large number of devices and the ability to operate with a significantly increasing data flow, AI, along with new technologies such as software-configurable networks (SDN), mobile edge computing (MEC) [19], device-to-device (D2D), network function virtualization (NFV) should be used to organize and manage data flows. Software-defined Network (SDN) technology is a new network theory that separates the control and data planes of a router/switch device to achieve better control of data flow and networks. A typical application of SDN to provide and manage network traffic is dynamically shown in [8]. However, for effective control of data flows, AI algorithms are also important in this area. To implement this algorithm, the AI is equipped with an SDN controller to provide network traffic control in the radio access network. In this work, AI is used on edge computing to solve one of the main problems of 5G networks. Thus, the paper has the goal of defining an AI algorithm and edge computing for the optimal organization and management of data flows. In addition, to model the flows of data traffic in the system, we use a mathematical model. After, we develop an effective system for the data flow of 5G networks with associated mathematical methods.

3 Problem Statement

Considering the described problems, technological capabilities of SDN/NFV networks, as well as new cloud architectural approaches, in particular Edge Computing in the introduction section, it is necessary to formulate a solution for predictive analytics of the Internet of Things. As part of the solution of this problem, it is proposed to use the capabilities of programmable networks to obtain data on flows for subsequent analytics. The developed analytical models, in order to ensure a high level of predictive analytics, as well as taking into account the structure of incoming analytical data, are developed on the basis of Deep Learning technology.

4 Proposed System and Algorithm

4.1 Architecture of Deep Learning Implementation Based on Edge-Computing

At the time of the research, in the scientific and technical world there are many works that are aimed at detecting traffic types, developing forecasting models, [3–5,16] both traffic and the load of telecommunication systems. These tasks are more interested in service providers and telecommunications.

Thus, as previously mentioned, this article proposes the use of Machine Learning algorithms with Deep Learning to generate forecasts of the activity of the Internet of Things traffic. These algorithms are a certain kind of analytical system that is deployed on a cloud structure - Edge Computing. Thus, this method of organizing analytics will allow monitoring traffic no longer entering the network, in addition, in conjunction with the forecasting system, a traffic detection system can work, which together will give a greater synergistic effect. In this paper, we consider the issue of predicting the traffic of the Internet of Things on Edge Computing nodes. The conceptual architecture of the solution is shown in Fig. 1.

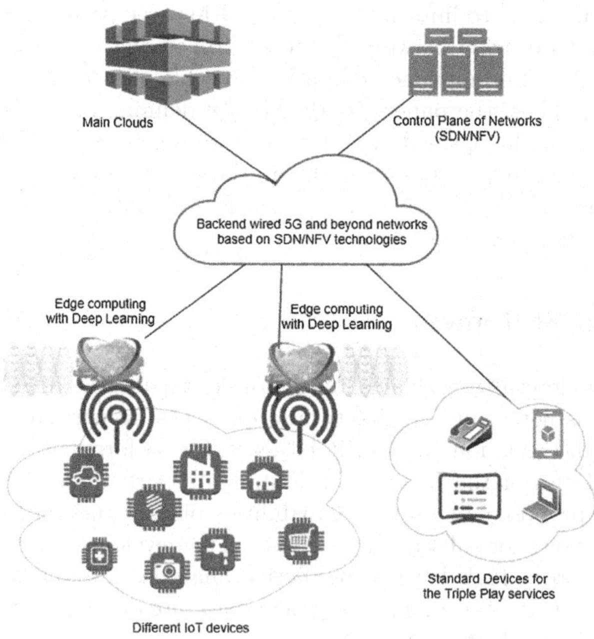

Fig. 1. The architecture of the concept

The proposed approach is implemented within the framework of a distributed analytical system for forecasting the development of infrastructure, where it is assumed to use metadata of flows that pass through programmable SDN switches. The principle of working with stream metadata for detecting the nature of data being transmitted (video, Internet of Things and others) was proposed in [3]. In [3], a method was proposed for compiling a meta-model of flows based on two global parts of the SDN stream table, namely: Match Field and Actions. Thus, since Edge Computing is located at the "edge of the network", that is, in fact the first object with high computing capabilities, after the infrastructure of mobile and other networks, there is the possibility of organizing an analytical module for forecasting flows, including as part of the Internet of Things.

Implementation of load forecasting algorithms in the form of analytical modules will allow for portability and flexible scaling of the system, independence of the data transmission medium. For the analytical system, all devices and streams are a "digital object" and, in fact, the system operates with "digital twins" of devices, traffic, and others, which are characterized by a set of parameters (variables for neural networks of the analytical module). This level of abstraction allows you to implement an analytical system that will work with data on flows (metadata) in the "On-the-fly" mode. Thus, this system allows not introducing additional delays into the traffic, and also does not in any way alter its activity and characteristics. The analytic module under consideration is part of a distributed analytical system, including one whose functional part was considered in [3–5]. In this paper, the hypothesis is tested about the possibility of traffic prediction on Edge computing structures; for this, the infrastructure of the Programmable Networks Laboratory (SDNLab) of the Department of Communication and Data Networks of St. Petersburg State Telecommunications University was used. prof. M.A. Bonch-Bruevich. To test the performance of the above proposed method, we used a model network with a working smart city service. To generate traffic, Internet of Things traffic generators were used, working with IoTDM (Internet of Things Data Management) system, built according to the specifications of the OneM2M international community. These traffic generators were developed as part of research work, the results of which are reflected in this article [1].

Based on this infrastructure, data were collected for their subsequent processing in a mathematical model of a neural network with deep learning. The mathematical method for processing "raw" data from SDN switches is given in [3], where the final data set is described by the following matrix (1).

$$DataSet_{ML} = \begin{matrix} [\text{TimeStamp}] & [\text{ByteCount}] & [\text{PacketCount}] \\ TS & BC_{delta_{12}} & PC_{delta_{13}} \\ TS & BC_{delta_{22}} & PC_{delta_{23}} \\ \cdots & \cdots & \cdots \\ TS & BC_{delta_{N2}} & PC_{delta_{N3}} \end{matrix}$$

Where TS is the value of the data acquisition period, in this work 1 s. And the parameters BC_{delta} and PC_{delta} indicate the number of Bits and Packets transmitted in the set time interval, according to the counters of the studied flows.

In the framework of this article, as already mentioned above, from the point of view of practical implementation, modeling of the developed Neural Network is carried out in order to test the hypothesis that it is possible to predict the Internet of Things traffic using Edge Computing.

4.2 Deep Learning Algorithm for IoT Traffic Prediction

Today, the Artificial Neural Network [14–16] is widely used for various tasks in various fields of life. For example, tasks such as speech recognition, text recognition, predicting the work and development of complex engineering systems, are largely solved with the help of Artificial Intelligence technologies, including artificial neural networks. The use of these algorithms is determined by a number of their advantages, for example, such as high recognition rates of the studied objects, prediction of complex systems within a limited time, as well as complexity, from the point of view of systematic objects.

A data stream from the generated $DataSet_{ML}$ is fed to the input layer of neurons (the so-called placeholders). Placeholders are connected to the first neural network layer by a fully connected architecture.

In this work, we perform IoT traffic prediction approaches using LSTM with deep learning. The prediction accuracy was evaluated using the RMSE as merit function and MAPE (mean absolute percent of error) and we tested the performance in three cases according to the number of hidden units in network. We made the prediction in cases of numbers of hidden units, 500, 100, 50, respectively.

Input and output time series depends on predicting the future value of one time-series is given another time-series. The past values of both series (for best accuracy), or only one of the series (for a simpler system) may be used to predict the target series.

5 Simulation Results

The dataset can be used to show how LSTM with deep learning can be trained to make predictions. The datasets are obtained from IoT traffic the model was simulated using method, which was described above. Then, the collection and preparation of the data set that was divided into 70%, 30% for training, and testing, respectively. The implementation of the feedback neural network to predict the performance accuracy of IoT traffic. Table 1 shows the prediction accuracy for the IoT throughput using RMSE and MAPE.

Table 1. The measure of prediction accuracy for the predicted model validation using RMSE and MAPE.

Hidden units	RMSE	MAPE
500	0.93432	2.4
100	2.4163	9.32
50	6.5104	16.75

Table 1 displays the prediction accuracy of IoT traffic in case of the number of hidden units in LSTM layer are 500, 100, 50, respectively in order to estimate the error of prediction we use the RMSE and another measure for performance accuracy MAPE.

From the tabulated results, the model predicted in the case of number hidden units 500 has the best prediction accuracy with RMSE value equal 0.93432 and MAPE equal 2.4% in comparison to its peers. The maximum average prediction accuracy improvement in the case of 500 hidden units is 14.35%. Also, the model predicted in the case of 100 hidden units dropped to has a prediction accuracy with RMSE value 2.4163 and the MAPE dropped to 9.32% and the maximum average of prediction accuracy in this case is 7.43%.

On the other hand, the model predicted in the cases of a number of hidden units 50 has the lowest prediction accuracy with RMSE value 6.5104 and MAPE dropped to 16.75% comparison of them.

Figures 2, 3 and 4 shows the prediction accuracy for the IoT throughput using RMSE.

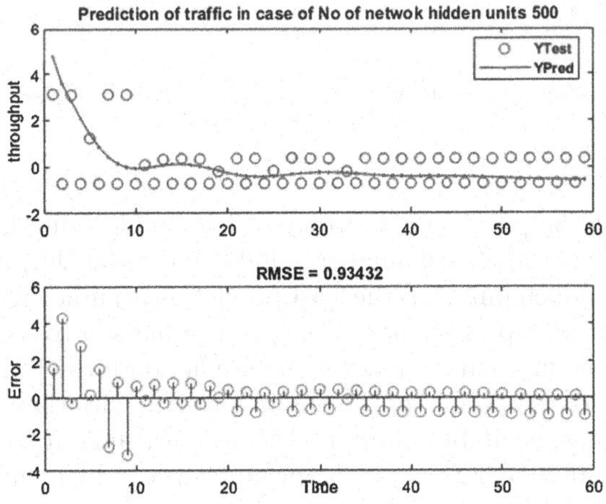

Fig. 2. The response of output element for time series in case of the number of hidden units 500

Figure 2 shows the prediction accuracy of IoT traffic using LSTM approach with deep learning in the case of number of hidden units 50. The model predicted in the case of 500 hidden units has the least prediction accuracy with RMSE value equal 0.93432 and the improvement is 14.35% in comparison to its peers. I Fig. 2 show two curves the first curve observes the prediction for IoT throughput with time and the second curve observes the prediction error with time. The prediction time 60 for verifying the ability of the LSTM deep learning in for optimizing the IoT traffic load prediction in the cases of the number of hidden units 500. As shown in Fig. 2 the throughput with time for the YTest and YPred (predicted model) models we notice that the predicted model increase at time 1 then decreases gradually until the time 9 then little increase then little decrease the time 59 which gives the best prediction accuracy in this case.

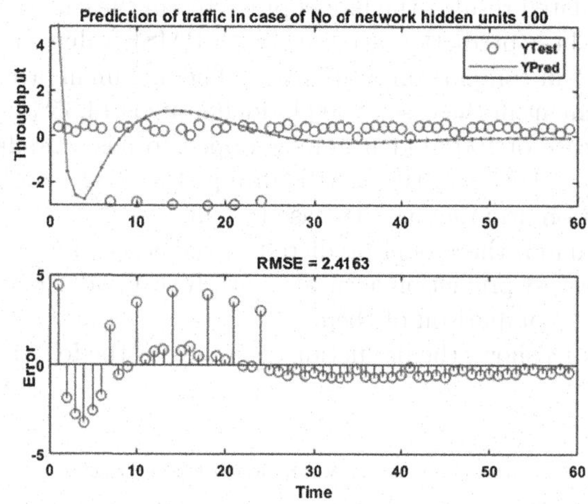

Fig. 3. The response of output element for time series in case of the number of hidden units 100

Figure 3 shows the prediction accuracy of IoT traffic using LSTM approach with deep learning in the case of number of hidden units 50. The model predicted in the case of 100 hidden units has the least prediction accuracy with RMSE value equal 2.4163 and the improvement is 7.43% in comparison to its peers. Figure 3 show two curves the first curve observes the prediction for IoT throughput with time and the second curve observes the prediction error with time. The prediction time 60 for verifying the ability of the LSTM deep learning in for optimizing the IoT traffic load prediction in the cases of the number of hidden units 50. As shown in Fig. 3 the throughput with time for the YTest and YPred (predicted model) models we notice that the predicted model increase at time 1 then decreases gradually until the time 5 then until time 14 increase then little decrease the time 60 which gives the best prediction accuracy in this case.

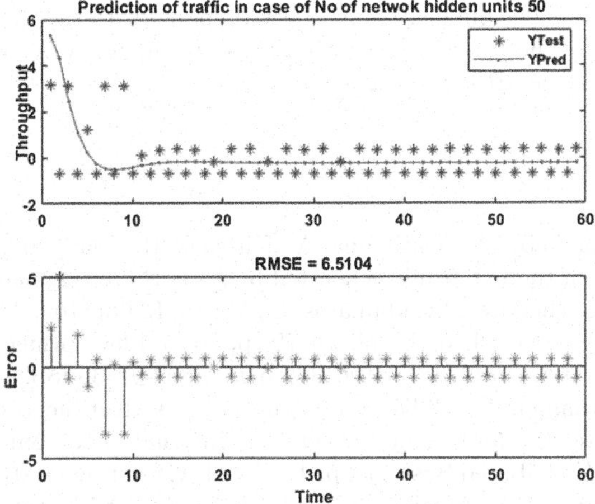

Fig. 4. The response of output element for time series in case of the number of hidden units 50

Figure 4 shows the prediction accuracy of IoT traffic using LSTM approach with deep learning in the case of number of hidden units 50. The model predicted in the case of 50 hidden units has the least prediction accuracy with RMSE value equal 6.5104 in comparison with its peers. Figure 4 show two curves the first curve observes the prediction for IoT throughput with time and the second curve observes the prediction error with time. The prediction time 60 for verifying the ability of the LSTM deep learning in for optimizing the IoT traffic load prediction in the cases of the number of hidden units 50. As shown in Fig. 4 the throughput with time for the YTest and YPred (predicted model) models we notice that the predicted model increase at time 1 then decreases gradually until the time 10 then little increase then become constant until the time 59 which gives the best prediction accuracy in this case.

6 Conclusion and Future Work

This paper has proposed a method for predicting the Internet of Things traffic in 5G/IMT-2020 communication networks with the possibility of its further implementation within the technological areas of IMT-2030 communication networks. In this paper, we discussed artificial intelligence methods to solve some problems in communication networks. In results was implemented the prediction of IoT traffic in time series using deep learning the prediction accuracy has been evaluated using the (RMSE) as a merit function and mean absolute percentage of error (MAPE). Our proposed deep learning algorithm for IoT prediction was developed at edge computing, it has performed IoT traffic prediction methods based on deep learning, the prediction accuracy was evaluated using the RMSE as a merit function and MAPE. Finally was tested the performance in three cases according to the number of hidden units in the network.

Acknowledgments. The publication has been prepared with the support of the grant from the President of the Russian Federation for state support of young russian scientists - doctors of science MD-2454.2020.9.

References

1. Volkov, A., Khakimov, A., Muthanna, A., Kirichek, R., Vladyko, A., Koucheryavy, A.: Interaction of the IoT traffic generated by a smart city segment with SDN core network. In: Koucheryavy, Y., Mamatas, L., Matta, I., Ometov, A., Papadimitriou, P. (eds.) WWIC 2017. LNCS, vol. 10372, pp. 115–126. Springer, Cham (2017). https://doi.org/10.1007/978-3-319-61382-6_10. 0302-9743 eISSN: 1611-3349
2. Ateya, A., Muthanna, A., Gudkova, I., Vybornova, A., Koucheryavy, A.: Intelligent core network for Tactile Internet system. In: International Conference on Future Networks and Distributed Systems, p. 15. ACM, Cambridge (2017)
3. Artem, V., Ateya, A.A., Muthanna, A., Koucheryavy, A.: Novel AI-based scheme for traffic detection and recognition in 5G based networks. In: Galinina, O., Andreev, S., Balandin, S., Koucheryavy, Y. (eds.) NEW2AN/ruSMART 2019. LNCS, vol. 11660, pp. 243–255. Springer, Cham (2019). https://doi.org/10.1007/978-3-030-30859-9_21
4. Muthanna, A., Volkov, A., Khakimov, A., Muhizi, S., Kirichek, R., Koucheryavy, A.: Framework of QoS management for time constraint services with requested network parameters based on SDN/NFV infrastructure. In: International Congress on Ultra Modern Telecommunications and Control Systems and Workshops, vol. 2018-November. IEEE Computer Society (2019). https://doi.org/10.1109/ICUMT.2018.8631274
5. Volkov, A., Proshutinskiy, K., Adam, A.B.M., Ateya, A.A., Muthanna, A., Koucheryavy, A.: SDN load prediction algorithm based on artificial intelligence. In: Vishnevskiy, V.M., Samouylov, K.E., Kozyrev, D.V. (eds.) DCCN 2019. CCIS, vol. 1141, pp. 27–40. Springer, Cham (2019). https://doi.org/10.1007/978-3-030-36625-4_3
6. Rathore, H., Agarwal, S., Sahay, S.K., Sewak, M.: Malware detection using machine learning and deep learning. In: Mondal, A., Gupta, H., Srivastava, J., Reddy, P.K., Somayajulu, D.V.L.N. (eds.) BDA 2018. LNCS, vol. 11297, pp. 402–411. Springer, Cham (2018). https://doi.org/10.1007/978-3-030-04780-1_28
7. Sewak, M., Sahay, S.K., Rathore, H.: An investigation of a deep learning based malware detection system. In: Proceedings of the 13th International Conference on Availability, Reliability and Security, p. 26. ACM (2018)
8. Sahay, S.K., Sewak, M., Rathore, H.: Comparison of deep learning and the classical machine learning algorithm for the malware detection. In: 19th IEEE/ACIS International Conference on Software Engineering, Artificial Intelligence, Networking and Parallel/Distributed Computing (SNPD). IEEE (2018)
9. Roul, R.K., Nanda, A., Patel, V., Sahay, S.K.: Extreme learning machines in the field of text classification. In: IEEE/ACIS 16th International Conference on Software Engineering, Artificial Intelligence, Networking and Parallel/Distributed Computing (SNPD). IEEE (2015)
10. Sharma, A., Sahay, S.K.: An effective approach for classification of advanced malware with high accuracy. Int. J. Secur. Appl. **10**(4), 249–266 (2016)
11. Boutaba, R., Salahuddin, M.A., Liman, N., et al.: J. Internet Serv. Appl. **9**, 16 (2018). https://doi.org/10.1186/s13174-018-0087-2

12. Roul, R.K., Sahay, S.K.: Categorizing text data using deep learning: a novel approach. In: Behera, H.S., Nayak, J., Naik, B., Abraham, A. (eds.) Computational Intelligence in Data Mining. AISC, vol. 711, pp. 793–805. Springer, Singapore (2019). https://doi.org/10.1007/978-981-10-8055-5_70
13. Evans, D.: The Internet of Things: how the next evolution of the internet is changing everything. CISCO White Paper 1, pp. 1–11 (2011)
14. Abdellah, A.R., Muthanna, A., Koucheryavy, A.: Robust estimation of VANET performance-based robust neural networks learning. In: Galinina, O., Andreev, S., Balandin, S., Koucheryavy, Y. (eds.) NEW2AN/ruSMART 2019. LNCS, vol. 11660, pp. 402–414. Springer, Cham (2019). https://doi.org/10.1007/978-3-030-30859-9_34
15. Abdellah, A.R., Muthanna, A., Koucheryavy, A.: Energy estimation for VANET performance based robust neural networks learning. In: Vishnevskiy, V.M., Samouylov, K.E., Kozyrev, D.V. (eds.) DCCN 2019. CCIS, vol. 1141, pp. 127–138. Springer, Cham (2019). https://doi.org/10.1007/978-3-030-36625-4_11
16. Abdellah, A.R., Mahmood, O.A.K., Paramonov, A., Koucheryavy, A.: IoT traffic prediction using multi-step ahead prediction with neural network. In: IEEE 11th International Congress on Ultra-Modern Telecommunications and Control Systems and Workshops (ICUMT) (2019)
17. Ellah, A.R.A., Essai, M.H., Yahya, A.: Robust backpropagation learning algorithm study for feed forward neural networks. Thesis, Al-Azhar University, Faculty of Engineering (2016)
18. Sopin, E., Samouylov, K., Shorgin, S.: The analysis of the computation offloading scheme with two-parameter offloading criterion in fog computing. In: Montella, R., Ciaramella, A., Fortino, G., Guerrieri, A., Liotta, A. (eds.) IDCS 2019. LNCS, vol. 11874, pp. 11–20. Springer, Cham (2019). https://doi.org/10.1007/978-3-030-34914-1_2
19. Daraseliya, A.V., Sopin, E.S., Samuylov, A.K., Shorgin, S.Y.: Comparative analysis of the mechanisms for energy efficiency improving in cloud computing systems. In: Galinina, O., Andreev, S., Balandin, S., Koucheryavy, Y. (eds.) NEW2AN/ruSMART 2018. LNCS, vol. 11118, pp. 268–276. Springer, Cham (2018). https://doi.org/10.1007/978-3-030-01168-0_25

On Optimal Placement of Base Stations in Wireless Broadband Networks to Control a Linear Section with End-to-End Delay Limited

Amir Mukhtarov[1(✉)], Oleg Pershin[2], Andrey Larionov[1], and V. M. Vishnevsky[1]

[1] V.A. Trapeznikov Institute of Control Sciences of RAS, 65 Profsoyuznaya street, 117997 Moscow, Russia
mukhtarov.amir.a@gmail.com, larioandr@gmail.com, vishn@inbox.ru

[2] Gubkin Russian State University of Oil and Gas (National Research University), 65 Leninsky Prospekt, 119991 Moscow, Russia
pershino@mail.ru

Abstract. The paper is dedicated to the design problem of a wireless communication network. The problem of base stations optimal placement along a linear section subject to control with end-to-end delay limited is formulated. The goal is to maximize the coverage area that comes to under the control of multiple deployed base stations while respecting technological conditions and budget constraints. In the paper, the features of the technological statement are analyzed. The formulation of the problem in the form of integer linear programming is proposed. When forming a mathematical model, a queuing network model with a Poisson arrival process is used.

Keywords: Wireless network · Integer linear programming · Poisson flow

1 Introduction

Wireless technologies are widely used in various areas of human life. Wireless broadband communication networks are used for operational control of industrial or civil objects, technological plants, smart moving vehicles. The use of wireless broadband technologies based on the IEEE 802.11 protocols family to organize such networks has several advantages over wired technologies. These include rapid deployment of communication networks, convenient modernization and scalability of the network architecture, and reduced installation and maintenance costs.

To improve the design efficiency of such a modern information transmission infrastructure, it is crucial to solving the problem of optimal placement of equipment, in our case, base stations of a wireless broadband communication network

The publication was supported in part by Russian Foundation for Basic Research (RFBR) according to the research project No. 19-07-00919.

© Springer Nature Switzerland AG 2020
V. M. Vishnevskiy et al. (Eds.): DCCN 2020, CCIS 1337, pp. 30–42, 2020.
https://doi.org/10.1007/978-3-030-66242-4_3

at various possible locations. A similar problem has been proposed and discussed in several works [1,4,8,9,11–13].

This work is a continuation of the researches [7] and [6], where the particular case of the problem is considered when the controlled area is a linear section, for example, the area along highways, the linear part of trunk pipelines, field communications. In the above papers, the formulation was given in the form of an integer linear programming model. The proof of NP – completeness was presented.

The problem considered in the previous work [6] it was necessary to place a given base station set formed at the previous network design stages. The present paper considers a more general case when solving an optimization problem is also determined by a set of placed stations from a given redundant set while respecting technical and economic constraints. This paper presents the preparation of main station characteristics, such as the coverage radius, link distance, and station service time. We need to prepare these characteristics before proceeding to the optimal placement problem. The paper proposes the problem in the form of an integer linear programming with the input of the above-calculated characteristics into the problem conditions with the end-to-end delay constraint. This restriction significantly impacts the mathematical model form of the problem.

2 Problem Statement

The problem is stated as follows. To control a given linear section, we need to place base transceiving stations (from now on referred to as stations) in such a way as to obtain maximum section coverage with restrictions on the total cost of the placed stations and end-to-end delay of signal transmission. It is essential to ensure the availability of communication of any station with gateways at the ends of the section through a system of placed stations.

A set of base stations $S = \{s_j\}$ is given. Each station has characteristics $s_j = \{r_j, \{R_{jq}\}, \mu_j, c_j\}$, $j = 1, \ldots, m; q = 1, \ldots, m; q \neq j$. Here r_j is a coverage radius of a station, R_{jq} is a link distance between stations s_j and s_q, μ_j is a service time rate and c_j is a cost.

Let we have a line segment of length L with the end points a_0 and a_{n+1}. Inside of the segment $[a_0, a_{n+1}]$ a finite set of arranged points $A = \{a_i\}, i = 1, \ldots, n$ is given; these points correspond to the set of vacant places where the stations can be placed. Each point a_i is defined with its one-dimensional coordinate l_i.

There is a special station type s_{m+1} which is gateway. These stations are already placed at the ends a_0 and a_{n+1} of the segment. For those stations $r_{m+1} = 0$. A link distance, a station service time, and a cost are not set.

It is required to place the stations in order to maximize a covered area of the segment by them, provided communication of each station with gateways through the system of placed stations and restrictions on the end-to-end delay T and cost C.

Let calculate characteristics of the station, such as link distance R_{jq}, coverage radius r_j, and service time rate μ_j before proceeding to integer programming.

3 Calculation of Link Distance and Coverage Radius of Stations

It is essential during deployment to provide maximum coverage of a given area and ensure communication between the placed base stations in the wireless broadband network.

Link Budget is a way of estimation of communication link's performance while accounting for the system's power, gains, and losses for both the transmitter and receiver. The complete equation can be written as follows:

$$P_{tr} - L_{tr} + G_{tr} - L_{fs} + G_{recv} - L_{recv} = SOM + P_{recv}, \tag{1}$$

where:

- P_{tr} is a transmitter output power, [dBm];
- L_{tr} is a transmitter losses, [dB];
- G_{tr} is a transmitter antenna gain, [dBi];
- L_{fs} is a free space path loss, [dB];
- G_{recv} is a receiver antenna gain, [dBi];
- L_{recv} is a receiver losses, [dB];
- SOM is a system operating margin, [dB];
- P_{recv} is a receiver sensitivity, [dBm].

The power received at the antenna is calculated by the Friis transmission equation:

$$\frac{P_{recv}}{P_{tr}} = G_{tr} G_{recv} \left(\frac{c}{4\pi R f} \right)^2,$$

where c is a speed of light, f is a frequency, R is a distance between transmit and receive antenna.

The Free Space Path Loss ($FSPL$) equation defines the propagation signal loss between two antennas through free space (air):

$$FSPL = \left(\frac{4\pi R f}{c} \right)^2. \tag{2}$$

The formula (2) expressed in decibels will be calculated as:

$$L_{fs} = 20 \lg F + 20 \lg R + K, \tag{3}$$

where F is a radio wave centre frequency of a communication link, R is a distance between transmit and receive antennas, and K is a constant.

Constant K depends on frequency and distance:

- for a frequency in GHz and a distance in km, constant K is equal to 92.45;
- for a frequency in MHz and distance in km, constant K is equal to 32.4;
- for a frequency in MHz and distance in m, constant K is equal to -27.55.

The loss L_{fs} is expressed from the formula (1) as follows:

$$L_{fs} = P_{tr} - L_{tr} + G_{tr} + G_{recv} - L_{recv} - SOM - P_{recv}. \quad (4)$$

Then the communication link equation is obtained from the formulas (3) and (4):

$$R = 10^{\left(\frac{L_{fs} - 20\lg F - K}{20}\right)}. \quad (5)$$

Using formulas (5) and (4), we can calculate the theoretical maximal communication link distance R_{jq} between base stations and the coverage radius r_j assuming the absence of obstacles, reflections, influence of terrain contours, etc. This is acceptable for our case of an open area.

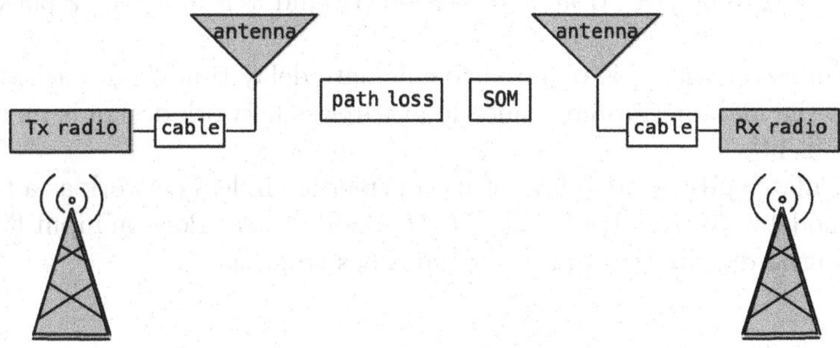

Fig. 1. The link between stations.

To calculate communication link distance R_{jq} (Fig. 1), base stations s_j and s_q will be considered as a *transmitter* and a *receiver* stations, with directional antenna for which gains G_{tr}^R and G_{recv}^R, respectively.

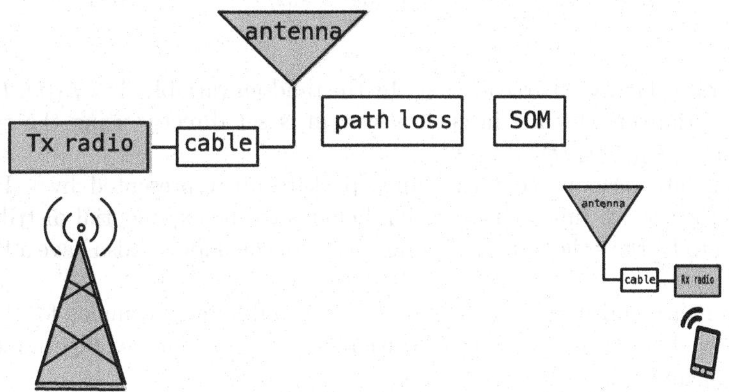

Fig. 2. Base station coverage.

Each base station is equipped with an omnidirectional antenna with given gain antenna G_{tr}^{r}. A station uses this antenna to cover a given area.

Calculating the coverage radius r_j (Fig. 2) base station will be considered a *transmitter* and a user device will be considered a *receiver*.

4 Tandem Queue Model for End-to-End Delay Evaluation

Each station is characterized by its throughput. Given a throughput and an average packet size, we can calculate the service time rate as follow:

$$\mu_j = \frac{p_j}{w},$$

where p_j is a throughput [Mbit/s] of station S_j, and w is an average packet size [Mbit].

This characteristic μ_j is required to calculate delay time T_j at each station.

One of the main performance metric in wireless network design is its end-to-end delay [2,14].

To calculate end-to-end delay, let us consider a wireless network as a tandem queuing model with cross-traffic and $M/M/1$ nodes, with Poisson input flow and an exponential distribution of the service times (Fig. 3).

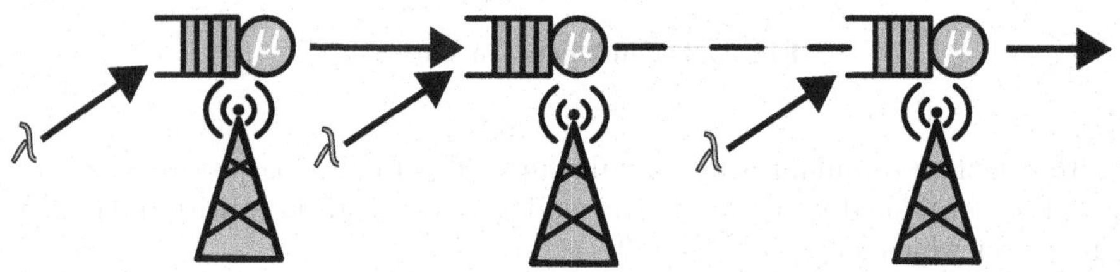

Fig. 3. Tandem queueing model $M/M/1 \to \ldots \to \cdot/M/1$.

The interval between arrivals is set by the random variable $A \sim f_A(t), f_A(t) = \lambda e^{-\lambda t}$ and the service time in such a system is set directly using the random variable $B \sim f_B(t), f_B(t) = \mu e^{-\mu t}$.

Taking into account that the input packets are represented by a Poisson flow and the service time at the station belongs to the exponential distribution, according to Burke's theorem [3], at the exit from the node we also have a Poisson flow.

To estimate end-to-end delays, we can use well-known formulas for $M/M/1$ systems [5]. For station s_j, the utilization is $\rho_j = \frac{\lambda}{\mu_j}$. The average number of packets on such a system is

$$\overline{N_j} = \frac{\rho_j}{1-\rho_j} = \frac{\lambda}{\mu_j - \lambda}.$$

By Little's law [10] the average time delay at each station is

$$\overline{T_j} = \frac{\overline{N_j}}{\lambda} = \frac{1}{\mu_j - \lambda}.$$

Then end-to-end delay is

$$\overline{T} = \sum_j \overline{T_j}, \quad \overline{T_j} = \frac{1}{\mu_j - \lambda}. \tag{6}$$

5 Integer Linear Programming Model

After estimating maximum communication link distances between stations R_{jq}, the maximum coverage radiuses r_j, and the delays at the network station $\overline{T_j}$, it is possible to move to the problem in the form of integer linear programming.

Let y_i^+ and y_i^-, $i = \overline{0, n+1}$ determine the size (the right and left, correspondingly) of stations covering at place a_i. The parameters y_i^+ and y_i^- can take non-negative integer values only.

Values of variables for gateways coverages $y_0^+, y_0^-, y_{n+1}^+, y_{n+1}^-$ are equal to 0.

Objective function will be presented as:

$$f = \sum_{i=1}^{n} (y_i^- + y_i^+) \rightarrow max \tag{7}$$

Let's also introduce binary variables x_{ij}. Then x_{ij} is equal to 1, if station s_j is placed at point a_i and $x_{ij} = 0$ otherwise; $i = \overline{1, n}$; $j = \overline{1, m}$.

Let us introduce binary variables e_i. Then e_i is equal to 1, if any station is placed at point a_i and e_i is equal to 0 otherwise; $i = \overline{1, n}$. For gateways placement points e_0 is equal to 1 and e_{n+1} is equal to 1.

Let us formulate the following system of the problem constraints.

By definition (8):

$$e_i = \sum_{j=1}^{m} x_{ij}, \quad i = \overline{1, n}. \tag{8}$$

Each station must be placed in only one point (9):

$$\sum_{j=1}^{n} x_{ij} \leq 1, \quad j = \overline{1, m}. \tag{9}$$

The values of coverages are no more than the coverage radius of the station placed at a_i, and equal to 0 if there is no station at a_i (10)–(11):

$$y_i^+ \leq \sum_{j=1}^{m} x_{ij} r_j, \quad i = \overline{1, n}; \tag{10}$$

$$y_i^- \leq \sum_{j=1}^{m} x_{ij} r_j, \quad i = \overline{1, n}. \tag{11}$$

The total coverage area between any two points a_i and a_k, where the stations are located cannot exceed the distance between these points (12)–(13).

$$y_i^+ + y_k^- \leq \frac{l_k - l_i}{2}(e_i + e_k) + (2 - e_i - e_k)L, \quad i = \overline{1, n}, \quad k = \overline{i+1, n+1}; \tag{12}$$

$$y_i^- + y_k^+ \leq \frac{l_i - l_k}{2}(e_i + e_k) + (2 - e_i - e_k)L, \quad i = \overline{1, n}, \quad k = \overline{i-1, 0}, \tag{13}$$

where l_k and l_i are the coordinates of the points a_i and a_k, respectively. This condition excludes the effect from intersections of station coverages when calculating the total coverage value for the entire segment.

According to the conditions of the problem, the station located at a_i must be connected with at least one station on the left and one station on the right, including stations at the end points a_0 and a_{n+1}.

We will introduce binary variables $z_{ijkq}, i = \overline{1, n}; j = \overline{1, m}; k = \overline{1, n}, k \neq i; q = \overline{1, m}, q \neq j$.

The variable z_{ijkq} is equal to 1, if there is a station s_j at point a_i and it is connected with a station s_q placed at the point a_k; and z_{ijkq} is equal to 0 otherwise.

The variable $z_{ij0(m+1)}$ is equal to 1, if here is a station s_j at point a_i and it is connected with a gateway s_{m+1} at the point a_0; $z_{ij0(m+1)}$ is equal to 0 otherwise.

The variable $z_{ij(n+1)(m+1)}$ is equal to 1, if here is a station s_j at point a_i and it is connected with a gateway s_{m+1} at the point a_{n+1}; $z_{ij0(m+1)}$ is equal to 0 otherwise.

Stations must be at both points a_i and a_k so that they can be connected (14)–(15):

$$z_{ijkq} \leq e_i, \quad i = \overline{1, n}; \quad j = \overline{1, m}; \quad k = \overline{1, n}, k \neq i; \quad q = \overline{1, m}, q \neq j; \tag{14}$$

$$z_{ijkq} \leq e_k, \quad k = \overline{1, n}; \quad j = \overline{1, m}; \quad i = \overline{1, n}, i \neq k; \quad q = \overline{1, m}, q \neq j. \tag{15}$$

It is necessary that station s_j at point a_i is connected to any one station located at point a_k, to the right of a_i $(k > i)$ or to the right gateway s_{m+1} (16)–(17).

$$\sum_{k=i+1}^{n} \sum_{\substack{q=1 \\ q \neq j}}^{m} z_{ijkq} + z_{ij(n+1)(m+1)} = x_{ij}, \quad i = \overline{1, n}, \quad j = \overline{1, m}. \tag{16}$$

Station s_j placed at a_n has only gateway s_{m+1} from the right at place a_{n+1} (17).

$$z_{nj(n+1)(m+1)} = x_{nj} \quad j = \overline{1, m}. \tag{17}$$

Also, at least, it is connected with any one station located at point a_k to the left of point a_i $(k < i)$ or with the left gateway s_{m+1} (18)–(19).

$$z_{1j0(m+1)} = x_{ij}, \quad j = \overline{1, m}; \tag{18}$$

Station s_j placed at a_1 has only gateway s_{m+1} from the left at place a_0 (18).

$$z_{ij0(m+1)} + \sum_{k=1}^{i-1} \sum_{\substack{q=1 \\ q \neq j}} z_{ijkq} = x_{ij}, \quad i = \overline{2, n}, \quad j = \overline{1, m}. \tag{19}$$

It is necessary that station s_q at point a_k is connected to any one station to the right located at point a_i (20).

$$\sum_{i=k+1}^{n} \sum_{\substack{j=1 \\ j \neq q}}^{m} z_{ijkq} = x_{kq}, \quad k = \overline{1, n-1}, \quad q = \overline{1, m}; \tag{20}$$

Also, station s_q at point a_k is connected to any one station to the left located at point a_i (21).

$$\sum_{i=1}^{k} \sum_{\substack{j=1 \\ j \neq q}}^{m} z_{ijkq} = x_{kq}, \quad k = \overline{2, n}, \quad q = \overline{1, m}; \tag{21}$$

Inequalities (14)–(15) and equalities (16)–(21) provide a condition for symmetry of communication between base stations located at points a_i and a_k for all i, k.

If station s_j and s_q are connected the maximal communication link distance of these placed stations must be no less than the distance between a_i and a_k, where s_i and s_q are located. Formally, this can be stated as (22)–(23).

For $i = \overline{1, n}$:

$$z_{ijkq}(R_{jq} - (a_i - a_k)) \geq 0, \quad k = \overline{0, i-1}; \quad j = \overline{1, m}; \quad q = \overline{1, m}, q \neq j; \tag{22}$$

$$z_{ijkq}(R_{jq} - (a_k - a_i)) \geq 0, \quad k = \overline{i+1, n+1}; \quad j = \overline{1, m}; \quad q = \overline{1, m}, q \neq j. \tag{23}$$

Let T be network end-to-end delay time limit. Using formula (6) to calculate the delay at each station, we write the inequality as:

$$\sum_{i=1}^{n} \sum_{j=1}^{m} x_{ij} \cdot \overline{T_j} \leq T. \tag{24}$$

And for cost limit C we have:

$$\sum_{i=1}^{n} \sum_{j=1}^{m} x_{ij} \cdot c_j \leq C. \tag{25}$$

6 Example

Let's look at one simple case of base stations placement problem.

Consider the section of length $L = 400$ with $n = 10$ placement points is given in Table 1:

Table 1. Placement points at the section of length $L = 400$.

a_i	a_1	a_2	a_3	a_4	a_5	a_6	a_7	a_8	a_9	a_{10}
Coordination	32	65	101	142	181	241	270	301	325	380

There are $m = 7$ base stations with parameters given in Table 2:

- P_{tr}^R is a transmit power for communication with base stations;
- G_{tr}^R is an antenna gain for communication with base stations;
- P_{recv}^R is a sensitivity for communication with base stations;
- P_{tr}^r is a transmit power for the coverage of section;
- G_{tr}^r is an antenna gain for the coverage of section;
- p is a throughput;
- c is a base station cost.

Table 2. Base station parameters.

BS	P_{tr}^R	G_{tr}^R	P_{recv}^R	P_{tr}^r	G_{tr}^r	p	c
No	[dBm]	[dBi]	[dBm]	[dBm]	[dBi]	Mbit/s	c.u
1	19	5	−69	20	2	54	2300
2	19	4	−80	19	3	54	1200
3	19	6	−69	18	2	54	4500
4	19	5	−83	18	3	54	6000
5	20	5	−85	20	2	54	3500
6	22	5	−69	18	2	54	4200
7	19	5	−69	18	2	54	4200

Finally, gateway stations of special type s_{m+1} placed on the ends of the segment are specified. Gateway parameters is given in Table 3:

Table 3. Gateway parameters.

Gateway	G_{tr}^R	P_{recv}^R
No	[dBi]	[dBm]
s_{m+1}	3	−69

6.1 Computation of the Communication Link Distance Between Base Stations

Base station is equipped with a directional antenna with a high gain to communicate with neighbouring stations. To calculate the losses between stations j and q, we use the formula (4):

$$L_{fs}^{jq} = P_{tr}^R(j) - L_{tr} + G_{tr}^R(j) + G_{tr}^R(q) - L_{recv} - SOM - P_{recv}^R(q).$$

The cable losses at the receiver L_{recv} and transmitter L_{tr} are equal to 1 dB. We will also provide system operating margin $SOM = 10\,\mathrm{dB}$.

Let us carry out an example of the calculation communication link between stations s_1 and s_2:

$$\begin{aligned} L_{fs}^{12} &= P_{tr}^R(1) - L_{tr} + G_{tr}^R(1) + G_{tr}^R(2) - L_{recv} - SOM - P_{recv}^R(2) \\ &= 19 - 1 + 5 + 4 - 1 - 10 - (-80) = 96(dB). \end{aligned} \tag{26}$$

To calculate the communication link, formula (5) must be used. The stations operate on 6th channel, carrier frequency $f = 2437\,\mathrm{MHz}$ and coefficient $K = -27.55$:

$$R_{jq} = 10^{\left(\frac{L_{fs}^{jq} - 20\lg F - K}{20}\right)} = 10^{\left(\frac{96 - 20\lg 2437 - (-27.55)}{20}\right)} = 617(m). \tag{27}$$

Table 4 summarizes the maximal communication link distances calculations between all stations s_j, $j = 1, \ldots, m$, and the gateway s_{m+1}.

6.2 Computation of the Coverage Radius

To cover a given section, the base station is equipped with an isotropic antenna with output power P_{tr}^r and gain G_{tr}^r is equal to 0. The cable loss L_{tr} is equal to 1.

A coverage area depends on a base station, as well as user device characteristics. Let us consider a user device with an antenna sensitivity $P_{RX} = -67\,\mathrm{dBm}$ and gain $G_{RX} = 0$. Loss L_{RX} is equal to 0.

Free space path loss between the j-th station and the user device

$$L_{fs}^{j} = P_{tr}^r(j) - L_{tr} - SOM - P_{RX}.$$

To calculate the coverage radius, must be used the formula (5). The stations operate on 6th channel, carrier frequency $f = 2437\,\mathrm{MHz}$ and coefficient $K = -27.55$

$$r_j = 10^{\left(\frac{L_{fs}^{j} - 20\lg F - K}{20}\right)}.$$

Table 4. The calculation of communication link distance between stations.

$R_{jq}, (m)$	s_1	s_2	s_3	s_4	s_5	s_6	s_7	s_{m+1}
s_1	–	617	219	978	1 232	195	195	123
s_2	174	–	195	872	1 098	174	174	109
s_3	219	692	–	1098	1 382	219	219	138
s_4	195	617	219	–	1 232	195	195	123
s_5	219	692	245	1 098	–	219	219	138
s_6	275	872	309	1 382	1 740	–	275	174
s_7	195	617	219	978	1 232	195	–	123

An example of calculating the coverage radius for the 1-st station:

$$r_1 = 10^{\left(\frac{20-1+2-10-(-67)-20\lg 2437-(-27.55)}{20}\right)} = 77(m)$$

Let's calculate the coverage radius for all stations s_j, $j = 1, \ldots, m$ (Table 5).

Table 5. Calculation of the coverage radius of stations.

STA	s_1	s_2	s_3	s_4	s_5	s_6	s_7
r_j	77	77	61	69	77	61	61

6.3 Time Delay Calculation

Let's calculate the delay for station s_1. The specified throughput is $p_1 =$ 54 Mbit/s. Let's assume that the average package size is $w =$ 2700 KByte (21.6 MBit). The arrival package rate is $\lambda = 0.5(s^{-1})$. Then the service rate according to the formula (4) will be

$$\mu_1 = \frac{54}{21.6} = 2.5(s^{-1}).$$

The utilization is equal to

$$\rho_1 = \frac{0.5}{2.5} = 0.2.$$

The average package size is

$$\overline{N}_1 = \frac{0.2}{1-0.2} = 0.25.$$

The average delay is

$$\overline{T_1} = \frac{0.25}{0.5} = 0.5(s).$$

Communication links between stations R_{jq}, the coverage radius of the station is r_j, the delays $\overline{T_j}$ are calculated, it is possible to search the optimal placement.

The problem formulated on the basis of (7)–(25) and given constraints on the cost $C = 18000$ and end-to-end delay $T = 3$ was solved by MATLAB Optimization Toolbox.

The optimal placement is presented in the Table 6.

Table 6. Solution result.

Placed station	s_6	s_7	–	–	s_2	–	s_5	–	s_1	–
Placement coordination	a_1	a_2	a_3	a_4	a_5	a_6	a_7	a_8	a_9	a_{10}

Obtained total coverage f is equal to 400 (m) with total cost c is equal to 15400 (c.u.), and end-to-end delay T is equal to 2.5 (s).

7 Conclusion

The paper considers the problem of finding an optimal placement of the given redundant set of base stations of wireless broadband communication network on a set of possible placement points to maximize the coverage area while respecting technological conditions and budget constraints.

To calculate a limit on the network delay time a network is considered as a tandem queue model with $M/M/1$ nodes.

The problem is formulated in the form of the integer linear programming model. Numerical example solution was presented.

It is planned to use the obtained model in practice in future work.

References

1. Ben Brahim, M., Drira, W., Filali, F.: Roadside units placement within city-scaled area in vehicular ad-hoc networks. In: Proceedings of the 2014 International Conference on Connected Vehicles and Expo, ICCVE 2014, pp. 1010–1016 (2014). https://doi.org/10.1109/ICCVE.2014.7297500
2. Bendel, D., Haviv, M.: Cooperation and sharing costs in a tandem queueing network. Eur. J. Oper. Res. **271**(3), 926–933 (2018). https://doi.org/10.1016/j.ejor.2018.04.049
3. Burke, P.J.: The output of a queuing system. Oper. Res. **4**(6), 699–704 (1956). https://doi.org/10.1287/opre.4.6.699
4. Chattopadhyay, A., Błaszczyszyn, B., Keeler, H.P.: Gibbsian on-line distributed content caching strategy for cellular networks. IEEE Trans. Wirel. Commun. **17**(2), 969–981 (2018). https://doi.org/10.1109/TWC.2017.2772911
5. Heyman, D.P.: Queueing Systems, Volume 1: Theory by Leonard Kleinrock. Wiley, New York (1975). $19.95, 417 p. Networks **6**(2), 189–190 (1976). https://doi.org/10.1002/net.3230060210

6. Ivanov, R., Mukhtarov, A., Pershin, O.: A problem of optimal location of given set of base stations in wireless networks with linear topology. In: Vishnevskiy, V.M., Samouylov, K.E., Kozyrev, D.V. (eds.) DCCN 2019. CCIS, vol. 1141, pp. 53–64. Springer, Cham (2019). https://doi.org/10.1007/978-3-030-36625-4_5
7. Ivanov, R., Pershin, O., Larionov, A., Vishnevsky, V.: On a problem of base stations optimal placement in wireless networks with linear topology. In: Vishnevskiy, V.M., Kozyrev, D.V. (eds.) DCCN 2018. CCIS, vol. 919, pp. 505–513. Springer, Cham (2018). https://doi.org/10.1007/978-3-319-99447-5_43
8. Khireddine, A., Amine, O.M.: Base station placement optimization using genetic algorithms approach. Int. J. Comput. Aided Eng. Technol. **12**(1), 1 (2020). https://doi.org/10.1504/ijcaet.2020.10006440
9. Kİzİloz, H.E.: On base station localization in wireless sensor networks. Balkan J. Electr. Comput. Eng. **8**, 57–61 (2020). https://doi.org/10.17694/bajece.613154
10. Little, J.D.C.: A proof for the queuing formula: L = λ W. Oper. Res. **9**(3), 383–387 (1961). https://doi.org/10.1287/opre.9.3.383
11. Liu, H.Q., Ding, S.J., Yang, L.C., Yang, T.: A connectivity-based strategy for roadside units placement in vehicular ad hoc networks. Int. J. Hybrid Inf. Technol. **7**(1), 91–108 (2014). https://doi.org/10.14257/ijhit.2014.7.1.08
12. Reis, A.B., Sargento, S., Neves, F., Tonguz, O.K.: Deploying roadside units in sparse vehicular networks: what really works and what does not. IEEE Trans. Veh. Technol. **63**(6), 2794–2806 (2014). https://doi.org/10.1109/TVT.2013.2292519
13. Shen, C., Yun, M., Arora, A., Choi, H.-A.: Efficient mobile base station placement for first responders in public safety networks. In: Arai, K., Bhatia, R. (eds.) FICC 2019. LNNS, vol. 70, pp. 634–644. Springer, Cham (2020). https://doi.org/10.1007/978-3-030-12385-7_46
14. Wu, K., Shen, Y., Zhao, N.: Analysis of tandem queues with finite buffer capacity. IISE Trans. **49**(11), 1001–1013 (2017). https://doi.org/10.1080/24725854.2017.1342055

CPU vs GPU Performance of MATLAB Clustering Algorithms

Andrey Ivanov[1(✉)], Ziazina Natalia[1,2], and Antonova Veronika[1,3]

[1] BMSTU, ul. Baumanskaya 2-ya, 5/1, Moscow, Russia
iam18u032@student.bmstu.ru, nataliacs@yandex.ru, ant_veronika@bmstu.ru
[2] ISP RAS, Alexander Solzhenitsyn St., 25, Moscow, Russia
[3] IRE RAS, Mokhovaya 11-7, Moscow, Russia

Abstract. Clustering is one of machine learning's tasks when given objects must be split into specific groups based on distance between them. Its applications include different fields such as pattern matching, data compression and image analysis. Many programing languages allow to create clustering algorithms, though using already implemented ones is much easier. MATLAB includes a few of them. Knowing the performance of MATLAB's cluster analysis algorithms may help choose the more optimal hardware for a given problem.

Keywords: MATLAB · Performance · Clustering · Machine learning · Inetl · AMD · Nvidia · GPU · CPU · K-means · DBSCAN · Hierarchical Clustering

1 Clustering Analysis

In mathematical notation, clustering problem is such: given is a set X: x_1, x_2, x_3, x_4, ... x_m, their labels Y: $y(x_1)$, $y(x_2)$, $y(x_3)$, $y(x_4)$, ..., $y(x_m)$. On the set X a metric $\rho(x, x')$ is given. It is necessary to group the sample into subsets (clusters), assign the label $y_i \in Y$ to each object $x_i \in X$, so that the objects inside each cluster are close relative to the metric ρ, and objects from different clusters are significantly farther. The clustering algorithm is a function F: X → Y, which associates the cluster identifier $y \in Y$ with any object $x \in X$. It is postulated that:

1. The clustering algorithm a is scale invariant.
2. The set of clustering results of algorithm a, depending on the change in the distance function ρ, must coincide with the set of all possible partitions of the set of objects X.
3. The clustering algorithm is consistent.

Clustering results vary between different algorithms. Also, some may require a predefined number of clusters, while others don't.

The work is partially supported by the Russian Foundation for Basic Research (project No. 19-07-00525 A – Developing flow-based models of routing problems in telecommunications networks).

© Springer Nature Switzerland AG 2020
V. M. Vishnevskiy et al. (Eds.): DCCN 2020, CCIS 1337, pp. 43–56, 2020.
https://doi.org/10.1007/978-3-030-66242-4_4

2 Clustering Algorithms in MATLAB

An overview of MATLAB's clustering algorithms (built-in):

1. Hierarchical Clustering – does not require number clusters, cannot detect anomalies. It creates a dendrogram, consisting of multiple levels of clusters. Resembles a tree of clusters [3,4].
2. k-Means – input should specify the number of groups. The shape of the cluster is spheroidal. Not useful for outlier detection. It is assumed that every object belongs to one of k classes, which are defined by a central vector. Classes are formed so that the square of distance from an object to the centroid is minimal [1–3].
3. Density-Based Spatial Clustering of Algorithms with Noise (DBSCAN) – does not need the number of clusters, can detect oddity in data. It takes the density of objects into account. The shape of clusters is arbitrary [3,6].
4. Gaussian Mixture Models (GMM) – the number of clusters is required. Can detect anomalies, since the algorithm works with distribution. The shape of clusters is clusters [3].
5. Nearest Neighbors – can work without specified number of groups. As they name states, the algorithm is distance-based, and because of that the shape of clusters is arbitrary [3,5].
6. Spectral clustering – can estimate the number of clusters, although works better if the value is predefined. The shape is arbitrary [3].

Because of the clustering analysis' nature, there is no right or wrong algorithm. It all depends on the data that should be split up. The performance of the algorithms is also dependent on the data. Some methods scale better with more cores and threads, some don't. Different algorithms require a different amount of calculations and some need specialized operations such as exponents, which means that performance is also dependent on supported hardware operations.

3 Benchmark Details

MATLAB has built-in tools to measure time. It is either *tic/toc* start timer/end timer or *timeit* function. This research will use *tic/toc* [11]. Time intervals of 500 clustering function calls will be recorded, their average and sample standard deviation will be calculated. In order to minimize distortion, the MATLAB process' priority will be set to Realtime before the benchmark is run. The following hardware will be used during tests:

1. AMD Ryzen 3900X, 12 cores @ 3.8 GHz base on x570 chipset and 32 GB of DDR4 3200 MHz RAM.
2. The processor above with Nvidia RTX 2080ti, 4352 CUDA cores & 11 GB of VRAM.
3. Intel core i7-3770k, 4 cores @ 3.5 GHz base, overclocked to ... with 16 GB of DDR3 ... MHz RAM.

4. The processor above with Nvidia GTX 660, 960 CUDA cores & 2 GB of VRAM.
5. Intel core i7-8750h, 6 cores @ 2.2 GHz base and 16 GB of DDR4 ... MHz RAM.
6. The processor above with Nvidia GTX 1070 (mobile), 2048 CUDA cores and 8 GB of VRAM.
7. AMD Ryzen 3700U, 4 cores @ 2.3 GHz base and 6 (effectively) GB of RAM.

For GPUs to be utilized, the Parallel Toolbox in MATLAB is installed. The dataset is NIPS Conference Papers 1987–2015 from UCI Machine Learning Repository [7]. Papers are referenced by "Vision", "Neural" and "Learning" values. All clustering algorithms use squared Euclidean distance as a metric:

$$|a-b|_2^2 = \sum_i (a_i - b_i)^2 \tag{1}$$

4 CPU Benchmark

The benchmark reads the dataset (transposed and optimized, available at [12]). k-Means, DBSCAN, hierarchy and nearest-neighbors clustering algorithms' implementations are run 500 times, with each iteration starting with setting timer with tic function and ending with reading time spent on iteration with toc function call. The data is then collected into an array (for each algorithm). The mean and sample standard deviation is calculated (Fig. 1 and Table 1):

$$\bar{t} = \frac{1}{N}\sum_{i=1}^{N} t_i \tag{2}$$

$$s = \sqrt{\frac{1}{N-1}\sum_{i=1}^{N}\left(t_i - \bar{t}\right)^2} \tag{3}$$

Since, probably, clustering will be performed once, first run results have been recorded as well (Fig. 2).

MATLAB can plot the results of clustering, so we can view the results in a more comprehensible format (Fig. 3).

Results of hierarchy clustering is usually represented by a dendrogram. The higher the level of a class on it, the broader it is (Figs. 4, 5 and 6).

Table 1. CPU-only benchmark

Algorithm	3900x			3700u			8750h			3770k		
	Average	Deviation	First run	Average	Deviation	First run	Average	Deviation	First run	Average	Deviation	First run
k_means	0.1485647	0.0294064	0.1733067	0.2369192	0.3332865	7.6030114	0.1607835	0.0688289	0.1874	0.197224	0.10638	0.302635
Hierarchy	0.6195843	0.1145432	0.7813761	1.005445	0.351111	3.366557	0.6148698	0.0583684	0.9526	0.986554	0.069673	1.412833
Neighbors	0.0240679	0.1134024	0.1685378	0.0428162	0.0239203	0.5702964	0.0268133	0.0119936	0.1322	0.031593	0.01869	0.23805
my_dbscan	0.2443005	0.0141887	0.3879294	0.1941701	0.0228209	0.5046445	0.1049712	0.0118161	0.1969	0.135808	0.015145	0.455906

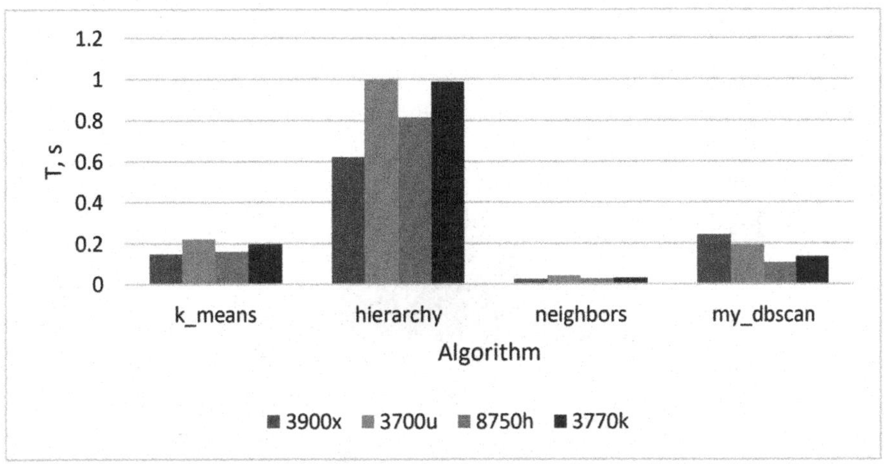

Fig. 1. CPU-only benchmark (average), lower – better

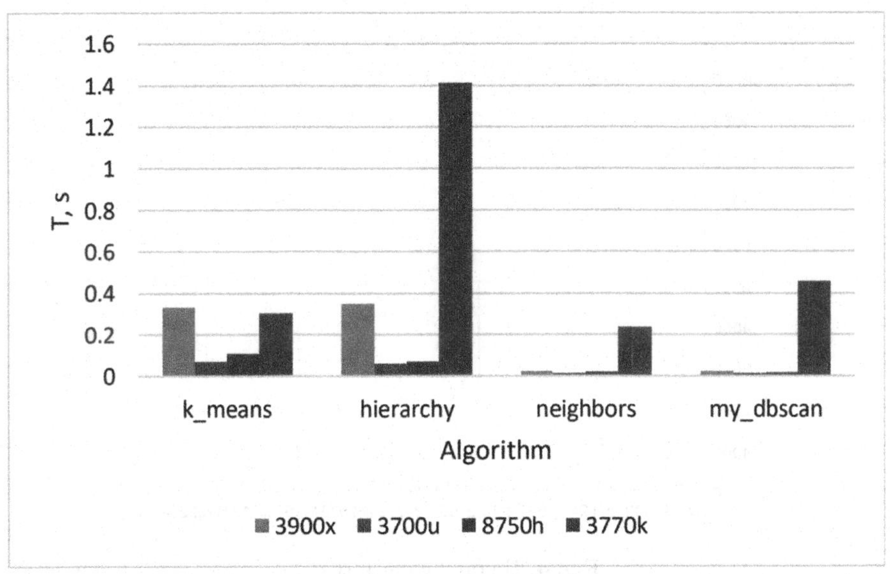

Fig. 2. CPU-only benchmark (first run), lower – better

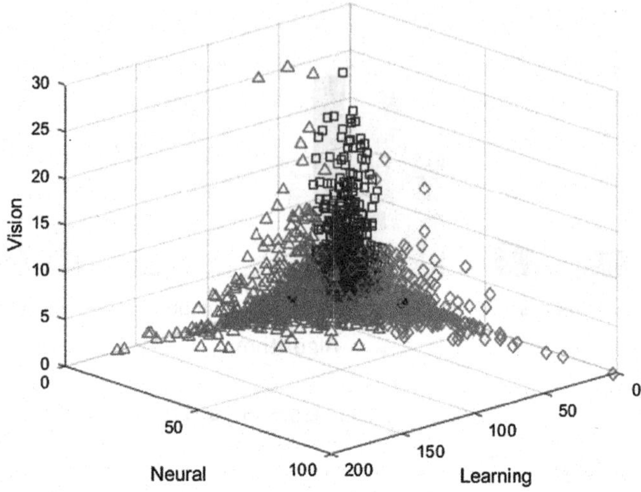

Fig. 3. k-means clustering

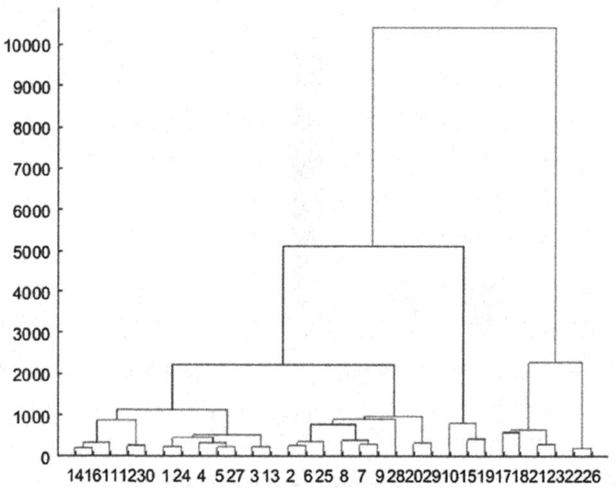

Fig. 4. Hierarchy clustering

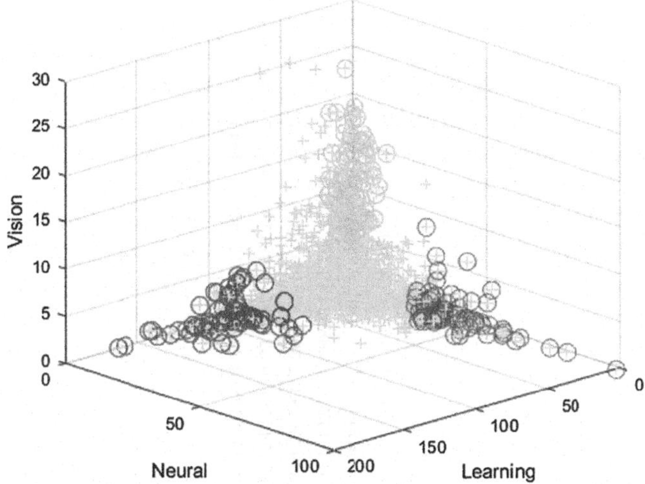

Fig. 5. Nearest-neighbor clustering

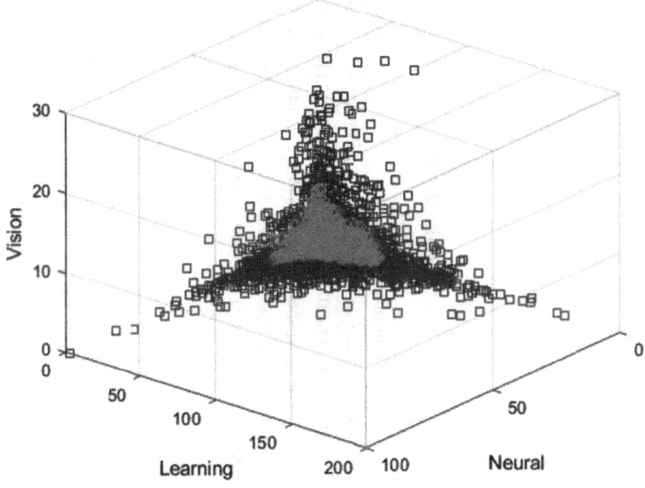

Fig. 6. DBSCAN clustering

5 GPU Benchmark

This benchmark uses the same dataset, as the one above. It only differs from the previous by utilizing GPUs. Parallel computing option is specified for k-means, so that the GPU is used for computing [8]. Also, 'IncludeTies', 'NSMethod', and 'SortIndices' name-value pair arguments are not used for knnsearch function [9], 'squaredeuclidean' Distance argument is supplied to pdist function for the same reason [10] (Fig. 7 and Table 2).

Table 2. GPU benchmark

Algorithm	2080ti			1070 mobile			660			3770k		
	Average	Deviation	First run	Average	Deviation	First run	Average	Deviation	First run	Average	Deviation	First run
k_means	0.101600341	0.022818093	0.3723852	0.193905003	0.559473012	12.6737349	0.263266735	1.013714704	22.88024	0.197224	0.10638	0.302635
Hierarchy	0.304797988	0.012092614	0.471679	0.539515566	0.044483149	1.4714957	0.598769451	0.089768183	2.596127	0.986554	0.069673	1.412833
Neighbors	0.029876952	0.010741517	0.2659459	0.038634962	0.014030987	0.3483474	0.043220559	0.021544788	0.521507	0.031593	0.01869	0.23805
my_dbscan	0.090224443	0.008407704	0.2682983	0.137256744	0.008500473	0.3158782	0.165918617	0.012611378	0.34114	0.135808	0.015145	0.455906

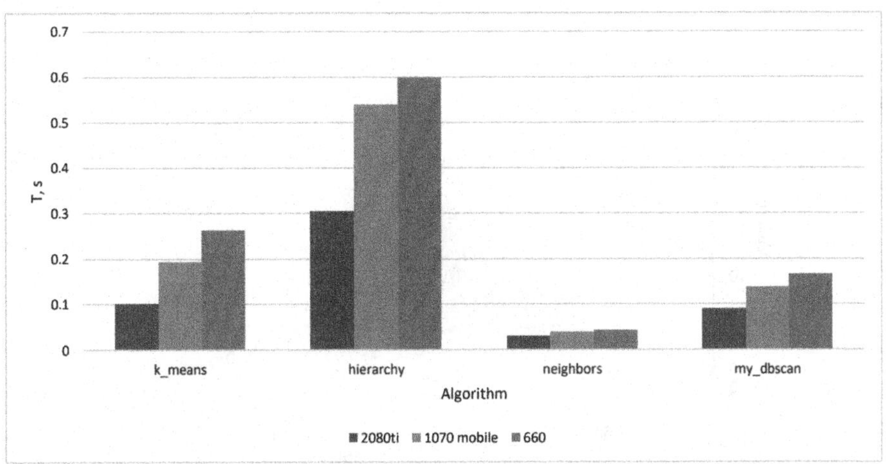

Fig. 7. GPU performance (average), lower – better

GPUs perform much better than CPUs in such tasks because of their architecture: they can perform the same step on different data in one tick (known as single instruction, multiple data, SIMD) (Fig. 8).

Nevertheless, data that is to be analyzed, must get from RAM to VRAM and that takes some time. That's why first run results are rather poor (especially noticeable for k-means algorithm) (Fig. 9).

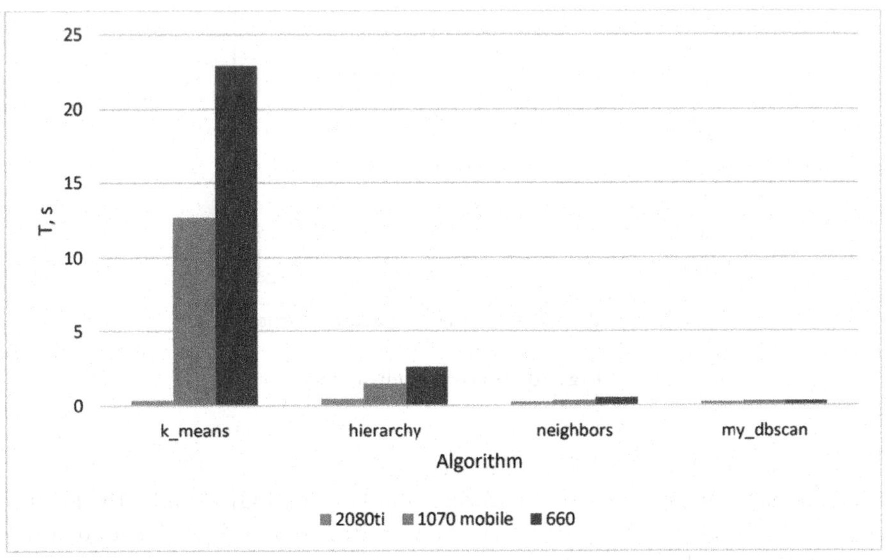

Fig. 8. GPU benchmark (first run), lower – better

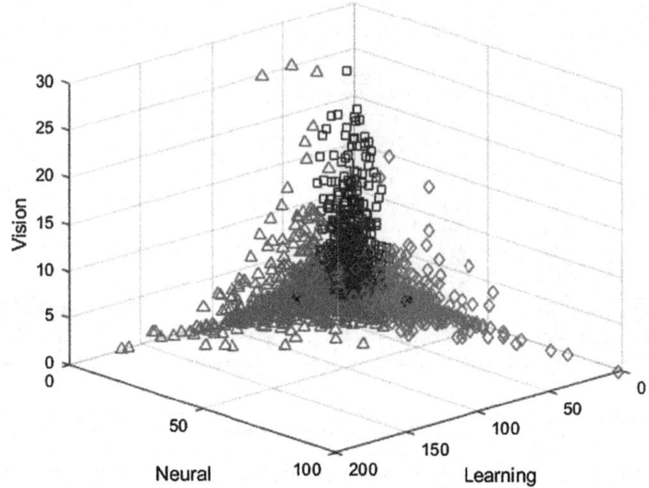

Fig. 9. k-means clustering

Results of k-means clustering performed on GPUs does not differ from results acquired on CPUs, which means that it is consistent (Fig. 10).

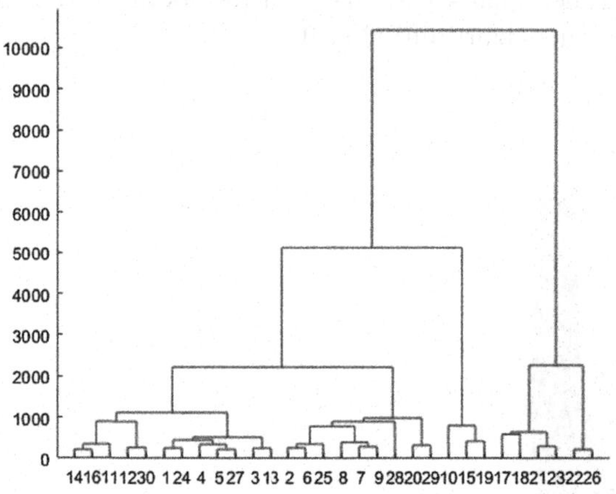

Fig. 10. Hierarchy clustering

Same for hierarchy clustering (Fig. 11).

Dissimilarities have been found during comparison of GPU and CPU results (only on the Nvidia RTX 2080ti). This may be due to mixed-precision architecture of GPU calculations (Fig. 12).

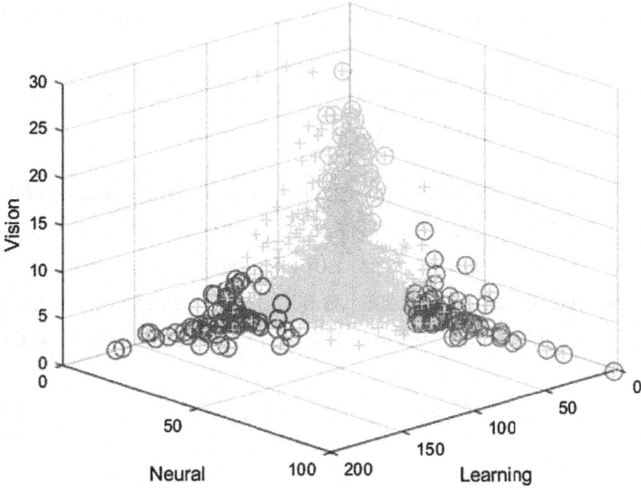

Fig. 11. Nearest-neighbor clustering

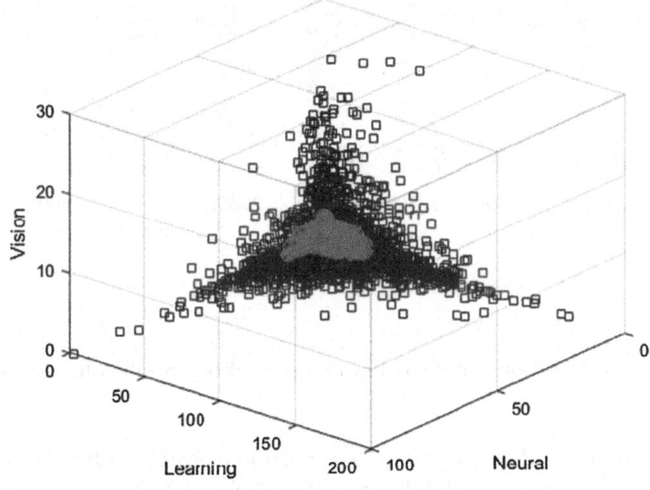

Fig. 12. DBSCAN clustering

6 Performance per Ruble

Performance in this section is defined as number of clustering function invocations per second, as in the formula:

$$P = \frac{1}{t}, \tag{4}$$

where P is performance (the higher the better), t is time, in seconds (Fig. 13 and Table 3).

Table 3. Average performance

Algorithm	3900x	3700u	8750h	3770k	2080ti	1070 mobile	660
k_means	6.731	4.221	6.2195	5.07	9.8425	5.157164522	3.798429
Hierarchy	1.614	0.995	1.2272	1.014	3.2809	1.85351464	1.670092
Neighbors	41.55	23.36	37.295	31.65	33.471	25.88329193	23.13714
my_dbscan	4.093	5.15	9.5264	7.363	11.083	7.28561653	6.027051

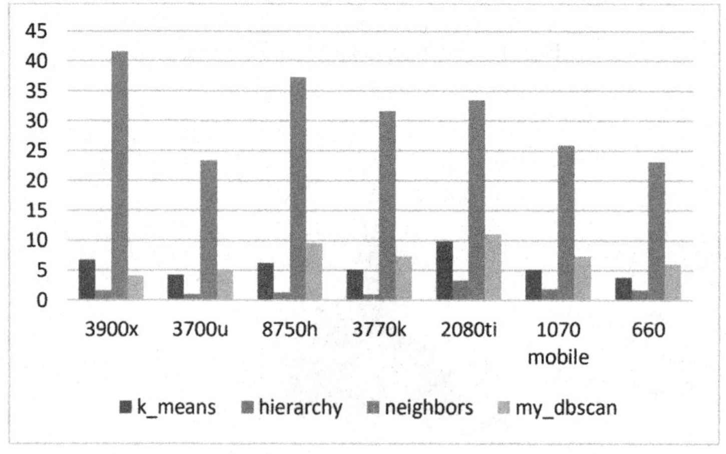

Fig. 13. Iterations of algorithm in a second on specified hardware

Performance per ruble, therefore, is performance divided by the price (CPU + RAM/ CPU + RAM + GPU/Laptop [i7-8750h & GTX 1070, AMD 3700U]) (Fig. 14, Tables 4 and 5):

Table 4. Price of hardware

Price	3900x	3700u	8750h	3770k	2080ti	1070 mobile	660
Rubles	51640	41400	117000	25027	143540	117000	39046.84

Table 5. Performance/ruble

Algorithm	3900x	3700u	8750h	3770k	2080ti	1070 mobile	660
k_means	0.000130346	0.000101953	5.31585E-05	0.000202596	6.85696E-05	4.40783E-05	9.72788E-05
Hierarchy	3.12546E-05	2.40238E-05	1.04888E-05	4.05014E-05	2.28568E-05	1.5842E-05	4.27715E-05
Neighbors	0.000804591	0.000564146	0.00031876	0.001264746	0.00023318	0.000221225	0.000592548
my_dbscan	7.92665E-05	0.000124399	8.14224E-05	0.000294217	7.72152E-05	6.22702E-05	0.000154354

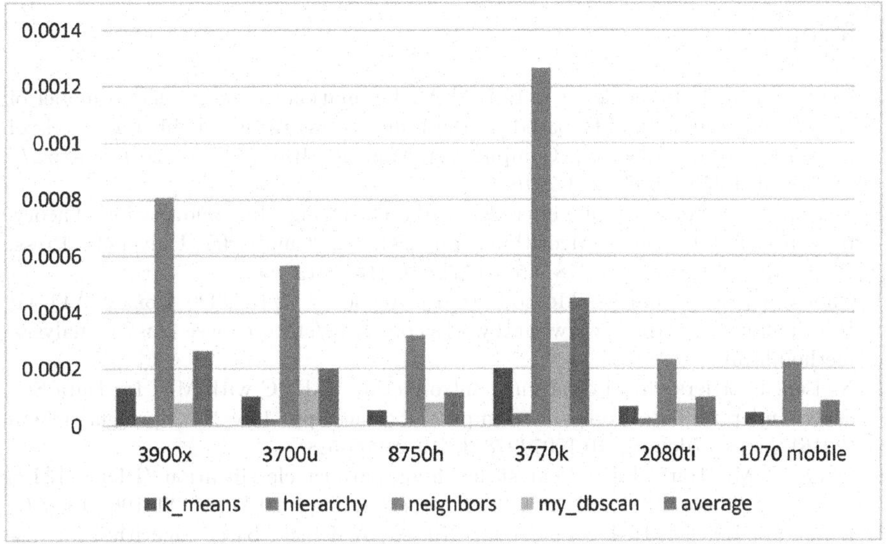

Fig. 14. Average performance/ruble of hardware

7 Conclusion

GPUs are usually considered much better in mathematical calculations because of their core count – they can perform way many more operations than most CPUs. Nevertheless, there is always a time lag for data synchronization – it must get from cache/RAM into VRAM and GPU's cache in order to be processed. Even though modern expansion card buses (PCIe 3.0 $\times$ 16) offer speeds of up to 16 GB/s, the transfer still takes some time, that is why first run on GPU takes more time than on the CPU. The most powerful piece of hardware, the Nvidia RTX 2080ti, outperformed almost all other pieces of equipment, as expected, although the AMD Ryzen 9 3900X can run 7 more iterations of the nearest neighbor clustering method that the graphics card. The best performance per ruble is delivered by Intel core i7-3770k, which is still a quite capable processor, despite its age. Also, being unlocked, it may be overclocked in order to improve its capabilities. However, the possibility and the extent of that depends on the quality of the die, in some cases higher potential may be achieved. CPU-only test has shown that the algorithms are more subject to single core performance, rather than core count. The parallel toolbox can unlock the processor's and the graphics card' potential by employing more threads on more cores. The CUDA hardware version also determines the performance of the GPU – fused multiply and add as well as mixed precision calculations give additional boost to its efficiency. The amount of RAM and VRAM can limit the effectiveness of an algorithm. It is usually stated that about 4 times more memory is required in order to perform a machine learning task that the size of the dataset.

References

1. Babichev, S., Lytvynenko, V., Taif, M.A.: Estimation of the inductive model of objects clustering stability based on the k-means algorithm for different levels of data noise. Radio Electron. Comput. Sci. Manage. **4**(39), 54–60 (2016). https://doi.org/10.15588/1607-3274-2016-4-7
2. MacKay, D.: An example inference task: clustering. In: Information Theory, Inference and Learning Algorithms, pp. 284–292. Cambridge University Press, Cambridge (2003). ISBN 978-0-521-64298-9. MR 2012999
3. Choose Cluster Analysis Method. (n.d.) Retrieved 5/13/2020 from MATLAB & Simulink. https://www.mathworks.com/help/stats/choose-cluster-analysis-method.html
4. Nielsen, F.: Hierarchical clustering. Introduction to HPC with MPI for Data Science. Undergraduate Topics in Computer Science, pp. 195–211. Springer, Cham (2016). https://doi.org/10.1007/978-3-319-21903-5_8
5. Cover, T.M., Hart, P.E.: Nearest neighbor pattern classification (PDF). IEEE Trans. Inf. Theor. **13**(1), 21–27 (1967). https://doi.org/10.1109/TIT.1967.1053964
6. Ester, M., Kriegel, H.-P., Sander, J., Xu, X.: A density-based algorithm for discovering clusters in large spatial databases with noise. In: Simoudis, E., Han, J., Fayyad, U.M. (eds.) Proceedings of the Second International Conference on Knowledge Discovery and Data Mining (KDD-96). AAAI Press, pp. 226–231 (1996). ArXiv:10.1.1.121.9220. ISBN 1-57735-004-9
7. Perrone, V., Jenkins, P.A., Spano, D., Teh, Y.W.: Poisson Random Fields for Dynamic Feature Models (2016). arXiv:1611.07460
8. k-means clustering - MATLAB kmeans (n.d.) Retrieved 5/13/2020 from MathWorks Help Center (2020). https://www.mathworks.com/help/stats/kmeans.html
9. Find k-nearest neighbors using input data - MATLAB knnsearch (n.d.) Retrieved 5/13/2020 from MathWorks Help Center (2020). https://www.mathworks.com/help/stats/knnsearch.html
10. Pairwise distance between pairs of observations - MATLAB pdist (n.d.) Retrieved 5/13/2020 from MathWorks Help Center (2020). https://www.mathworks.com/help/stats/pdist.html
11. Measure the Performance of Your Code - MATLAB & Simulink (n.d.) Retrieved 5/13/2020 from MathWorks Help Center (2020). https://www.mathworks.com/help/matlab/matlab_prog/measure-performance-of-your-program.html
12. https://github.com/berkut126/MatlabPerformance/blob/master/NIPS.csv (2020)

High-Capacity Photon Switching Systems Based on the Two-Stage 256 × 256 Switch

E. A. Barabanova[1,2(✉)], K. A. Vytovtov[1,2], V. S. Podlazov[1], and V. M. Vishnevsky[1]

[1] V.A. Trapeznikov Institute of Control Sciences RAS, Profsoyuznaya 65 str., Moscow, Russia

[2] Astrakhan State Technical University, Tatischeva 16 str., Astrakhan, Russia

elizavetaalexb@yandex.ru

Abstract. The new type of high-capacity photon switching systems based on the two-stage 256 × 256 switch are proposed in this work for the first time. The main advantage of this type of photon systems is the constant period between optical signals that required for non-blocking property of the switch. This period is equal to four bits for the systems with any number of inputs. The method of invariant scaling of photon switching systems based on the two-stage 256 × 256 switch is offered for the first time also. It provides non-blocking of the switching scheme and the constant period between optical signals. Additionally the method of circuit and fiber complexities calculation of the proposed high-capacity photon switching systems is developed in this work for the first time. The numerical calculations of the circuit and fiber complexities and comparison with the crossbar and tree-type switching systems are carried out. The results show the gain in circuit complexity of the presented schemes in comparison with the strictly non-blocking crossbar scheme and the gain in fiber complexity in comparison with the tree-type one.

Keywords: Photon switch · Two-stage switch · Multiplexer · Demultiplexer · Complexity

1 Introduction

Today, stream data processing technology is widely used in electronics, opto-electronic and all-optical "Big data" processing systems [1,2]. Parallel data processing switches with the large number of inputs should be used in such systems for high performance interconnections [3]. Electronic and optoelectronic switching systems with the large number of inputs has been widely described in literature [4,5]. As for all-optical switching systems, there are only small capacity switches based on well-known crossbar, Banyan or schemes [6,7] as rule. Recently, the new type of all-optical high-capacity strictly non-blocking switching systems has been presented by the authors in [8–10]. The main element of those photon switching systems is the 4 × 4 photon switch [10]. Next, the 16 × 16 switching

The reported study was funded by RFBR, project number 19-29-06043.

© Springer Nature Switzerland AG 2020

V. M. Vishnevskiy et al. (Eds.): DCCN 2020, CCIS 1337, pp. 57–69, 2020.
https://doi.org/10.1007/978-3-030-66242-4_5

system is based on these 4×4 photon switches has been presented [8]. One of the advantages of those systems is the low complexity which is approximately equal to one of blocking schemes [7]. However, their throughput is linearly decreased with increasing in the number of inputs. Indeed in this case the period T that necessary for strictly non-blocking functioning must be increased. For example, this period is equal to four bits ($T = 4$) for the 4×4 switching system, it is equal to sixteen ($T = 16$) for the 16×16 switching system, and $T = 49$ for the non-blocking functioning 256×256 one. Thus we can say that the throughput of the 256×256 switching system is 12 times less than the throughput of the 4×4 switching system and 3 times less than the throughput of the 16×16 one. This paper is devoted to developing the design principles of non-blocking high-capacity switches. First of all we propose the new type of the 256×256 base element (Sect. 3) allowing us to design the strictly non-blocking high-capacity photon switching systems. Actually, it is the complex multi-stage scheme contains the new quasi-complete photon 16×16 switches with constant period $T = 4$ (Sect. 2). Such the approach allows us to design, describe, and calculate arbitrary high-capacity switching systems with constant period $T = 4$ from the same point of view. It also is offered the method of the switch scaling (Sect. 4), and the new complexity calculation method (Sect. 5).

2 The Quasi-Complete Photon 16 × 16 Switch

In this section we present the new scheme of the 16×16 switch (Fig. 1) that is used as the cell of the 256×256 base element (Fig. 2). Note as the advantage that the period T of the offered 16×16 quasi-complete photon cell is 4 times smaller then the one of the 16×16 switch described in [8]. At the same time, high throughput of the new type of switch is achieved by increasing circuit complexity in six time in comparison with the photon one [10].

The 16×16 cell is based on the sixteen 4×4 photon switches in the central stage, the sixteen 1×4 photon demultiplexers in the input stage and the sixteen 4×1 photon multiplexers in the output stage(Fig. 1). These elements connect to each other in accordance to the quasi-complete graph principal [11]. For this purpose all input 1×4 photon demultiplexers are divided into four groups of four 1×4 photon demultiplexers. And all central 4×4 photon switches are divided into four groups of four 4×4 photon switches also. The input demultiplexers of each group are connected with the central stage photon switches of the corresponding group in accordance with the tree-type topology [7]. The central stage photon switches are connected with the output 4×1 photon multiplexers in accordance with the following rule: the outputs of the first photon switch are connected with the inputs of the next numbers of output 4×1 photon multiplexers: 1, 5, 9 and 13; the outputs of the second photon switch are connected with the inputs of the next numbers of output 4×1 photon multiplexers: 2, 6, 10 and 14,..., the outputs of the fourth photon switch are connected with the inputs of the numbers of multiplexers: 4, 8, 12 and 16. Analogously for the fifth-eighth photon switches it can be written the following sequence of the output multiplexer numbers: 5,

9, 13, 1; 6, 10, 14, 2; ... 8, 12, 16, 4. Finally for the ninth-thirteenth photon switches and for fourteens-sixteenth ones the sequence of the output multiplexer numbers will be following: 9, 13, 1, 5;...12, 16, 4, 9; and 13, 1, 5, 9;...16, 4, 8, 12. The quasi-complete graph principal provides the strictly non-blocking of the switching system and does not increase the period of optical signals with an increase in the number of inputs.

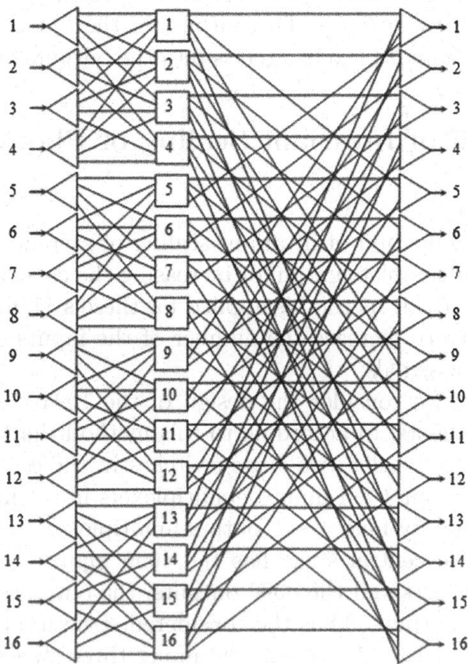

Fig. 1. The design of photon quasi-complete 16 × 16 switch

3 The Two-Stage Photon 256 × 256 Switch as the Basic Element of High-Capacity Photon Switching Systems

The proposed 256 × 256 base element consists of the two stages of the quasi-complete photon 16 × 16 cells (Fig. 2). Here $N_1 = 16$ is the number of the inputs of the 16×16 switch, $N_2 = 256$ is the number of the inputs of the 256×256 switch, the trapezoids $C_{2,1}$ denote the 16×16 switch schemes (Fig. 1) without the output multiplexers, the squares $C_{2,2}$ denote the schemes of the complete switches with the $N_1 = 16$ inputs and the $N_1 = 16$ outputs (Fig. 1). The triangles $C_{2,3}$ denote the schemes of the 4 × 1 multiplexers. The set $C_1 = \{1...1024\}$ corresponds to the outputs of the circuits $C_{2,1}$ and consists of three subsets I, J, k, where the elements of the subset $I = \{1...16\}$ are numbers of the circuits $C_{2,1}$, the elements of the subsets $J = \{1...16\}$ are the groups numbers of the four outputs of the circuits $C_{2,1}$, and the elements of the subset k ={1...4} are the output numbers in the groups. It also is seen in Fig. 2 that there are the $p = 4$ copies of the sets C_2 = {1...256}. Each set C_2 corresponds to the inputs of the circuits $C_{2,2}$. It consists of the subsets I and J, where the elements of the subsets I = {1...16} are numbers of the circuits $C_{2,2}$, and the elements of the subsets $J = \{1...N_1\}$ are the input numbers of the circuits $C_{2,2}$. There are interconnections between the outputs of the circuits $C_{2,1}$ and the inputs of the circuits $C_{2,2}$ in each of the four copies of C_2. Here the outputs I, J, k are connected to the inputs J, I of the k-th copy of the sets C_2. The outputs of the k copy of the set C_2 are combined by the 4×1 multiplexers circuits into the outputs of 256×256 switch.

4 The Method of Invariant Scaling of Photon Switching Systems

Now let us apply the method of invariant scaling of system networks based on the principal of quasi-complete graph for increasing the capacity of the proposed switches [10]. However this method is applied for these systems here for the first time, and it allows us to increase the number of the inputs without increasing the period T between signals.

As example, here we consider the design of the 1024 × 1024 switch based on the 256 × 256 elements. In according to this method the 1024 × 1024 switch contains the sixteen 256 × 256 two-stage switches in the central stage, the one thousand and twenty-four 1 × 4 input demultiplexers in the input stage and the one thousand and twenty-four 4 × 1 output multiplexers in the output stage (Fig. 3). Therefore the central stage has the 4096 inputs (outputs), the input stage has the 1024 inputs and the 4096 outputs, the output stage has the 4096 inputs and the 1024 outputs. Thus the considered structure is analogous to the quasi-complete photon 16 × 16 switch containing three stages (Fig. 1). According to the principal of the quasi-complete graph, each of the sixteen 256 × 256 switches is divided into the sixty-four zones containing the four inputs. Every two hundred and fifty-six inputs of the 1024 × 1024 switch are connected with the inputs of the central stage zones through the 1 × 4 input demultiplexers analogous to the quasi-complete photon 16 × 16 scheme (Fig. 1). And the sixty-four outputs of the zones must be connected with the 1024 × 1024 system outputs through the 4 × 1 multiplexers, analogous to the outputs of the central stage of the 16 × 16 switch connect with the 4 × 1 multiplexers (Fig. 4). In such the switch the signal from any input transmits to one of the copies of the 256 × 256 switch. Therefore, the 1024 × 1024 switch is non-blocking as well as the 256 × 256 one. The analogous principal can be used for next scaling of photon switches for designing 4096 × 4096,16384 × 16384, 65536 × 65536 photon switches etc.

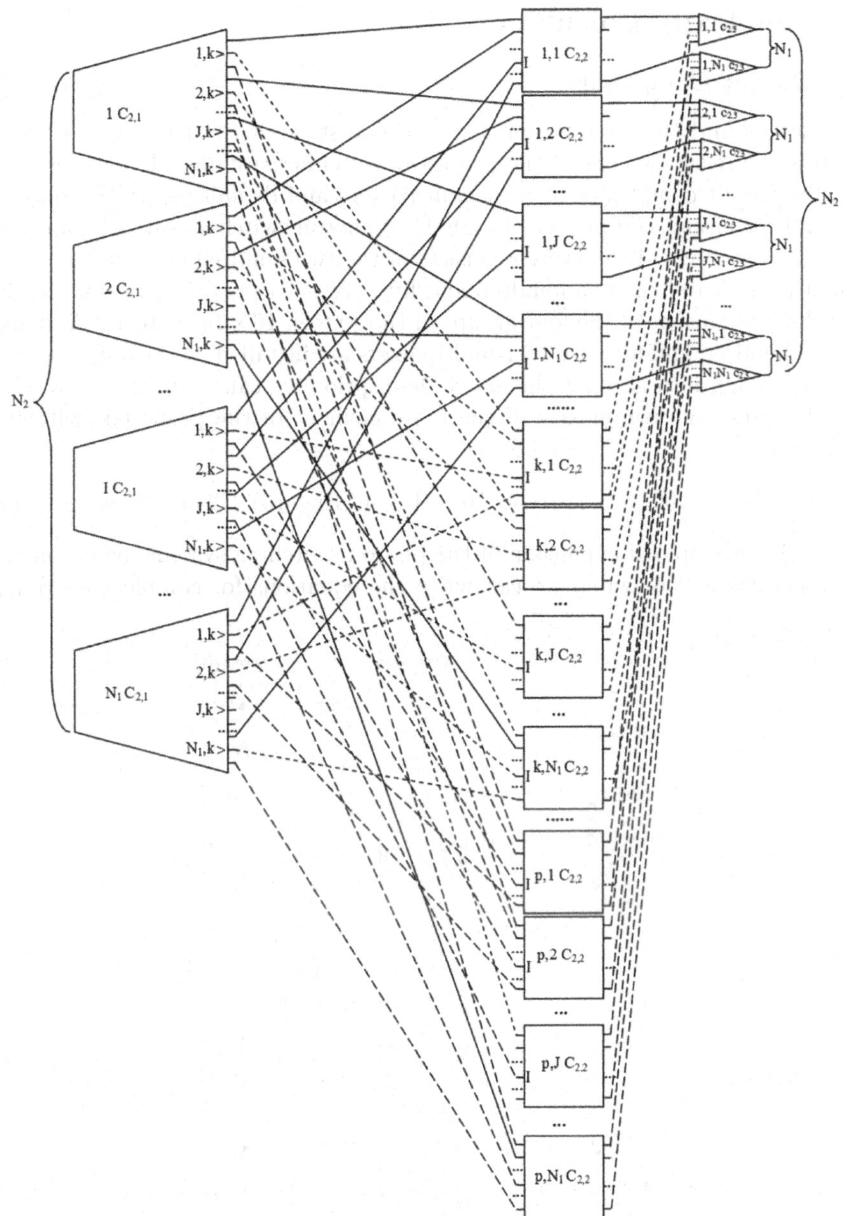

Fig. 2. The design of two-stage photon 256 × 256 switch

5 Complexity Calculation

5.1 Circuit Complexity

First, let us find the circuit complexity of two-stage 256 × 256 switch (Fig. 2) which is the basic element of the proposed switching systems. Here we assume that the circuit complexity of each multiplexers and demultiplexers is equal to unit and the circuit complexity of each 4 × 4 photon switch is equal to four [10]. Here the first stage of the switch consists of the two hundred and fifty-six 1 × 4 demultiplexers and the two hundred and fifty-six 4 × 4 photon switches and the second stage consists of the four groups of the sixteen 16 × 16 switches. And also the two hundred and fifty-six 4 × 1-multiplexers are installed on the output of the system. Taking into account the above description we can write the expression for calculating the complexity of the basic element of the proposed switching system as

$$S_{256} = 256 + 4 \cdot 256 + 4 \cdot 16 \cdot (16 + 4 \cdot 16 + 16) + 256 = 7680 \tag{1}$$

To use the principle of expansion of the photon switching systems based on the two-stage 256 × 256 switch we can write the equations for complexities S_{1024},

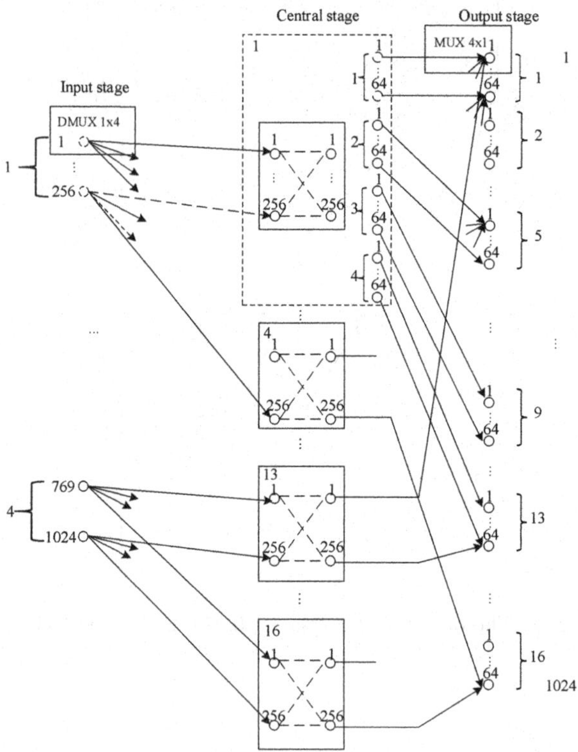

Fig. 3. The design of photon quasi-complete 1024 × 1024 switch

S_{4096}, S_{16384}, S_{65536} of the 1024×1024, 4096×4096, 16384×16384, 65536×65536 switches correspondingly:

$$S_{1024} = 16 \cdot S_{256} + 2N \cdot 4^0 = 124928 \tag{2}$$

$$S_{4096} = 16^2 \cdot S_{256} + 4^1 \cdot 2N + 2N \cdot 4^0 = 2007040 \tag{3}$$

$$S_{16384} = 16^3 \cdot S_{256} + 4^2 \cdot 2N + 4^1 \cdot 2N + 2N \cdot 4^0 = 32145408 \tag{4}$$

$$S_{65536} = 16^4 \cdot S_{256} + 4^3 \cdot 2N + 4^2 \cdot 2N + 4^1 \cdot 2N + 2N \cdot 4^0 = 514457600 \tag{5}$$

Based on this regularity (3)–(5) and the method of mathematical induction, the formula for calculating the complexity of the proposed photon switches for any N inputs can be obtained in the following form

$$\begin{aligned} S_n &= 16^n \cdot S_{256} + 4^{n-1} \cdot 2N + 4^{n-2} \cdot 2N + ... \\ &+ 4^1 \cdot 2N + 4^0 \cdot 2N = 16^n \cdot S_{256} + 2N \sum_{i=0}^{n-1} 4^i \end{aligned} \tag{6}$$

where n is the circuit serial number which is associated with the number of the inputs as

$$N = 2^{2n+8} \tag{7}$$

Expressing n from (7) we get

$$n = \log_2 \sqrt{N} - 4 \tag{8}$$

The results of numerical calculations are presented in Table 1. Taking into account the fact that $S_{256} = 7680$ and inserting (8) in (6) we get the final expression for calculating the complexity of the proposed photon switching systems

$$S(N) = \frac{15}{128} N^2 + 2N \sum_{i=0}^{\log_2 \sqrt{N} - 5} 2^{2i} \tag{9}$$

for the case $N > 256$. And since (1) we have $S(N) = 7680$ for $N = 256$. Therefore it can be written finally

$$S(N) = \begin{cases} 7680 & \text{for } N = 256 \\ \frac{15}{128} N^2 + 2N \sum_{i=0}^{\log_2 \sqrt{N} - 5} 2^{2i} & \text{for } N > 256 \end{cases} \tag{10}$$

5.2 Fiber Complexity

Fiber complexity is the total number of fibers required for connecting each input with each output through the switching elements. First of all let us calculate the fiber complexity of the 256 × 256 two-stage switch, and then the fiber complexity of $N \times N$ switch with arbitrary N. As the 256 × 256 two-stage switch, consists

Table 1. The values of the circuit complexity of the high-capacity photon switching systems based on the two-stage 256 × 256 switch for the different number of inputs

n	N	$S(N)$	n	N	$S(N)$
0	$2^8 = 256$	7680	5	$2^{18} = 262144$	8231845888
1	$2^{10} = 1024$	124928	6	$2^{20} = 1048576$	131711631400
2	$2^{12} = 4096$	2007040			...
3	$2^{14} = 16384$	32145408			...
4	$2^{16} = 65536$	514457600	n	2^{2n+8}	$S(2^{2n+8})$

of the two stages of the quasi-complete photon 16 × 16 cells we must calculate internal fiber lines inside each switching cell and external fiber lines between stages. The 16 × 16 cell without the output multiplexers of the input stage contains 16 × 4 = 64 internal fibers (Fig. 1) and each complete the 16 × 16 cell of the second stage contains 2 × 16 × 4 = 128 internal fibers. Also there are 16 × 4 × 16 = 1024 external fiber lines between the first and the second stages and there are 16 × 4 × 16 = 1024 external fiber lines between the second and the third stages. (Fig. 1) So the expression for calculating the fiber complexity of the 256×256 two-stage switch is $F(256) = 64\times16+128\times16\times4+1024+1024 = 11264$. Considering the method of invariant scaling of photon switching systems (Sect. 4) we can calculate the fiber complexity $F(1024)$, $F(4096)$, $F(16384)$, $F(65536)$ of the proposed schemes for 1024, 4096, 16384, 65536 inputs correspondingly

$$F(1024) = 8 \times 1024 + 16 \times 11264 = 188416 \quad (11)$$

$$F(4096) = 8 \times 4096 + 16 \times 188416 = 3047424 \quad (12)$$

$$F(16384) = 8 \times 16384 + 16 \times 3047424 = 48889856 \quad (13)$$

$$F(65536) = 8 \times 65536 + 16 \times 48889856 = 782761984 \quad (14)$$

And analogously, for N number of inputs we can write

$$F(N) = 8 \times N + 16 \times F(N/4) \quad (15)$$

Now let us write the expressions (10)–(14) as

$$F_1 = F(2^{10}) = 8N + 2^4 \cdot F_{256} = 188416 \quad (16)$$

$$F_2 = F(2^{12}) = 8N + 2^{17} + 2^8 \cdot F_{256} = 3047424 \quad (17)$$

$$F_3 = F(2^{14}) = 8N + 2^{19} + 2^{21} + 2^{12} \cdot F_{256} = 48889856 \quad (18)$$

$$F_4 = F(2^{16}) = 8N + 2^{21} + 2^{23} + 2^{25} + 2^{16} \cdot F_{256} = 78761984 \quad (19)$$

In general case for n-number scheme we obtain

$$F_n = F(2^{2n+8}) = 8N + 2^{17} \sum_{i=0}^{n} 2^{2(i+n-4)} - 2^{2n+11} - 2^{2n+9} + 2^{4n} \cdot 11264 \quad (20)$$

Substituting (8) into expression (20) we obtain the resulting expression of the fiber complexity for high-capacity photon switches based on the 256 × 256 two-stage switch for the first time

$$F(N) = N\left[\sum_{i=0}^{\log_2\sqrt{N}-4} 2^{2i+1} - 2 + \frac{11}{64}N\right] \tag{21}$$

The results of numerical calculations are presented in Table 2. The results are equal to the ones obtained in simple recalculation of the fiber connections in the corresponding schemes.

Table 2. The values of the fiber complexity of the high-capacity photon switching systems based on the two-stage 256 × 256 switch for the different number of inputs

n	N	$F(N)$	n	N	$F(N)$
0	$2^8 = 256$	11264	5	$2^{18} = 262144$	12526288900
1	$2^{10} = 1024$	188416	6	$2^{20} = 1048576$	200429011000
2	$2^{12} = 4096$	3047424			...
3	$2^{14} = 16384$	48889856			...
4	$2^{16} = 65536$	782761984	n	2^{2n+8}	$F(2^{2n+8})$

6 The Comparative Analysis

At first, we present the comparative analysis results of the circuit complexities of the three switch type: the crossbar switch [7], the photon switches [8–10] and the new type of high-capacity switching systems (Fig. 4).

The calculation results show that the circuit complexity of the new type of the high-capacity switching systems is more than the complexity of the photon switches [8,10] and significantly less than the complexity of the crossbar ones [5]. For example, the circuit complexity of crossbar switch [7] with the input number $N = 4096$ is $S_c(N) = N^2 = 16777216$; the circuit complexity of the photon switch [6] is $S_d(N) = N/2 \cdot log_2\, N = 24576$ and the circuit complexity of the proposed high-capacity photon switch is equal to 2007040. Thus, in this case the circuit complexity of the proposed type of photon switches in 82 times more than the complexity of the photon switches [6] and in 8 times less than the complexity of crossbar switches [5].

Now we present the comparative analysis results of the fiber complexities of the three switch type: the tree-type switch [7], the photon switches [8,10] and the new type of high-capacity switching systems. As the fiber complexity of photon switches [10] has not been calculated earlier we present derivation of its formula for the first time also. From the design of the photon switches [5]

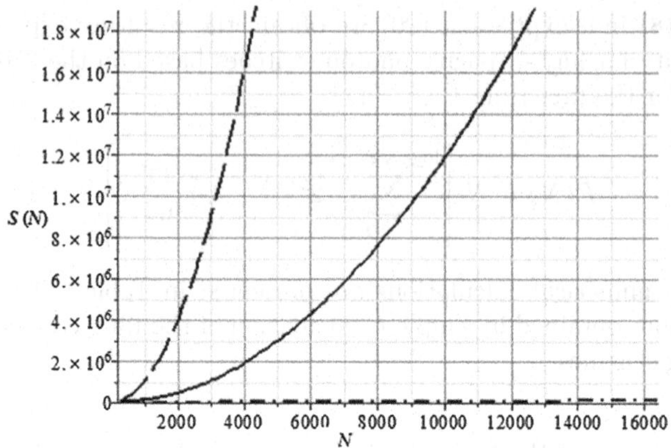

Fig. 4. The comparison of the circuit complexity. The solid line corresponds to the photon switching system based on two-stage 256 × 256 switch, the long-dash line corresponds to the crossbar switch, the dash-dot line corresponds to the photon switch

it follows that the fiber complexity of the 16 × 16, 256 × 256, 65536 × 65536 and 4294967296 × 4294967296 photon switches can be calculated by using the expressions:

$$F_{d0} = F(2^4) = F(16) = 4 \cdot 4 = 16 \tag{22}$$

$$F_{d1} = F(2^8) = F(256) = N + 2 \cdot 16 \cdot F_{d0} = 256 + 2 \cdot 16 \cdot 16 = 768 \tag{23}$$

$$F_{d2} = F(2^{16}) = F(65536) = N + 2 \cdot 256 \cdot F_{d1} = 65536 + 512 \cdot 768 = 458752 \tag{24}$$

$$\begin{aligned} F_{d3} &= F(2^{32}) = F(4294967296) = N + 2 \cdot 65536 \cdot F_{d2} \\ &= 4294967296 + 2 \cdot 65536 \cdot 458752 = 64424509440 \end{aligned} \tag{25}$$

Let us write the expressions (22)–(25) in the form:

$$F_{d0} = F(2^4) = 2^4 \cdot 2^0 = 16 \tag{26}$$

$$F_{d1} = F(2^8) = 2^8 + 2^9 = 2^8 \cdot (2^0 + 2^1) = 768 \tag{27}$$

$$\begin{aligned} F_{d2} &= F(2^{16}) = 2^{16} + 2 \cdot 2^8 \cdot (2^8 + 2^9) = 2^{16} + 2^{17} + 2^{18} \\ &= 2^{16} \cdot (2^0 + 2^1 + 2^2) = 458752 \end{aligned} \tag{28}$$

$$\begin{aligned} F_{d3} &= F(2^{32}) = 2^{32} + 2 \cdot 2^{16} \cdot (2^{16} + 2^{17} + 2^{18}) = 2^{32} + 2^{33} + 2^{34} + 2^{35} \\ &= 2^{32} \cdot (2^0 + 2^1 + 2^2 + 2^3) = 64424509440 \end{aligned} \tag{29}$$

where F_{d0} is the fiber complexity of 256 × 256 switch, F_{d1} is the fiber complexity of 1024 × 1024 switch, etc. After analyzing the above dependences, we write down the expression

$$F_d(N) = N \sum_{i=0}^{n-1} 2^i \tag{30}$$

for the $N \times N$ fiber complexity of the photon switch. Here n is the serial number of the scheme

$$n = \log_2 \log_2 N - 1 \tag{31}$$

The numerical calculation results (Fig. 5) show that the fiber complexity of the new type of the high-capacity switching systems is more than the fiber complexity of the photon switches [6] and significantly less than the complexity of the tree-type ones [5]. For example, the fiber complexity of the tree-type switch [5] with the input number $N = 65536$ is $F_t(N) = N^2 = 4294967296$; the fiber complexity of the photon switch is (30) $F_d = 458752$ and the fiber complexity of the proposed high-capacity photon switch is equal to 782761984. Thus in this case the fiber complexity of the proposed type of photon switches in 1706 times more than the complexity of the photon switches [6] and in 5.5 times less than the complexity of tree-type switches [5].

Significantly lower circuit and fiber complexities are the advantages of the proposed high-capacity switches in comparison with the tree-type scheme, and in comparison with the photon switches, the proposed systems have higher throughput. Increasing the throughput of the proposed schemes is provided by the decreasing a period of signals. A period of signals for proposed switches is in 4 times less than the period of the 16×16 systems and in 12 times less than the period of the 256×256 photon systems [10].

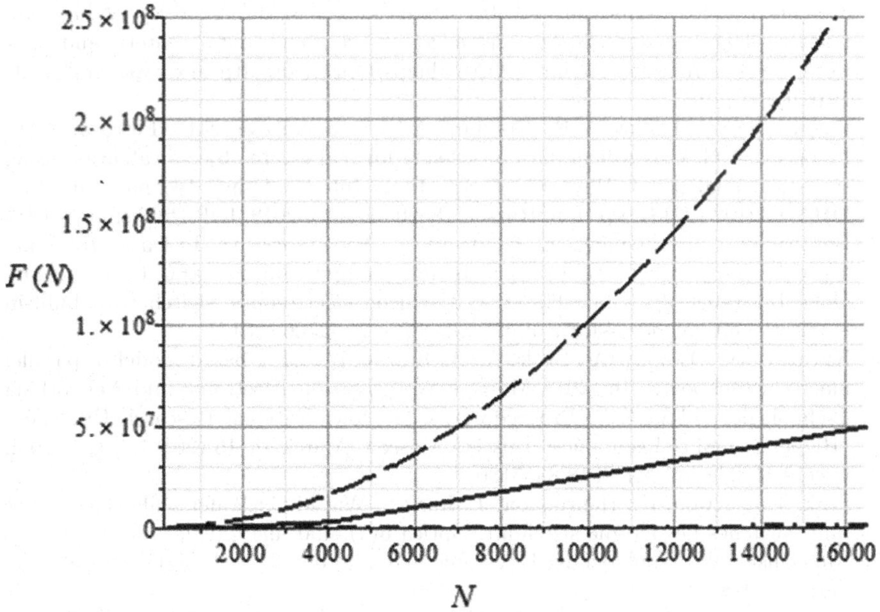

Fig. 5. The comparison of the fiber complexity. The solid line corresponds to the photon switching system based on two-stage 256×256 switch, the long-dash line corresponds to the tree-type switch, the dash-dot line corresponds to the photon switch

7 Conclusion

In this paper the new type of high-capacity photon switching systems based on the two-stage switch 256×256 is presented for the first time. The designing principals of such the 256×256 element are described in detail by using the set-theoretic representation of the switch. The method of invariant scaling of the proposed photon switching systems is considered also. As the example the 1024×1024 system bipartite graph is presented and described in detail.

The main advantages of the proposed systems are strictly non-blocking, scalability, low circuit and fiber complexities in comparison with strictly non-blocking cross-bar and tree-type switches [5], and high throughput in comparison with the photon switches [6]. The throughput of the proposed switches is in 4 times higher than the throughput of the 16×16 photon switching systems [10] and in 12 times higher than the throughput of 256×256 ones.

The analytical models for calculating the circuit and fiber complexities of the proposed switching systems are presented here for the first time. The numerical calculations showed that the difference in the circuit and fiber complexities of the crossbar and tree-type switches and the proposed switching systems increases with the increase in the number of inputs.

References

1. Mohamed S.H., El-Gorashi, T.E.H., Elmirghani, J.M.H.: A survey of big data machine learning applications optimization in cloud data centers and networks. Netw. Internet Archit. (2020). https://arxiv.org/ftp/arxiv/papers/1910/1910.00731.pdf
2. Parygin, D.S., Malikov, V.P., Golubev, A.V., Sadovnikova, N.P., Petrova, T.M., Finogeev, A.G. Categorical data processing for real estate objects valuation using statistical analysis. J. Phys. Conf. Ser. Proc. Int. Conf. Inf. Technol. Bus. Ind. **1015**, 032102 (2018). pp. 1–6, Russia, 18–20 January 2018. IOP Publishing (2018)
3. Barabanov, I., Barabanova, E., Maltseva, N., Kvyatkovskaya, I.: Data processing algorithm for parallel computing. Commun. Comput. Inf. Sci. **466**, 61–69 (2014)
4. Qiao, L., Tang, W., Chu, T.: 32×32 silicon electro-optic switch with built-in monitors and balanced-status units. Sci. Rep. **7**, 42306 (2017)
5. Kutuzov, D., Osovsky, A., Stukach, O., Starov, D.: CPN-based model of parallel matrix switchboard. In: 2018 Moscow Workshop on Electronic and Networking Technologies (MWENT). Proceedings. - Moscow: National Research University "Higher School of Economics" Russia.- Moscow, March 14–16 (2018). https://doi.org/10.1109/MWENT.2018.8337180
6. Seok, T.J., Kwon, K., Henriksson, J., Luo, J., Wu, M.C.: Wafer-scale silicon photonic switches beyond die size limit. Optica **6**(4), 490–494 (2019)
7. Kabacinski, W.: Nonblocking Electronic and Photonic Switching Fabrics, p. 282. Springer, USA (2005)
8. Barabanova, E.A., Vytovtov, K.A., Maltseva, N.S., Kravchenko, O.V., Kravchenko, V.F.: Models and algorithms of optical switching systems with decentralized control. In: 2019 IEEE Conference of Russian Young Researchers in Electrical and Electronic Engineering, pp. 64–68 (2019). https://doi.org/10.1109/EIConRus.2019.8657063

9. Vytovtov, K.A., Barabanova, E.A.: Optical switching cell based on metamaterials and ferrite Films. In: 12th International Congress on Artificial Materials for Novel Wave Phenomena - Metamaterials 2018 Espoo, Finland, Aug. 27th–Sept. 1st. 2018, pp. 424–426
10. Barabanova, E.A., Vytovtov, K.A., Vishnevskiy, V.M.: Novyj princip postroeniya opticheskih ustrojstv obrabotki informacii dlya informacionno-izmeritel'nyh sistem. Datchiki i sistemy 9, 3–9 (2019, in Russian)
11. Karavai, M.F., Parkhomenko, P.P., Podlazov, V.S.: Combinatorial methods for constructing bipartite uniform minimal quasicomplete graphs (symmetrical block designs). Autom. Remote Control **30**, 312–327 (2009). Pleiades Publishing, Ltd

Investigation of the Guaranteed Traffic Rate in Enterprise WLAN

M. Rudenkova, H. Khayou, and L. I. Abrosimov(✉)

National Research University "Moscow Power Engineering Institute",
Krasnokazarmennaya 14, 111250 Moscow, Russia
{RudenkovaMA,AbrosimovLI}@mpei.ru,
hussein.khayou@gmail.com

Abstract. Enterprise WLAN uses an IEEE 802.11 wireless channel to transfer data from a variety of network applications. The predominant types of traffic are the various types of real-time traffic that currently play an important role in business operations. Network applications with real-time traffic have different network requirements: packet bandwidth, jitter, packet delay, and loss tolerance. WLANs have variable packet throughput depending on the bandwidth, the number of connected wireless stations, the traffic intensity of the network applications, and completely different media access protocols. Real-time traffic creates challenges and demands on wireless resource management. The purpose of this paper is to improve end-user experience of real-time traffic transmission through the wireless channel. To achieve this goal, we developed a testbed to obtain the mean and variance of wireless channel service time using event-driven simulator. We also proposed an analytical model of enterprise WLAN and equations to compute the guaranteed traffic intensity through wireless channel. An example showing how to obtain the guaranteed traffic intensity using the analytical model and event-driven simulation results is provided.

Keywords: WLAN · Performance · DCF · PCF

1 Introduction

Wireless local area network (WLAN) is the most popular technology for providing access to local services or/and Internet services in the enterprise. Recently, various types of voice and video network applications, such as real-time interactive video applications, streaming video, telepresense and video conference applications are becoming increasingly important to the enterprise. Real-time traffic requires guaranteed delivery time of network packet.

Wireless channel media access protocols provide: mechanism to place one network packet in the wireless medium (contention or contention-free protocol), delivery control and network packet acknowledgment. If the network packet is placed in the wireless medium without collision, the wireless channel deliver

© Springer Nature Switzerland AG 2020
V. M. Vishnevskiy et al. (Eds.): DCCN 2020, CCIS 1337, pp. 70–81, 2020.
https://doi.org/10.1007/978-3-030-66242-4_6

this packet to the receiver. When the intensity of real-time traffic increases, the wireless channel queue increases and delivery time also increases. If the waiting time is greater than the guaranteed delivery time, the packet is removed from queue and dropped.

The guaranteed traffic rate is one of the criteria for QoS estimation [1]. The fundamental purpose of QoS is to provide the required level of service in current network conditions. The basic characteristics of QoS are bitrate, jitter, packet delay and loss tolerance. But for real-time traffic the more important is End-to-end QoS [2] which is very difficult to estimate.

There is a problem in determining the numerical characteristics of End-to-End QoS especially the packet delivery rate or the actual time to service the network packet (delay) which depend on the following: wireless channel throughput, number of connected wireless stations, intensity of inbound traffic, wireless channel media access protocol and its parameters. There are many publications on estimating the delay, throughput or performance of the IEEE 802.11 wireless channel for a set of configuration parameters and WLAN characteristics [3–7]. The authors developed $M/M/1/\infty$ [3–5] or $M/G/1/\infty$ [6,7] model for WLAN, which is unsuitable for real-time traffic in enterprise WLAN. The authors of [8] [9] examine the performance and delay of the wireless channel for a set of configuration parameters and different WLAN characteristics, but it is very important to develop a methodology to obtain guaranteed traffic intensity for enterprise WLAN specificity.

Therefore, we developed an event-driven simulation testbed to obtain the wireless channel service time and the variance for a set of WLAN parameters and characteristics. We modified $M/G/1/s$ model to describe the enterprise WLAN model and obtain the equations to compute the wireless channel actual service time to service real-time traffic, the guaranteed delivery rate and the wireless channel efficiency of packet delivery. We proposed a methodology and an example to compute the guaranteed delivery rate for the specified guaranteed delivery time of real-time network application.

2 Characteristics of Wireless Local Area Network

We study enterprise WLAN in infrastructure mode which contains an access point **AP**, wireless stations **STA** $(k = \overline{1, K})$ and an internal corporate LAN as **Switch** and corporate LAN resources as **Server** (Fig. 1). Each **STA** has its own set of real-time network applications. **AP** has a IEEE 802.11 wireless channel with media access control protocol CF $(CF = [DCF, PCF])$.

The WLAN under study has the following characteristics.

The real-time network applications utilize the wireless channel by sending packets with an average length l [bit/packet]. The intensity of real-time network application traffic determines the time between network packets θ_k which is assumed to have an exponential distribution.

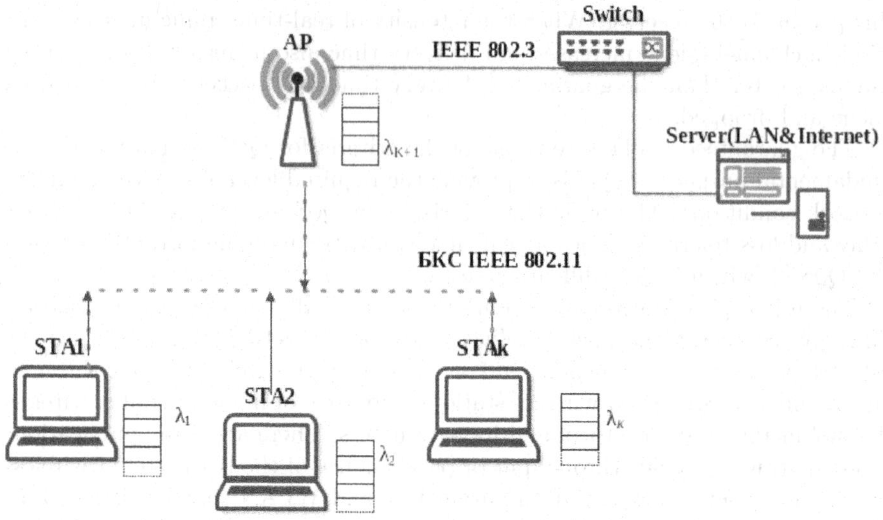

Fig. 1. The wireless local area network

Total traffic intensity of real-time network application Λ is:

$$\Lambda = \sum_{k=1}^{K+1} \lambda_k \tag{1}$$

The main feature of real-time network application traffic is following. Inbound traffic intensity is λ_k (where $k = K + 1$ is index for **AP**), the outbound wireless channel delivery intensity is u_k. The part of the traffic is dropped with intensity $\Delta\lambda_k$ if a waiting time is higher than an specified guaranteed delivery time T_g.

$$\lambda_k = u_k + \Delta\lambda_k \tag{2}$$

The total network application traffic intensity Λ can be separated into two parts: the flow of network packets which is delivered with an intensity u and the flow of network packets which is dropped with an intensity $\Delta\Lambda$:

$$\Lambda = u + \Delta\Lambda \tag{3}$$

The intensity of dropped network packets $\Delta\Lambda$ depends on the denial of service probabilities p_d and the total network application traffic intensity Λ:

$$\Delta\Lambda = p_d \cdot \Lambda \tag{4}$$

3 Performance Parameters of Wireless Local Area Network

The wireless channel is used to service network packets. The main resource in the wireless channel is throughput C [bit/s]. However, a more important parameter is the throughput in network packets C_p[packet/s] with average length l_p [packet/s] which can be sent through the wireless channel. The time τ^1 to transfer one network packet is:

$$\tau^1 = \frac{l_p}{C} \tag{5}$$

Let's assume that μ^1 is the intensity of network packet service in a wireless channel without a specific wireless channel media access protocol. We have:

$$\tau^1 = \frac{1}{\mu^1} \tag{6}$$

If the wireless channel uses a specific wireless channel media access protocol CF, let the service time be T_{CF} , which is increased by the time t_{CF} because of the introduced overhead for the specific functions of CF. Then

$$T_{CF} = \tau^1 + t_{CF} \tag{7}$$

The intensity of network packet service in a wireless channel with a specific CF is:

$$\mu_{CF} = \frac{1}{T_{CF}} \tag{8}$$

To estimate the functioning capacity of the wireless channel we introduce a coefficient α which depends on the total intensity of network application traffic and the service intensity:

$$\alpha = \frac{\Lambda}{\mu_{CF}} \tag{9}$$

The the wireless channel load ρ depends on the intensity of serviced network packets u and the specific CF, therefore:

$$\rho = \frac{u}{\mu_{CF}} = u{\cdot}T_{CF} \tag{10}$$

4 Investigation of Wireless Channel Media Access Protocol

To obtain numerical wireless channel characteristics we used event-driven Network Simulator NS-3 which is a very popular tool and is used in a lot of research works [11,12]. We developed a program for NS-3 [13] to obtain the service time value for a specific scenario such as set of WLAN characteristic (Table 1) for different K. Number of experiments $N = 1000000$.

Scenario 1. To obtain the service time value $\overline{T_{DCF}}(\Lambda)$ and variance $\sigma^2_{DCF}(\Lambda)$ for $CF = DCF$ the WLAN topology Fig. 1 is used, let K = 4. WLAN characteristic (Table 1). The results of experiments are shown on Figs. 2 and 3.

Scenario 2. To obtain the service time value $\overline{T_{PCF}}(\Lambda)$ and variance $\sigma^2_{PCF}(\Lambda)$ for $CF = PCF$ the WLAN topology Fig. 1 is used, let K = 4. WLAN characteristic (Table 1). The results of experiments are shown on Figs. 4 and 5.

Table 1. The configuration parameters and WLAN characteristics

σ_c	9 us	CW_{max}	1023	CW_{min}	15
t_{SIFS}	16 us	t_{DIFS}	34 us	l_{ack}	14 B
C	54 Mbit/s	Λ	250 - 12000 packet/s	K	4
$\bar{l}$	1500 B	$l_{CF-POLL}$	20 B		

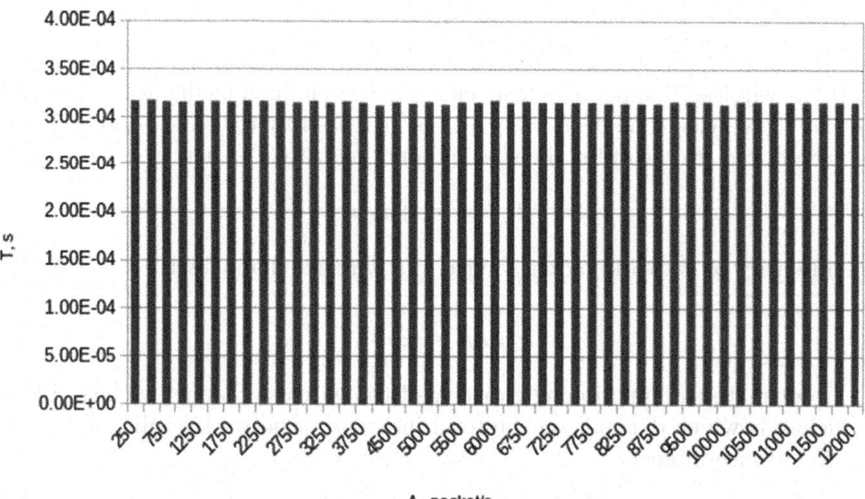

Fig. 2. Evaluation of service time $\overline{T_{DCF}}(\Lambda)$ $(CF = DCF)$

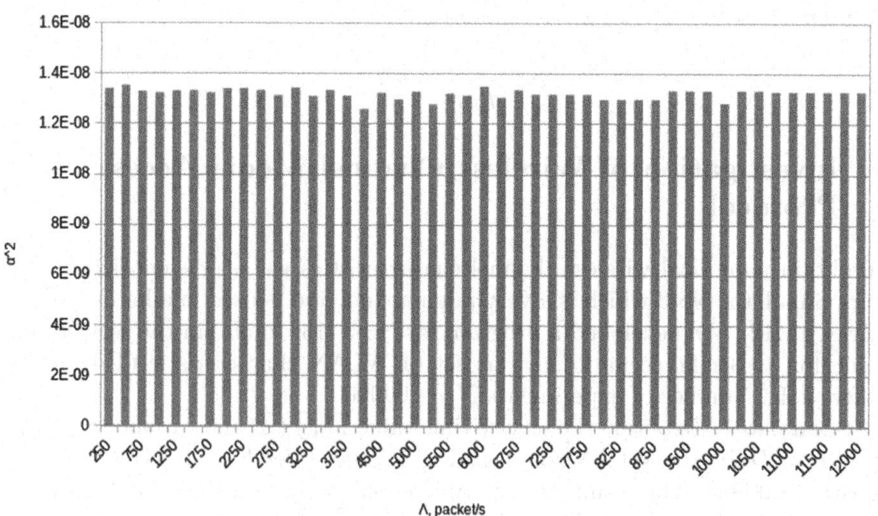

Fig. 3. Evaluation of variance $\sigma^2_{DCF}(\Lambda)$ $(CF = DCF)$

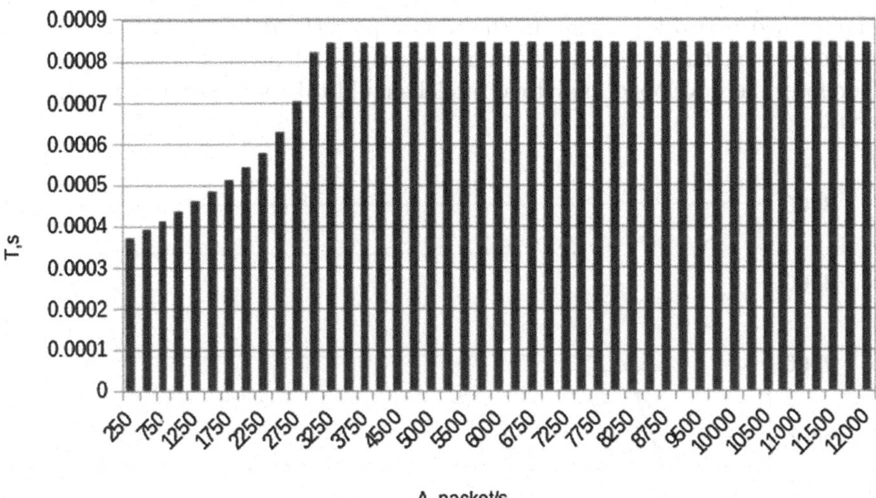

Fig. 4. Evaluation of service time $\overline{T_{PCF}}(\Lambda)$ $(CF = PCF)$

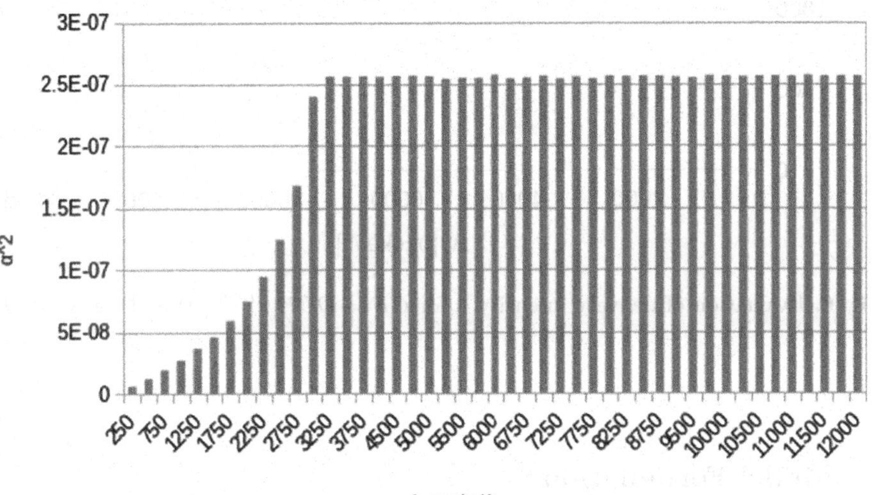

Fig. 5. Evaluation of variance $\sigma^2_{PCF}(\Lambda)$ $(CF = PCF)$

The results show that $\sigma^2_{DCF}(\Lambda) < 1.0$ and $\sigma^2_{PCF}(\Lambda) < 1.0$. Therefore, in the developed analytical model, the wireless channel service time intervals could be described by an arbitrary distribution law.

Let's obtain dependence of delivery intensity $u_C F$ and total intensity of traffic Λ for CF = DCF and CF = PCF Fig. 6.

The Fig. 6 shows that dependencies have two parts. In the first part the delivery intensity u increases linearly with traffic intensity increasing Λ. There is no queued network packets. In the second part the delivery intensity u does not depend on the intensity of traffic Λ because of queued network packets. Therefore, the developed analytical model should take into account the limited queue.

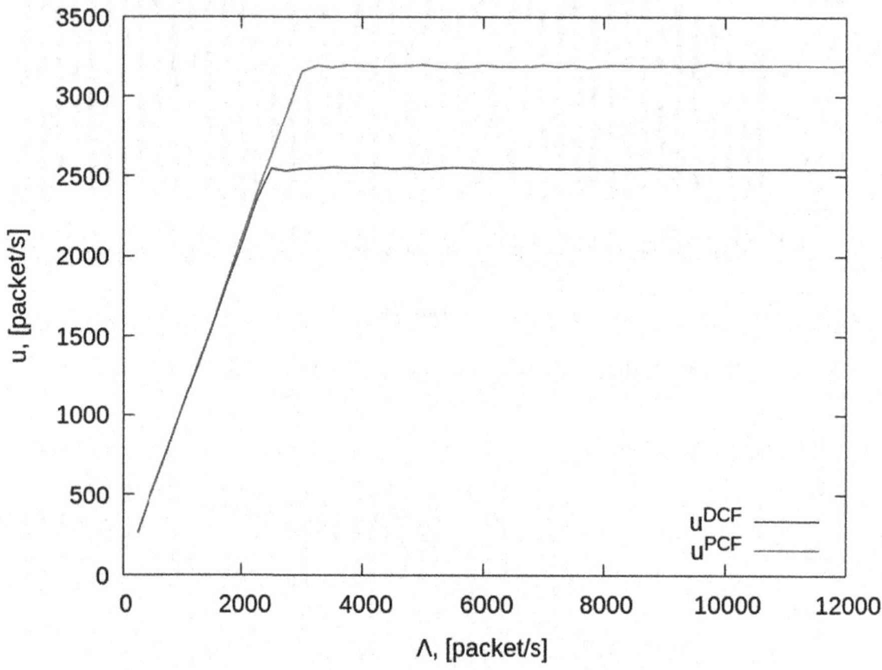

Fig. 6. Evaluation of delivery intensity u for $CF = DCF$ $u^{DCF}(\Lambda)$ and $CF = PCF$ $u^{PCF}(\Lambda)$

5 Model Formulation

For enterprise WLAN (Fig. 1) with wireless channel characteristics (Sect. 4) $\overline{T_{DCF}}(\Lambda)$, $\sigma^2_{DCF}(\Lambda)$ $(CF = DCF)$ and $\overline{T_{PCF}}(\Lambda)$, $\sigma^2_{PCF}(\Lambda)$ $(CF = PCF)$ let's develop modification of M/G/1/s model to determine guaranteed delivery intensity of real-time traffic u^a_{CF} for specified guaranteed delivery time T_g.

First, let's assume $M/M/1/s$ model as the basic model of wireless channel where s is the queue size. We assume the number of places in the queue s depends on the number of wireless stations K, the guaranteed delivery time T_g and the CF protocol service time:

$$s = K \cdot (T_g/T_{CF}) \tag{11}$$

The system has i states $(i = \overline{0, s+1})$:

when $i = 0$ - the wireless channel is empty, there are 0 packets waiting the service;

when $i = 1$ - the wireless channel services 1 packet, there are 0 packets waiting the service;

when $i = 2$ - the wireless channel services 1 packet, there is 1 packet waiting the service;

when $i = x$ - the wireless channel services 1 packet, there are $x - 1$ packets waiting the service;

when $i = s + 1$ - the wireless channel services 1 packet, there are s packets waiting the service.

The probabilities $P_i(i = \overline{0, S+1})$ of these states are:

$$\begin{aligned} P_0 &= \frac{1}{1+\alpha+\alpha\cdot\sum_{k=1}^{K+1}\alpha^k} \\ P_1 &= \frac{\alpha}{1+\alpha+\alpha\cdot\sum_{k=1}^{K+1}\alpha^k} \\ &\cdots \\ P_{s+1} &= \frac{\alpha\cdot\alpha^s}{1+\alpha+\alpha\cdot\sum_{k=1}^{K+1}\alpha^k} \end{aligned} \tag{12}$$

To switch to M/G/1/s model let's apply the Khintchine-Pollaczek rule. Let the multiplier coefficient be γ:

$$\gamma = \frac{\left(1 + \frac{\sigma_{CF}^2}{\left(\overline{T_{CF}}\right)^2}\right)}{2} \tag{13}$$

where $\overline{T_{CF}}$ is the average service time for the specific CF. σ_{CF}^2 is the variance of service time for the specific CF.

When the new probabilities of system states for M/G/1/s model of wireless channel are:

$$\begin{aligned} \widehat{P_{i\in 1I}} &= \frac{\alpha^i\cdot\gamma^{\frac{i-1}{2}}}{\sum_{i\in 1I}\alpha^i\cdot\gamma^{\frac{i-1}{2}} + \sum_{i\in 2I}\alpha^i\cdot\gamma^{\frac{i}{2}}} \\ \widehat{P_{i\in 2I}} &= \frac{\alpha^i\cdot\gamma^{\frac{i}{2}}}{\sum_{i\in 1I}\alpha^i\cdot\gamma^{\frac{i-1}{2}} + \sum_{i\in 2I}\alpha^i\cdot\gamma^{\frac{i}{2}}} \end{aligned} \tag{14}$$

where $I = 1I + 2I$ is the set of states, 1I – is the subset of odd states, 2I – is the subset of even states.

The average number $\overline{n}$ of packets in the system is:

$$\overline{n} = \sum_{i=1}^{s+1} i \cdot \widehat{P_i} \tag{15}$$

The guaranteed delivery rate of network applications traffic u_{CF}^g passing through the wireless channel with a specific CF:

$$u_{CF}^g = \Lambda \cdot (1 - P_{s+1}) \tag{16}$$

or

$$u^g_{CF} = \Lambda \cdot \left(1 - \frac{\alpha^i \cdot \gamma^{\frac{i-1}{2}}}{\sum_{i \in 1I} \alpha^i \cdot \gamma^{\frac{i-1}{2}} + \sum_{i \in 2I} \alpha^i \cdot \gamma^{\frac{i}{2}}}\right) + \Lambda \left(1 - \frac{\alpha^i \cdot \gamma^{\frac{i}{2}}}{\sum_{i \in 1I} \alpha^i \cdot \gamma^{\frac{i-1}{2}} + \sum_{i \in 2I} \alpha^i \cdot \gamma^{\frac{i}{2}}}\right) \tag{17}$$

The actual service time T^a_{CF} is determined as:

$$T^a_{CF} = \frac{\overline{n}}{u_{CF}} \tag{18}$$

The efficiency of packet delivery using the specific CF is determined as:

$$U_{CF} = \frac{1}{T^a_{CF}} \tag{19}$$

Let's obtain the actual service time $T^a_{CF}(\Lambda)$ and the efficiency of packet delivery $U_{CF}(\Lambda)$ for WLAN (Fig. 1) with characteristics and parameters (Table 1) $CF = DCF, PCF$.

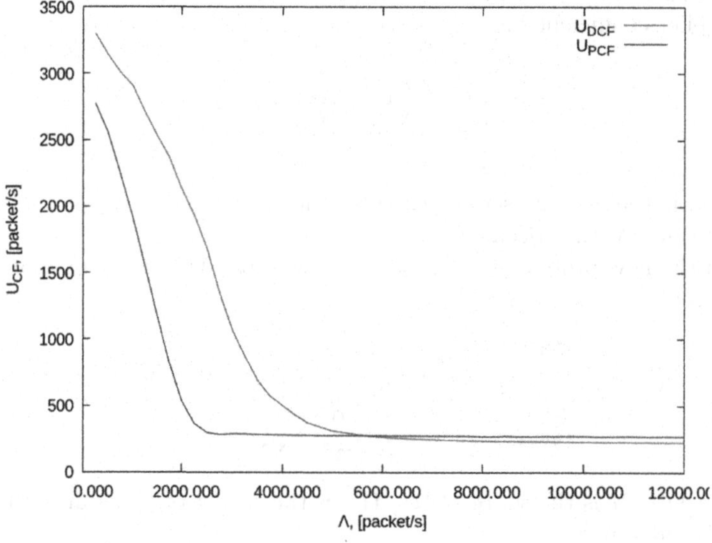

Fig. 7. The efficiency of packet delivery U_{CF} $CF = [DCF, PCF]$

6 Methodology for Guaranteed Delivery Rate Estimation

For enterprise WLAN with structure (Fig. 1), parameters (σ_c, CW_{max}, CW_{min}, t_{SIFS}, t_{DIFS}, l_{ack}, C, Λ, K, $\bar{l}$, $l_{CF-POLL}$) and specified CF we provide methodology to determine guaranteed delivery rate of traffic for specified guaranteed delivery time. The methodology includes the following steps:

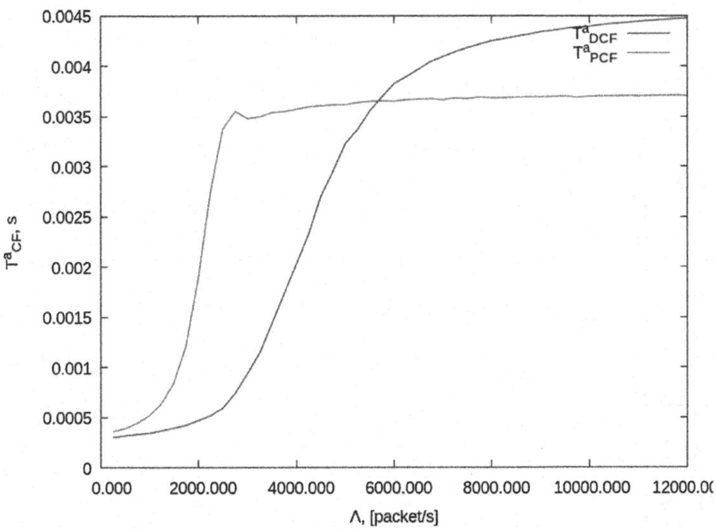

Fig. 8. Actual service time T^a_{CF} $CF = [DCF, PCF]$

Step 1. Determine the configuration parameters of the wireless channel, specific CF and determine the range of WLAN characteristics and parameters set.

Example: WLAN Fig. 1, $K = 4$, $CF = DCF$, WLAN characteristics and parameters set Table 1.

Step 2. Using the ns-3 program which describes the topology of the WLAN of interest, the set of characteristics and configuration parameters, obtain dependencies $\overline{T_{CF}}(\Lambda)$, $\sigma^2_{CF}(\Lambda)$, $u^{CF}(\Lambda)$ and plot the numerical results.

Example: See results on Figs. 2, 3 and 6.

Step 3. Using equations (14) from Sect. 5, obtain $P_{s+1}(\alpha)$

Example: The denial of service probability p_d for $CF = DCF$ and $K = 4$ vs. α (Fig. 9).

Step 4. For specified T_g:

(1) using obtained analytical results from Sect. 5. (Figs. 7 and 8), obtain Λ for $T^a_{CF} = T_g$;

(2) using results of simulation for $\overline{T_{CF}}(\Lambda)$, $\sigma^2_{CF}(\Lambda)$ and equation (13) obtain γ;

(3) using Eqs. (8) and (9) obtain α;

(5) using α and Fig. 9 obtain P_{s+1};

(6) using Eq. (16) obtain $u^a_C F$;

Example:

(1) let $T_g = 0,001s$, using Figs. 7 and 8 $\Lambda \approx 3100 packet/s$;

(2) using results of simulation Figs. 2 and 3 and Eq. (13) $\gamma \approx 0.56$;

(3) using Eqs. (8) and (9) $\alpha \approx 1.0$;

(5) using α and Fig. 9 $P_{s+1} \approx 0.05$;

(6) using Eq. (16) obtain $u^a_{DCF} \approx 2945 packet/s$;

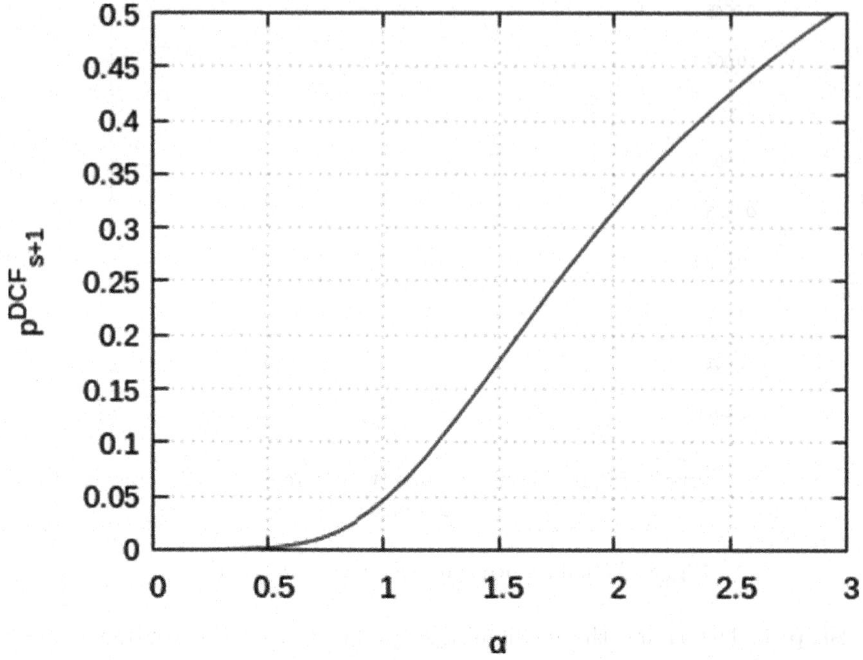

Fig. 9. Denial of service probability p_d (DCF)

7 Conclusion

The main contributions in this research are as follows: The wireless channel service time and variance for DCF and PCF wireless channel media access protocols under certain conditions (WLAN parameters and characteristics) are obtained using NS-3. The analytical model of enterprise WLAN is developed. Using the equations and the simulation results, the dependencies on actual service time and intensity of traffic are obtained. A methodology to obtain the guaranteed intensity of traffic is proposed. An Example to obtain the guaranteed intensity of traffic is provided.

References

1. Szigeti, T., Barton, R., Hattingh, C., Briley Jr., K.: End-to-End QoS Network Design, 2nd edn. Cisco Press, Indianapolis (2014)
2. RFC 1633 - Integrated Services In The Internet Architecture: An Overview. Tools. Ietf. Org, 2020, https://tools.ietf.org/html/rfc1633. Accessed 20 Sept 2020
3. Bianchi, G.: Performance analysis of the IEEE 802.11 distributed coordination function. IEEE J. Sel. Areas Commun. **18**, 535–547 (2000)
4. Ziouva, E., Antonakopoulos, T.: CSMA/CA performance under hightraffic conditions: throughput and delay analysis. Comput. Commun. **25**, 313–321 (2002)

5. Yang, X.: Performance analysis of priority schemes for IEEE 802.11 and IEEE 802.11e wireless LANs. IEEE Trans. Wirel. Commun. **4**, 1506–1515 (2005)
6. Malone, D., Duffy, K., Leith, D.: Modeling the 802.11 distributed coordination function in nonsaturated heterogeneous conditions. IEEE/ACM Trans. Netw. **15**(1), 172 (2007)
7. Daneshgaran, F., Laddomada, M., Mesiti, F., Mondin, M.: Unsaturated throughput analysis of IEEE 802.11 in presence of non ideal transmission channel and capture effects. IEEE Trans. Wirel. Commun. **7**, 1276–1286 (2008)
8. Shaaban, S., et al.: Performance evaluation of the IEEE 802.11 wireless LAN standards. World Congress Eng. I (2008)
9. Ali, Q.: Performance evaluation of WLAN internet sharing using DCF & PCF modes. Int. Arab J. e-Technol. (2009)
10. Tinnirello, I., Bianchi, G., Xiao, Y.: Refinements on IEEE 802.11 distributed coordination function modeling approaches. IEEE Trans. Veh. Technol. **59**, 1055–1067 (2010)
11. Yin, Y., Gao, Y., Hei, X.: Performance evaluation of a unified IEEE 802.11 DCF model in NS-3. In: Song, H., Jiang, D. (eds.) Simulation Tools and Techniques. SIMUtools2019. Lecture Notes of the Institute for Computer Sciences, Social Informatics and Telecommunications Engineering, vol. 295, pp. 395–406. Springer, Cham (2019)
12. Deutsch, P., Veyster, L., Cheng, B.-N.: LL SimpleWireless: A Controlled MAC/PHY Wireless Model to Enable Network Protocol Research
13. Nsnam. n.d., ns-3: a discrete-event network simulator for internet systems. Available on https://www.nsnam.org/
14. IEEE Standard for Information technology - Telecommunications and information exchange between systems Local and metropolitan area networks-Specific requirements - Part 11: Wireless LAN Medium Access Control (MAC) and Physical Layer (PHY) Specifications. IEEE Std 802.11-2016 (Revision of IEEE Std 802.11-2012), 14 Dec 2016

SARSA Based Method for WSN Transmission Power Management

Alexander Alexandrov(✉) and Vladimir Monov

Institute of Information and Communication Technologies - Bulgarian Academy of Sciences, Akad. G.Bonchev 1113, Sofia, Bulgaria
{akalexandrov,vmonov}@iit.bas.bg
http://www.iict.bas.bg

Abstract. The scope of this research is to propose an adaptive machine learning approach which can help the WSN's nodes to manage their transmission power and to improve the internode wireless communications. The optimized transmission power has benefits in terms of WSN energy consumption and RF interlink interference. The paper proposes an adaptive method of a wireless sensor node based on Multi-Layer Perceptron (MLP) network representation and machine learning. The presented in the paper approach uses the SARSA (State-Action-Reward-State-Action) algorithm which is a form of reinforcement machine learning. The aim of the new method is to improve the sensor nodes Transmission Power Management (TPM) process. This inspires many practical solutions that max-imize resource utilization and prolong the shelf life of the battery-powered wireless sensor networks.

Keywords: ANN · MLP · Transmission power control

1 Introduction

Internode communications are usually the most energy consuming event in Wireless Sensor Networks (WSNs). One way to significantly reduce energy consumption is by applying an adaptive transmission power management techniques. This approach dynamically adjusts the transmission power in which depends on factors as wireless link Quality of Service (QoS) and the wireless node Received Signal Strength (RSS) value. As is illustrated on the Fig. 1 the reliable connection between sensor nodes depends on the distance between nodes, received signal strength and the level of the existing RF noise. In the real environment, the deviation between the needed transmission power for reliable communication between WSN nodes at one and the same distance can reach dramatically high values because of the mentioned above factors.[1]

The task of the WSN power management (WSN-PM) stays more complex when the propagation of the RF transmission signal is influenced by the factors which are time changeable as cyclic sources of RF noise, interference RF sources with variable sizes and etc. Therefore, one of the possible ways to solve the

© Springer Nature Switzerland AG 2020
V. M. Vishnevskiy et al. (Eds.): DCCN 2020, CCIS 1337, pp. 82–93, 2020.
https://doi.org/10.1007/978-3-030-66242-4_7

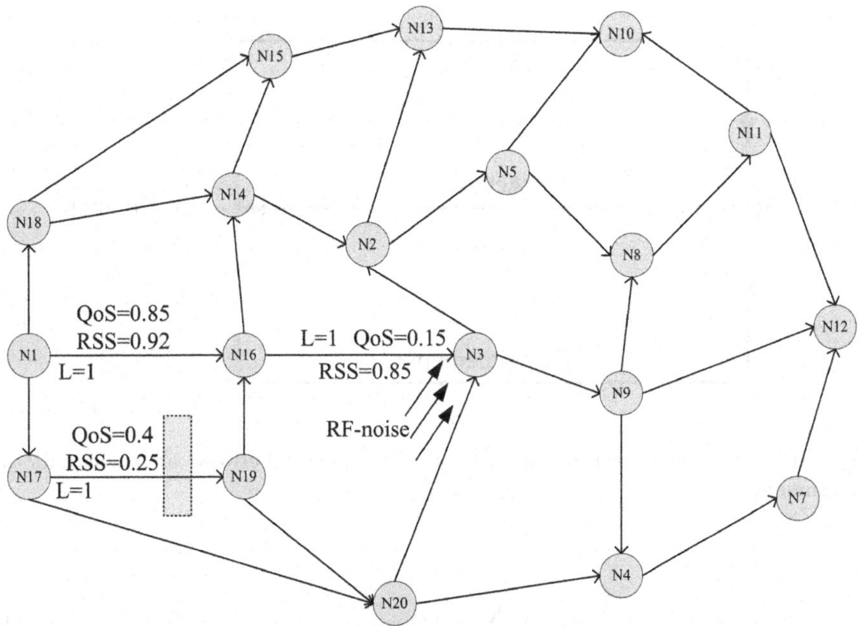

Fig. 1. Wireless sensor network with RF barrier between Note 17 and Node 19 and RF noise around Node 3.

complexity problem is to use an adaptive method of a wireless sensor node based on a self-learning Artificial Neural Network (ANN). The term machine learning is a set of algorithms and statistical models that software application use to perform a specific task without using explicit instructions, relying on patterns and inference instead. It is seen as a subset of artificial intelligence. The method takes place when the problem is too complicated to be solved in real time, or in case that is not impossible the problem to be solved in a classical way. One of the machine learning methods is the Reinforcement Learning (RL) [2]. The method of Reinforcement Learning uses an agent executor - environment approach and is based on the concept of reward. Reinforcement Learning involves two main entities: an agent and the environment (Fig. 2). The agent plays as a learner and decision-maker at the same time, while the environment is unpredictable and unknown which influences the agent's performance.

Where:

S_k - represents the status of the environment;
a_k – actions – decisions of the Agent. It is noted by default that the agent can choose among a predefined list of possible actions.
r_k - feedback called reward which evaluates the effect of the actions a_k

The objective function of the reinforcement learning approach is as follows: In our case the target of the objective function for the reinforcement learning approach is to maximize the cumulative reward rk as follows:

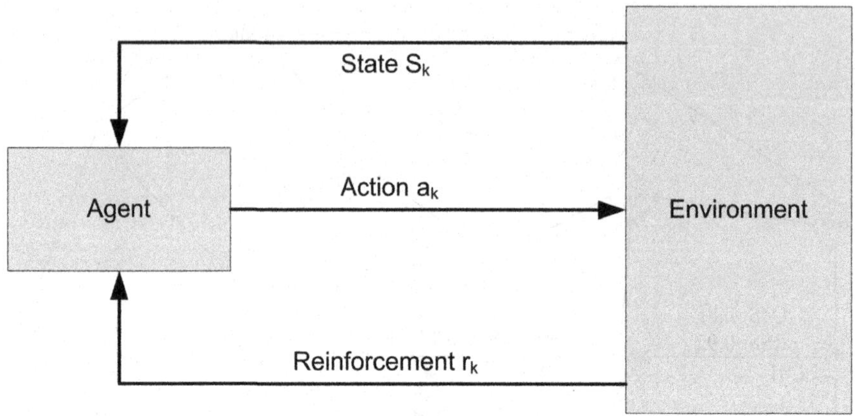

Fig. 2. Reinforcement Learning model - the interaction between an agent and the environment.

$$MAX \sum_{k=0}^{k=\infty} \lambda^k \mathbf{E}[r_k(a_k, s_k)] \tag{1}$$

where:

λ^k - is the probability distribution of the k reward

A possible implementation of the reinforcement learning methods is the Temporal Difference (TD) Learning approach. The TD approach refers to a class of the model-free reinforcement methods which learns by the state of the current estimate of the value function. The possible options of the TD methods are the Q-Learning, SARSA, Rescorla-Wagner, PVLV and etc. In the current research, we are focused mainly on the SARSA method and the proposed algorithm as relatively the most adaptive and flexible for the needs of the WSN power management. SARSA is part of the group of Temporal Difference (TD) algorithms used in Reinforcement Learning and it was proposed in 1994 by Rummery & Niranjan [3,4]

2 Related Work

To overcome the disadvantages of the proactive and reactive techniques, machine learning represents an attractive solution [5] to reach a defined goal by learning the dynamics of the WSNs [6,7], predicting and adapting the transmission power values in different conditions. The objective is making WSNs autonomous without the intervention of developers and users to set the transmission power.

To the best of our knowledge, only a few contributions have applied machine learning in TPC, mainly Reinforcement Learning (SARSA, Q-Learning and etc.) and fuzzy logic [8].

Q-Learning in WSNs has been used as WSN management approach in the literature but mainly for path selection in routing protocols and sleeping techniques, maintaining constant learning factors [9]. The static values would either bring the system slowly to convergence or make the system too reactive if the learning factor is constantly low or high respectively. The reinforcement learning [10]method is similar to Q-Learning but has some additional benefits. In [11] the reinforcement learning approach is used to calculate and update the weights in the Neural Networks. In [12]the reinforcement learning is used for adaptive routing in WSNs. Other approach used reinforcement learning is described in [13] for task scheduling mobile nodes communication in WSns. The authors from [14,15] propose different variations of Task Scheduling Algorithm for adaptive power management in WSNs SARSA [16–18] as a reinforcement learning method is similar to Q-Learning. The main difference between SARSA and Q-Learning is that SARSA uses an on-policy algorithm which means that SARSA calculates the Q-value based on the action executed by the current policy and is contrary to the off-policy used in Q-Learning.

3 Proposed Model and SARSA Based Method

3.1 Implemented Core Technologies – Modified Multi-layer Perceptron

In the current development, we represent the WSN as a set of Multi-Layer Perceptron's (MLPs). Every wireless node can be represented as a perceptron consisting of four components, i.e. inputs, weights, activation function, and output. The generalized architecture of an MLP perceptron related to the sensor node power management is given in Fig. 3.

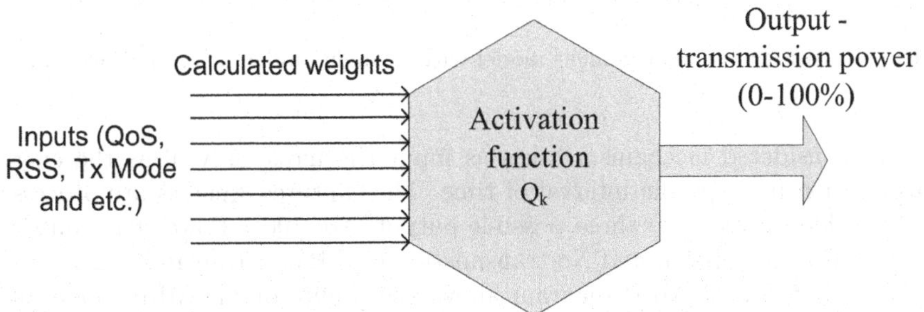

Fig. 3. MLP based model of WSN node.

A is shown on the diagram above, different inputs with different weights are captured, and in our implementation if some weights match threshold values, it may activate the function to generate an output. The proposed MLP architecture consists of three types of layers such as input, hidden and output similar to

[12,13]. Our approach is utilizing unsupervised back propagation based learning in which threshold values are used as the activation function. If output values do not matches, it can be adjusted.

The proposed MLP model of the wireless sensor node is shown in Fig. 4. The proposed relatively low number of the nodes (only 4) in the hidden layer is related to the hardware limitations of the simulated sensor node embedded microcontroller and memory. A bigger number of nodes in the hidden layer will increase drastically the needed resources and time for calculation and will consume more power which is an real issue in the battery powered sensor nodes.

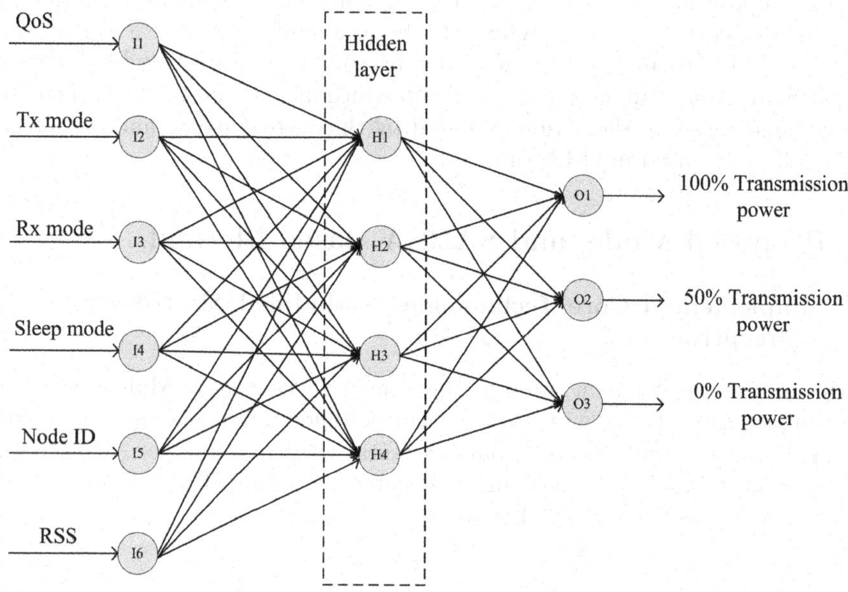

Fig. 4. MLP power management model with a single hidden layer of WSN node.

The considered mechanism takes as input the mode as well as the energy consumption in a specific interval of time. The captured weights are processed in the hidden layer. The three possible outputs are 100% Power transmission, 50% Power transmission and No transmission (0% Power transmission).

According to the MLP diagram above the inputs of the MLP model of a sample WSN node are as follows:

- QoS – Quality of Service. This parameter is calculated by the wireless sensor node and depends on the bandwidth, packet delay, and packet loss real-time measurement;
- Tx mode – the mode when the wireless sensor device transmits RF packets. The transmission mode refers to the mechanism of transferring data between two devices connected over the network. In the current research is considered

only the Tx mode parameter transmission power which can vary from 0 to 100% of the existing RF device transmission capacity;
- R_x mode – mode of the receiving data packets from the sensor node;
- Sleep mode – mode when the sensor node doesn't transmit or receive any data;
- Node ID – the unique ID assigned of the node during the WSN forming and configuration.
- RSS – Received Signal Strength. Also referred to as RSSI (Received Signal Strength Indicator) is the parameter calculated on the basis of the RF power presented in the received radio signal. RSS is measured in dB and typically vary between 0dBm (excellent) and -110 dBm (very poor).

The output of the proposed MLP model is related to the level of power transmission as the main energy utilization parameter and the key factor for transmission power management.

In the active mode (Tx or Rx mode), the node fall in the following active patterns: 100% transmit power, 50% transmit power and 0% transmit power.

The proposed method, considered in this paper consists of three main phases – data collection, learning, and results in phases.

Data collection – in this phase, every wireless sensor node keeps track of the RF packets received from neighbors, the packets forwarded to neighbors and the average energy consumption in a specific period of time. This phase may take time to get a few optimal values of the system usage over a specific period of time.

The following parameters are collected and calculated during the initial data collection phase:

- the average value of the RSS signal for the last 10 received RF packets;
- the average value of the QoS parameter calculated for the last 10 received RF packets;
- Tx mode is set to 100% RF transmission power;
- Rx mode is set to state 0 (wait);
- Sleep mode is set to state 0(wait);
- NodeID is assigned and fixed during the WSN configuration;

Learning phase – in this phase, the modified MLP is trained to identify different communication sources which are located in its environment.

In this stage, the sensor node learns the parameters of packets which the neighbor wireless node receives and sending over a specific time and the related energy consumption.

The learning process is based on the SARSA algorithm as a form of implementation of Reinforcement Learning.

Based on the SARSA based Reinforcement Learning function definition described in details in [7], we have:

$$Q_k = r_k(s_k, a_k) + \gamma \, maxQ(S_k, a_k, w_k) \quad Q_k \in (0, 0.5, 1) \tag{2}$$

where: r_k – reinforcement value

s_k - state of the environment in stage k
a_k – action state
γ - discount factor $0 \leq \gamma \leq 1$
w_k – weights in stage k
$maxQ(s_k, a_k, w)$ - is the function value state/action

A diagram which describes the practical implementation of the proposed method is shown on Fig. 5.

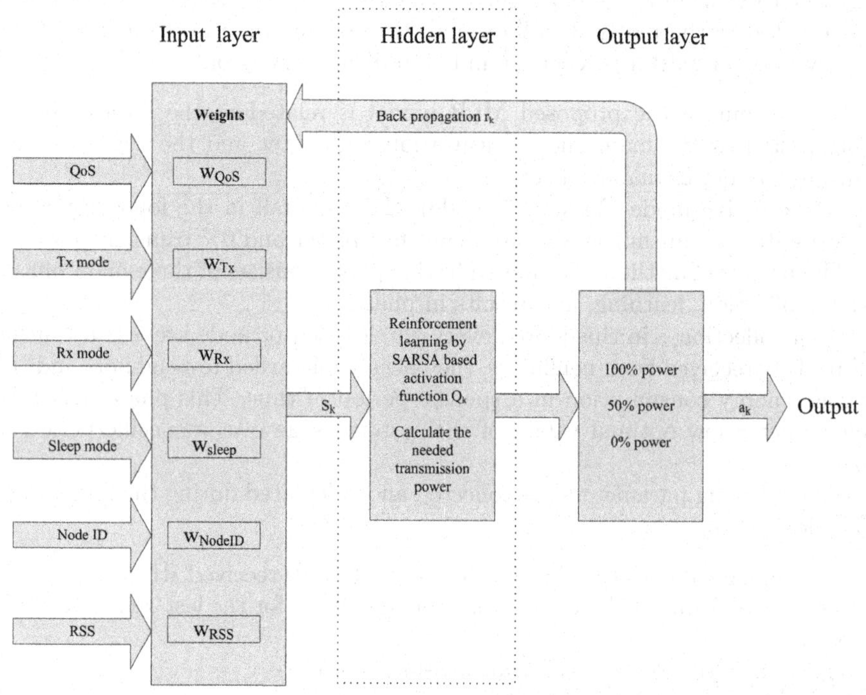

Fig. 5. The working mechanism of the proposed method.

In the proposed method the learning process starts with a preliminary fixed combination of weights (w_k), $s_k = 0$ and $a_k = 0$, forming the initial function value state/action $Q(s_k, a_k, w)$.

During the iterations stages of the learning process, Q-function generates a corrective signal and send it to the input of the system.

The reinforcement value r_k provides for every new state S a signal that can be reward or punishment. In our case, r_k takes values -1, 0 or 1.

The discount factor γ role is to mark the importance of future reward. In our case,γ takes only two values: 0 or 1.

The captured weights (w_{QoS}, w_{Tx}, w_{Rx}, w_{sleep}, w_{NodeID} and w_{RSS}) are multiplied with the input values and processed in the hidden layer for an activation function generation.

As part of the reinforcement learning process the SARSA algorithm follows the main actions shown on Fig. 2:

- the environment sends his state to the agent;
- the agent takes action in response;
- the environment sends a pair of next state and rewards back to the agent;
- the agent updates its knowledge with the feedback from the environment to evaluate its last action;
- the cycle continues until the environment sends a termination signal.

The updated equation, related to the modified SARSA algorithm is:

$$QS_t, a_t \leftarrow QS_t, a_t + \alpha[r_{t+1} + \gamma QS_t, a_{t+1} - QS_t, a_t] \tag{3}$$

where Q – action value which refers to state S and action a in moment t and t+1;

S_t – State of the environment in moment t;

a_t – action of the agent in moment t;

r_{t+1} - reinforcement value in moment t+1;

α – learning rate level $\alpha \in [0, 1]$;

γ – discount factor $\gamma \in [0, 1]$;

The NS2 simulation of the proposed method is based on the following pseudo code which was used in the software implementation of the proposed SARSA based method:

```
Prerequisites:
State G = 1, n;
Actions A = 1, n;
Reward R = G x A
Learning rate Alpha = 0,1;
Discount factor Gamma = 0, 1;
Function: SARSA_Learning (G. A, R, Alpha,Gamma)
{
Repeat {
Initialize S;
Calculate A from S using defined policy;
Repeat {
Take action A;
Observe S;
Q(s,a) ← (s,a) + Alpha(R + Gamma, Q(S',A') − Q(s,a));
S= S';
A= A';
Until S is terminal;
}
Until Q is not converged;
}
}
```

The learning phase takes a time which depends on the needed accuracy of the system.

Results – In this phase, the accuracy of the learning process is reviewed. The system is capable to calculate the needed transmission energy amount depend on the current topology of the neighbor nodes and their current status.

Based on the described above method were prepared simulations based on the SARSA algorithm and Q-Learning algorithm.

The executed simulation experiments based on NS2 network simulator shows that the learning phase has a typical length between 1 and 10 hours for 1000 sensor node based WSN and depends on the needed accuracy of the wireless system.

The experimental results from the NS2 simulations if the Q-Learning and SARSA algorithm are shown on Tables 1 and 2 accordingly.

Table 1. Results from the NS2 based Q-Learning simulation

Power level	A	R	Iteration steps
0 %	0,187	0,194	1
10 %	0,073	0,072	200
20 %	0,148	0,147	1000
30 %	0,036	0,036	2300
40 %	0,479	0,476	3000
50 %	0,57	0,062	11200

Table 2. Results from the NS2 based SARSA simulation

Power level	A	R	Iteration steps
0 %	0,112	0,121	1
10 %	0,341	0,339	100
20 %	0,348	0,337	250
30 %	0,215	0,287	500
40 %	0,421	0,429	700
50 %	0,036	0,036	6000

Using the NS2 network simulator we did multiple simulations with implemented Q-Learning and the proposed modification of SARSA algorithms to see the performance and the reliability of the calculated results.

The diagram from Fig. 6 shows that the proposed modification of SARSA algorithm is sensitively faster and reach the level of the predefined power level of the simulated sensor node. As is shown from the diagram after approximately 6000 iterations the SARSA algorithm achieved the predefined level of 50 percent power level compared to the 11000 iterations of the Q-Learning algorithm.

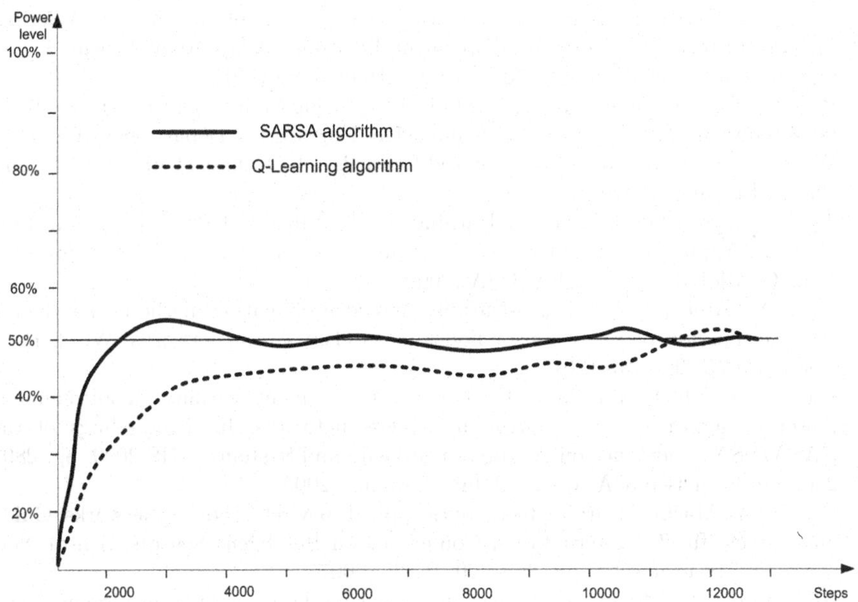

Fig. 6. The performance of SARSA compared to the Q-Learning algorithm

4 Conclusions

We have proposed a new model and adaptive method for wireless sensor node power management based on the SARSA algorithm which uses an AI Reinforcement Learning approach. The proposed sensor node model is developed on the basis of Multi-Layer Perceptron (MLP) network representation and machine learning.

The simulation results show that the implemented in a wireless sensor node SARSA algorithm has sensitively better performance compared to the related Q-Learning algorithm.

At the same time, the software implementation of the SARSA algorithm is more compact and uses less sensor node microcontroller resources compared to the Q-Learning algorithm.

This inspires many practical solutions that can maximize resource utilization and prolong the shelf life of the battery-powered wireless sensor networks.

References

1. Gummeson, J., Ganesan, D., Corner, M.D., Shenoy, P.: An adaptive link layer for heterogeneous multi-radio mobile sensor networks. IEEE J. Sel. Areas Commun. **28**, 1094–1104 (2010)
2. Lange, S., Gabel, T., Riedmiller, M.: Batch reinforcement learning. In: Wiering, M., van Otterlo, M. (eds.) Reinforcement Learning. Adaptation, Learning, and Optimization, vol. 12, pp. 45–73. Springer, Heidelberg (2012)
3. Rummery, G., Niranjan, M.: On-line Q-learning using Connectionist systems, Technical report no.166, University of Cambridge, Engineering Department (1994)
4. Watkins, J.C.H.: Learning from Delayed Rewards, Ph.D. thesis, University of Cambridge, England (1989)
5. Torrey, L., Shavlik, J.: Transfer learning. In: Handbook of Research on Machine Learning Applications and Trends: Algorithms, Methods, and Techniques, pp. 242–264. IGI Global, Hershey, PA, USA (2009)
6. Sung, Y., Ahn, E., Cho, K.: Q-learning reward propagation method for reducing the transmission power of sensor nodes in wireless sensor networks. Wirel. Pers. Commun. **73**, 257–273 (2013)
7. Udenze, A., McDonald-Maier, K.: Direct reinforcement learning for autonomous power configuration and control in wireless networks. In: Proceedings of the NASA/ESA Conference on Adaptive Hardware and Systems, AHS 2009, pp. 289–296, San Francisco, CA, USA, 29 July-1 August 2009
8. Forster, A.: Machine learning techniques applied to wireless ad-hoc networks: Guide and survey. In: 3rd International Conference on Intelligent Sensors, Sensor Networks and In-formation, pp. 365–370. IEEE (2007)
9. Alexandrov, A., Monov, V.: Q-learning based model of node transmission power management in WSN. In: Proceedings of International Conference on Big Data, Knowledge and Control Systems Engineering BdKCSE 2018, pp. 15–112, 21–22 November 2018
10. Kaelbling, L.P., Littman, M.L., Moore, A.P.: Reinforcement learning: a survey. J. Artif. Intell. Res. **4**, 237–285 (1996)
11. Hocenski, Z., Antunovic, M., Filko, D.: Accelerated gradient learning algorithm for neural network weights update. In: Lovrek, I., Howlett, R.J., Jain, L.C. (eds.) KES 2008, Part I. LNCS (LNAI), vol. 5177, pp. 49–56. Springer, Heidelberg (2008)
12. Kadam, K., Srivastava, N.: Application of machine learning (reinforcement learning) for routing in Wireless Sensor Networks (WSNs). In: 1st International Symposium on Physics and Technology of Sensors (ISPTS-1), Vol. 2012, pp. 349–352, Pune (2012). https://doi.org/10.1109/ISPTS.2012.6260967
13. Cirstea, C., Davidescu, R., Gontean, A.: A reinforcement learning strategy for task scheduling of WSNs with mobile nodes. In: 2013 36th International Conference on Telecommunications and Signal Processing (TSP), pp. 348–353, Rome (2013). https://doi.org/10.1109/TSP.2013.6613950
14. Zhang, B., Wu, W., Bi, X., Wang, Y.: A task scheduling algorithm based on Q-learning for WSNs. In: Liu, X., Cheng, D., Jinfeng, L. (eds.) ChinaCom 2018. LNICST, vol. 262, pp. 521–530. Springer, Cham (2019). https://doi.org/10.1007/978-3-030-06161-6_51
15. Wei, Z., Zhang, Y., Xiangwei, X., Shi, L., Feng, L.: A task scheduling algorithm based on Q-learning and shared value function for WSNs. Comput. Netw. **126**, 141–149 (2017)

16. Van Seijen, H., Van Hasselt, H., Whiteson, S., Wiering, M.: A theoretical and empirical analysis of expected Sarsa. In: 2009 IEEE Symposium on Adaptive Dynamic Programming and Reinforcement Learning, Nashville, 30 March-2 April 2009, pp. 177–184 (2009). https://doi.org/10.1109/ADPRL.2009.4927542
17. Yu, S., Zhou, J., Li, B., Mabu, S., Hirasawa, K.: Q value-based Dynamic Programming with SARSA Learning for real time route guidance in large scale road networks. In: Proceedings of the International Joint Conference on Neural Networks, pp. 1–7 (2012)
18. Wen, F., Wang, X.: Sarsa learning based route guidance system with global and local parameter strategy. IEICE Trans. Fundam. Electron. Commun. Comput. Sci. **E98A**(12), 2686–2693 (2015)

Method of Frequency Coding in Microwave RFID

Sergey Suchkov, Viktor Nikolaevtsev(✉), and Dmitry Suchkov

Saratov State University named after N.G. Chernyshevsky,
Astrakhanskaya str. 83, 410012 Saratov, Russia
nikolaevcev@yandex.ru

Abstract. The concept of constructing of a radio frequency identification system at frequencies 6 GHz based on frequency coding using radio frequency identification tags on resonators on bulk acoustic waves with high Q factor is presented. The axially symmetric 3D structure of the film piezoelectric resonator with the quasi-single-crystal AlN film on the acoustic Bragg reflector which is formed from the alternating molybdenum and silicon oxide layers was considered. Optimization of the acousto-electronic FBAR resonator design for the RFID tag in the frequency band 10–13 GHz was accomplished. The dependence of the top electrode thickness upon the required resonant frequency of the FBAR resonator was obtained. Design of the RFID tag in the frequency band 10–13 GHz was proposed. There are real technical possibilities for creating of the RFID system in the frequency band 6–30 GHz with the FBAR resonators. Such systems will be protected to electromagnetic fields and ionizing radiation and they will have the high temperature stability characteristic of the FBAR resonators.

Keywords: Radio frequency identification · RFID tag · Anticollision problem · Frequency coding · FBAR resonators.

1 Introduction

Currently, radio frequency identification (RFID) systems are widely applied all around the world for building of automatic systems for accounting and control of the large flows of goods, passengers, vehicles and other objects. These systems meet the demands of the operative control and logistic operations and reduce the negative influence of the "human factor".

The major part of RFID systems are based on radio frequency identification tags (RFID tags) on integrated circuits (ICs) which operate in the frequency bands "130 kHz", "13.5 MHz", "900 MHz", and most recently, in the "2.4 GHz" band. Depending upon the frequency band, they have the distance of the objects identification 10 cm to 5 m. The advantage of such tags is the low price, and the

This work was supported by the Ministry of Education and Science of Russia in the framework of project No. FSRR-2020-0005.

© Springer Nature Switzerland AG 2020
V. M. Vishnevskiy et al. (Eds.): DCCN 2020, CCIS 1337, pp. 94–104, 2020.
https://doi.org/10.1007/978-3-030-66242-4_8

disadvantages are associated with the semiconductor chip of the IC, which is sensitive to the external electromagnetic fields, the ionizing radiation, and also to the temperature changes. Under these influences, the code in the IC RFID tag can be changed or even destroyed. Therefore, the past 15 years, RFID systems based on tags on surface acoustic waves (SAWs) reflective delay lines has been developed. Such RFID systems are increasingly being realized to account for particularly valuable and critical objects [1–3]. RFID tags on SAW are resistant to the indicated influences and changes, and they also require less energy of the interrogation impulse for the same identification frequency bands than the RFID tags on IC. In particular, the system of accounting for cargo turnover at the International Space Station was built using such RFID tags [4]. SAW RFID tags were developed for the open frequency bands "433 MHz", "900 MHz", "2.45 GHz" and "6 GHz" [5,6].

One of the difficulties in using of the RFID systems is the presence of domestic and industrial electromagnetic interferences. They reduce the distance of the identification of objects and they are significant for all the indicated frequency bands, except for the "6 GHz" frequency band. However, in the coming years the mobile communication systems operation is also planned to expand on this frequency band.

Therefore, it is advisable to consider those frequency bands in which the domestic and industrial electromagnetic interferences are not expected in the near future. As a consequence, we can consider the frequency band 6–30 GHz. In this band, the creation of digital ICs and, therefore, RFID tags on ICs are not expected in the near future. Therefore, the element base for RFID tags for this band could be the analog acoustoelectronic devices. It is obvious that SAW RFID tags are not applicable in the indicated band due to the high propagation losses [6]. However, if one switch from the time domain of the formation of code signals to the frequency domain, in which the signal is not the pulse delay time, but frequency of the signal which is excited by the tag, then one can use the thin film resonators on the bulk acoustic waves (FBAR), which have relatively small losses on resonant frequency (2–3 dB) and high Q factor (about 1000) [7,8] in the considered frequency band.

Another important problem of RFID systems is the collision of code signals from several RFID tags at the input of a code reader, which arises in the case of simultaneous identification of many RFID tags. In the RFID systems with the SAW tags the different methods are used: the spatial discretization method [1] the correlation method [3,4,9], the multilateration method [10,11]. For the frequency coding, the spatial discretization method becomes very convenient due to the high directivity of the electromagnetic signal propagation with a small reader antenna size. In addition the application of the unlimited anticollision method based on the multilateration method using the multiband tags and the multi-antenna reader system is also simplified [11].

2 Frequency Coding

For frequency coding, the reader must generate and radiate enough long in time (up to 10 s) interrogation impulses, which are sequences of short radio pulses in which the carrier frequency changes consequentially in discrete in accordance with the frequency code zones. When a certain frequency of the interrogation impulse coincides with the resonant frequency of the coded FBAR, oscillations occur in it and they decay exponentially after the exposure of the interrogation impulse of the reader according to the law

$$e^{-\frac{\omega_0}{2Q}t}, \tag{1}$$

where $\omega_0 = 2\pi f_0$, f_0 is the resonant frequency of the FBAR and Q is its Q factor.

Then the e times decrease time of the oscillation amplitude is

$$t_e = \frac{Q}{\pi f_0}. \tag{2}$$

The resonator is connected to the RFID tag antenna. The damped current oscillations in the resonator radiate the corresponding signal into space. The signal should be detected by the reader and it should be associated with the code frequency zone of the interrogation impulse. Thus, if the RFID tag consists of the several FBARs with different resonant frequencies, their signals form a number of frequencies in the reader, which is the code of this tag.

We consider the RFID system in the frequency band 10–13 GHz with 30 frequency code zones Δf_i = 100 MHz (i = 1.2, ..., 30). In such a system, for example, if there are eight FBARs in the RFID tag, the number of possible codes will be

$$N_K = C_{30}^8 = 5\ 825\ 925. \tag{3}$$

This number of codes is sufficient for the most possible applications of very expensive RFID tags on FBAR. For the code frequency f_i the interrogation pulse duration T_0 is determined by the frequency code zone bandwidth $T_0 = 1/\Delta fi =$ $= 10$ ns.

In this case, the interrogation impulse code zones will contain from 100 to 130 oscillation periods, and their quantity depends upon the code zone number. This pulse duration determines the distance of the "dead zone", in which the code signal of the RFID tag will be suppressed by the reader signal. For this case the "dead zone" distance is 1.5 m.

In order to avoid signal spectral leakage into the neighboring code areas the required Q factor of the FBAR code resonators is also determined by the size of the code area and it must provide the emitted signal bandwidth significantly less than the code area bandwidth. Therefore, it will be enough to have a resonator bandwidth BR = 20 MHz, then the loaded Q factor of the resonators Q_i should be at least 500 at i = 1 and at least 650 at i = 30. Resonators with such Q factors of are really feasible in the frequency band 10–13 GHz [7,8].

It follows from (1) that the resonator decay time is $t_e = 16$ ns for all of the code zones. If the distance between the RFID tag is, for example, $R = 10$ m, the code pulse delay time is $T_D = T_0 + R/c$=76 ns, and the total time of one code signal receiving is $T_r = T_D + t_e$=92 ns.

Thus, the time of interrogation of one code zone could be taken equal to 100 ns. Then the time of interrogation of all 30 code zones would be about 3 μs. Such operating speed is sufficient to control fast moving objects. So for the velocity of about 300 km/h, during the system code reading, the displacement of the object will be less than 1 mm.

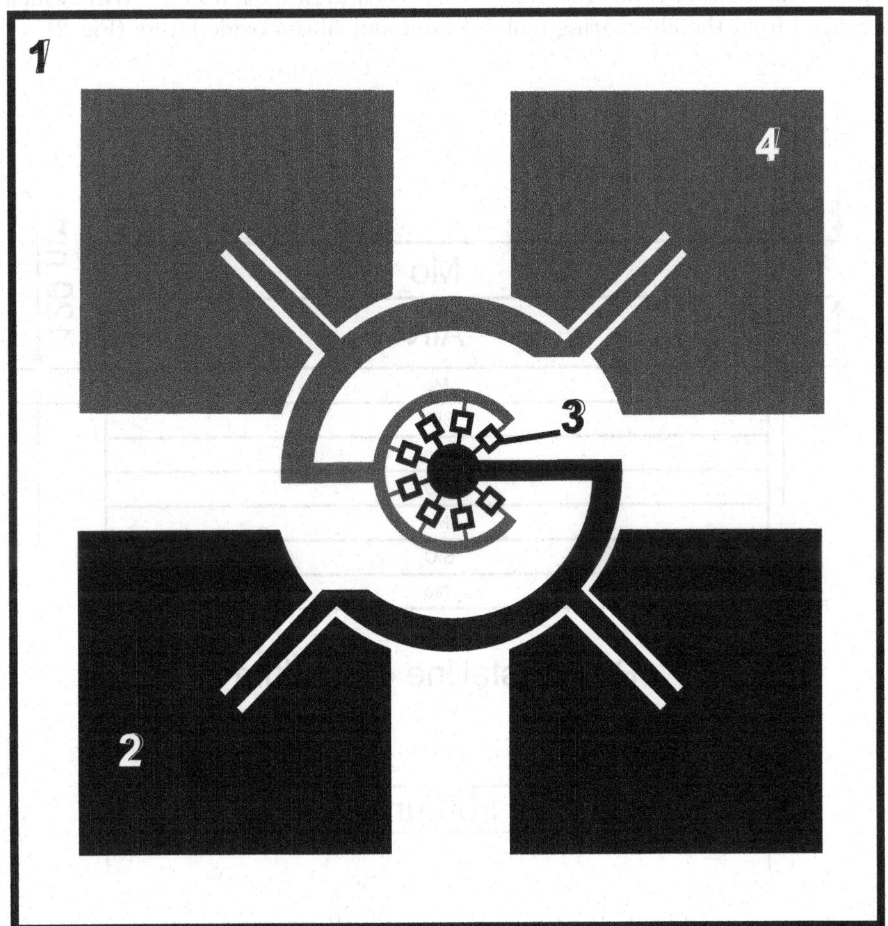

Fig. 1. The construction of the RFID tag on FBAR in the frequency band 10–13 GHz.

Shown in Fig. 1 is the construction of the proposed RFID tag. All the RFID tag including the hybrid antenna is fabricated on the polycrystalline corundum substrate (1). One of the elements of the antenna (2) is connected with the bottom electrodes of the FBAR resonators (3) and another element (4) is connected with the top electrodes. The construction of the hybrid antenna with the coplanar elements allows to obtain the antenna efficiency not less than 90% in the frequency band 10–13 GHz. The size of the antenna is 14 mm for the used frequency band. Microstrips feed the signal from the antenna elements to the FBARs and they are also the matching elements for the resonators with the antenna.

In order to analyze of the FBAR frequency responses, we consider the axially symmetric 3D structure of the film piezoelectric resonator with the quasi-single-crystal aluminum nitride (AlN) film on the acoustic Bragg reflector (ABR) which is formed from the alternating molybdenum and silicon oxide layers (Fig. 2).

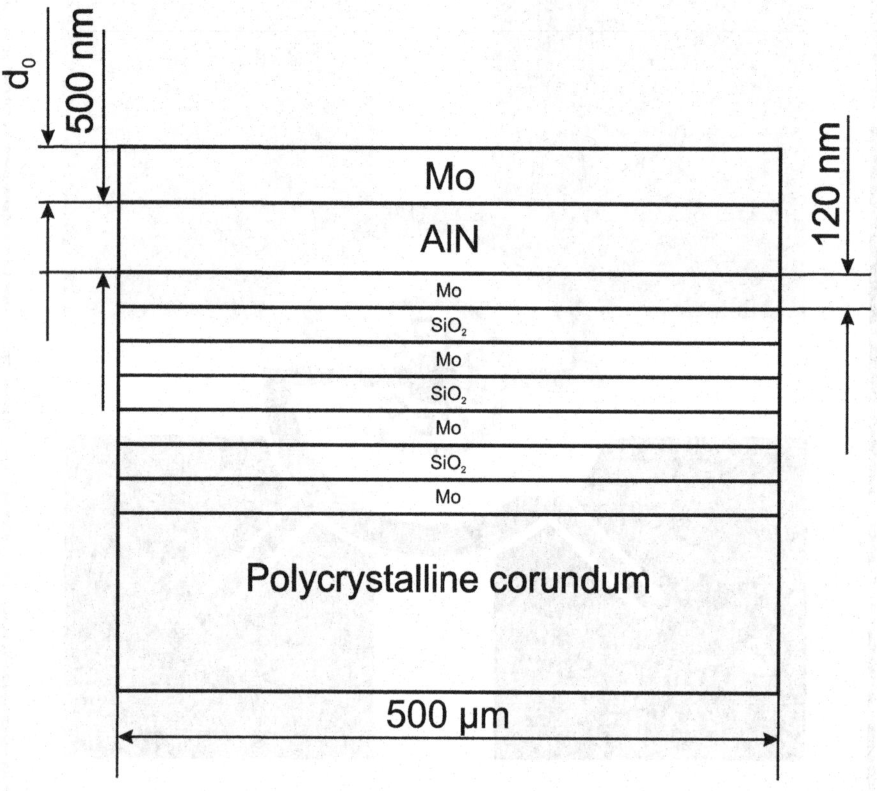

Fig. 2. The structure of the FBAR resonator.

We simulated the impedance real part frequency response of the piezoelectric resonator in the frequency band 10–13 GHz by means of the software COMSOL Multiphysics®.

As a result of maximum Q factor optimization of FBAR in considered frequency band the following sized of the FBAR structure were determined, which could be used for the RFID tags creation: the thickness of the AlN piezoelectric film is 500 nm, the thickness of alternating layers of molybdenum and silicon oxide is 120 nm, and the number of layers of the Bragg mirror is 7. Results of the simulation show that for the thickness of the ABR layers which is indicated above, the ABR reflection coefficient is not less than 0.95 in all the frequency band of 10–13 GHz. Shown in Fig. 3 is the simulated impedance real part ReZ_R frequency response of the FBAR for the thickness of the top electrode d_0=200 nm. The loaded Q factor of the resonator is about 1000, which is in the agreement with the results of the works [7,8].

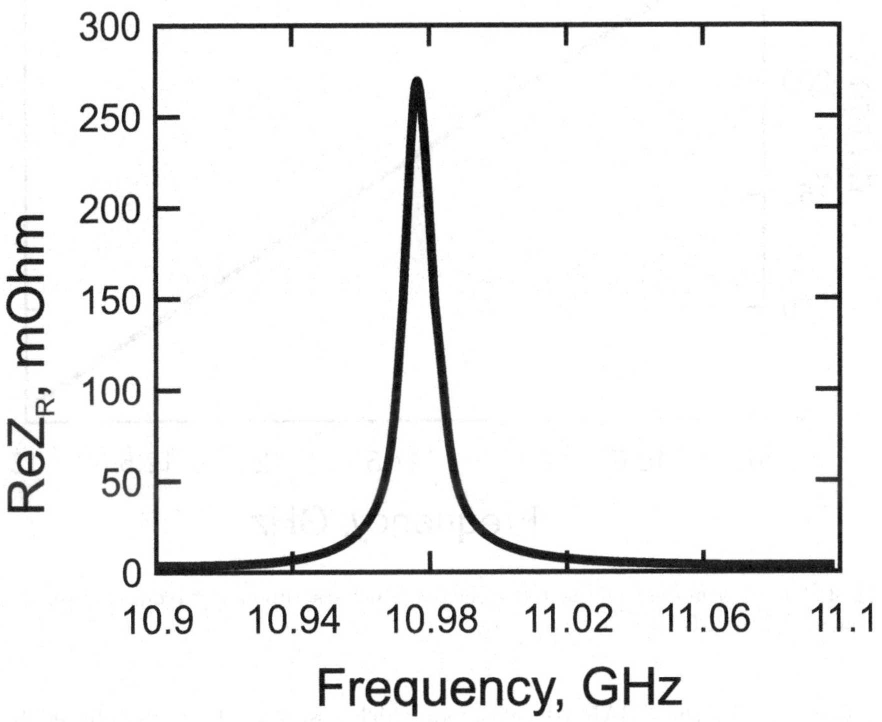

Fig. 3. The FBAR impedance real part frequency response.

The application of molybdenum piezoelectric resonator electrodes allows to obtain the desirable adhesion of the AlN film to the electrodes. The choice of silicon oxide as the second material for the ABR is associated with the value of the longitudinal BAW velocity, which is close to that of molybdenum, and, at the same time, a significant difference in acoustic impedances of these two materials.

In order to tune the FBAR resonant frequencies to the required frequency code positions (frequency bands of the code zones), the strong dependence of the FBAR resonance frequency upon the top electrode thickness is used. Shown in Fig. 4 is this dependence. Thus, in order to construct an RFID tag on an FBAR in the frequency band 10–13 GHz, it is required to vary the thickness of the top electrode in the range from 50 to 300 nm.

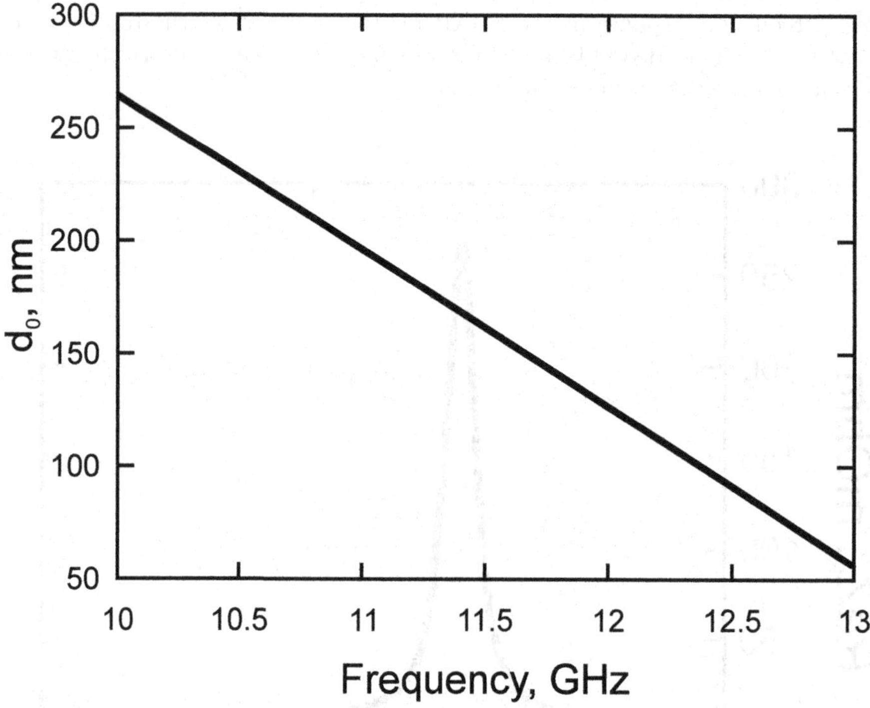

Fig. 4. The dependence of the top electrode thickness upon the required resonant frequency of the FBAR.

Changing of the FBAR top electrode thickness could be accomplished by means of local ion-beam etching technology [12]. As it follows from Fig. 4, the top electrode thickness d_0 of the FBAR for the i-th code zone, should be determined as

$$d_0 = 52 + 7i \pm 2 \ (nm). \tag{4}$$

So, for the resonator of the first code zone, d_0 could vary from 57 to 61 nm, and for the resonator of the 30-th code zone d_0 could vary from 260 to 264 nm. In this case, the bandwidth of the code zones is not less than 57 MHz, and the bandwidth of the bandgaps is not less than 43 MHz. It is sufficient to suppress the code signal in neighboring code zones for a frequency bandwidth 20 MHz for the loaded resonator.

The technology of FBARs fabrication with the noted parameters and RFID as a whole is quite feasible using modern photolithographic equipment, vacuum deposition and ion-beam etching installations, which make it possible to create the layered nanoscale structures [12] with an accuracy of $\pm$ 2 *nm*.

3 Suppression of False Signals

Since RFID systems operate with electromagnetic waves propagation in the open space then the surrounding objects reflect the electromagnetic waves and the spurious signals appear at the input of the reader. In order to eliminate such signals, an initial delay (about 1 μs) was used in the SAW RFID tags, after which the reception of code signals began. In such case the spurious signals from objects at a distance up 150 m are removed from the reader.

For frequency coding, one could introduce an additional delay line into the tag construction, but this greatly complicates the design of the tag which becomes very expensive. Another way to suppress the spurious reflected signals is the ability of the creation of a narrow radiation pattern of the reader antenna, which is not the technical problem in the frequency band 6–30 GHz. However, in such case it is necessary to orient the reader antenna to the object, which can be done manually or automatically, for example, by means of the phased antenna array (PAA).

Using the RFID system in stationary conditions allows one to determine the parameters of all the spurious signals previously and to suppress them during the software processing of the signals which are received by the reader.

The correlation method for suppress of the spurious signals based on differences in the amplitude-time characteristics of reflected and code signals is possible. So, the reflected signals have an almost rectangular amplitude-time characteristics which coincides with the characteristic of the code frequency in the interrogation impulse and the code pulses which are radiated by the RFID tag have the exponentially reducing amplitude-time characteristics. Thus, each received pulse should be analyzed on the presence of an exponential amplitude change, which can be done by the known correlation methods. This method also could be used for moving objects identification.

4 Solution of Anticollision Problem

The problem of collision of code signals emerges while the several tags respond to the interrogation impulse. In such case, the code pulses at the input of the reader without anticollision protection form an unreadable set of pulses.

As it was already indicated in the previous section, in the frequency band 6–30 GHz, one can apply the spatial separation of the code signals receiving from the RFID tags using a narrow radiation pattern of the reader antenna. Such a method of anticollision protection was proposed in 2002 [1], but its application was limited due to the difficulties of a narrow diagram obtaining in the used frequency bands of the SAW tags.

A lot of works including those implemented in practice [3,4,9], are devoted to the correlation method of anticollision protection of RFID systems with SAW tags. In brief, it consists in the fact that the reader radiates not one interrogation impulse, but it sequentially radiates the various sets of the time-position impulses with the enumeration of all the used time-position codes. If the time-position code is the same as the time response which is formed by a particular SAW tag, this tag radiates the autocorrelation signal of a large amplitude in its response. If there is no coincidence, then each of the tags which are located in the identification area of the reader antenna radiates a cross-correlation signal which is much smaller in amplitude. The correlation method of anticollision protection is accomplished, for example, in RFID system at the ISS [4], where up to 30 objects are identified simultaneously. However, this method requires a lot of time to sort through all possible codes for systems with a large number of codes and it is limited by the number of simultaneously recognized objects Nmax by Shannon–Hartley theorem [13]

$$N_{max} = BT, \tag{5}$$

where B is the system bandwidth, T is the signal time delay in the system.

The system which is realized at the ISS is close to the maximum number of codes which are identifying simultaneously in the RFID systems with the SAW tags.

The limitation which follows from (5) is explained by the fact that with an increase in the number of tags which are identifying simultaneously, the total level of cross-correlation signals begins to exceed the level of the autocorrelation signal.

For frequency coding, the correlation method application is also possible if impulses with carrier frequencies are simultaneously radiated from the given set of the frequency code combinations. But the limitations of its using which were indicated above also take a place.

Multiband tags are used in some applications of RFID systems with SAW tags [11]. In such systems only one code pulse is generated in each subband and there are 100 time code zones in each subband. Thus, in a triband system, $100^3 = 1$ million codes could be used, in a quadband system, $100^4 = 100$ million codes could be used, etc. In such the RFID system, the limited anticollision protection could be accomplished in one subband with the number of simultaneously identified objects equal to the number of code zones in this subband, in the condition that the group of monitored objects is selected with the same codes in the other subbands. In the RFID system with multiband tags the absolute anticollision protection is also possible with the simultaneous identification of any number of codes generated by pulses from all subbands. But it requires the multilateration signal processing method in the multi-antenna reader [14]. For frequency coding, a multiband method for limited anticollision protection could also be used. So, if one use four subbands: 10–13.3 GHz, 13.3–16.6 GHz,

16.6–19.9 GHz and 19.9–23.2 GHz with 32 code zones then each total number of codes will be 1 048 576 with simultaneous identification of up to 32 codes.

5 Conclusion

These estimates show that there are real technical possibilities for creating of the RFID system in the frequency band 6–30 GHz with acoustoelectronic FBAR resonators. In addition to the advantages already described of the low level of industrial and domestic noise for the secure of the RFID tag code exposed to electromagnetic fields and ionizing radiation such systems also have the high temperature stability characteristic of FBAR resonators. Therefore, in the near future we could expect the creation of such RFID systems.

References

1. Hartmann, C.: A Global SAW ID Tag With Large Data Capacity. In: IEEE Ultrasonics Symposium Proceedings, pp. 65–69. IEEE, Munich, Germany (2010)
2. Harma, S., Plessky, V., Li., X.: Feasibility of ultra-wideband SAW tags. In: IEEE International Ultrasonics Symposium Proceedings, pp. 1944–1947. IEEE, Beijing, China (2008)
3. Kozlovski, N., Malocha, D.: SAW noise-like anti-collision code study. In: IEEE International Frequency Control Symposium Proceedings, pp. 616—621. IEEE, Besancon, France (2009)
4. Brown, P., et al.: Asset tracking on the international space station using global SAW tag RFID technology. In: IEEE International Ultrasonics Symposium Proceedings, pp. 72—75. IEEE, New York, NY, USA (2007)
5. Plessky, V., Reindl, L.: Review on SAW RFID tags. IEEE Trans. UFFC **57**(3), 654–668 (2010)
6. Plessky, V., Lamothe, M., Davis, Z., Suchkov, S.: SAW tags for the 6-GHz range. IEEE Trans. UFFC **61**(12), 2149–2152 (2014)
7. Umeda, K., Kawamura, H., Takeuchi, M., Yoshino, Y.: Characteristics of an AlN based bulk acoustic wave resonator in the super high frequency range. Vacuum **83**(3), 672–674 (2009)
8. Nor, N., et al.: Film bulk acoustic wave resonator in 10–20 GHz Frequency Range. In: 3rd International Conference on Electronic Design (ICED) Proceedings, pp. 481—485. IEEE, Phuket, Thailand (2016)
9. Kojgerov, A., Dmitriev, V., Noskov, A.: Korrelyacionnye svojstva radioidentifikatorov na PAV s fazomanipulirovannymi kodami. M.: Voprosy radioelektroniki. Ceriya OT (4) 5—14 (2010) (in Russian)
10. Bechteler, T., Yenigun, H.: Localization and identification based on SAW ID-tags at 2.5 GHz. IEEE Trans. Microw. Theory Techn. **51**(5), 1584–1590 (2003)
11. Vishnevskiy, V.M., Kozyrev, D.V. (eds.): DCCN 2018. CCIS, vol. 919. Springer, Cham (2018). https://doi.org/10.1007/978-3-319-99447-5
12. Pochon, S., Ion-Beam Nano-patterning: experimental results with chemically-assisted beam. In: SPIE 10589, Advanced Etch Technology for Nano-patterning VII Proceedings, p. 105890. SPIE, San Jose, California, USA (2018)

13. Shannon, C.: The Mathematical Theory of Communication. University of Illinois Press, Urbana, Illinois, USA, IL (1949)
14. Suchkov, S., et al.: Multiband SAW tag for RFID systems with unlimited anti-collision. In: Surface Acoustic Wave Sensors Symposium Proceedings, SAWLab Saxony, Dresden, Germany (2016)

On Synchronisation of ISS-OFDM Signals

Semyon Dorokhin(✉)

Moscow Institute of Physics and Technology, 9 Institutskiy per., Dolgoprudny, Moscow Region 141701, Russian Federation
dorohin.sv@phystech.edu

Abstract. Interleaved Spread Spectrum OFDM (ISS-OFDM) is a new spread-spectrum modulation method which is similar to conventional OFDM. Despite properties of the signal itself are quite well studied, the problem of synchronisation remains unsolved. This study is focused on ISS-OFDM baseline synchronisation design. A novel sampling clock offset estimation algorithm is proposed and compared with an approach which is classical in OFDM systems. Simulation of ISS-OFDM communication system shows that the novel SCO estimator is more robust. The entire communication system is modeled as well. It can operate at negative signal-to-noise ratios with acceptable bit error rate.

Keywords: ISS-OFDM · SCO · Synchronisation · OFDM · Spread spectrum

1 Introduction

Interleaved Spread Spectrum OFDM (ISS-OFDM) is an OFDM-like spread spectrum method which was introduced by Pingzhou Tu et al. [8] in 2006. ISS-OFDM has diversity both in frequency and time domain, which was exploited to adapt ISS-OFDM for cognitive radio. Pingzhou Tu et al. [4] also managed to demonstrate that ISS-OFDM modulation can be used as PARP (peak-to-average-ratio) reduction technique. The same research group [3] suggested parallel FFT demodulation method, which reduced demodulation time. The subband-like spectral structure of ISS-OFDM was exploited in works featuring adaptive subband filtering. Pingzhou Tu et al. emphasised that it is possible to avoid information loss even if some subbands are not present in the spectrum [3]. The above-mentioned researchers suggested [9] flexible adaptive filtering scheme which can be used by several radiodevices to share the same frequency band. This direction was further explored by Qin Danyang et al. [5], who studied ISS-OFDM assuming fading channel model.

So far ISS-OFDM was studied primary as a technique for cognitive radio. However, ISS-OFDM has an interesting feature: increasing the number of subcarriers four times allows to decrease the minimal acceptable SNR (signal-to-noise ratio) by 3 dB. With sufficient number of subcarriers it is possible to operate at extremely low SNR and the main limiting factor is computational complexity.

© Springer Nature Switzerland AG 2020
V. M. Vishnevskiy et al. (Eds.): DCCN 2020, CCIS 1337, pp. 105–116, 2020.
https://doi.org/10.1007/978-3-030-66242-4_9

The available computational power increased significantly since the introduction of ISS-OFDM. Nowadays one can afford to greatly increase the number of subcarriers, thus achieving reliant communication at negative SNR.

Despite the fact that ISS-OFDM did not receive much attention in cognitive radio applications, it can be revived as low-SNR communication technique. It can be used in applications where power consumption is a parameter to be minimised (IoT networks).

1.1 ISS-OFDM Properties

Since ISS-OFDM is quite an uncommon type of modulation, a brief insight into its properties may be useful. Let us denote the number of constellation points in one symbol as N_c, and for simplicity assume that the signal is not oversampled, in other words, its spectrum occupies frequencies from zero to sampling rate.

Time-Domain. In baseband each ISS-OFDM symbol carrying N_c constellation points $a_i, i = \overline{0, N_c - 1}$ consists of N_c^2 samples $y_m, m = \overline{0, N_c^2 - 1}$:

$$y_m = y_i(n) = a_i \exp(2\pi j \frac{in}{N_c}), \quad i = m \bmod N_c, \quad n = \lfloor \frac{m}{N_c} \rfloor \tag{1}$$

where j is a complex unity. The process of interleaving may bring more clarity to the structure of OFDM (Fig. 1). Every constellation point modulates the corresponding subcarrier. Every subcarrier is upsampled with the factor of N_c, so that there are $N_c - 1$ points between every sample. After that, the i-th subcarrier is shifted by $i - 1$ sample and all the subcarriers are summed up thus giving the resulting ISS-OFDM signal.

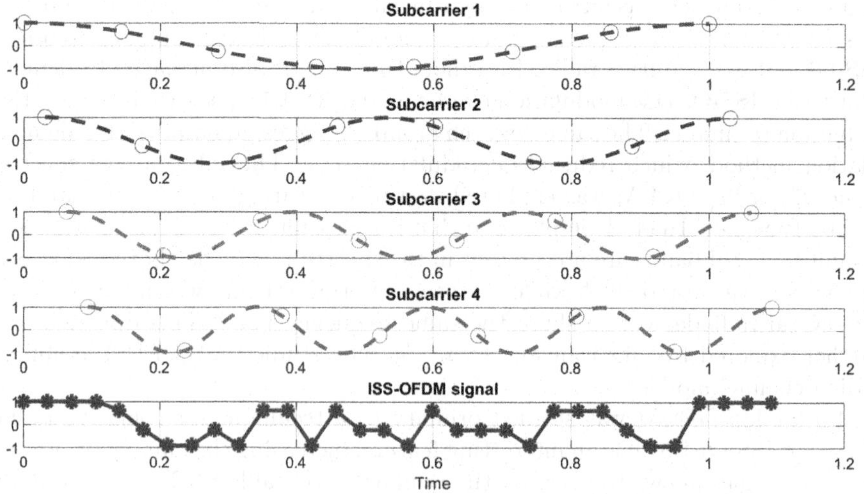

Fig. 1. Signal interleaving

Frequency Domain. Discrete Fourier Transform of one ISS-OFDM symbol is

$$Y_k = Na_i e^{-2\pi j \frac{(nN_c+i)i}{N_c^2}}, i = k \bmod N_c, n = \lfloor \frac{k}{N_c} \rfloor, k \in \overline{0, N_c^2 - 1} \qquad (2)$$

Fig. 2 illustrates this equation. The subcarriers of the same colour correspond

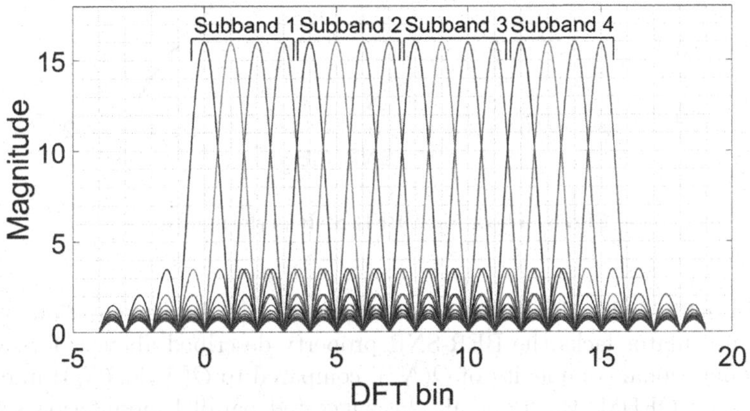

Fig. 2. Structure of spectrum

to the same constellation point. It can be seen that ISS-OFDM also has diversity in frequency domain: every constellation point is spread on N_c subcarriers. This diversity can be exploited to perform adaptive subband filtering, as it was done by P. Tu et al. [9] and Q. Danyang et al. [5].

BER-SNR Performance. It is important to emphasize that ISS-OFDM allows theoretically unlimited potential for BER-SNR performance improvement. As it was derived in [3], ISS-OFDM BER-SNR dependency in case of additive white Gaussian noise (AWGN) model could be described by the following equation:

$$BER(dB) = 10 \log Q(\sqrt{N_c^2 \cdot SNR}), \qquad (3)$$

where $Q(x)$ is Q-function: $Q(x) = \frac{1}{\sqrt{2\pi}} \int_x^\infty \exp(-\frac{u^2}{2}) du$

If the parameter N_c is doubled, the number of subcarriers increases be the factor of 4, which results in an improvement of $3dB$, as it can be seen in Fig. 3.

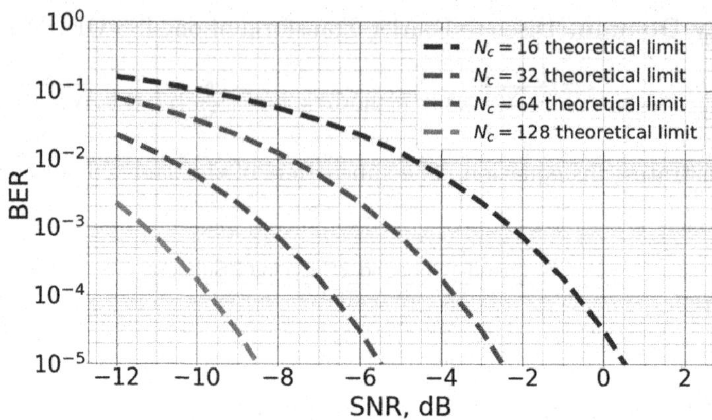

Fig. 3. Theoretical BER-SNR performance.

Parallel FFT Algorithm. A straightforward [8] implementation of ISS-OFDM modulator lacks the BER-SNR property described above. Moreover, it has computational complexity of $O(N_c^2)$, compared to $O(N_c \log(N_c))$ in case of conventional OFDM. Pu T. et al. [3] suggested parallel modulation scheme, which not only reduced the computational complexity to $O(N_c \log(N_c))$, but also improved robustness to Gaussian noise.

In the parallel IFFT-modulation scheme subcarrier corresponding to every constellation point is modulated separately and the result is then combined. For constellation point $a_i, i \in \overline{0, N_c - 1}$ an array of N_c complex numbers is generated, where the constellation point is placed as the i-th element. The other elements are zeros. Next, N_c-point IFFT is performed. Result of the i-th IFFT block is:

$$y_i(n) = \sum_{k=0}^{N_c-1} a_k e^{2\pi j \frac{nk}{N_c}} \delta_{k,i} = a_i e^{2\pi j \frac{in}{N_c}}, \tag{4}$$

which exactly matches the expression (1). Demodulation can be easily done using the inverse version of this algorithm.

Cyclic Shift. Using cyclic prefix is a common technique in OFDM systems. It allows to avoid inter-symbol interference and is usually used for coarse time synchronisation, as well as coarse CFO estimation. Therefore it is essential to understand how ISS-OFDM signals behave when the cyclic shift is present. In conventional OFDM cyclic shift results in rotation of the entire constellation and can be easily compensated with the help of pilot tones. In case of ISS-OFDM it is not true. A shift even in one time sample will result in wrong demodulation and further compensation will be needed. Since the DFT is known (2), one can represent ISS-OFDM signal in time domain using Fourier series:

$$y(t) = \sum_{k=0}^{N_c^2-1} Y_k e^{2\pi j \frac{kt}{N_c^2}} \sum_{k=0}^{N_c^2-1} a_p e^{-2\pi j \frac{kp}{N_c^2}} e^{2\pi j \frac{kt}{N_c^2}}$$
$$= \sum_{p=0}^{N_c-1} a_p e^{-2\pi j \frac{p}{N_c^2}(p-t)} \sum_{q=0}^{N_c-1} e^{-2\pi j \frac{q}{N_c}(p-t)} =$$
$$= \sum_{p=0}^{N_c-1} a_p e^{-2\pi j \frac{p}{N_c^2}(p-t)} e^{-\pi j(p-t)} \frac{sin(\pi(p-t))}{sin(\frac{\pi(p-t)}{N_c})}, \quad (5)$$

where $p = k\text{mod}N_c$. If the shift expressed in time samples is an integer, it is quite easy to compensate. Let us suppose that the aforementioned parallel FFT algorithm is used and that there is a cyclic shift of δ ($t = i - \delta$ in 5). Let us denote $\delta_i = \delta\text{mod}N_c$, $\delta_n = \lfloor\frac{\delta}{N_c}\rfloor$. Then

$$y(t = m - \delta) = a_{i-\delta_k} e^{-2\pi j \frac{(i-\delta_k)\delta_n}{N_c}} \quad (6)$$

In that case impact of cyclic shift can be expressed as following: constellation point a_i will be seen at the output of FFT-block number $(i - \delta_k)\text{mod}N_c$ and it will be multiplied by $exp(-2\pi j \frac{((i-\delta_k)\text{mod}N_c)\delta_n}{N_c})$.

If the shift is not an integer of time sample, the situation is much more complicated. Let us suppose that the integer (in terms of time samples) part of the cyclic shift was compensated. It can be seen from (5) that fractional cyclic shift will result in interference between the subcarriers in case of parallel FFT demodulation: the energy of different constellation points will be mixed.

The main drawback of parallel FFT demodulation method arises: it separates time samples correspond to different constellation points into different FFT blocks. When there is fractional time shift, every FFT block contains some energy from other constellation points, which is not compensated by demodulation algorithm.

That effect will definitely degrade the performance of the systems, the impact of fractional time shift should be studied more carefully. However, this impact and the severeness of degradation will be estimated later via simulations.

1.2 Scope of the Paper

As it was said before, ISS-OFDM received some attention in context of cognitive radio only. As far as the author is concerned, ISS-OFDM was studied in simulations and the question of ISS-OFDM synchronisation is still untouched. This paper is focused on designing the synchronisation algorithm which will take into account effects inherent in real-world systems. Such effects include Carrier Frequency Offset (CFO) and Sampling Clock Offset (SCO). In OFDM systems CFO and SCO are usually coupled [2] and should be estimated jointly. However, CFO is left out of scope of this paper, since it is the initial study on ISS-OFDM synchronisation. After the baseline model is studied, it is possible to move on and design a joint CFO and SCO estimator. This paper focuses on time synchronisation and SCO estimation. The main contributions of this paper are:

- Fine time synchronisation and tracking algorithm
- Novel SCO estimation algorithm
- Simulation of the proposed synchronisation method

Moreover, the influence of SCO was studied in terms of BER performance.

The rest of the paper is organised as follows: Sect. 2 gives a brief overview of existing synchronisation algorithms, Sect. 3 describes the proposed tracker and estimator, Sect. 4 presents the simulation results followed by a concise conclusion.

2 State of the Art

As it was mentioned above, the area of ISS-OFDM synchronisation is unexplored. Therefore, there is no ISS-OFDM synchronisation algorithms to compare with. However, since ISS-OFDM inherited many properties of conventional OFDM, it is tempting to adapt existing OFDM synchronisation techniques to ISS-OFDM. The existing algorithms were selected to be tested on ISS-OFDM according to the following criterion:

- Popularity and simplicity. Since it is the first study on ISS-OFDM synchronisation, only classical OFDM synchronisation algorithms are considered.
- Robustness. ISS-OFDM offers acceptable performance at negative SNRs (refer to Fig. 3), so algorithms must comply with this feature.
- Scalability in the number of subcarriers. The main method to improve ISS-OFDM robustness is to increase the number of subcarriers. That is why this parameter must be explicit in the algorithm's structure.

2.1 Time-Domain Synchronisation

Generally, there are three approaches to initial synchronisation in OFDM systems. The first one is to use cyclic prefix and correlation alongside with estimator (commonly Maximum Likelihood [10]). Another approach is to perform acquisition based on training symbols [6] or 2D-pilot map [1], without utilising guard interval. The third approach, which Chinese standard DTMB-A relies on, implies using PN-sequences as guard interval [11].

Both PN-sequence and pilot-aided approaches appear to be difficult to scale in the number of subcarriers and to keep robust at the same time. Additional study has to be carried out in order to determine optimal parameters (length of PN-sequence, number of scattered pilots etc.) for low-SNR applications. On the contrary, correlation-based approach is easy to scale as the length of guard interval depends on the number of subcarriers. Averaging over consequent symbols can improve robustness. Moreover, cyclic prefix will be needed to avoid inter-symbol interference when fading is taken into account. Therefore, cyclic prefix and correlation approach is chosen for coarse time synchronisation.

2.2 SCO Estimation

For continuous transmission it is essential to track sampling clock offset (SCO). In OFDM systems there is a classical approach based on pilot correlation between subsequent symbols [7]. It was also shown [2] that SCO and carrier frequency offset (CFO) are coupled in OFDM systems and should be estimated jointly. Since it is the first study on ISS-OFDM synchronisation, CFO is left out of scope. In this paper a novel robust SCO tracking scheme is proposed and compared with this classical approach. The new tracker exploits ISS-OFDM signal structure in modulation domain, yielding promising results. It is also possible to adapt the classical OFDM approach for ISS-OFDM, since the DFT of ISS-OFDM symbol is known. The aforementioned adapted classical approach will be compared with the novel one.

3 Proposed Method

The first step in synchronisation process is typically time synchronisation. For coarse time synchronisation cyclic prefix was added to the modulated signal and the start of the symbol was estimated using correlation. Due to SCO the beginning of symbol slightly drifts away from the observation window. Taking into account the cyclic shift properties of ISS-OFDM described in Sect. 1, a fine time tracker is needed to compensate that drift.

3.1 Time-Domain Tracker

ISS-OFDM symbols carrying N_c constellation points consists of N_c^2 points and a guard interval (typically $\frac{1}{8}$ of bare symbol length). For the following analysis it is assumed that correlation-based initial sychronisation algorithm provides the estimation of the next symbol start with an error of no more than $\frac{N_c}{2}$.

If the signal is upsampled with the factor of k, the total length of a symbol is $\frac{9}{8}N_c^2k$. From (1) it can be seen that $y_0(n) = a_0 \ \forall n = \overline{0, N_c - 1}$, so every N_c sample of non-upsampled signal is the same. The new time tracking algorithm uses this periodicity. The simplified time tracking algorithm consists of the following steps (refer to Fig. 4): 1. jump to the estimated middle of guard interval; 2. Step $kN_c/2$ samples back; 3. Sum up N_c subsequent samples with a distance of kN_c between them; 4. Step k samples forward; 5. Sum up again, until N_c steps are performed

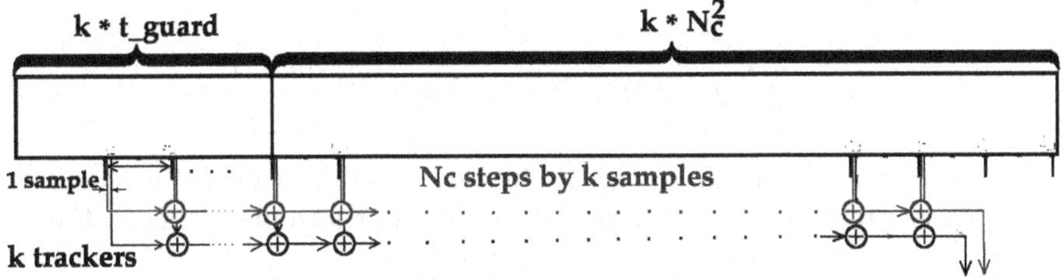

Fig. 4. The proposed clock frequency offset tracker

Let us explain the idea behind this tracker, assuming interpolation by DFT-zero padding for simplicity of analysis. The normalisation factor is set to 1 as it does not affect the results. The signal, interpolated by zero-padding with interpolation factor k (see (2)):

$$y_p = \sum_{l=0}^{N_c^2-1} \sum_{m=0}^{N_c^2-1} s_m e^{-2\pi j \frac{ml}{N_c^2}} e^{2\pi j \frac{pl}{kN_c^2}} = \sum_{m=0}^{N_c^2-1} s_m \sum_{l=0}^{N_c^2-1} e^{-2\pi j \frac{l(km-p)}{kN_c^2}} \tag{7}$$

$$= \sum_{m=0}^{N_c^2-1} s_m e^{-\pi j l(m-\frac{p}{k})(1-\frac{1}{N_c})} \frac{sin(\pi k(m-\frac{p}{k}))}{sin(\frac{\pi k(m-\frac{p}{k})}{N_c^2})} = \sum_{m=0}^{N_c^2-1} s_m e^{-\pi j l(m-\frac{p}{k})(1-\frac{1}{N_c})} \delta_{m,\frac{p}{k}},$$

where $\delta_{i,j}$ is the Kronecker delta. Tracker $l, l = \overline{0, k-1}$ has starting offset of l samples and provides maximum location estimation m_l from following N_c variables:

$$S_i^l = \sum_{\substack{p=l \\ \text{step } k}}^{kN_c-1} y_p = \sum_{\substack{p=l \\ \text{step } k}}^{kN_c^2-1} \sum_{m=0}^{N^2-1} s_m e^{-\pi j l(m-\frac{p}{k})(1-\frac{1}{N})} \frac{sin(\pi k(m-\frac{p}{k}))}{sin(\frac{\pi k(m-\frac{p}{k})}{N^2})} \tag{8}$$

It can be seen from this formula that for a tracker which has a shift of t from the closet a_0 sample the abscissa of maximum is not an integer, as it is shifted by $\frac{l}{k}$. Interpolation is used to determine a precise location of maximum for every tracker. Taking into account upsampling and start offset for every tracker yields the following formula for time offset m_{offs} measured in samples:

$$m_{offs} = \frac{\sum_{l=0}^{k-1} (m_l - \frac{N_c}{2})k + l}{k} \tag{9}$$

It is important to emphasis that the proposed algorithm gives time offset estimation by modulo N_c, and therefore the error of initial estimation must not exceed $\frac{N_c}{2}$.

3.2 SCO Estimator

If time synchronisation is performed correctly, one can track SCO in modulation domain based. As it was mentioned in Sect. 1, cyclic shift which is not an integer of one time sample introduces ICI in ISS-OFDM modulation domain, However, for small fractional shifts Δ ICI can be omitted. In that case the demodulation via parallel FFT yields:

$$\tilde{a}_k = \sum_{n=0}^{N_c-1} y_{i-\Delta}(n) e^{-2\pi j \frac{kn}{N_c}} = N_c a_k e^{\pi j \Delta} \frac{sin(\pi \Delta)}{sin(\frac{\pi \Delta}{N_c})} \tag{10}$$

If the time tracker is applied, the cyclic shift is caused by SCO ξ. When demodulation via parallel FFT is performed, the cyclic shift introduces phase offset:

$$arg(\tilde{a}_k^l) = \pi \Delta = \pi \xi (N_c^2 + N_g) l, \tag{11}$$

where l is the number of symbol since perfect time-domain synchronisation and N_g is the length of the guard interval.

The phase difference of pilots in two subsequent symbols $\Delta\phi$ is proportional to SCO:

$$\xi = \frac{\Delta\phi}{\pi(N_c^2 + N_g)}, \tag{12}$$

where $\Delta\phi$ is the difference between pilots' phases in subsequent symbols. If one constellation point a_0 is reserved for continual pilot tone, a_0 remains the same for every symbol and the SCO estimation can be averaged over many symbols, thus improving robustness.

3.3 Simulation Setup

Aforementioned synchronisation algorithms were simulated in Matlab first separately and then as a united system. Gaussian channel model was chosen as the simpliest one to start with. In following simulations only QPSK modulation was used, as the impact of different modulation schemes on synchronisation performance is out of scope of this study. Random data bytes were generated and then scrambled with a PN-sequence of length $2^{14} - 1$. Guard interval of $\frac{1}{8}$ symbol length was chosen. The general structure of baseband model is presented in Fig. 5.

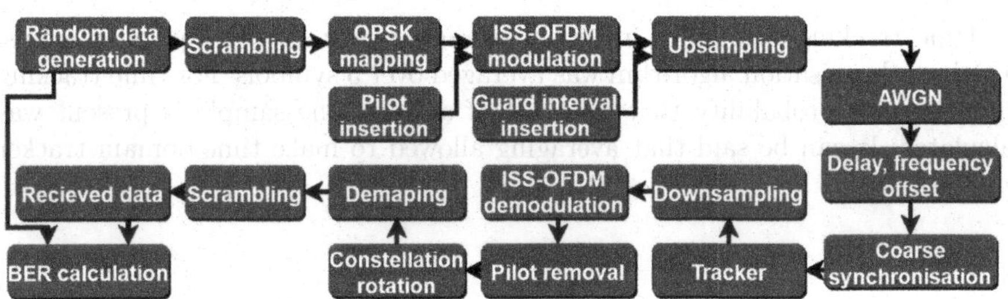

Fig. 5. Simulated baseband communication system

Since it is the first study on ISS-OFDM SCO tracking, the CFO has not been taken into account. One of ISS-OFDM subcarriers was reserved for a constant pilot tone to compensate constellation rotation caused by remained frequency offset. SCO was simulated via resampling. Every plot data point was averaged over 4980 symbols, with random time delay being introduced to every group of 60 symbols. Initial synchronisation and tracking algorithm were applied to every group and resulting BER was measured for further averaging.

4 Results and Discussion

SCO and time tracking algorithms described in Sect. 4 were simulated separately as well. Because the algorithms are required to operate at very low SNR, the estimation based on one symbol is not enough. Algorithms must average their estimations over several subsequent symbols. The number of symbols for averaging is a parameter to be optimised. If it is too small, there will be too many errors and it will be impossible to approach theoretical performance limit. If this parameter it too large, the theoretical limit can be approached, but the algorithms will have redundant computational complexity. The optimal value was chosen experimentally by running several simulations and the final results are presented in Fig. 6.

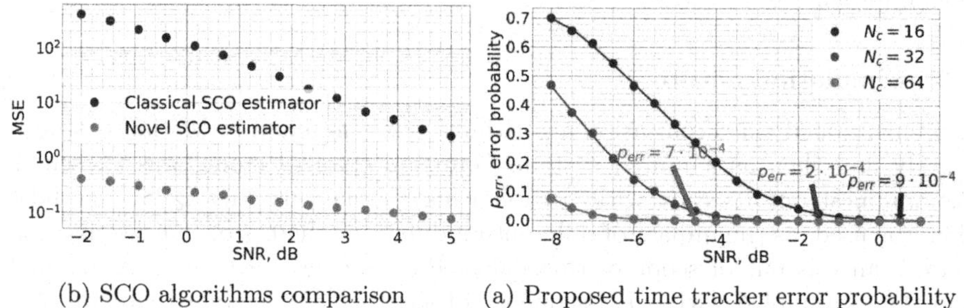

(b) SCO algorithms comparison (a) Proposed time tracker error probability

Fig. 6. Synchronisation algorithms simulation

Time tracking algorithm included averaging over 15 consequent symbols, initial synchronisation algorithm was averaged over 3 symbols. For time tracking algorithm the probability that an error of at least one sample is present was calculated. It can be said that averaging allowed to make time-domain tracker quite reliant.

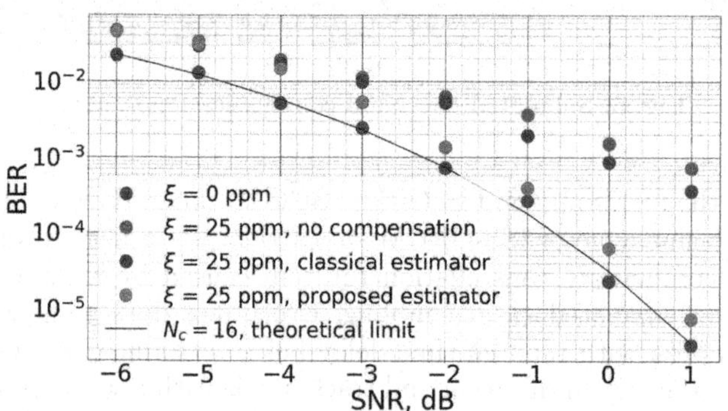

Fig. 7. Performance of the modeled communication system

SCO tracker was compared with the one proposed in [7] and both SCO trackers were averaged over 80 symbols to obtain reasonable MSE. It can be seen that the proposed SCO estimator outperforms the classical approach. However, in OFDM-scenario MSE is usually lower by several orders of magnitude. It can be explained by ICI introduced by fractional time shift described in Sect. 1.

The optimal parameters of averaging were used in a simulation to evaluate the performance of the entire system. The results of final simulation together with theoretical curves are presented in Fig. 7.

Simulation data lay closer to theoretical boundary as SNR increases. It is explained by high error probability of synchronisation algorithms at low SNRs, as the averaging parameter was optimsed for SNRs corresponding to acceptable BER only. It can be seen that the system with the proposed SCO estimator performs better than the one with classical estimator, as expected. It means that time and frequency diversity present in ISS-OFDM system give a potential for robust synchronisation algorithms.

5 Conclusion

In this article synchronisation algorithms for ISS-OFDM signals were discussed for the first time. The effect of sampling clock offset on demodulated constellations was studied. A novel robust sampling clock offset tracking scheme was proposed and compared with an approach which is classical in OFDM systems. Simulation proved the novel SCO estimator to be more robust to Gaussian noise.

It turned out that cyclic shift compensation is quite a difficult task in ISS-OFDM systems. The shift causes interference between subcarriers if it is not an integer of time sample. Parallel FFT-demodulation scheme should be modified to perform compensation of arbitrary cyclic shift.

Robust fine time-domain ISS-OFDM tracker was introduced to track the symbol "drifting" caused by SCO and to estimate the part of cyclic shift which is an integer of time sample.

The entire ISS-OFDM communication system model was simulated. In particular, it was shown that with 16^2 subcarriers presented synchronisations algorithms can operate at SNR as low as 1 dB, with the system keeping the bit error rate at a value of 10^{-5}. The proposed algorithms scale in the number of subcarriers and the scaled system can perform even at lower SNR values.

The major drawback of the study is that it does not consider CFO. CFO and SCO should be estimated jointly and the coupling between them should be studied. CFO may also degrade the performance of time-domain synchronisation and SCO tracker.

Moreover, only AWGN model was used without any consideration of fading. The robustness is expected to deteriorate in fading environment. Furthermore, it is not evident how to perform efficient channel estimation in case of ISS-OFDM. Additional research is needed to determine whether the proposed alogrithms are suitable for fading channels.

References

1. Fernandez-Getino Garcia, M.J., Paez-Borrallo, J.M., Zazo, S.: DFT-based channel estimation in 2D-pilot-symbol-aided OFDM wireless systems. In: Proceedings of IEEE VTS 53rd Vehicular Technology Conference, Spring 2001. (Cat. No.01CH37202), Vol. 2, pp. 810–814 (2001)
2. Jung, Y., Kim, J., You, Y.: Complexity efficient least squares estimation of frequency offsets for DVB-C2 OFDM systems. IEEE Access **6**, 35165–35170 (2018)
3. Tu, P., Huang, X., Dutkiewicz, E.: Diversity performance of interleaved spread spectrum OFDM signals over frequency selective multipath fading channels. In: 2007 International Symposium on Communications and Information Technologies, pp. 184–189 (2007)
4. Tu, P., Huang, X., Dutkiewicz, E.: Peak-to-average power ratio performance of interleaved spread spectrum OFDM signals. In: 2007 International Symposium on Communications and Information Technologies, pp. 82–86 (2007)
5. Qin, D., Ma, L,, Wang, E., Ma, H., Ding, Q.: An interference suppression mechanism for WSN. In: Proceedings of 2013 International Conference on Sensor Network Security Technology and Privacy Communication System, pp. 28–33 (2013)
6. Schmidl, T.M., Cox, D.C.: Robust frequency and timing synchronization for OFDM. IEEE Trans. Commun. **45**(12), 1613–1621 (1997)
7. Speth, M., Fechtel, S., Fock, G., Meyr, H.: Optimum receiver design for OFDM-based broadband transmission. ii. a case study. IEEE Trans. Commun. **49**(4), 571–578 (2001)
8. Tu, P., Huang, X., Dutkiewicz, E.: A novel approach of spreading spectrum in OFDM systems. In: 2006 International Symposium on Communications and Information Technologies, pp. 487–491 (2006)
9. Tu, P., Huang, X., Dutkiewicz, E.: Subband adaptive filtering for efficient spectrum utilization in cognitive radios. In: 2008 3rd International Conference on Cognitive Radio Oriented Wireless Networks and Communications (CrownCom 2008), pp. 1–4 (2008)
10. van de Beek, J., Sandell, M., Isaksson, M., Ola Borjesson, P.: Low-complex frame synchronization in OFDM systems. In: Proceedings of ICUPC 1995 - 4th IEEE International Conference on Universal Personal Communications, pp. 982–986 (1995)
11. Wu, J., Chen, Y., Zeng, X., Min, H.: Robust timing and frequency synchronization scheme for DTMB system. IEEE Trans. Consum. Electron. **53**(4), 1348–1352 (2007)

Reserve Navigation System of Tether Powered Unmanned Aerial Platform in Conditions of Turbulent Atmosphere

V. M. Vishnevsky[1(✉)], E. A. Mikhailov[2], and Nguyen Duy Phuong[3]

[1] V. A. Trapeznikov Institute of Control Sciences of Russian Academy of Sciences, 65 Profsoyuznaya Street, Moscow 117997, Russia
vishn@inbox.ru

[2] M. V. Lomonosov Moscow State University, 1 Leninskie gori, Moscow 119991, Russia
ea.mikhajlov@physics.msu.ru

[3] Moscow Institute of Physics and Technology (National Research University), 9 Institutskiy per., Dolgoprudny, Moscow Region 141701, Russia
ndphuong2207@gmail.com

Abstract. Unmanned aerial vehicles play a very important role in different technical problems. They can take photographs of the area, track forest fires, control offenses and be useful in another fields. Nowadays tethered high-altitude platforms have a particular interest. They have a number of advantages over autonomous devices. For example, they can work for a sufficiently long time and do not depend on the limited resources of batteries. This allows them to be used as transmitters for mobile phones and for other purposes where it is impossible to use autonomous unmanned platforms. In addition, they can carry a much higher load and have a higher power. However, there are some problems connected with navigation of the platform. The wind can change the location of the platform, which should be corrected. The coordinates of unmanned aerial vehicles are usually found using GPS or GLONASS systems. Unfortunately, the signal can be quite weak in some areas, so it is necessary to take the reserve navigation system, which will be connected with beacons located on the ground. In this work we present the idea of system of navigation of the tethered platform, which takes into account its main features and can give us the opportunity to stabilize the location of the vehicle. We give the main formulaes for the coordinates and estimate typical mistakes while measuring them.

Keywords: Unmanned aerial vehicles · Navigation system · Tethered platform

1 Introduction

Unmanned aerial vehicles have recently become widespread in a variety of industries. They are used for the purpose of photographing and filming, to control

The reported study was funded by RFBR, project number 19-29-06043.

© Springer Nature Switzerland AG 2020
V. M. Vishnevskiy et al. (Eds.): DCCN 2020, CCIS 1337, pp. 117–128, 2020.
https://doi.org/10.1007/978-3-030-66242-4_10

various offenses. Such devices can help receive information about forest fires and other emergencies. As a rule, unmanned devices are autonomous, and the engines located on board are powered by batteries. This approach has several advantages, but at the same time creates a number of problems. So, the flight time of the drone turns out to be significantly limited due to the not very large battery life. It also turns out that the transported cargo also has significant weight restrictions - engines powered by batteries cannot have a significant mass. A possible increase in battery capacity will naturally lead to an increase in the mass of the copter and therefore is not a solution to the problem. All this makes it impossible to use autonomous unmanned platforms to perform a number of tasks that require long-term stay in the air and the use of complex and heavy equipment on board.

One of the possible methods for solving emerging problems is associated with the use of tethered high-altitude unmanned platforms [1–4]. They work by using a cable-rope that connects them to the ground station. The ground station can be stationary or located on a car (in the event that the mobility of the used complex is required). It contains the equipment that is associated with the control of the copter. The cable must be sufficiently strong and serve both for signal transmission and for stabilizing the position of the drone. Existing developments make it possible to transmit power up to 15 kW by cable. This makes it possible to place on the unmanned platform both powerful enough engines (this allows to lift a large payload), and serious attachments that consume a large amount of electricity. Usually, the operating time of such a device is practically unlimited. Thus, tethered high-rise platforms can be used for a large number of tasks, where their autonomous counterparts are not applicable. So, they can be used to provide mobile communication in the event of a breakdown of towers with transmitters. The tethered platform can also be used to "distribute" high-speed internet at public events. Also, tethered platforms can be more effectively used to control various offenses. (The experience of the coronavirus epidemic has shown that in some cases strict control over the movement of citizens under quarantine conditions is necessary.) In addition, tethered high-altitude platforms can be used to control the border.

At the same time, despite a number of obvious advantages, tethered platforms also have a number of problems. The first of these is their stability under the influence of wind loads. In case of calm weather, the aircraft is at a given point above the base station. At the same time, the presence of wind can significantly shift its position relative to the required point. This process should be carefully considered. This requires careful consideration of the equations of motion for different parts of the cable. The equations that can be obtained in this case turn out to be rather complicated for an analytical solution, and require the use of numerical methods. In this paper, we briefly describe the main methods for studying the corresponding models.

Another, no less significant problem, is that almost any unmanned platform is oriented in space using a satellite navigation system (GPS or GLONASS). A tethered platform is no exception in this case - in modern conditions it turns out to be quite effective, especially in the case of densely populated areas, where we can talk about a fairly good signal reception quality. At the same time, it should be noted that the guaranteed error in determining the coordinates using only data received from the satellite is within a few meters [5]. Typically, it can be improved with ground base stations. They can improve the accuracy of determining coordinates a few tens of kilometers from their location. For this reason, they are located in places where navigation system services are in high demand. At the same time, if we are talking about sparsely populated regions, where the signal level is noticeably lower, the accuracy of determining the coordinates also deteriorates significantly. In addition, deliberate attempts to jam the signal of the satellite navigation system cannot be ruled out. All this puts us in front of the need to create a duplicate navigation system that would allow the device to navigate in space even in the absence of an acceptable signal quality.

If we talk about the existing experience in creating backup navigation systems for unmanned vehicles, then we can distinguish a number of approaches that were used primarily for the needs of autonomous devices. Nevertheless, we can say that the fundamental points in this case will differ little from what is the case for tethered unmanned platforms. Basically, all existing methods, one way or another, are reduced to one of four approaches [6].

The first approach refers to the so-called inertial navigation system [6]. The device has IMU sensors that can measure the acceleration of the device, as well as the change in the direction of its movement. Thus, it is possible to obtain the dependence of the acceleration components of the quadcopter on time. They can be integrated - and then we get the dependence of the velocity components on time. After integrating the velocity, it is possible to obtain the dependence of the coordinates on time. This method is quite simple to realize, which is its obvious advantage. At the same time, a significant drawback is that the integration must be performed numerically using the equipment on board the quadcopter. Unfortunately, any of the numerical methods has a certain error, which will only grow with time. Thus, if we are talking about tethered high-altitude unmanned platforms, which should work for a sufficiently long time, this method can hardly be considered suitable. During the operation of the device, the error will surely reach unacceptable values.

The second approach uses what is called a visual navigation system [6]. It is connected with the placement of a video camera on board the high-altitude platform. It takes pictures of the surrounding area, fixes the location of the main buildings, features of the relief and objects. After that, using machine learning methods and a special computer program, a map of the area over which the device is moving is built. After the device has been trained, it can independently move over a given area and navigate using the camera mentioned above. This method is quite promising for autonomous platforms that move over large enough territories. At the same time, if we are talking about tethered platforms, then their “habitat” is rather small. For this reason, there is no need to implement a rather complex visual navigation system, and it makes sense to use much simpler methods and approaches.

The third approach involves the use of a radar navigation system [6]. A radar is placed on board the unmanned flying platform, which can both emit and receive radio waves. By the time that elapsed from the moment the waves were emitted to the reception of the reflected wave, it is possible to determine the distance from the radar to a certain object. As such items, you can use special marks that can be placed on the earth's surface. One of the most obvious advantages of this method is the fact that it is practically independent of weather conditions. It seems to us that such an approach is quite promising from the point of view of application for unmanned high-altitude platforms.

The fourth approach is related to the so-called laser navigation [6]. Lasers can be installed on the unmanned high-altitude platform, which will emit corresponding beams. On the ground, in the same way as in the case of the "radar" approach, it is possible to set marks (where mirrors can play this role). By measuring the distances from the device to them, it is possible to determine mathematically the distance to the unmanned platform. It seems to us that this method also has a number of significant advantages from the standpoint of application for tethered unmanned platforms.

It makes sense to mention the experience of predecessors in this area. Thus, Tordesillas proposes the joint use of both inertial navigation and visual navigation [16]. The authors propose the FASTER system, which uses mathematical methods that make it possible to improve the accuracy of calculating coordinates. Their work presents the simulation of the movement of a quadcopter in the forest. In addition, the authors of this work carried out experiments on the operation of the quadrocopter in a room where the researchers placed artificial obstacles that the device learned to fly around.

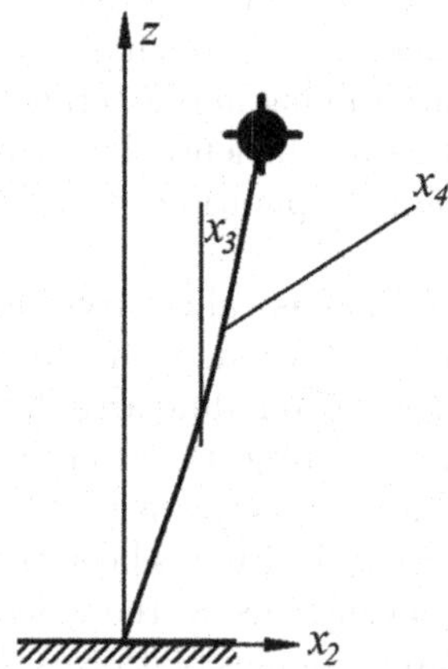

Fig. 1. A typical scheme of the unmanned aerial vehicle.

Oleinikova et al. [7] discusses the use of a visual navigation system for quadrocopters. At the beginning of operation, the device takes pictures of the earth's surface and creates a map of the territory in its memory. After that, he is quite successful in using it. The authors conducted a study of the efficiency of the apparatus in a variety of landscapes, from urban development to garbage dumps. The device is also successfully oriented when flying indoors.

Kumar [8] proposes the use of laser navigation in combination with inertial sensors on unmanned aerial vehicles. This makes it possible to efficiently calculate its coordinates in various situations.

Different authors describe the advantages of using radars for navigation of unmanned platforms. Quist and Beard [9,10] show the robustness of such methods both using the mathematical modeling of the flight and the real tests. Lindstrom with co-authors [11] compare the usage of radars and the visual navigation. They come to the conclusion that the radar navigation is much more effective. The navigation based on using synthetic aperture radars has also been described in detail in works of Christensen et al. [12], Reid and Ash [13] and other groups.

The works listed above are for autonomous unmanned aerial platforms. If we talk about tethered unmanned platforms, then we can mention the work of Dowling [14]. It is associated with devices that move indoors. So, they can be used, for example, in the machine rooms of industrial enterprises. They are navigated by a laser system. Laser beams are reflected from walls and objects, due to which a room map is built in the device's memory. Then, according to these data, the device successfully flies around obstacles. Similar methods were also used in Zeng's work [15], where the results were refined using visual navigation methods.

Summarizing the results given above, we can say that for tethered high-altitude platforms, the use of a laser navigation system is most effective. We discuss this approach in detail below.

2 Main Laws for the Motion of the Tethered Aerial Vehicle

Here we describe the main approaches of the modeling the tethered vehicle (Fig. 1). Here we can introduce the following equations [3]:

$$\frac{dx_1}{dz} = \rho g;$$

$$\frac{dx_2}{dz} = x_3;$$

$$\frac{dx_3}{dz} = -\frac{\rho g x_3\left(1+x_3^2\right) + AV^2\left(1+x_3^2\right)^{1/2}}{x_1};$$

$$\frac{dx_4}{dz} = \left(1+x_3^2\right)^{1/2};$$

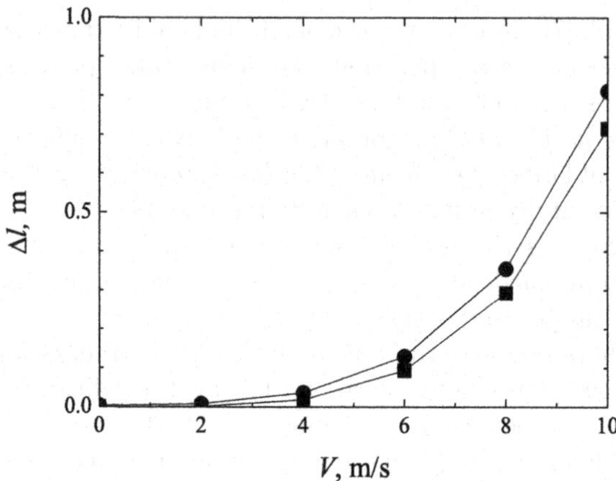

Fig. 2. Elongation in connection with velocity of the wind. Rectangles show $V = 0$ and rounds show $V = 10$ m/s.

where z is the distance from the earth surface, x_1 is the force pulling force, x_2 is the distance from the vertical line, x_3 is the tangent of the angle of inclination, x_4 is the length of the cable. Also we should say that ρ is the linear density of the cable, g is the acceleration of gravity, A is the drag coefficient and V is the velocity of the wind (Fig. 1).

Table 1. Elongation of the cable for different velocity of the wind.

V, m/s	$x_2(z_{max}) = 0$	$x_2(z_{max}) = 5$ m	$x_2(z_{max}) = 10$ m	$x_2(z_{max}) = 15$ m
0	0	0	0	0
2	0.001	0.004	0.008	0.014
4	0.018	0.027	0.036	0.048
6	0.093	0.111	0.129	0.150
8	0.293	0.324	0.355	0.389
10	0.716	0.763	0.812	0.861

The equations are solved numerically using Runge – Kutta fourth order method [3]. This method is quite stable and gives us an opportunity to solve the equations quite precisely.

One of the main questions is connected with the full length of the cable:

$$L = x_4(h_{max}).$$

Also it is interesting to count the value of the distance from the ground station and the unmanned aerial vehicle. If the distance from the vertical line at the highest point z_{max} is $x_2(z_{max})$, we can find it as:

$$l_1 = \left(z_{max}^2 + x_2^2(z_{max})\right)^{1/2}.$$

To find the difference between these values we can introduce the parameter:

$$\delta l = L - l_1.$$

For example, we can calculate the values for Δl for linear density $\rho = 0.04$ kg/m and drag coefficient $A = 0.004$ kg/m^2, $L = 150$ m. It could be interesting to model this process for different velocity of the wind. We have found the elongation for different cases (Table 1 and Fig. 2) .

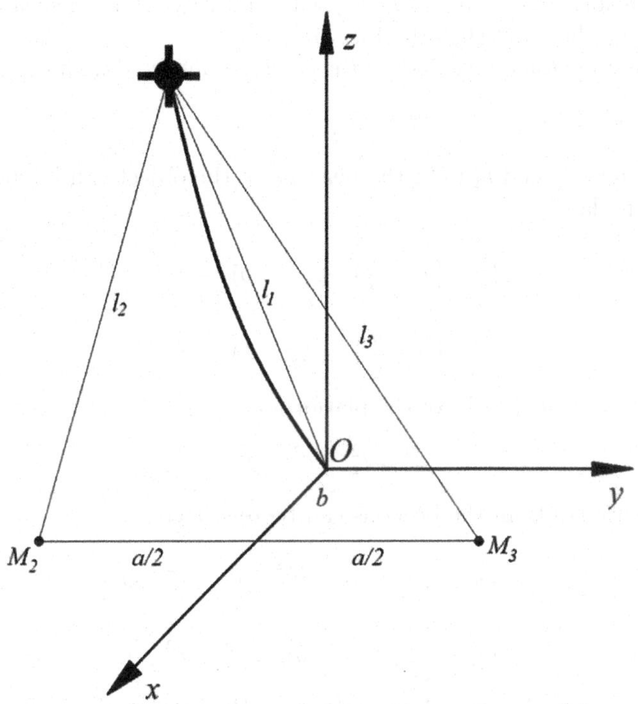

Fig. 3. Location of the radio beacons.

The connection between velocity and the elongation can be found as:

$$\delta l \approx \alpha V^4,$$

where the coefficient α depends on the parameter x_2 (Table 2). It can be seen that the value of the elongation is much smaller than the length of the cable. So, we can approximately assume that the distance from the basic station is equal to the length of the cable.

Table 2. Coefficient α for different x_2.

x_2, m	α, s^4/m^3
0	0.092
5	0.095
10	0.096
15	0.098

3 Coordinates of the Vehicle

The coordinates of the tethered unmanned aerial vehicle can be found using the radio beacons located in points $M_2(-\frac{a}{2}, b, 0)$ and $M_3(\frac{a}{2}, b, 0)$. The third point is connected with the point $(0, 0, 0)$ (Fig. 3).

We can assume that the length of the cable is connected with the coordinates of the vehicle:

$$l_1^2 = x^2 + y^2 + z^2.$$

The distances l_2 and l_3 from the beacons to the object can be found by the following formulaes:

$$l_2^2 = \left(y + \frac{a}{2}\right)^2 + (x - b)^2 + z^2;$$

$$l_3^2 = \left(y - \frac{a}{2}\right)^2 + (x - b)^2 + z^2.$$

We can obtain the height of the platform as:

$$z^2 = l_1^2 - x^2 - y^2.$$

So the distances from the beacons can be rewritten:

$$l_2^2 = \left(y + \frac{a}{2}\right)^2 + (x - b)^2 + l_1^2 - x^2 - y^2;$$

$$l_3^2 = \left(y - \frac{a}{2}\right)^2 + (x - b)^2 + l_1^2 - x^2 - y^2.$$

We can subtract the second equation from the first one:

$$l_2^2 - l_3^2 = \left(y + \frac{a}{2}\right)^2 - \left(y - \frac{a}{2}\right)^2.$$

This leads us to the expression for the y-coordinate:

$$y = \frac{l_2^2 - l_3^2}{2a}.$$

The equation for l_2 can be rewritten, if we expand the brackets:

$$-2xb + b^2 = l_2^2 - l_1^2 - ay - \frac{a^2}{4}.$$

If we take into account the formulae for y, we shall obtain that:

$$-2xb + b^2 = \frac{l_2^2 - 2l_1^2 + l_3^2}{2} - \frac{a^2}{4}.$$

For coordinate x we can take:

$$x = \frac{b}{2} + \frac{a^2}{8b} - \frac{l_2^2 - 2l_1^2 + l_3^2}{2b}.$$

The equation for l_1, it can be rewritten:

$$l_1^2 = \left(\frac{b}{2} + \frac{a^2}{8b} - \frac{l_2^2 - 2l_1^2 + l_3^2}{2b}\right)^2 + \left(\frac{l_2^2 - l_3^2}{2a}\right)^2 + z^2.$$

For the height of the unmanned aerial vehicle we shall obtain the following expression:

$$z = \sqrt{l_1^2 - \left(\frac{b}{2} + \frac{a^2}{8b} - \frac{l_2^2 - 2l_1^2 + l_3^2}{2b}\right)^2 - \left(\frac{l_2^2 - l_3^2}{2a}\right)^2}.$$

As it was said in previous paragraph, the length if the cable can differ from the distance from the ground station. So it is necessary to estimate the mistake connected with the uncertainty for the distances.

If we assume the mistakes independent, the mistake for x will be the following:

$$\Delta x = \sqrt{\left(\frac{\partial x}{\partial l_1}\right)^2 \Delta l_1^2 + \left(\frac{\partial x}{\partial l_2}\right)^2 \Delta l_2^2 + \left(\frac{\partial x}{\partial l_3}\right)^2 \Delta l_3^2}.$$

The partial derivatives can be found as:

$$\frac{\partial x}{\partial l_1} = \frac{2l_1}{b};$$

$$\frac{\partial x}{\partial l_2} = -\frac{l_2}{b};$$

$$\frac{\partial x}{\partial l_3} = -\frac{l_3}{b}.$$

So the mistake for the parameter is:

$$\Delta x = \sqrt{\frac{4l_1^2}{b^2}\Delta l_1^2 + \frac{l_2^2}{b^2}\Delta l_2^2 + \frac{l_3^2}{b^2}\Delta l_3^2}.$$

In a similar way, we can write the formulae for the mistake for y:

$$\Delta y = \sqrt{\left(\frac{\partial y}{\partial l_1}\right)^2 \Delta l_1^2 + \left(\frac{\partial y}{\partial l_2}\right)^2 \Delta l_2^2 + \left(\frac{\partial y}{\partial l_3}\right)^2 \Delta l_3^2}.$$

The partial derivatives will be:

$$\frac{\partial y}{\partial l_1} = 0;$$

$$\frac{\partial y}{\partial l_2} = \frac{l_2}{a};$$

$$\frac{\partial y}{\partial l_3} = -\frac{l_3}{a}.$$

For the mistake we shall have:

$$\Delta y = \sqrt{\frac{l_2^2}{a^2}\Delta l_2^2 + \frac{l_3^2}{a^2}\Delta l_3^2}.$$

For the z-coordinate we will have:

$$\Delta z = \sqrt{\left(\frac{\partial z}{\partial l_1}\right)^2 \Delta l_1^2 + \left(\frac{\partial z}{\partial l_2}\right)^2 \Delta l_2^2 + \left(\frac{\partial z}{\partial l_3}\right)^2 \Delta l_3^2}.$$

As for the derivatives, we have:

$$\frac{\partial z}{\partial l_1} = \frac{1}{2}\frac{\frac{l_1}{b^2}\left(\frac{a^2}{2} + 2l_2^2 - 4l_1^2 + 2l_3^2\right)}{\sqrt{l_1^2 - \left(\frac{b}{2} + \frac{a^2}{8b} - \frac{l_2^2 - 2l_1^2 + l_3^2}{2b}\right)^2 - \left(\frac{l_2^2 - l_3^2}{2a}\right)^2}};$$

$$\frac{\partial z}{\partial l_2} = \frac{1}{2}\frac{l_2\left(1 + \frac{1}{b^2}\left(\frac{a^2}{4} - l_2^2 + 2l_1^2 - l_3^2\right) - \frac{1}{a^2}\left(l_2^2 - l_3^2\right)\right)}{\sqrt{l_1^2 - \left(\frac{b}{2} + \frac{a^2}{8b} - \frac{l_2^2 - 2l_1^2 + l_3^2}{2b}\right)^2 - \left(\frac{l_2^2 - l_3^2}{2a}\right)^2}};$$

$$\frac{\partial z}{\partial l_3} = \frac{1}{2}\frac{l_2\left(1 + \frac{1}{b^2}\left(\frac{a^2}{4} - l_2^2 + 2l_1^2 - l_3^2\right) + \frac{1}{a^2}\left(l_2^2 - l_3^2\right)\right)}{\sqrt{l_1^2 - \left(\frac{b}{2} + \frac{a^2}{8b} - \frac{l_2^2 - 2l_1^2 + l_3^2}{2b}\right)^2 - \left(\frac{l_2^2 - l_3^2}{2a}\right)^2}}.$$

For the value of the mistake we will have:

$$\Delta z = \frac{1}{2\sqrt{l_1^2 - \left(\frac{b}{2} + \frac{a^2}{8b} - \frac{l_2^2 - 2l_1^2 + l_3^2}{2b}\right)^2 - \left(\frac{l_2^2 - l_3^2}{2a}\right)^2}}$$

$$\times(\frac{l_1^2}{b^4}\left(\frac{a^2}{2} + 2l_2^2 - 4l_1^2 + 2l_3^2\right)^2$$

$$+l_2^2\left(1 + \frac{1}{b^2}\left(\frac{a^2}{4} - l_2^2 + 2l_1^2 - l_3^2\right) - \frac{1}{a^2}\left(l_2^2 - l_3^2\right)\right)^2 \Delta l_2^2$$

$$+l_3^2\left(1 + \frac{1}{b^2}\left(\frac{a^2}{4} - l_2^2 + 2l_1^2 - l_3^2\right) + \frac{1}{a^2}\left(l_2^2 - l_3^2\right)\right)^2 \Delta l_3^2)^{1/2}.$$

The expressions are quite complicated, so it is necessary to take some simplifications. For example, to estimate the mistakes, we can take the maximum values of the distances and the mistakes:

$$\Delta l = \max(\Delta l_1, \Delta l_2, \Delta l_3);$$

$$l = \max(l_1, l_2, l_3).$$

So, the mistakes can be estimated by formulaes:

$$\Delta x = \frac{l\sqrt{6}}{b}\Delta l;$$

$$\Delta y = \frac{l\sqrt{2}}{a}\Delta l;$$

$$\Delta z = \frac{l\Delta l}{2z}\sqrt{\left(\frac{a^2+8l^2}{2b^2}\right)^2 + 2\left(1+\frac{1}{2}\left(\frac{a^2+8l^2}{2b^2}\right)+\frac{l^2}{a^2}\right)}.$$

It can be assumed that the mistake is mainly connected with the mistake in measuring the length of the cable. So, we can estimate it as:

$$\Delta l \approx \delta l = \alpha V^4.$$

For wind velocity $V = 10$ km/s we obtain $\Delta l \approx 0.7$ m. If we take the values $z = 150$ m, $l = 150$ m, $a = 400$ m and $b = 300$ m :

$$\Delta x = \frac{150 \text{ m} \cdot 2.45}{300 \text{ m}} \cdot 0.7 \text{ m} = 0.9 \text{ m};$$

$$\Delta y = \frac{150 \text{ m} \cdot 1.41}{400 \text{ m}} \cdot 0.7 \text{ m} = 0.4 \text{ m};$$

$$\Delta z = \frac{150 \text{ m}}{2 \cdot 150 \text{ m}}$$

$$\times(\left(\frac{(400 \text{ m})^2 + 8(150 \text{ m})^2}{2(300 \text{ m})^2}\right)^2$$

$$2\left(1+\frac{1}{2}\left(\frac{(400 \text{ m})^2 + 8(150 \text{ m})^2}{2(300 \text{ m})^2}\right)+\frac{(150 \text{ m})^2}{(400 \text{ m})^2}\right))^{1/2} \cdot 0.7 \text{ m} = 1 \text{ m}$$

So, it can be seen, that the mistake is quite moderate and it is comparable with the one for the GPS/GLONASS.

4 Conclusion

We have described the reserve navigation system for the tethered unmanned aerial vehicles. It can be based on radio beacons or on the laser navigation. We have shown that here we can take only two ground beacons and use the length of the cable as a third distance. We have estimated possible mistakes connected with the wind and another factors. It has been shown that the mistakes are quite comparable with the ones for the space navigation system.

References

1. Kiribayashi, S., Ashizawa, J., Nagatani, K.: Modeling and design of tether powered multicopter. In: 2015 IEEE International Symposium on Safety, Security, and Rescue Robotics (2015)
2. Vishnevsky, V., Tereschenko, B., Tumchenok, D., Shirvanyan, A.: Optimal method for uplink transfer of power and the design of high-voltage cable for tethered high-altitude unmanned telecommunication platforms. In: Vishnevskiy, V.M., Samouylov, K.E., Kozyrev, D.V. (eds.) DCCN 2017. CCIS, vol. 700, pp. 240–247. Springer, Cham (2017). https://doi.org/10.1007/978-3-319-66836-9_20
3. Vishnevsky, V.M., Mikhailov, E.A., Tumchenok, D.A., Shirvanyan, A.M.: Mathematical model of the operation of a tethered unmanned platform under wind loading. Math.l Models Comput. Simul. **12**, 492–502 (2020). https://doi.org/10.1134/S2070048220040201
4. Kozyrev, D.V., Phuong, N.D., Houankpo, H.G.K., Sokolov, A.: Reliability evaluation of a hexacopter-based flight module of a tethered unmanned high-altitude platform. In: Vishnevskiy, V.M., Samouylov, K.E., Kozyrev, D.V. (eds.) DCCN 2019. CCIS, vol. 1141, pp. 646–656. Springer, Cham (2019). https://doi.org/10.1007/978-3-030-36625-4_52
5. Grimes, J.G.: Global positioning system standard positioning service performance standard (2008)
6. Mohamed, S.A.S., Haghbayan, M.-H., Westerlund, T., Heikkonen, J., Tenhuhen, H., Plosila, J.: A survey on odometry for autonomous navigation systems. IEEE Access **7**, 97466 (2019)
7. Oleinikova, H., et al.: An open-source system for vision-based micro-aerial vehicle mapping, planning, and flight in cluttered environments. arXiv:1812.03892 (2018)
8. Kumar, G.A., Patil, A.K., Patil, R., Park, S.S., Chai, Y.H.: A LiDAR and IMU integrated indoor navigation system for UAVs and its application in real-time pipeline classification. Sensors (Basel) **17**, 1268 (2017)
9. Quist, E.B., Beard, R.W.: Radar odometry on small unmanned aircraft. In: AIAA Guidance, Navigation, and Control (GNC) Conference, 4698 (2013)
10. Quist, E.B., Beard, R.: Radar odometry on fixed-wing small unmanned aircraft. IEEE Trans. Aerosp. Electron. Syst. **52**, 396 (2016)
11. Lindstrom, C., Christensen, R.S., Gunther, J.H.: An investigation of GPS-denied navigation using airborne radar telemetry. In: IEEE/ION Position, Location and Navigation Symposium, vol. 168 (2020)
12. Christensen, R. S., Gunther, J., Long, D.: Toward GPS-denied navigation utilizing back projection-based synthetic aperture radar imagery. In: Proceedings of the ION 2019 Pacific PNT Meeting, p. 108 (2019)
13. Reid, Z., Ash, J.N.: Leveraging 3D models for SAR-based navigation in GPS-denied environments. In: Algorithms for Synthetic Aperture Radar Imagery XXV, vol. 10647, 106470H (2018)
14. Dowling, L., et al.: Accurate indoor mapping using an autonomous unmanned aerial vehicle. arXiv: 1808.01940 (2018)
15. Zeng, Q., Wang, Y., Liu, J., Chen, R., Deng, X.: Integrating vision and laser pont for outdoor UAV SLAM. In: Ubiquintous Positioning Indoor Navigation Based Service, vol. 170 (2014)
16. Tordesillas, J., Lopez, B.T., Everett, M., How, J.P.: FASTER: Fast and Safe Trajectory Planner for Flights in Unknown Environments arXiv: 2001.04420 (2020)

Teleportation of the Bell States on IBM Q Computers Under Their Hardware Errors

V. P. Gerdt[1,2,3] and E. A. Kotkova[1,3(✉)]

[1] Joint Institute for Nuclear Research, Dubna 141980, Russian Federation
gerdt@jinr.ru, ekaterina.a.kotkova@gmail.com
[2] Peoples' Friendship University of Russia, Moscow 117198, Russian Federation
[3] Dubna State University, Dubna 141982, Russian Federation

Abstract. We present and analyze our experimental results on teleportation of two-qubit maximally entangled Bell states on the NISQ (Noisy Intermediate-Scale Quantum) five-qubit processors IBM Q Burlington, Essex, London, Ourense, Rome, Santiago, Vigo and Yorktown. The main obstacle in practical implementation of quantum algorithms on the NISQ computers is caused by hardware errors which depend on the depth of the underlying circuit and its gates. We suggest several modifications of the original teleportation protocol to optimize the depths of its circuit and the connectivity of hardware qubits. In addition, we compare the dynamics of the output probabilities on the processor IBM Q Yorktown within one and a half years of our use of this processor. They clearly demonstrate the significant progress made in the hardware of quantum computers.

Keywords: Quantum information · Quantum communication · Quantum teleportation · Bell states · Qiskit

1 Introduction

The superposition and entanglement of quantum states are the main resources of quantum computation, information and communication [1]. Computational cost of quantum algorithms that do not rest on entanglement appears to scale exponentially with the size of the input problem and often allow rather efficient simulation on classical higher performance computers. An example of promising entangled - based algorithm oriented to NISQ computers is the variational quantum eigensolver (VQE) proposed in [2] and intensively studied in computational quantum chemistry (see the review [3] and its bibliography). The VQE is a hybrid quantum-classical algorithm whose quantum part serves for a state preparation and measurement, whereas a classical computer provides processing the measurement results. The key part of VQE are "entanglers", two-qubit entangling gates (cf. [4]).

The following circuit (Fig. 1), taken from [5], is an example of such entangler.

This work was supported by the Grant from the Ministry of Science and Higher Education of the Russian Federation (Agreement No. 075-10-2020-117).

© Springer Nature Switzerland AG 2020
V. M. Vishnevskiy et al. (Eds.): DCCN 2020, CCIS 1337, pp. 129–143, 2020.
https://doi.org/10.1007/978-3-030-66242-4_11

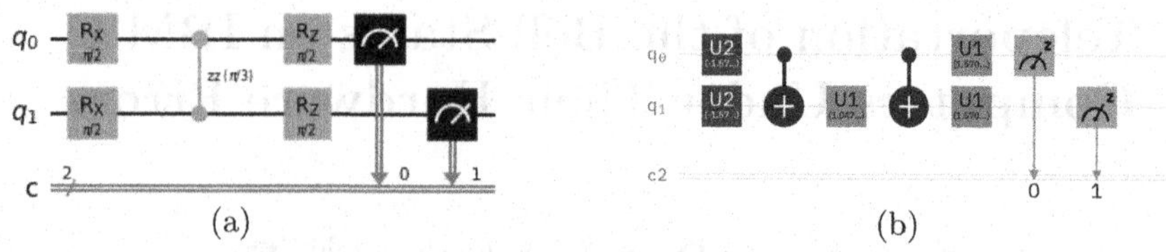

Fig. 1. Entangler gate. a) Initial circuit, b) transpiled into quantum computer gates one

For the input 2-qubit state $|00\rangle$ and entangler parameters $\theta_1, \theta_2, \theta_4, \theta_5 = \pi/4$, $\theta_3 = \pi/6$ its measurements results obtained on IBM Q Santiago and those on the "ideal" (i.e., computed with Qasm Simulator [6]) quantum computer (cf. Sect. 2.2) are given below (Fig. 2). As one can see, in spite of small depth of the entangler circuit the errors are quite notable, and their main source is the two-qubit CNOT gates.

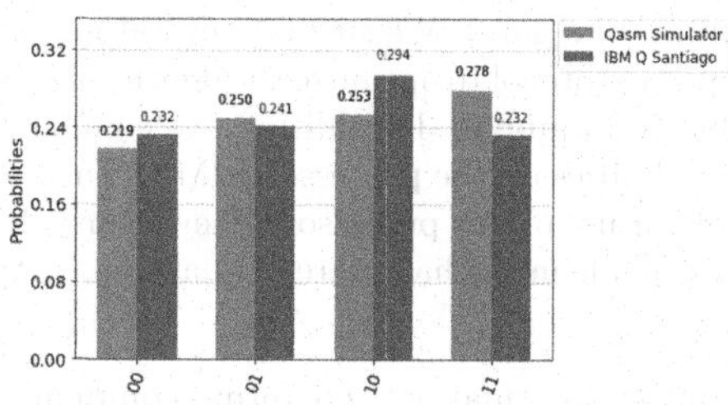

Fig. 2. Results for entangler gate implementation on Qasm Simulator and IBM Q Santiago.

To investigate dependence of the resulting hardware errors on the structure of a circuit and connectivity of the qubits it uses, we consider the circuit (protocol) suggested in [7] for teleportation of the maximally entangled Bell states (Fig. 3).

One-qubit quantum teleportation was first proposed by Bennett et al. in [8]. To teleport a two-qubit state of a quantum channel should consist of three or more qubits. One of the quantum states $|GHZ\rangle$ [9]

$$|GHZ\rangle = \frac{1}{\sqrt{2}}(|000\rangle + |111\rangle) \tag{1}$$

or $|W\rangle$ [10]

$$|W\rangle = \frac{1}{\sqrt{3}}(|001\rangle + |010\rangle + |100\rangle), \tag{2}$$

belonging to different equivalence classes, is used for creating three-qubit quantum channels, depending on the chosen protocol. In [11,12], circuits using the state $|W\rangle$ are presented. The states belonging to the same equivalence class as

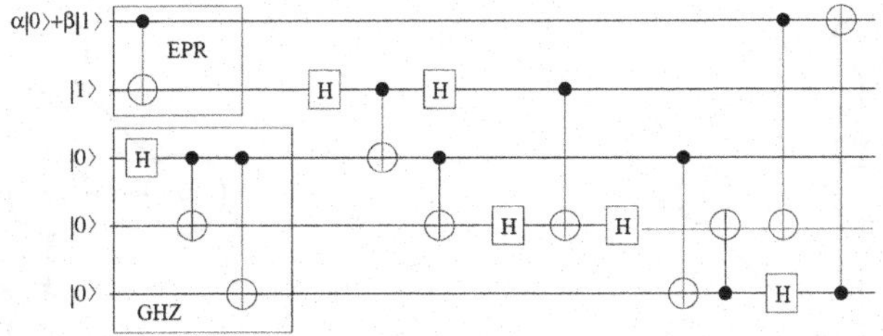

Fig. 3. Initial circuit of the two-qubit teleportation.

$|GHZ\rangle$ are used in [13,14]. The four-qubit cluster state is used to teleport an arbitrary two-qubit state in IBM Quantum Experience [15] in [16]. Protocols for teleportation were also developed using other cluster states [17,18].

Below we present our results on implementation of the protocol [7] (see Fig. 3) of two-qubit quantum teleportation using $|GHZ\rangle$ state on the 5-qubit IBM computers accessible via the cloud platform IBM Quantum Experience [15] and Qiskit framework [19]. All calculations were made within circuit model of computations [1].

We teleported the maximally entangled 2-qubit states known as the Bell (EPR) states (3) [20].

$$|q_0\rangle = |x\rangle\,, |q_1\rangle = |y\rangle \longrightarrow |\beta_{xy}\rangle = \frac{1}{\sqrt{2}}\left(|0,y\rangle + (-1)^x\,|1,1-y\rangle\right), \;\; x,y \in \{0,1\}. \tag{3}$$

The composition of the paper is as follows. Section 2 provides an overview of 5-qubit IBM quantum computers used in our work. We discuss CNOT-gates construction between physically disconnected qubits in Sect. 3. The experimental realization of the teleportation protocol is also demonstrated there.

2 IBM Quantum Computers

Let's briefly overview IBM quantum computers used in this paper. All IBM Q quantum computers are built using superconducting qubit technology [21].

IBM Q Yorktown has been accessible to users since 2017. It underwent a modification about a year ago: its error rates decreased and connections between qubits changed from unidirectional (Fig. 4a) to bi-directional (Fig. 4b).

IBM Q Burlington, IBM Q Essex, IBM Q London, IBM Q Ourense and IBM Q Vigo have T-shaped architecture (Fig. 5).

The third architecture (Fig. 6) is represented by IBM Q Rome and IBM Q Santiago.

Fig. 4. The connectivity of IBM Q Yorktown (a) before its modification, (b) actual version.

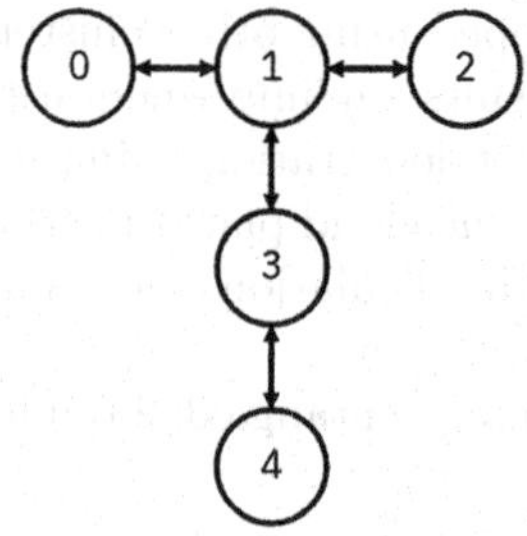

Fig. 5. T-shaped connection graph.

Fig. 6. Connection graph of IBM Q Rome and IBM Q Santiago.

Quantum volume [22] is a measure of quantum computer performance. It is equal to 8 for IBM Q Burlington, IBM Q Essex, IBM Q Ourense, and IBM Q Yorktown. IBM Q London and IBM Q Vigo have the quantum volume 16. Both IBM Q Rome and IBM Q Santiago have the quantum volume 32 [23]. But quantum volume of IBM Q Rome was calculated after the implementations under consideration.

2.1 Gates

The one-qubit gates used in the teleportation protocol are

$$X = \oplus = \begin{pmatrix} 0 & 1 \\ 1 & 0 \end{pmatrix}, \quad H = \frac{1}{\sqrt{2}} \begin{pmatrix} 1 & 1 \\ 1 & -1 \end{pmatrix}, \tag{4}$$

and the 2-qubit control-$\oplus$ (CNOT) gate. Before runnung a circuit on a device, they are transpiled by IBM into the set of the physical gates, which includes two-qubit CNOT gate and one-qubit parametrical ones

$$U3(\theta,\phi,\lambda) = \begin{pmatrix} \cos\left(\frac{\theta}{2}\right) & -\sin\left(\frac{\theta}{2}\right)e^{i\lambda} \\ \sin\left(\frac{\theta}{2}\right)e^{i\phi} & \cos\left(\frac{\theta}{2}\right)e^{i(\lambda+\phi)} \end{pmatrix}, \quad \theta,\lambda,\phi \in [0,2\pi], \tag{5}$$

$$U2(\phi,\lambda) = U3(\frac{\pi}{2},\phi,\lambda) = \begin{pmatrix} \frac{1}{\sqrt{2}} & -\frac{\exp i\lambda}{\sqrt{2}} \\ \frac{\exp i\phi}{\sqrt{2}} & \frac{\exp i(\lambda+\phi)}{\sqrt{2}} \end{pmatrix}, \tag{6}$$

$$U1(\lambda) = U3(0,0,\lambda) = \begin{pmatrix} 1 & 0 \\ 0 & \exp i\lambda \end{pmatrix}. \tag{7}$$

We also used the SWAP gate

$$SWAP(q_0,q_1) = CNOT(q_0,q_1)CNOT(q_1,q_0)CNOT(q_0,q_1), \tag{8}$$

which exchanges the states of qubits q_0 and q_1, for providing CNOT gate between physically unconnected qubits, as discussed in the next section.

2.2 Sources of Errors

Consider the causes of errors in running quantum circuits on real quantum computers. One type of such errors occurs from "noise" in gates and the measurement operations. Due to peculiarities of the physical implementation of one-qubit gates, the value of the U3 gate error is twice larger than the U2 error. U1, which changes the phase of the qubit, is performed in software, which affects the physical implementation of subsequent gates. Despite of the fact that a software implementation itself without physical effects on qubits should not lead to errors in quantum computations, errors in its application are still possible because of decoherence.

Decoherence includes two processes: energy relaxation and dephasing. Energy relaxation means spontaneous transition of a qubit state from $|1\rangle$ to $|0\rangle$ and is characterized by time T_1. Dephasing is the destruction of the superposition state like $\frac{|0\rangle+|1\rangle}{\sqrt{2}}$. Time T_2 reflects both effects of energy relaxation and dephasing.

Since the teleported state is a superposition with equal weights, when it is measured in the classical basis, one of the states included in it is to be obtained with probability 0.5. A quantum computer runs a quantum circuit a finite number of times. So even with ideal computations the resulting probabilities will differ from the expected values. Figure 7a shows the resulting state of the qubits, to which $|\beta_{00}\rangle$ is teleported, calculated using Statevector Simulator [24] without measurement implementation. Statevector Simulator performs single ideal run of a quantum circuit and returns the resulting state vector of the simulator. Results for 1024 shots of ideal implementation of the same circuit on Qasm Simulator [6] are shown in Fig. 7b.

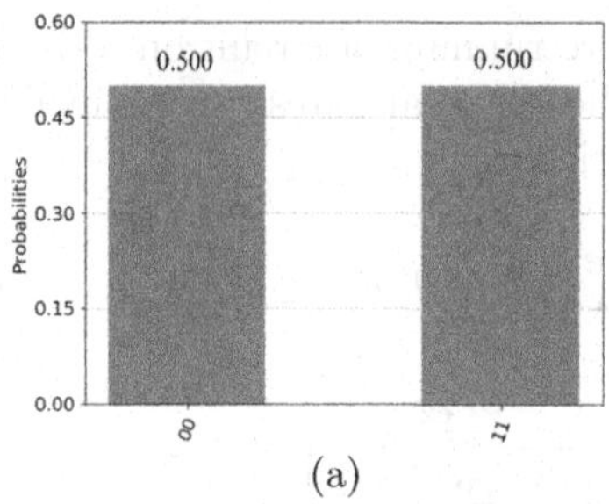

(a)

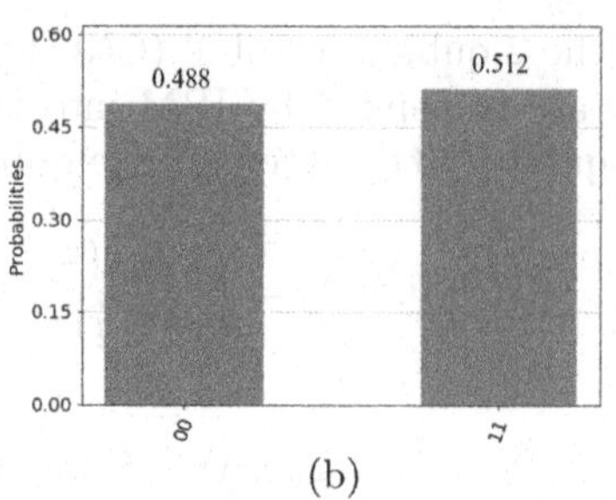

(b)

Fig. 7. Results for $|\beta_{00}\rangle$ teleportation on simulators. (a) One shot of ideal simulation without measurement, (b) ideal simulation on Qasm Simulator with measurement in classical basis. Shots = 1024.

3 Experimental Realization

3.1 CNOT Gates Implementation

The scheme of qubit interconnections in the quantum teleportation protocol is presented in Fig. 8. As one can see from comparing it with any of the architectures from the previous section, some connections that are necessary for implementation are not available among the devices. All connections in IBM Q Yorktown before its modification were unidirectional, so we used a circuit with additional Hadamard gates (Fig. 9a) to change the direction of CNOT gates when necessary. Another problem is implementation of CNOT gates between physically unconnected qubits. In the IBM transpiler, this problem is solved by adding a SWAP gate between an auxiliary qubit and one of the pair of unconnected qubits (Fig. 9b). In some cases, this can lead to the fact that, as a result of applying the teleportation protocol, the state is sent to and received from the same qubit, which negates the very idea of the teleportation. To prevent this and minimize the number of auxiliary gates as much as possible, we combine this approach and the usage of circuits with the same number of CNOT gates but without swap of qubit states (Fig. 9c,d).

3.2 IBM Q Yorktown

In [25] we presented three implementations of the protocol on IBM Q Yorktown before modification of this computer. The correspondence of qubit numbers in the initial circuit to quantum computer qubits is shown in Table 1.

After a certain modification of IBM Q Yorktown when all connections between qubits had become bidirectional, a reduction of implemented circuits depths became possible, since a program change of directions in some CNOT gates was no longer necessary. The depth (see [26]) of all circuits (without the measurement gates) decreased from 25 to 17 for $|\beta_{00}\rangle$ and $|\beta_{01}\rangle$ and from 26 to 18 for $|\beta_{10}\rangle$ and $|\beta_{11}\rangle$. The initial and reduced versions of Circuit 1 are presented in Fig. 10.

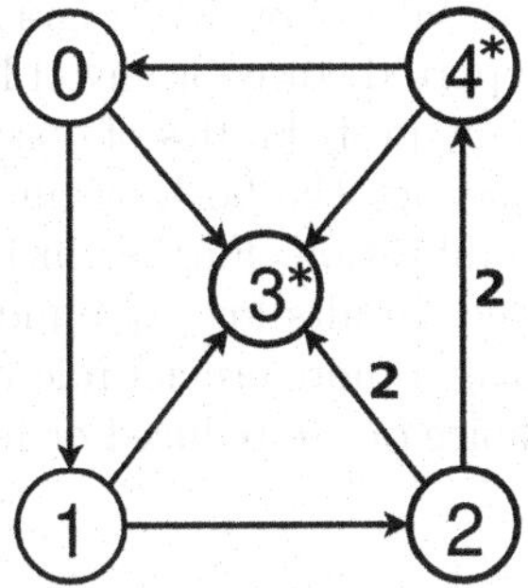

Fig. 8. Qubit connections in the teleportation protocol

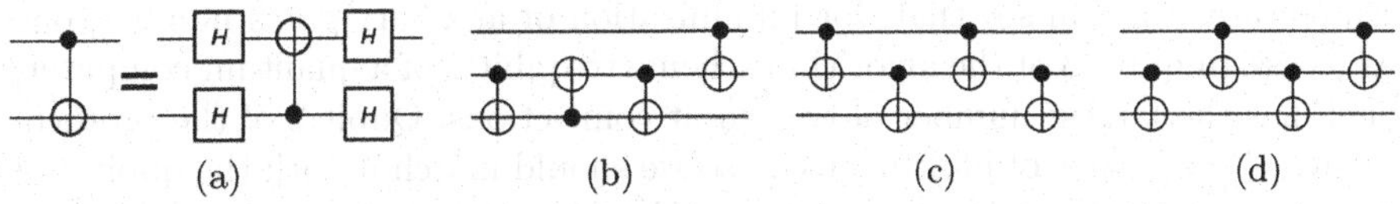

Fig. 9. CNOT implementation: (a) changing the direction; (b,c,d) $CNOT(q_0, q_2)$ with SWAP gate usage (b) and without it (c,d)

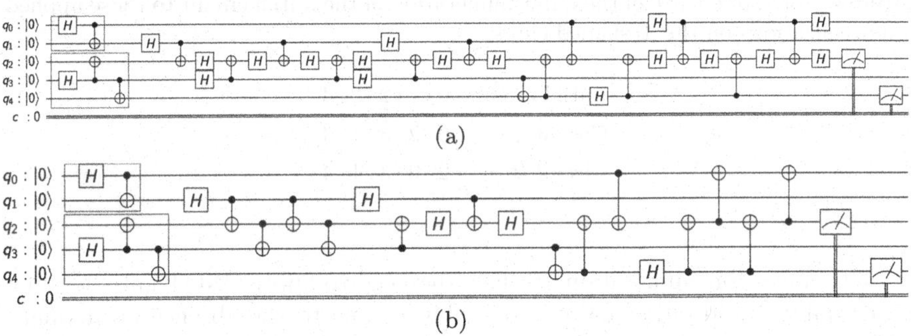

Fig. 10. Quantum Circuit 1 for teleportation of $|\beta_{00}\rangle$ on IBM Q Yorktown a) initial version, b) reduced after the device modification

Table 1. Matchings of the quantum algorithm qubits to the qubits of the IBM Q Yorktown.

Qubit of the protocol	Circuit 1	Circuit 2	Circuit 3
0	0	0	3
1	1	1	4
2	3	4	1
3	2	2	2
4	4	3	0

As one can see from the results shown in Fig. 11, the achieved error rates didn't allow to detect the teleported state for any of Bell ones before the modification of IBM Q Yorktown. The probabilities of observing the "correct" results were higher than incorrect ones for the both original and modified versions of Circuit 1 after the device modification. Further increase of the gate and measurement errors led to incorrect results for $|\beta_{00}\rangle$ and $|\beta_{10}\rangle$ states. The changes in error rates of the device play a more crucial role in the quality of the implementation rather than the choice of the reduced or initial version of the circuit.

3.3 T-Shaped IBM Quantum Computers

When comparing the protocol qubit interconnections scheme with the devices' architecture, one can see that, for minimization of auxiliary gates number, one has to assign qubit 3 of the algorithm scheme to qubit 1 of a quantum computer, which has the largest number of two-qubit connections. Qubit 2 of the protocol has two CNOT gates of multiplicity 2, so we should match it with the qubit 3 of the device and qubit 4 of the protocol with qubit 4 of the quantum computer. Thus, there are only two ordering options, presented in Table 2.

Table 2. Correspondence of the qubit numeration in the initial circuit to those applied in its realizations on the T-shaped chips.

Initial circuit	0	1	2	3	4
Circuit 1	0	2	3	1	4
Circuit 2	2	0	3	1	4

With any qubit numeration, we face the need to apply CNOT gate between device qubit 4 and either qubit 0 or qubit 2. Due to the absence of auxiliary qubit connected to both of them, we combine two approaches implementing SWAP gate between qubits 3 and 4 of the device. That is, $SWAP(q_2, q_4)$, where q_2, q_4 are qubits of the protocol, is used at the place which is marked with the red line in the protocol circuit (Fig. 12). Since $SWAP(q_3, q_4)$ between qubits q_3 and q_4 of the computer is not repeated for the reverse exchange of states between the q_3 and q_4 qubits, as a result of the implementation of the quantum circuit, the state is transferred not to the q_1 and q_4 qubits of the quantum computer, but to the q_1 and q_3, which corresponds to the q_3 and q_2 qubits of the algorithm.

The obtained circuits with a depth (without measurement gates) of 20 for $|\beta_{00}\rangle$ and $|\beta_{01}\rangle$ teleportation and 21 for teleportation of $|\beta_{10}\rangle$ and $|\beta_{11}\rangle$ are presented in Fig. 13.

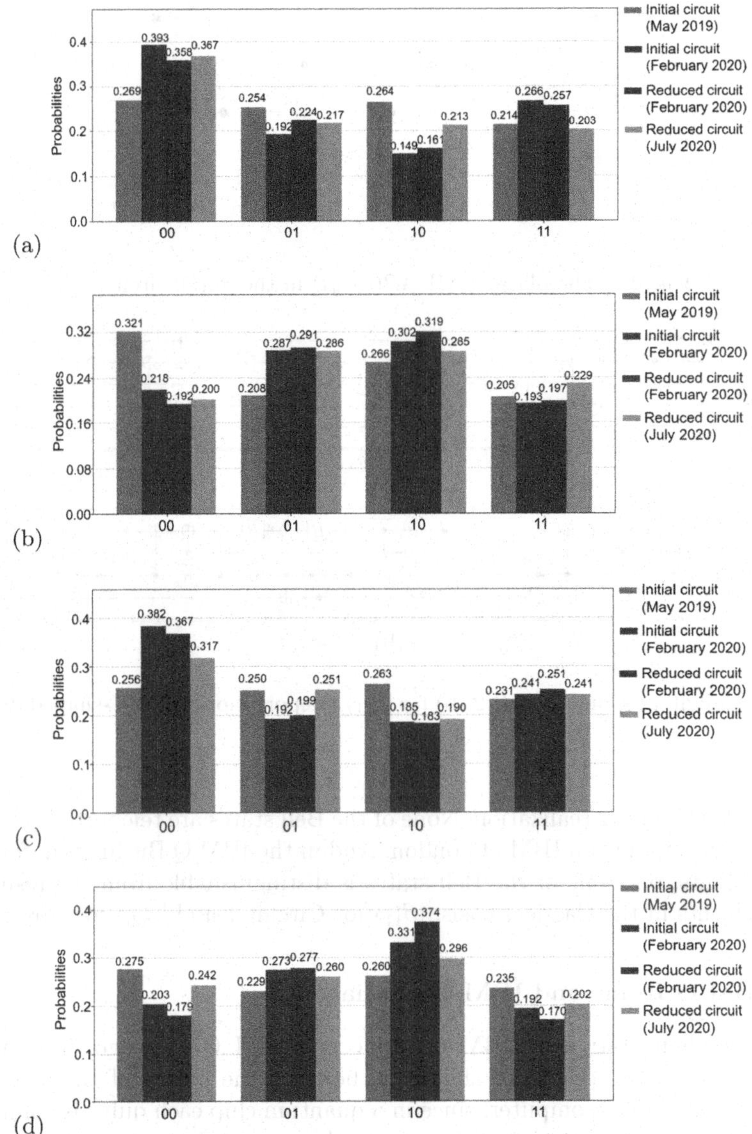

Fig. 11. Results for Circuit 1 for teleportation of the Bell states on the IBM Q Yorktown with readout in the classical basis. a) state $|\beta_{00}\rangle$, b) state $|\beta_{01}\rangle$, c) state $|\beta_{10}\rangle$, d) state $|\beta_{11}\rangle$.

As one can see from Fig. 14, the probability of obtaining "correct" results is higher than incorrect ones for all four Bell states for Circuit 1 implemented on the quantum computers IBM Q Essex, IBM Q Ourense and IBM Q Vigo. Similar, but slightly better for IBM Q Vigo and slightly worse for IBM Q Ourense results are

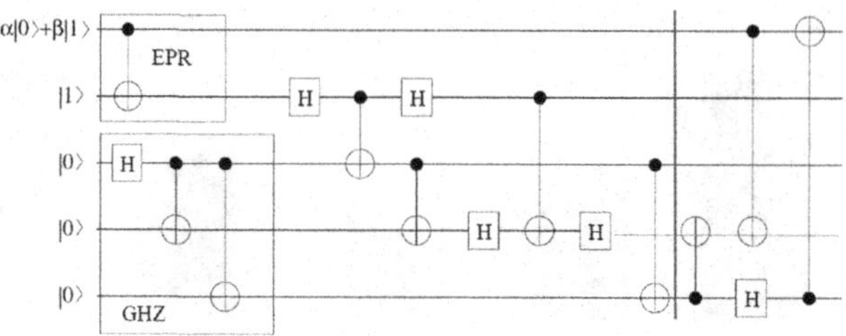

Fig. 12. The place of $SWAP(q_2, q_4)$ in the initial circuit.

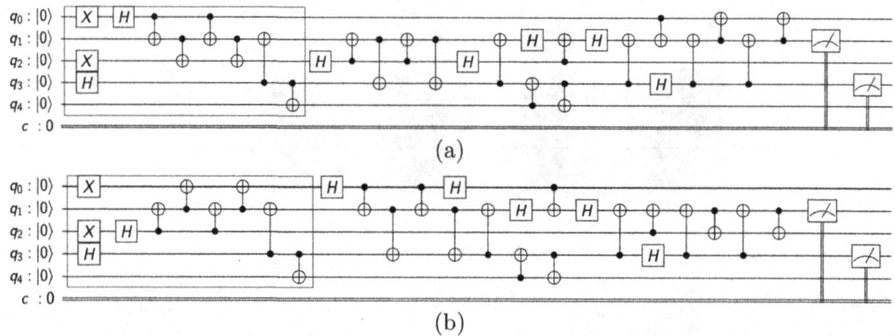

Fig. 13. Circuit 1 (a) and Circuit 2 (b) for $|\beta_{11}\rangle$ teleportation on the T-shaped devices.

achieved for Circuit 2 realization. None of the Bell states are teleported correctly by any of the circuits on IBM Q London. And in the IBM Q Burlington, for each of the circuits, only one of the Bell states is distinguishable from the results of measurements in the classical basis: $|\beta_{11}\rangle$ for Circuit 1 and $|\beta_{00}\rangle$ for Circuit 2.

3.4 IBM Q Rome and IBM Q Santiago

For the implementation on IBM Q Rome and IBM Q Santiago, it is not so obvious which correspondence should be between the qubits of the algorithm and the qubits of the computer, since in a quantum chip each qubit is connected to no more than two others (see Fig. 6). In this regard, let us consider for IBM Q Rome an implementation in which a protocol qubit corresponds to a quantum computer qubit with the same number. The same circuit is applied upside down on IBM Q Santiago, i.e., the protocol qubit q_0 corresponds to the quantum computer qubit q_4, the algorithm qubit q_1 to the device qubit q_3 etc.

Let's take a look at the resulting circuit (Fig. 15). The Roman numerals indicate places that are needed to be given special attention:

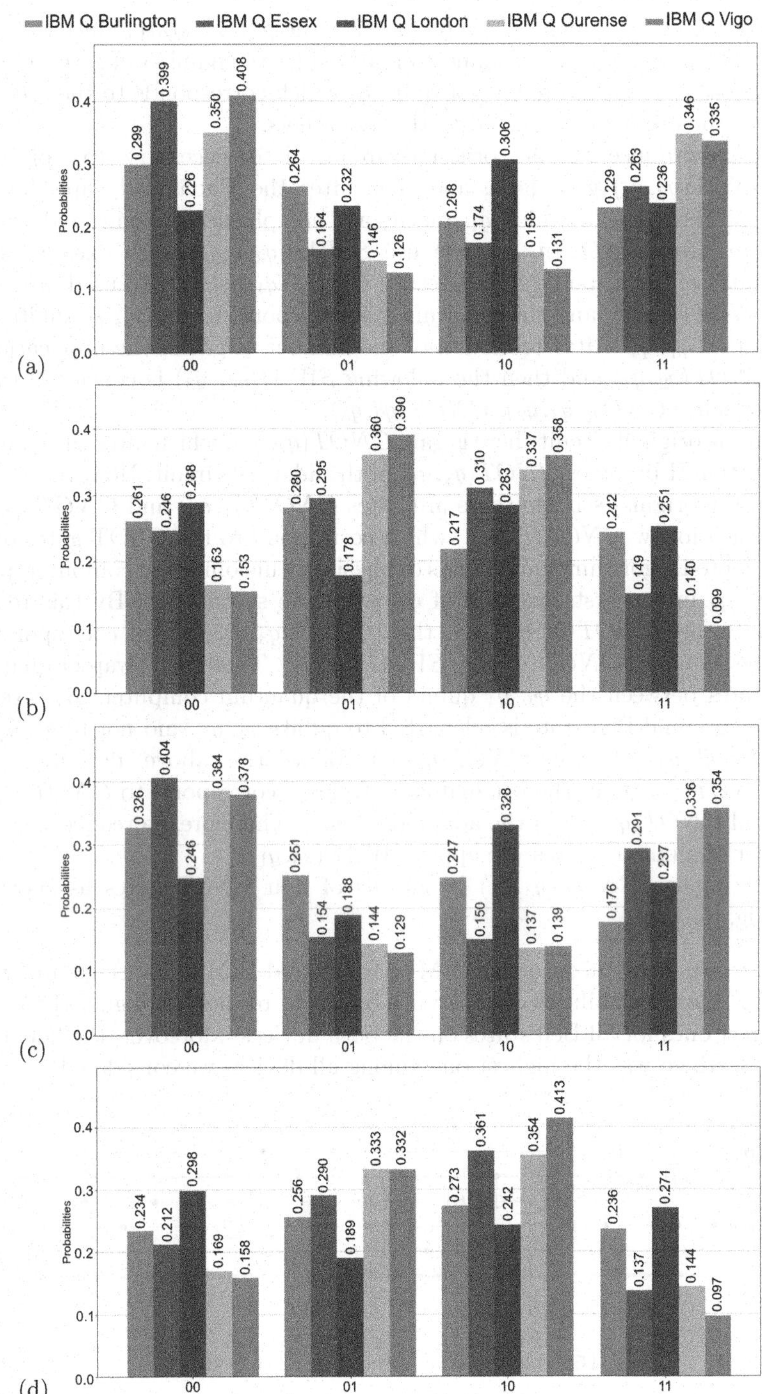

Fig. 14. Results of Circuit 1 for teleportation of the Bell states on the T-shaped devices with readout in the classical basis. a) state $|\beta_{00}\rangle$, b) state $|\beta_{01}\rangle$, c) state $|\beta_{10}\rangle$, d) state $|\beta_{11}\rangle$.

I. The qubits q_2, q_3 and q_4 are prepared in the $|GHZ\rangle$ state. Since the q_2, q_4 qubits have no physical connection, to reduce the number of gates used, we choose as a control qubit not qubit q_2 which corresponds to the ordering, but the qubit q_3 connected to the two others.

II. In accordance to the original circuit, it is necessary to apply the $CNOT(q_2, q_3)$ gate here, and then after the Hadamard gate the gate $CNOT(q_1, q_3)$ in which q_1 and q_3 are not physically connected. Therefore, after $CNOT(q_2, q_3)$, we use $SWAP(q_2, q_3)$, which exchanges the states of the q_2 and q_3. As a result, $CNOT(q_1, q_2)$ is then used instead of $CNOT(q_1, q_3)$, and the quantum state teleports to the q_2, q_4 qubits, and not to q_3, q_4. After reduction of the number of gates, the application of $CNOT(q_2, q_3)$ and then the following $SWAP(q_2, q_3)$ is converted to the sequence $CNOT(q_3, q_2)$, $CNOT(q_2, q_3)$.

III. In the original circuit there is gate $CNOT(q_2, q_4)$, which after the transformation II becomes $CNOT(q_3, q_4)$ of the adapted circuit. However, further in the original scheme there are gates $CNOT(q_4, q_3)$ and $CNOT(q_4, q_0)$, which follow $CNOT(q_2, q_4)$, which correspond to the CNOT gates of the adapted algorithm and consists of physically unconnected computer qubits q_4, q_2 in the first case and of q_4, q_0 in the second one. By this reason, we apply $CNOT(q_4, q_3)$ and then $CNOT(q_3, q_4)$, which corresponds to the sequence $CNOT(q_3, q_4)$, $SWAP(q_3, q_4)$. There is a swap of quantum states between the q_3, q_4 qubits of the quantum computer. As a result, the original Bell state is teleported to qubits q_2, q_3, and not to q_2, q_4.

IV. $CNOT(q_0, q_3)$ and $CNOT(q_4, q_0)$ follow the above described gate $CNOT(q_4, q_3)$ in the original circuit. They correspond to $CNOT(q_0, q_2)$ and $CNOT(q_3, q_0)$ in the adapted scheme. Therefore, to reduce the number of auxiliary gates, we apply $SWAP(q_0, q_1)$.

V. We realize $CNOT(q_3, q_1)$ by means of four CNOT gates as shown in Fig. 15.

The results for realization on IBM Q Rome and IBM Q Santiago are shown in Fig. 16. The probabilities of observing "correct" results are higher than those of incorrect ones for all Bell states on the both devices. Moreover, the fidelity on IBM Q Santiago was the highest one among all devices we considered.

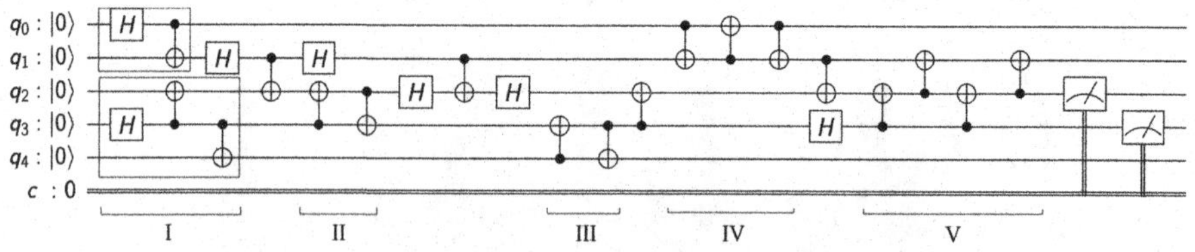

Fig. 15. Teleportation circuit on IBM Q Rome.

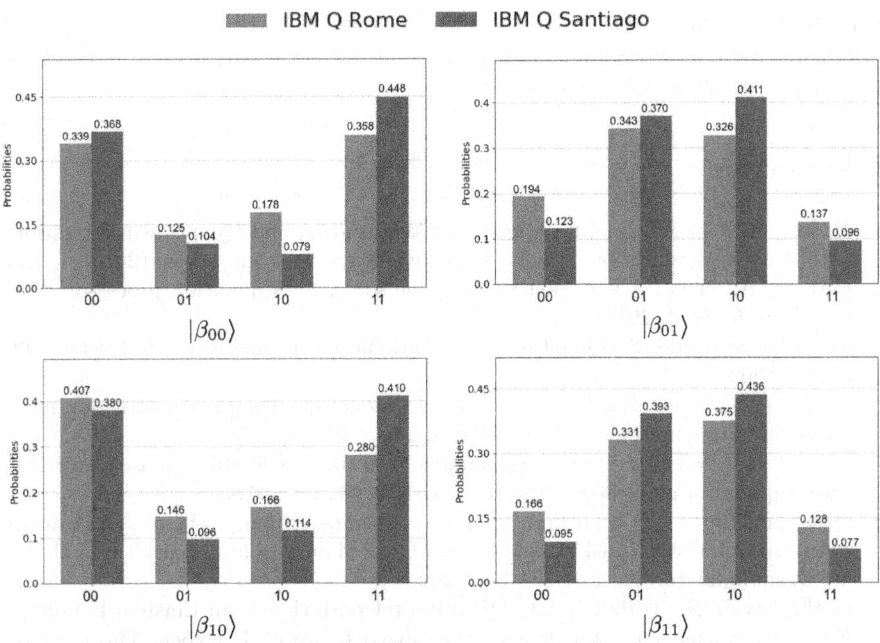

Fig. 16. Results for IBM Q Rome and IBM Q Santiago.

4 Conclusion

The results for implementations of the teleportation of Bell states on five-qubit IBM Q quantum computers with three different connection graphs are presented in Fig. 11, 14, 16. The fact that probabilities of detecting "correct" results for all Bell states were higher than incorrect ones on the devices including IBM Q Santiago, IBM Q Rome, IBM Q Vigo, IBM Q Ourense, and IBM Q Essex, which became publically accessible in the last years, shows the technical improvement achieved by IBM. Nevertheless, errors make recognition of transported states impossible for teleportation of all or the majority of Bell states on the T-shaped computers IBM Q Burlington and IBM Q London. The results for IBM Q Yorktown became better than those obtained in our previous study [25], but some of the Bell states may still be indistinguishable from the measurement results depending on the current device's calibrations. Thus, further decreasing of error rates is needed. It can be done using the error correction codes [27] and the quantum control [28] to suppress errors [29].

As part of this work, a set of programs in Python was written using the Qiskit framework to implement the quantum teleportation protocol on specific quantum computers.

Acknowledgements. The authors are deeply grateful to Michael Biercuk, Michael Hush and Andre Carvalho for informing us about error suppression research at Q-CTRL (https://docs.q-ctrl.com/).

References

1. Nielsen, M.A., Chuang, I.L.: Quantum Computation and Quantum Information. 10th Anniversary Edition. Cambridge University Press, Cambridge (2010)
2. Peruzzo, A., et al.: A variational eigenvalue solver on a quantum processor. Nat. Commun. **5**, 4213 (2014)
3. McArdle, S., Endo, S.: Quantum computational chemistry. Rev. Mod. Phys. **92**, 015003 (2020)
4. Kandala, A., et al.: Hardware-efficient variational quantum eigensolver for small molecules and quantum magnets. Nature **549**, 242–246 (2017)
5. Uvarov, A.V., Kardashin, A.S., Biamonte, J.D.: Mashine learning phase transitions with a quantum processor. Phys. Rev. A. **102**, 012415 (2020)
6. QasmSimulator – Qiskit 0.21.0 documentation, https://qiskit.org/documentation/stubs/qiskit.providers.aer.QasmSimulator.html#qiskit.providers.aer.QasmSimulator. Accessed 15 Sept 2020
7. Gorbachev, V.N., Trubilko, A.I.: Quantum teleportation of an Einstein-Podolsky-Rosen pair using an entanglement three-particle state. J. Exper. Theor. Phys. **118**(5), 1036–1040 (2000)
8. Bennett, C.H., Brassard, G., Crepeau, C., et al.: Teleporting an unknown quantum state via dual classical and Einstein-Podolsky-Rosen channels. Phys. Rev. Lett. **70**, 1895–1899 (1993)
9. Greenberger, D., Horne, M., Zeilinger, A.: Similarities and differences between two-particle and three-particle interference. Fortschr. Phys. **48**(4), 243–252 (2000)
10. Dür, W., Vidal, G., Cirac, J.I.: Three qubits can be entangled in two inequivalent ways. Phys. Rev. A **62**, 062314–062325 (2000)
11. Cao, Z.-L., Song, W.: Teleportation of a two-particle entangled state via W class states. Physica A: Stat. Mechanics Appl. **347**, 177–183 (2005)
12. Joo, J., Park, Y.-J., Oh, S., Kim, J.: Quantum teleportation via a W state. New J. Phys. **5**, 136.1–136.9 (2003)
13. Ghosh, S., Kar, G., Roy, A., Sarkar, D. et al.: Entanglement teleportation through GHZ-class states. New J. Phys. **4**, 48.1–48.9. (2002)
14. Tsai, C., Hwang, T.: Teleportation of a Pure EPR State via GHZ-like State. Int. J. Theor. Phys. **49**, 1969–1975 (2010)
15. IBM Quantum Experience, https://www.ibm.com/quantum-computing/experience/. Accessed 20 Sept 2020
16. Rajiuddin, S., Baishya, A., Behera, B.K., Panigrahi, P.K.: Experimental realization of quantum teleportation of an arbitrary two-qubit state using a four-qubit cluster state. Quantum Inf. Process. **19**(3), 1–13 (2020). https://doi.org/10.1007/s11128-020-2586-x
17. Li, D.-C., Cao, Z.-L.: Teleportation of two-particle entangled state via cluster state. Commun. Theor. Phys. **47**(3), 464–466 (2007)
18. Liu, Z., Zhou, L.: Quantum teleportation of a three-qubit state using a five-qubit cluster state. Int. J. Theor. Phys. **53**(12), 4079–4082 (2014). https://doi.org/10.1007/s10773-014-2158-x
19. Qiskit, https://qiskit.org/. Accessed 20 Sept 2020

20. Sutor, R.S.: Dancing with Qubits: How quantum computing works and how it can change the world. Packt (2019)
21. Kjaergaard, M., et al.: Superconducting qubits: current state of play. Ann. Rev. Condensed Matter Phys. **11**, 369–395 (2020)
22. Cross, A.W., et al.: Validating quantum computers using randomized model circuits. Phys. Rev. A. **100**(3), 032328 (2019)
23. IBM Quantum Experience - Docs and Resources, https://quantum-computing.ibm.com/docs/manage/backends/. Accessed 15 Sept 2020
24. StatevectorSimulator – Qiskit 0.21.0 documentation, https://qiskit.org/documentation/stubs/qiskit.providers.aer.StatevectorSimulator.html#qiskit.providers.aer.StatevectorSimulator. Accessed 15 Sept 2020
25. Gerdt, V.P., Kotkova, E.A., Vorob'ev, V.V.: The teleportation of the Bell states has been carried out on the five-qubit quantum IBM computer. Phys. Particles Nuclei Lett. **16**(6), 975–984 (2019). https://doi.org/10.1134/S1547477119060153
26. Preskill, J.: Quantum computing in the NISQ era and beyond. Quantum **2**, 79 (2018)
27. Harper, R., Flammia, S.T.: Fault-tolerant logical gates in the IBM quantum experience. Phys. Rev. Lett. **122**, 080504 (2019)
28. Ball, H., Biercuk, M.J., Carvalho, A., et al.: Software tools for quantum control: Improving quantum computer performance through noise and error suppression. arXiv:2001.04060 (2020)
29. Superconducting qubits: improving the performance of single qubit gates, https://docs.q-ctrl.com/boulder-opal/application-notes/superconducting-qubits-improving-the-performance-of-single-qubit-gates. Accessed 18 Sept 2020

Management of Risks for Complex Computer Network

A. O. Kalashnikov(✉) and E. V. Anikina(✉)

V.A. Trapeznikov Institute of Control Sciences Russian Academy of Sciences, Moscow, Russia
aokalash@ipu.ru, janet0584@mail.ru
https://www.ipu.ru/

Abstract. The paper is concerned with a general model of complex computer network, within which a risk manager exercises efficient management of risks of a complex system through distribution of available resource among its elements (units of computer networks). A tasks of risks management are considered in a context of uncertainty and mutual influence of system elements on each other and methods for solving them are proposed.

Keywords: Computer network · Management of risk · Risk manager · Arbitral solution · Cognitive game

1 Introduction

The modern stage of Russia's development may be described just in one word – digitalization, as the sphere and volumes of application of digital and information technologies are each day expanding and increasing. This fact is reflected among other in the programs for implementation of National Projects adopted in Russia in 2018–19, and not only those that, to some extent, directly relate to introduction of digital and information technologies in various areas of public and economic activity, for example, "Digital Economy", "Science" or "Education", but also those that involve a significant improvement in the quality of life and development of citizens: "Healthcare", "Culture", "Safe and High-Quality Roads", "Housing and Urban Environment", "Ecology" and some others.

Within these projects a great number of large-scale, distributed systems are created, often having no analogues both in their complexity, on the one hand, and in size of potential threats that arise in the event of their failure or incorrect operation, on the other hand [1,2]. Examples of such systems are Internet segments, a set of objects that form a critical information infrastructure of Russia, distributed automated control systems for technological production, segments of the Internet of things and industrial Internet, distributed corporate information systems, telecommunications systems, fixed and mobile communication systems, and a number of others.

© Springer Nature Switzerland AG 2020
V. M. Vishnevskiy et al. (Eds.): DCCN 2020, CCIS 1337, pp. 144–157, 2020.
https://doi.org/10.1007/978-3-030-66242-4_12

Meanwhile, performed analysis shows that with few exceptions approaches to solution of risk management challenges and safety of complex systems are exercised exclusively based on traditional positions, when a set of independent, autonomous chains of a following type is considered as management mechanisms: standard risk – standard scenario – standard resources. This approach contemplates that events, when standard risks are brought about, occur independently of each other, and as far as probability estimation of the said events is minor, expectation of simultaneous or successive occurrence of such several events may be disregarded. It therefore follows from the foregoing assumption that as only one risk-generating event is exercised during some period of time, a single standard scenario may be implemented for its detecting, preventing and relieving the consequences and a set of standard resources may be applied as determined by such scenario.

Under present-day conditions this approach can't be unfortunately considered as satisfactory, whereas as evidenced by the analysis, during several dozens of years occurrence of one risk-generating events may directly or indirectly instigate occurrence of other risk-generating events, such events may in exchange bring about further risk-generating events and so forth. And we can't exclude possibility of influence of further events on initial ones, both towards strengthening and weakening of any related risks.

Thus, for current efficient resolution of risk management tasks and safety of complex systems it is necessary to use approaches allowing to consider risk-generating events and related risks not from the "point" in some phase space but rather as a dynamic network, which units significantly affect the state of each other.

2 General Model

Let's consider a complex system consisting a variety of elements:
$S = \{s_1, \ldots, s_i, \ldots, s_n\}, i \in N = \{1, \ldots, n\}$. Let's assume within the framework of a model $s_i \in S, i \in N$, are elements of the system S, that are not independent and can have a certain effect on each other's state.

Suppose there is a subject that we will call RM (risk manager). Let's assume RM has a certain amount of resource $X \geqslant 0$, that can be arbitrarily allocated between elements of the system $S : x = (x_1, \ldots, x_n), x_i \geqslant 0, i \in N, \sum_{i=1}^{n} x_i \leqslant X$.

Let us denote the set of a admissible distributions of resource X between the elements of the system S by
$RM : X(X) = \{x = (x_1, \ldots, x_n) \in \mathbb{R}^n : x_i \geqslant 0, i \in N, \sum_{i=1}^{n} x_i \leqslant X\}$.

Since elements of the system S are not independent and may have some effect on the state of each other within the model, local risk shall mean some local characteristic of a separate element $s_i \in S, i \in N$, depending not only on amount of resource $x_i \geqslant 0$, reserved by RM for this element, but also on distribution of resources for remaining elements of the system S, i.e. on the vector $x = (x_1, \ldots, x_n) \in \mathbb{R}^n$. Integral risk shall similarly mean some integral characteristic of the overall system S as a whole, also depending on the vector contingent on $x = (x_1, \ldots, x_n) \in \mathbb{R}^n$.

Let's define the function of local risk: $\rho_i(x) : \mathbb{R}^n \rightarrow \mathbb{R}^1$ for each element $s_i \in S, i \in N$ of the system S and assume the vector-function of risk $(\rho_1(x), \dots, \rho_n(x))$ uniquely describes the state of the system S. Similarly, we define the function of integral risk: $\rho(x) = \rho(\rho_1(x), \dots, \rho_n(x))$ for the system S as a whole. In the simplest form the sum of local risks of all elements of the system S can be chosen as the integral risk function: $\rho(x) = \sum_{i=1}^{n} \rho_i(x)$.

Let us denote vector as: $z = (z_1, \dots, z_n) \in \mathbb{R}^n$, then: $z = (z_1, \dots, z_n) \geqslant 0$, if $z_i \geqslant 0, i \in N$.

Let us assume the functions of local risk $\rho_i(\cdot), i \in N$, are continuous, differentiable everywhere, and have the following properties within the model:
C1 (the nonnegativity of risk): for any $i \in N, x \geqslant 0 : \rho_i(x) \geqslant 0$;
C2 (the monotony of risk): for any $i \in N, k \in N : \dfrac{\partial \rho_i(x)}{\partial x_k} < 0$;
C3 (the limited of risk): for any $i \in N, x \geqslant 0$: there is $\rho_i^{\infty} > 0$ such that $\rho_i(x) > \rho_i^{\infty}$.

Property C1 means potential damage connected with implementation of the local risk for any element $s_i \in S, i \in N$ of the system S which can't be negative.

Property C2 means for any element $s_i \in S, i \in N$ of the system S that additional allocation of the resource by RM subject shall lead to reduction of local risk for all elements of the system S.

Property C3 means that for any element $s_i \in S, i \in N$ of the system S no additional allocation of the resource by RM subject is incapable to reduce local risk for this element "to zero". In other words, notwithstanding the volumes of resources spent by RM subject for any element $s_i \in S, i \in N$ of the system S a positive residual risk always subsists.

Thus, the functions of local risk $\rho_i(\cdot), i \in N$ are a family of non-negative, bounded, and strictly monotonic functions that decrease in all arguments.

The following tuple sets the model of risk management for a complex system:

$$\langle RM, X, S = \{s_i\}, \{p_i(\cdot)\}, p(\cdot), i \in N \rangle \tag{1}$$

We assume that within the General Model the target of RM subject is the following: to achieve maximum possible reduction in value of the integral risk $\rho(x)$, using the available resource X and distributing it among elements of the system S. The RM's goal can be formally written as:

$$\inf_{x \in X} p(x) = \inf_{x \in X} p(p_1(x)), \dots, p_n(x)) \tag{2}$$

Handling of task (2) is a search for the global minimum of the function $\rho(x)$ in a bounded domain $x \in X$ and can be obtained by traditional numerical methods (see, for example, [3]) for known functions of local risk $\rho_i(\cdot), i \in N$.

Let us assume we do not know the specific type of functions local risk $\rho_i(\cdot), i \in N$ for any element $s_i \in S, i \in N$, of the system S, it is usually the case in practice (see , for example, [4]), but we know they satisfy the properties C1, C2, and C3. In this case, the methods for handling of task (2) become significantly more complex.

3 The Arbitral Solution

Let us consider the case when the elements $s_i \in S, i \in N$ of the system S are independent and do not affect each other within the model presented in Sect. 2. An approach with an unknown specific type of functions of local risk to the handling of this task was first outlined, although in a slightly different formulation, in the article [5], and most fully described in the monographs [1,2]. This approach is based on the following considerations (for more details, see [1,2]). Since we do not know the specific form of the functions of local risk, let's pass from the "global" task of minimizing the integral risk (2) to the "local" task of reducing the maximum of local risks $\rho_i(\cdot), i \in N$:

$$\inf_{x \in X} \sup_{i \in N} p_i(x) \tag{3}$$

Then the following tuple sets the model of information risk management for a complex system:

$$\langle RM, X, S = \{s_i\}, \{p_i(\cdot)\}, i \in N \rangle$$

The "good" resource distributions $\widehat{x}(X) \in X(X)$ are a handling to task (3) such that:

$$\widehat{x}(X) = arg \inf_{x \in X} \sup_{i \in N} p_i(x)$$

Let us denote a subset of "good" resource distributions as $\widehat{X}(X) \subseteq X(X)$.

The following statement "on equalization of local risks" (see [1,2]), in the notation of the model considered above, is true for the task (3).

Statement 1. *Let $\rho_i(\cdot), i \in N$ satisfy the properties C1, C2 and C3 and there is a resource distribution $(\widetilde{x}_1, \ldots, \widetilde{x}_n) \in X(X)$ such that: $\sum_{i=1}^{n} \widetilde{x}_i = X$ and $\rho_1(\widetilde{x}_1) = \ldots = \rho_n(\widetilde{x}_n) = c = const$, then $(\widetilde{x}_1, \ldots, \widetilde{x}_n)$ is the only handling of task 3.*

The essence of this statement is that, if it is impossible to implement the "global" criterion of risks reduction, the proper strategy of RM subject will be striving to ensure aligning of local risks for all elements of the system S through corresponding allocation of resources.

Before proceeding to further review of we should note that in the absence of knowledge about specific type of local risk functions for any element $s_i \in S, i \in N$ of the system S it is reasonable to assume that all elements of the system S are similar to each other. Then in a certain sense the local risk functions themselves will be "similar": $\rho_1(\widetilde{x}_1), \ldots, \rho_n(\widetilde{x}_n)$.

Suppose RM knows the current local risk values for each element $s_i \in S, i \in N$, of system S, before any resources are allocated to them. Let us denote the specified values $\rho_i(0), i \in N$ and order them in descending order: $\rho_{(1)}(0) \geqslant \ldots \geqslant \rho_{(n)}(0)$. It follows that different amounts of resources will be spent, if RM's goal is to equalize local risks, then to achieve this goal for different

elements of
$s_i \in S, i \in N$, of system S, and: if $\rho_{(i)}(0) \geqslant \rho_{(j)}(0)$, then $x_{(i)} \geqslant x_{(j)}$ must be performed. In this case, the values $\widetilde{p}_i = \rho_i(0)$ can be considered as a kind of "requests" of the elements $s_i \in S, i \in N$, of the system S for the provision of a resource from RM. Let denote the vector of "requests" of the elements $s_i \in S, i \in N$, of the system S for the provision of a resource from RM as $\widetilde{p} = (\widetilde{p}_1, \ldots, \widetilde{p}_n)$.

Suppose RM has a resource X, which is a function of the "requests" of elements of the system S, such that $X = X(\widetilde{p}_1, \ldots, \widetilde{p}_n)$ is symmetric, continuous, strictly monotonic, and $X(0, \ldots, 0) = 0$. The indicated properties are quite natural and reflect the following features of the defender's behavior: 1) not to allocate a resource without the need; 2) increase the amount of allocated resource in case of increasing risks; 3) consider all elements of the system S homogeneous in the absence of additional information.

From the point of view of risk management, in [1,2] formulated and justified a number of "reasonable" requirements that must be satisfied with a "good" allocation of resources.

T1 (Pareto optimality): for any $\widehat{x}(X) \in \widehat{X}(X) : \sum_{i=1}^{n} \widehat{x}_i(X) = X$.

T2 (monotone): for any $X_1 > X_2 \geqslant 0$ and $\widehat{x}(X) \in \widehat{X}(X) : \widehat{x}(X_1) > \widehat{x}(X_2)$, that is $\widehat{x}_i(X_1) \geqslant \widehat{x}_i(X_2), i \in N$ and there is $j \in N$ such that $\widehat{x}_j(X_1) > \widehat{x}_j(X_2)$.

T3 (parity): for any $\widehat{x}(X) \in \widehat{X}(X) : if \rho_{(1)}(0) \geqslant \ldots \geqslant \rho_{(n)}(0)$, then $\widehat{x}_{(1)}(X) \geqslant \ldots \geqslant \widehat{x}_{(n)}(X)$.

The subset of "good" distributions of the resource $\widehat{X}(X) \subseteq X(X)$ satisfying the requirements of T1, T2, and T3 is not empty, since, the uniform distribution of the resource satisfies these requirements : $\widehat{e}(X) \in \widehat{X}(X) : \widehat{e}_i(X) = X/n$ it was shown in [1,2].

Given assumptions allow using for finding of efficient distribution of resource a game-theoretic approach on the basis of the arbitration scheme, based on the principles of stimulation and non-suppression (for further information refer to [6,7]).

Let us briefly recall the main results, adhering to the terminology and designations defined in the General Model presented above.

We will consider elements $s_i \in S, i \in N$ of the system S as "players" of a game $\Gamma(\widetilde{p})$, where $\widetilde{p} = (\widetilde{p}_1, \ldots, \widetilde{p}_n)$ is a vector of "requests" from elements of the system S for resource provision by RM, acting as a "arbiter".

Let us define $X(\widetilde{p}) = X(\widetilde{p}_1, \ldots, \widetilde{p}_n)$ as an RM resource available for distribution and the set of admissible distributions of the resource X between elements of the system S:

$$X(\widetilde{p}) = \{(x_1, \ldots, x_n) \in \mathbb{R}^n : x_i \geqslant 0, i \in N, \textstyle\sum_{i=1}^{n} x_i \leqslant X(\widetilde{p})\}.$$

Let's denote $\widehat{X}(\widetilde{p}) \subseteq X(\widetilde{p})$ — a subset of "good" resource distributions that satisfying the requirements of T1, T2, and T3.

Definition 1. *Let* $\widetilde{p}_{(1)} \geqslant \ldots \geqslant \widetilde{p}_{(n)}$, *if:*

1) $\widehat{\pi}(\widetilde{p}) \in \widehat{X}(\widetilde{p})$;

2) $\widehat{\pi}_{(1)}(\widetilde{p}) = \sup\limits_{\widehat{x}(\widetilde{p}) \in \widehat{X}(\widetilde{p})} \widehat{x}_{(1)}(\widetilde{p});$

$\widehat{\pi}_{(2)}(\widetilde{p}) = \sup\limits_{\widehat{x}(\widetilde{p}) \in \widehat{X}^{(1)}(\widetilde{p})} \widehat{x}_{(2)}(\widetilde{p});$

...

$\widehat{\pi}_{(n-1)}(\widetilde{p}) = \sup\limits_{\widehat{x}(\widetilde{p}) \in \widehat{X}^{(1)(2)\dots(n-2)}(\widetilde{p})} \widehat{x}_{(n-1)}(\widetilde{p});$

where
$\widehat{X}^{(1)(2)\dots(k)}(\widetilde{p}) = \{\widehat{x}(\widetilde{p}) \in \widehat{X}(\widetilde{p}) : \widehat{x}_{(1)}(\widetilde{p}) = \widehat{\pi}_{(1)}(\widetilde{p}), \dots, \widehat{x}_{(k)}(\widetilde{p}) = \widehat{\pi}_{(k)}\widetilde{(p)}\}$
and $k = 1, 2, \dots, n-2$,
then $\widehat{\pi}(\widetilde{p}) = (\widehat{\pi}_{(1)}(\widetilde{p}), \dots, \widehat{\pi}_{(n)}(\widetilde{p}))$ *– is the resource allocation by the "maximum stimulating solution" (MS-solution).*

As a result, MS-solution will mean such distribution of resource X among elements of the system S when: firstly, specifications T1, T2 and T3 are met, and, secondly, the amount of resource as much as possible for number (1) among all such distributions is assigned for the element with number (1) with a maximum "request"; the amount of resource as much as possible for number (2) among distributions, where amount of resource for number (1) has already fixed and equal to $\widehat{\pi}_{(1)}(\widetilde{p})$ so forth, is assigned for the element with number (2) with second major "request".

The existence of an MS solution is not obvious as follows from the above definition. However, the following statement turns out to be true (see [1,2,5–7]):

Statement 2. *Let* $\widetilde{p}_{(1)} \geqslant \dots \geqslant \widetilde{p}_{(n)}$*, then MS-solution* $\widehat{\pi}(\widetilde{p}) = (\widehat{\pi}_{(1)}(\widetilde{p}), \dots, \widehat{\pi}_{(n)}(\widetilde{p}))$ *exists and is unique.*

The aforesaid assumption enables within solution of task (3) to arrange efficient management of risks of a complex system, when maximum risk reduces first, then next by significance risk reduces, and so forth. Unfortunately, the proof of Statement 2 is not constructive and does not determine the MS-solution in an analytical form. Nevertheless, this can be done for a number of frequent cases, which allows applying the MS-solution in practice.

Suppose the function $X(\widetilde{p})$ *has the form:* $X(\widetilde{p}) = X(\widetilde{p}_1 + \dots + \widetilde{p}_n)$*, that is, depending only on the amount of "requests" all the elements* $s_i \in S, i \in N$ *of system S, which is quite common in practice, and* $\dfrac{\partial X(\widetilde{p}_1 + \dots + \widetilde{p}_n)}{\partial \widetilde{p}_i} > 0$*, for* $i \in N$*. The following statements [1,2] turn out to be true for the cases when* $X(\widetilde{p})$ *is convex, concave, or linear.*

Statement 3. *Let* $\widetilde{p}_{(1)} \geqslant \dots \geqslant \widetilde{p}_{(n)}$ *(we assume* $\widetilde{p}_1 \geqslant \dots \geqslant \widetilde{p}_n$ *for simplicity) and* $X(\widetilde{p})$ *is concave, that is:* $\dfrac{\partial^2 X(\widetilde{p}_1 + \dots + \widetilde{p}_n)}{\partial \widetilde{p}_i^2} \geqslant 0$*, for* $i \in N$*. Then the MS-solution has the form:*

$\mu_n^+(\widetilde{p}) = \dfrac{1}{n} X(n\widetilde{p}_n);$

$\mu_n^+(\widetilde{p}) = \dfrac{1}{k}(X(k\widetilde{p}_k + \sum_{i=k+1}^{n} \widetilde{p}_i) - \sum_{i=k+1}^{n} \mu_i^+(\widetilde{p})), k = 1, 2, \dots, n-1.$

Statement 4. *Let $\widetilde{p}_{(1)} \geqslant \ldots \geqslant \widetilde{p}_{(n)}$ (we assume $\widetilde{p}_1 \geqslant \ldots \geqslant \widetilde{p}_n$ for simplicity) and $X(\widetilde{p})$ is convex, that is: $\dfrac{\partial^2 X(\widetilde{p}_1 + \ldots + \widetilde{p}_n)}{\partial \widetilde{p}_i^2} \leqslant 0$, for $i \in N$. Then the MS-solution has the form:*

$\mu_n^-(\widetilde{p}) = \dfrac{1}{n} X(n\widetilde{p}_1)$;

$\mu_n^-(\widetilde{p}) = \dfrac{1}{n-(k-1)}(X(\sum_{i=1}^{k-1}\widetilde{p}_i + (n-(k-1))\widetilde{p}_k) - \sum_{i=1}^{k-1}\mu_i^-(\widetilde{p})), k = 2, \ldots, n.$

Statement 5. *Let $\widetilde{p}_{(1)} \geqslant \ldots \geqslant$ (we assume $\widetilde{p}_1 \geqslant \ldots \geqslant \widetilde{p}_n$ for simplicity) and $X(\widetilde{p}) = \alpha(\sum_{i=1}^{n}\widetilde{p}_i) + \beta$ (linear function), that is: $\dfrac{\partial^2 X(\widetilde{p}_1 + \ldots + \widetilde{p}_n)}{\partial \widetilde{p}_i^2} \equiv 0$, for $i \in N$. Then the MS-solution has the form:*

$\mu_k(\widetilde{p}) = \alpha \widetilde{p}_k = 1, \ldots, n.$

Let us conclude from the analysis of the given MS-solution that reliable information about the values of "requests" of elements $s_i \in S, i \in N$, of the system S for the provision of a resource by player RM: $\widetilde{p} = \rho_i(0)$ is key for the implementation of this approach since:

- firstly, it is an opportunity to order the elements $s_i \in S, i \in N$ of the system S, taking into account their potential risk, which determines the priorities when allocating resources in accordance with the ordered vector of "requests" $\widetilde{p}_{(1)} \geqslant \ldots \geqslant \widetilde{p}_{(n)}$;
- secondly, the ability to determine the corresponding amount of the resource $X(\widetilde{p}) = X(\widetilde{p}_1, \ldots, \widetilde{p}_n)$, as a function of the indicated "requests";
- third, if RM chooses $X(\widetilde{p}) = X(\widetilde{p}_1 + \ldots + \widetilde{p}_n)$, as a function for calculating the resource volume , then it is possible to calculate the values of the MS-solution directly in accordance with statements 4–6.

Thus, if we want to use some analogue of the MS-solution to handling task (3) within the framework of the model presented in Sect. 2, we need to develop a method that will allow us to evaluate the "requests" of the elements $s_i \in S, i \in N$, of the system S on the provision of a resource by RM: $\widetilde{p} = \rho_i(0)$, and their ranking $\widetilde{p}_{(1)} \geqslant \ldots \geqslant \widetilde{p}_{(n)}$ taking into account the mutual influence on each other.

Please, consider a few examples.

Let for any element $s_i \in S, i \in N$ of the system S the local risk function is of the following form: $\rho_i(x) = r_i \times p_i(x) + \rho_i^\infty, x \in \chi(X)$, where $r_i > 0$ is maximum possible damage from destructive effect on the element $s_i \in S, i \in N, \rho_i^\infty \geqslant 0$ is a residual risk, and $p_i(x)$ is the function meeting the following properties:

C4: for any $i \in N, x \geqslant 0 : p_i(x) \in [0,1]$;

C5: for any $i \in N, k \in N, \dfrac{\partial \rho_i(x)}{\partial x_k} < 0$;

In fact, we may assume that $p_i(x), i \in N$ is the probability of implementing a destructive effect on the element $s_i \in S, i \in N$, the system S. An example of such function is, for example, the function $p_i(x) = e^{-x_i}$.

It is rather clear that for so defined functions of the local risk $\rho_i(x), i \in N$ the properties C1–C3 are met.

Let's assume for simplicity:
$r_1 = \ldots = r_n = r, \rho_1^\infty = \ldots = \rho_n^\infty = 0, p_1(x) = \ldots = p_n(x) = p, 0 < p < 1$. It is obvious that if the elements $s_i \in S, i \in N$ of the system S are independent and autonomous, then "requests" are such that $\overline{p}_1 = \ldots = \overline{p}_n$, and as MS-solution equal allocation of the resources should be chosen: $e_i(X) = X/n$.

Example 1. If the graph G(S,W) is a "chain".
Case A. Let's assume that basing on the structure of the graph G(S,W) some destructive influence on the system S may be performed sequentially n times, and the "entry point" for the influence is the element s_1, and distribution of this influence under the system S may only be carried out along the "lines of interrelation" between the elements. Thus (see Fig. 1A) the element s_1 is affected first, and then if successful, the element s_2 and so on up to the element s_n. Let's denote $P_k, k \in N$ as the probability of "successful destructive influence" on the element s_k, as it is easy to see that $P_k = p_1 p_2 \ldots p_k = p^k$. Therefore we have: $P_1 > P_2 > \ldots > P_n$, and thus the following shall be met: $\widetilde{p}_1 > \widetilde{p}_2 > \ldots > \widetilde{p}_n$, and MS-solution $\widehat{\pi}(\widetilde{p}) = (\widehat{\pi}_{(1)}(\widetilde{p}), \ldots, \widehat{\pi}_{(n)}(\widetilde{p}))$ shall be such that $\widehat{\pi}_1(\widetilde{p}) > \ldots > \widehat{\pi}_n(\widetilde{p})$.

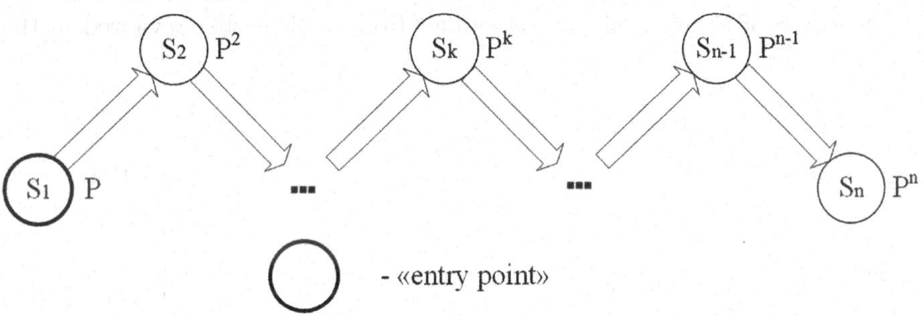

Fig. 1. 1A. For example 1A.

Case B. It is obvious, that if the - "entry point" of destructive influence is the element s_n, the situation will change strictly into the opposite:
$P_k = p_n p_{n-1} \cdots p_{n-(k-1)} = p^k, P_1 < P_2 < \ldots < P_n$, so the following shall be performed: $\widetilde{p}_1 < \widetilde{p}_2 < \ldots < \widetilde{p}_n$, and MS-solution $\widehat{\pi}(\widetilde{p}) = (\widehat{\pi}_{(1)}(\widetilde{p}), \ldots, \widehat{\pi}_{(n)}(\widetilde{p}))$ shall be such that $\widehat{\pi}_1(\widetilde{p}) < \ldots < \widehat{\pi}_n(\widetilde{p})$
Case C. If the "entry point" is the element $s_k, k \in N, 1 < k < n$, then we have (see Fig. 1C): $P_k = p, P_{k-1} = P_{k+1} = p^2$ and so on, where:
$P_k > P_{k-1} = P_{k+1} > \ldots, \widetilde{p}_k > \widetilde{p}_{k-1} = \widetilde{p}_{k+1} > \ldots$ and MS-solution shall be such that $\widehat{\pi}_k(\widetilde{p}) > \widehat{\pi}_{k-1}(\widetilde{p}) = \widehat{\pi}_{k+1}(\widetilde{p}) > \ldots$ (Fig. 2).

Finally, if "entry points" for destructive influences are all elements $s_i \in S, i \in N$ of the system S, these influences are applied in parallel (i.e., simultaneously on all elements of the system S), we have:

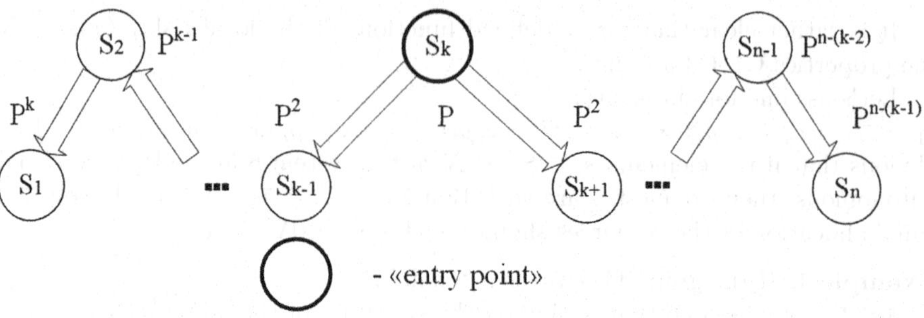

Fig. 2. 1C. For example 1C.

$P_1 = P_2 = \ldots = P_n = p, \widetilde{p}_1 = \widetilde{p}_2 = \ldots = \widetilde{p}_n$ with equal allocation of resources as MS-solution to be chosen: $e_i(X) = X/n$, which coincides with MS-solution in the case, when elements of the system S are independent and autonomous.
Example 2. If the graph G(S,W) is a "ring".

If we assume that as in the Case A (Example 1), "entry point" is the element s_1 and so on, destructive influence is sequentially applied l-times, $l \in N$. We will review two cases below.
Case A. If $l \leqslant n$, then, as you can clearly note, situations for the elements $s_i \in S, i \in N$ of the system S do not differ from those earlier reviewed in the Example 1 (Fig. 3).

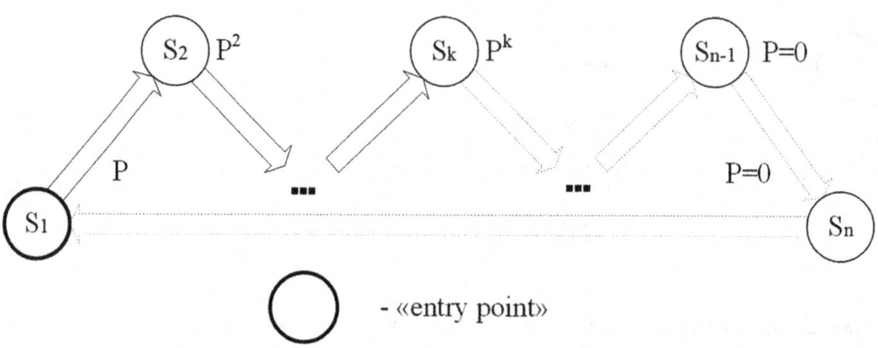

Fig. 3. 2A. For example 2A.

Case B. In case of $l > n$. To be more definite let's assume that $l = n + 1$, then for all $s_k \in S$ such that $2 \leqslant k \leqslant n$ the following is met: $P_k = p_1 p_2 \ldots p_k = p^k$ and $P_2 > \ldots > P_n$. However, for the element s_1 probability of successful destructive influence, in contrast to the Case A, if of the following form:
$P_1 = p + p^{n+1}(1 - p) > p$. In other words, more resources than in the Case A of this example, shall be allocated for the element s_1 (Fig. 4).

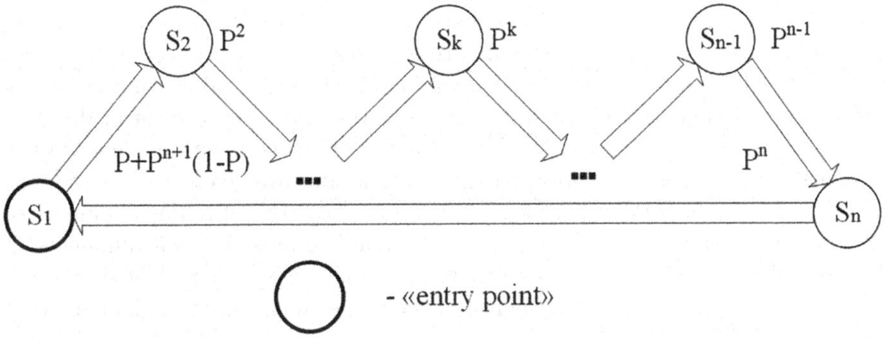

Fig. 4. 2B. For example 2B.

Analysis of the examples discussed above shows that in the case, when elements of the system S may influence each other, an effective solution of the "defender's task":

- Firstly, significantly depends on the structure and "weights" of relations between elements of the system S, set by the graph G(S,W), as well as knowledge about possible ways of having destructive influence on the system S;

- Secondary, local risks of the elements of the system S, as a result of the said interaction, may both increase or decrease.

- Thirdly: finding of an efficient solution for resource allocation, provided that the graph G(S,W) has a rather complex structure, cannot be obtained in analytical form.

4 Cognitive Game

Either expert or expert-formalized or finally a formalized approach may be obviously used to solve the task (3).

Let's discuss this issue in more detail.

As part of the expert approach, all "responsibility" for recording of effects that arise as a result of mutual influence of risks associated with the elements of the system S on each other is fully assigned to experts, being specialists in the field of information risk assessment. For more information about methods used in implementing this approach, please, see, for example, [8–11].

Within the expert-formalized approach the experts use formalized procedures that allow assessing, among other, multi-factorial risks. An example of such approach is a mechanism of comprehensive assessment, which is based on the apparatus of sequential binary convolutions. More information about this approach can be found, for example, in [12, see detailed bibliography ibidem], and in relation to information risk assessment - in [13–16].

We now proceed to consideration of the formalized approach.

Let us consider the model in more detail described in Sect. 2. Let us assume that there is a complex system consisting of elements:

$S = s_1, \ldots, s_i, \ldots, s_n, i \in N$, which have a certain impact on each other, and transfer the risks associated with them to each other. Let us describe this effect by a weighted oriented graph G(S,W), where S is the set of nodes that coincides with the set of elements of the system S , and $W \subseteq S \times S$ is the set of directed arcs $w_i j = (s_i, s_j) \in W, i \in N, j \in N$, which reflect the connections between the elements of the system S.. Suppose that two functions are given: $\rho : S \to \mathbb{R}^1$ and $\sigma : W \to \mathbb{R}^1$, on G(S,W) where $\rho_i, i \in N$ is the "weight" of nodes (elements of the system S), and $\sigma_{ij}, i \in N, j \in N$ — "weight" of arcs (mutual influences of elements of the system S on each other). The matrix $\sum = \|\sigma_{ij}\|$ of size $n \times n$ reflects the "intensity" of the influence of the i-th element on the j-th element of the system S.

Let us assume the mutual influence is carried out at discrete moments of time and the initial state of the system corresponds to the zero moment of time and the values of the weights of the elements of the system S: $\rho_i(t = 0) = \overline{p}_i$. Let us describe the change in the values of the weights of the elements $s_i \in S, i \in N$, of the system S as a result of their mutual influence on each other by the following expressions:

$$p_i(t+1) = p_i(t) + \sum\nolimits_{k=1}^{n} \sigma_{ik}(p_i(t)) - p_i(t-1)), t = 0, 1, \ldots, p_i(t=0) = \overline{p}_i \quad (4)$$

System (4) uniquely describes the dynamics of changes in the values of the "weights" of elements of system S as a result of their mutual influence on each other in the case when there is no information about a specific type of local risk functions $\rho_i(\cdot), i \in N$ for any element $s_i \in S, i \in N$, system S, but it is known that they satisfy the properties C1, C2 and C3 (see the description of the corresponding model in Sect. 2).

Let $\triangle\rho_i(t) = \rho_i(t) - \rho_i(t-1)$ is the "rate" of change in the values of the "weights" of the elements $s_i \in S, i \in N$, of the system S (in the terminology [17,18], "momentum"), then expressions (4) can be represented as follows:

$$\triangle p_i(t+1) = \sum\nolimits_{k=1}^{n} \sigma_{ik} \triangle p_i(t), t = 0, 1, \ldots, \triangle p_i(t=o) = \overline{p}_i \quad (5)$$

The description of the graph G(S,W) together with the dynamics of changes in the values of the "weights" of elements of the system S up to the time t defined by expressions (4) or (5) will be called (for more details, see, for example, [17–22] "cognitive map"(the term was first introduced in [19])).

Let $\overline{p}_i(t) = (\rho_i(0), \rho_i(1), \ldots, \rho_i(t))$ is the vector of values of the "weights" of the element $s_i \in S, i \in N$, of the system S up to the time
$t, \overline{P}(t) = (\rho_1(t), \rho_2(t), \ldots, \rho_n(t))$ is the vector of values of the "weights" of elements of the system S at time $t, \mathcal{P}(t) = (\overline{P}(0), \overline{P}(1), \ldots, \overline{P}(t))$ — is the matrix of the dynamics of changes in the values of the "weights" of the elements of the system S up to the time t (in the terminology [17,18] — "trajectories"), $t = 0, 1, \ldots$.

The most important feature of "cognitive maps" is the ability to take into account the indirect influence of elements of the S system on each other, when one of the elements affects the other, through several intermediate ones.

Let us denote a unit matrix of size $n \times n$ as E_n and consider, following [18] the expression:

$$B^t = E_n + \sum + \sum^2 + \ldots + \sum^t, t = 1, 2, \ldots \tag{6}$$

The dynamics of changes in the values of the "weights" of elements of the system S up to the time t, following (4) and (6) can be described (taking into account the above notation) by the expression:

$$\overline{P}(t) = B^t \overline{P}(0) + (E_n - B^t)\overline{P}(-1), t = 1, 2, \ldots \tag{7}$$

Then if we assume that $\rho_i(t) \equiv 0 fort < 0$ (for more information, see [18]), then expression (7) takes the form:

$$\overline{P}(t) = B^t \overline{P}(0), t = 1, 2, \ldots \tag{8}$$

If matrix $\sum$ is such that all its eigen values are contained inside unit circle in complex plane, performance of this requirement is sufficient for provision of precision of a sum of natural degrees of matrix $\sum^t$ at $t \to \infty$ (for more information, see [18]). Let us denote the value of sum (6) as B^∞ as $t \to \infty(B^\infty \approx (E_n - \sum)^{-1}$, see [18]), then as $t \to \infty$ expression (8) takes the form:

$$\overline{P}(\infty) = B^\infty \overline{P}(0) \tag{9}$$

where $\overline{P}(\infty) = (\rho_1(\infty), \rho_2(\infty), \ldots, \rho_n(\infty))$ are the "steady-state" values of the local risks of the element $s_i \in S, i \in N$, of the system S, which we will consider as "requests" in the model described in Sect. 2. By ordering them in descending order $\rho_{(1)}(\infty) \geqslant \rho_{(2)}(\infty) \geqslant \ldots \geqslant \rho_{(n)}(\infty)$ and applying the approach presented above, we can choose as an effective solution problem (3) the corresponding MS-solution.

5 Conclusion

The paper is concerned with a general model of complex computer network, within which a risk manager exercises efficient management of risks of a complex system through distribution of available resource among its elements (units of computer networks). Local risk functions, meeting some specified requirements, are applied for estimation of the state of system elements, and for estimation of a state of system as a whole – a function of integral risk.

A task of risks management is considered in a context of uncertainty, when there is no information about certain type of functions of a local risk of system elements.

It is shown that in case of independence (absence of mutual influence on each other) of system elements to find efficient distribution of resource a game-theoretic approach on the basis of an arbitration network under principles of stimulation and non-suppression (MS-solution) may be used.

It is also demonstrated that if system elements may have certain influence on the state of each other, another game-theoretical approach may be applied with the use of a game on cognitive map (a cognitive game).

As further directions of study it seems appropriate to consider a model of information confrontation within a complex computer network, where except for RM, playing as a "defender" of complex system, there is also a subject, fulfilling a role of "attacker".

References

1. Kalashnikov, A.O.: Modeli i metodi organizacionnogo upravleniya informacionnimi riskami korporacii, p. 312. Egves, Moscow (2011). ISBN: 978-5-91450-078-5
2. Kalashnikov, A.O.: Organizacionnie mehanizmi upravleniya informacionnimi riskami korporacii, p. 175. PMSOFT (2008). ISBN: 978-5-9900281-9-7
3. Matthias, E.: Multicriteria Optimization, p. 382. Springer, Heidelberg (2010)
4. Kozlov, A.D., Noga, N.L.: Riski informacionnoi bezopasnosti korporativnih informacionnih system pri ispolzovanii oblachnih tehnologii. Upravlenie riskom **3**, 31–46 (2019)
5. Kalashnikov, A.O.: Upravlenie informacionnimi riskami s ispolzovaniem arbitrajnih shem. Sistemi upravleniya i informacionnie tehnologii **4**(16), 57–61 (2004)
6. Rotar, V.I.: O principe stimulyacii v arbitragnoy sheme. Ekonomika i matematicheskie metodi, t. XVII, i. 4, pp. 751–764 (1984)
7. Rotar, V.I., Kalashnikov, A.O.: O maksimalno stimuliruyshem reshenii zadachi raspredeleniya dohodov. Tezisi dokladov soobshenii Vsesouznogo simpoziuma "Sovremennie problem matematicheskoy ekonomiki", pp. 48–49. Vilnys: IMK AN Litovskoy SSR (1984)
8. Petrenko, S.A., Simonov, S.V.: Upravlenie informacionnimi riskami. Ekonomicheski opravdannaya bezopasnost, p. 384. Kompaniya AiTi, DMK Press (2004). ISBN: 5-98453-001-5 (AiTi), ISBN: 5-94074-246-7 (DMK Press)
9. Astahov, A.M.: Iskusstvo upravleniya informacionnimi riskami, p. 312. DMK Press, Bremen (2010)
10. Damodaran, A.: Strategicheskiy risk-manadjement: principi i metodiki. Per.s angl. OOO "I.D. Viliams", p. 496 (2017). ISBN 978-5-8459-1453-8 (rus.)
11. Markova, A.S., Barabanov, A.V., Dorofeev, A.V., Markov, A.S., Cirlov, V.L.: Sem bezopasnih informacionih tehnologii. In: Pod red, p. 224. DMK Press (2017). ISBN: 978-5-97060-494-6
12. Novikov, D.A.: Teoriya upravleniya organizacionnimi sistemami, 3-e izd, p. 604. Izdatelstvo fiziko-matematicheskoi literaturi (2012). ISBN/ISSN: 978-5-94052-222-5
13. Kalashikov, A.O.: Upravlenie informacionnimi riskami organizacionnih system: obshaya postanovka zadachi. Informaciya i bezopasnost, Tom 19. No **1**(4), 36–45 (2016)
14. Kalashikov, A.O.: Upravlenie informacionnimi riskami organizacionnih system: mehanizmi kompleksnogo ocenivaniya. Informaciya i bezopasnost, Tom 19. No **3**(4), 315–322 (2016)

15. Kalasnikov, A.O., Anikina, E.V.: Model upravleniya informacionnoi bezopasnostiyu kriticheskoi informacionnoi infrstructuri na osnove viyavleniya anomalnih sostoyanii (Chast 1). Informaciya I bezopasnost, Tom 21. No **2**(4), 145–154 (2018)
16. Kalashnikov, A.O., Anikina, E.V.: Model upravleniya informacionnoi bezopasnostiyu kriticheskoi informacionnoi infrstructuri na osnove viyavleniya anomalnih sostoyanii (Chast 2). Informaciya I bezopasnost, Tom 21. No **2**(4), 155–164 (2018)
17. Roberts, F.S.: Discretnie matematicheskie modeli s prilojeniyami k socialnim, biologicheskim i ekologicheskim zadacham. Nauka, p. 496 (1986)
18. Novikov, D.A.: "Kognitivnie igri": lineinaya impulsnaya model. Problemi upravleniya **3**, 14–22 (2008)
19. Tolman, E.: Cognitive maps in rats and men. Psychol. Rev. **55**, 189–208 (1948)
20. Kulinich, A.A.: Sistematizaciya kognitivnih kart i metodov ih analiza. Tr. VII-i mejdunar. konf. "Kognitivnii analiz i upravlenie razvitiem situacii", pp. 50–56. IPU RAN (2007)
21. Maksimov, V.I.: Structurno-celevoi analiz razvitiya socialno-ekonomicheskih situacii. Problemi upravleniya **3**, 30–38 (2005)
22. Avdeeva, Z.K., Kovriga, S.V., Makarenko, D.I., Maksimov, V.I.: Kognitivnii podhod v upravlenii. Problemi upravleniya **3**, 2–8 (2007)

Prerequisites and Methodology for Digital Transformation of 4G Networks into 5G Ecosystem

V. L. Shirokov(✉)

Department of Network Integration (LLC "LANIT-Integration"), LANIT Inc. (LAboratory of New Information Technologies), Murmansk passage, 14-1, Moscow, Russia
ShirokofVL@mail.ru

Abstract. The concept, architecture, operating modes, deployment and transformation scenarios, models, methods and stages of the evolutionary transition from 4G mobile networks to 5G ecosystems are considered.

Existing 4G mobile networks are conditionally divided into the passive part, or the radio access network (RAN), and the active part or the communication system (CS). Moreover, the RAN is the mini heterogeneous part of the network, as well as the active part, is more complex, according to the functionality of the entire network.

The models, methods of the modernization paradigm, as well as infrastructure and resources management of the existing 4G mobile network, ways and scenarios of evolutionary transformations into 5G are proposed.

Keywords: Concept · Models · Methods · Mobile networks · RAN · Clusters · Communication systems · Transformation scenarios · Accelerating · 5G ecosystems

1 Introduction

Existing mobile communication networks require continuous improvement, modernization, and scaling to meet the growing requirements of users [1].

The volume of traffic is growing on average by 26% per year. The requirements of users to the speed of network interaction and quality service are also increasing. Multimedia traffic components are growing. Video traffic already occupies 75% of total volume, and it forecasts to grow to 82% in 2022. In addition, the structure of mobile communications is complex, hybrid, and its capabilities are quickly exhausted.

The network-oriented aggregate model used in the analysis of the parameters of existing 3G and 4G/LTE mobile networks and the network as a whole have the following limitations:

– Basic communication systems CSs of the mobile network are focused on limited number of manufacturers, what increases as the capital cost (CAPEX) both operating cost (OPEX) of this network;

© Springer Nature Switzerland AG 2020
V. M. Vishnevskiy et al. (Eds.): DCCN 2020, CCIS 1337, pp. 158–168, 2020.
https://doi.org/10.1007/978-3-030-66242-4_13

- Radio access networks RANs are the least heterogeneous part of the mobile networks than CSs;
- The communication networks CSs are more heterogeneous and determine the basic functionality of the mobile network, which laid down by equipment manufacturers.

The above limitations of aggregate models are constraining factors in the improvement of 4G mobile communications and the transition to 5G. Overcoming limitations is possible through the evolutionary transformation of 4G networks into a 5G ecosystem.

Note. A mobile network consists of radio access network (RAN), transport communication network, and core, or service systems in general. Hereinafter, mobile networks maybe named as the general term of RAN, unless otherwise not specified.

2 Statement of the Problem

The basic parameters of 5G ecosystem (compared to 3G and 4G mobile networks) are as follows:

- time delay reducing to 1 ms;
- increasing 100-fold data transfer rate;
- guaranteed for five nines (99.999%) connection;
- service of up to 1 million devices and users per km2.

The best way to transition to 5G is to implement models of an open virtual infrastructure [2]:

- from a monolithic network of one manufacturer to a flexible, disaggregated, multivendor;
- from a closed, inefficient network to an open, modular IP-based and intelligent automation;
- from a restricted CS and RAN systems to self-configurable, self-organizing, self-optimizing networks based on SON, software-defined operating modes via types and loading of services.

Therefore, common concept, models, methods, and 4G network architecture scenarios will be need:

- decomposition, i.e. division of a network into elements, modules, systems, clusters;
- disaggregation, i.e. separation of aggregate network model into software and hardware;
- migration to virtual open slicing infrastructure from aggregated model of network.

Thus, transition willingness of 4G network to the 5G ecosystem is determined by digital transformation into a new infrastructure. Next, we consider the concept, models, methods, scenarios of accelerated evolutionary transition from existing mobile networks 4G/LTE to 5G ecosystem.

3 The Concept of 5G Ecosystem

The better way to 5G is software-defined systems and control decomposed mobile networks:

- from monolithic, proprietary and single-vendor systems to flexible, disaggregated, multi-vendor systems;
- from closed, inefficient, limited choices to open, modular, applying intelligent automation and e2e IP;
- from network defined (and constrained) by RAN to network defined by the services and desired operational model

So Cloud-RAN is an essential step to realize the potential of 5G networks.

The readiness of existing 4G networks for transition to 5G ecosystem must go through evolutionary transformation into an open distributed communication cloud edge infrastructure [2–4], as well as show on Fig. 1.

Integration of the O-RAN Alliance in the project of Telecommunication infrastructure project (TIP) is based on the principles of interaction and specifications of open radio access networks and radio communications (Open RAN, or O-RAN). TIP focuses on infrastructure modularity and interoperability via open source systems and plug-fests. And also the readiness into 5G transition is driving by distributed edge infrastructure.

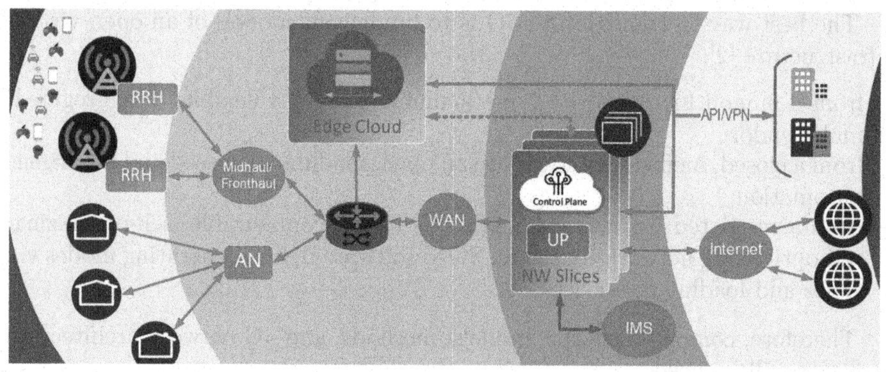

Fig. 1. 5G ecosystem transformation

The main pathway to 5G is shown on Fig. 2.

Where LTE is Long Term Evolution, NSA – Non-Stand Alone, SA – Stand Alone systems.

The TIP project includes the following key provisions [2]:

- multi-vendor construction;
- modularity of the network structure, clusters, CS and RAN systems;

- dividing the network into software and hardware, thanks to the disaggregation method;
- standard interfaces of RAN and CS modules, providing decomposition of network systems;
- interoperability in the interaction of elements, modules, CS and RAN systems;
- compatibility of systems providing open source software.

Fig. 2. The main pathway to 5G

RAN (Cloud O-RAN) offers the following features:

- a greater selection of equipment due to multi-vendor;
- accelerated market entry due to a greater variety of functionality;
- reduction in capital costs due to multi-vendor equipment selection;
- reduction of operating costs due to smart automation and control;
- due to cloud cover, great infrastructure flexibility, increasing network efficiency;
- increased revenue through support for open API and rapid introduction of new services.

5G are software-defined open radio access networks, or SD O-RANs. In addition, an important step in creating an end-to-end network (from beginning to end) is the creation of a mobile C-RAN network for more efficient use of resources and the entire network.

The basis of the approach to the new software architecture is that 5G is a common telecommunications distributed platform, which includes a virtual implementation, the following principles:

- virtualization of radio access network, or vRAN;
- peripheral data processing and distributed management, or Edge computing;
- flexible access to user applications thanks to the user plane function (UPF), and MEC applications;
- centralized processing, polices control, and operations services.

In addition, the vBranch, or virtual branch layer, is specified here, which includes:

- a common part of the infrastructure of the 5G distributed telecommunications platform;
- end-to-end automatic resource dispatching, or orchestration.

The level of through orchestration is divided into several sublevels:

- about mobility;
  - data transfer;
  - edge resources;
- about centralized processing.

In general, distributed 5G telecommunications platform includes the following hardware components:

- base stations;
- channels-fronthauls (these are new elements on 5G edges, or RANs);
- edge switches;
- edge processing nodes (new subsystems on 5G edges, or RANs);
- channels-middlehauls;
- core switches;
- channels-backhauls;
- core routers;
- core's data centers.

Thus, a new structure, topology and entire architecture of 5G network is formed.

So the features of subsystems and structures of traditional models of RANs, CSs and 5G network models are discussed below.

4 Subsystems and Structure of RAN and CS Models

Conventional CS and RAN systems are hybrid, aggregated models:

- were built, as rule, on equipment of one manufacturer;
- in terms of the functionality were not effective;
- by cost were not optimal;
- were poorly scalable;
- limited innovation.

In contrast, open O-RANs and cloud-based CSs are flexible NaaS "network as a service" models that have the following advantages:

- equipped with equipment from several vendor firms;
- ready for innovation and transformation to 5G;
- flexible in functionality;
- cost-effective;
- well scalable.

Thus, the structure of the new virtualized 5G network model includes the following logical systems:

1. software-defined radio access network (SD-RAN);
2. virtualized cloud systems and cloud (Cloud);
3. intelligent support systems (or intellectual, ISS) for automation, support of network operations (SON, BSS, HSS, OSS, IP, and transport).

Thus all elements, modules and subsystems of the common 5G infrastructure are shown on Fig. 3.

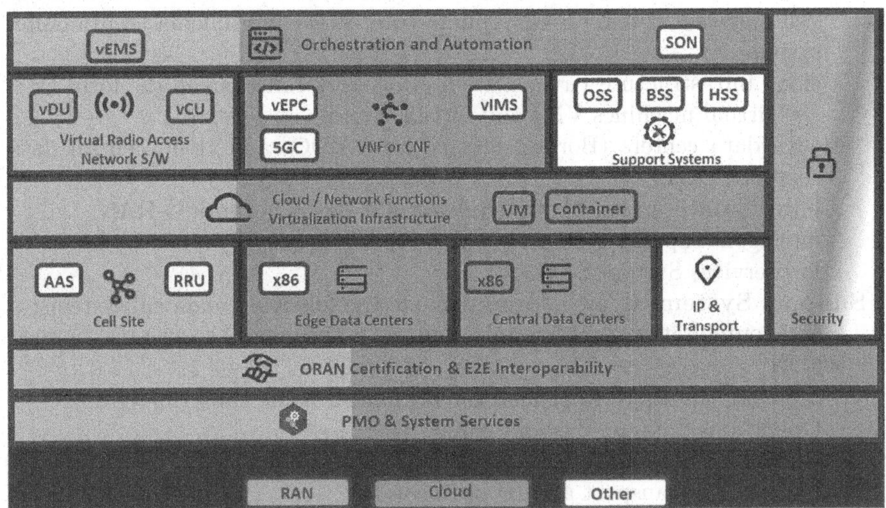

Fig. 3. 5G infrastructure elements

v – virtual, EMS – Element Management System, AAS – Adaptive Antenna System, RRU – Remote Radio Unit, EPC – Evolving Packet Core Such a stratification of the network into a subsystem allows to build fully open 5G generation software-defined cloud virtual network.

Now then:

1. **Radio Access Network (RAN)** is open virtualized SD-vRAN, which implies a slicing decomposition of vRAN into following modules and systems:
   - virtual element management system vEMS of virtual radio access network vRAN;
   - software (SoftWare, S/W) of the vRAN virtual radio access network, consisting of:
     - virtualized distributed units vDUs and
     - virtualized centralized control units vCUs;
   - part of the cloud infrastructure that is responsible for virtualizing network functions for the RAN;

   - base stations, i.e. cell sites Cell Sites, consisting of:
     - remote radio units RRUs;
     - adaptive antenna systems AASs;
   - Open RAN (O-RAN), implying interoperability and compatibility for RAN devices and systems from different manufacturers;
   - project management office (PMO).
2. **Cloud** is an open part of the 5G O-RAN ecosystem, which includes the following modules and subsystems:
   - centralized automation system SON and dispatching resources, or orchestration SON & Orchestration;
   - centralized CNF and virtualized VNF network functions of the evolving vEPC packet core [6], 5GC system core, vIMS multimedia transmission system [5];
   - cloud infrastructure virtualization of network functions in the form of:
     - virtual machines VMs and virtual containers VCs;
   - edge data centers (Border data centers) EDCs and cloud central data centers CDCs;
   - fully certified, i.e. compatible open radio access network O-RAN;
   - intelligent support systems (OSS, BSS, HSS);
   - low services System Services.
3. **Support Systems** of 5G network include the following main subsystems:
   - Self-configuration, self-organization, self-optimization system, named SON;
   - Support of subscriber profiles server, named home subscriber HSS;
   - Operations Support System, named OSS;
   - Business Support System, named BSS;
   - Support of Transport and IP protocols.

5 RAN Deployment Models

First model is shown on Fig. 4.

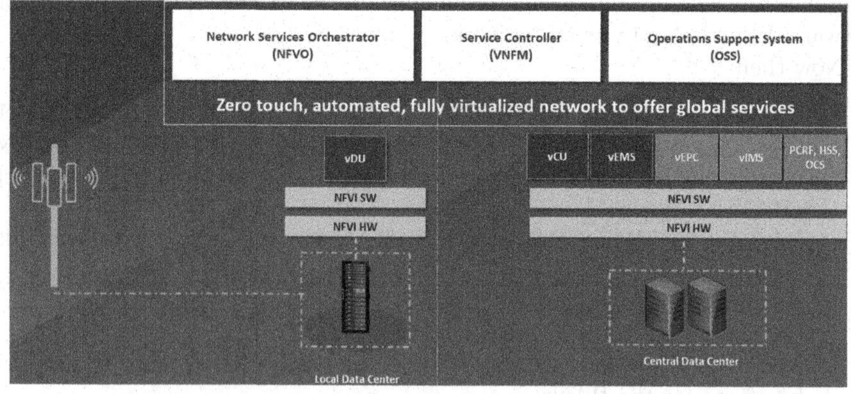

Fig. 4. RAN deployment model (First Model)

Where NFVI – Network Functions Virtualization Infrastructure, SW – SoftWare, HW – HardWare, EMS – Element Management System, IMS – IP Multimedia Subsystem, v – virtualized, DU – Distributed Unit, CU – Control Unit (of control plane), EPC – Evolved Packet Core (or vEPC of 5G Core), HSS – Home Subscriber System, OCS – Operating Control System, PCRF – Policy and Charging Rules Function.

It is centrally distributed, which requires good optical transport and includes the following blocks:

- NFV orchestrator (NFVO) of network functions virtualization, which manages resources and network services, as well as a number of other functions;
- Service Controller (VNFM, or VNF Manager);
- Operation Support System (OSS).

This is a fully automated virtualized service delivery network (Zero Touch, without operator intervention, i.e. automatically, except in emergency cases). Provides a fully dynamic allocation of resources between 4G and 5G networks without any predefined levels (presets).

It includes following vRAN modules and subsystems:

- on the edge side NFVI HW, NFVI SW, local data center, vDU, vCU, vEMS;
- on the central side – vEPC, vIMS, central data center, PCRF, HSS, OCS.

Second model is shown on Fig. 5.

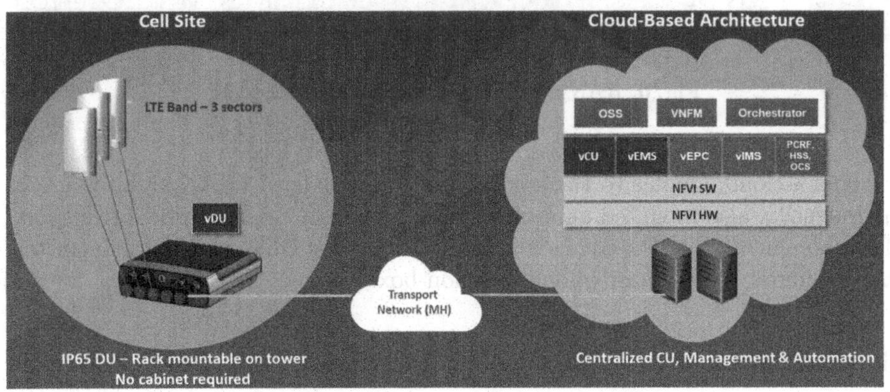

Fig. 5. RAN deployment model (Second Model)

Here LTE – Long Term Evolution, MH – Middle Haul

It is a low-density cloud-based virtualized vRAN architecture mounted on a mast, which does not require optical transport, includes the following blocks:

- vDU - virtualized DU, mounted on a mast in its own box;
- centralized CU, Management & Automation, including:
  - OSS, VNFM and orchestrator;
  - vCU, vEMS; vEPC, vIMS, PCRF, HSS, OCS;
  - NFVI HW, NFVI SW and centralized data center CDC;
- Transport network (MH, or Middle Haul) between vDU and CDC.

Third model is shown on Fig. 6.

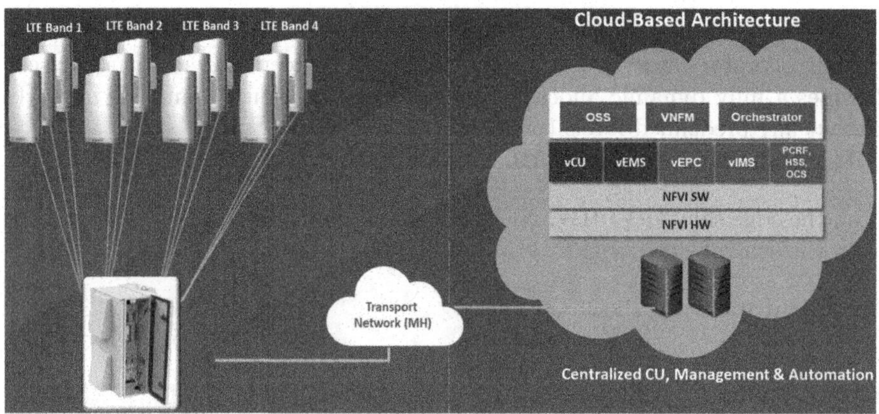

Fig. 6. RAN deployment model (Third Model)

It is a cloud, similar to the second model, virtualized vRAN architecture of high density, also mounted on a mast, which does not require optical transport. The difference is that the edge vDU consists of several DUs, which form a cluster, also mounted on the mast in one common box.

6 Stages of Transition to the Open O-RAN 5G Network

Three consecutive steps are proposed for moving from O-RAN to the 5G ecosystem:

- Improving 4G/LTE by increasing the speed, quality of multimedia data transmission, introducing intelligent automation;
- 5G NSA (Non-Stand Alone) - joint use of existing 4G/LTE base stations and new 5G with improved parameters of mobile communications in speed, capacity, reliability;

– 5G SA (Stand Alone) - the use of only 5G base stations providing real-time communications, mass telecommunications, ultra-reliable machine communications, IoT, etc.

Along with orchestration, it is necessary to monitor, analyze and manage data transmission. So, for the use of machine learning and artificial intelligence methods in network analytics, a new paradigm about the location of subscribers, as well as the statistics of traffic of devices and users, is gaining interest. In the future, also the transition to a hybrid network architecture is the key to the development of artificial intelligence models. Therefore, the proposed concept, models and analysis methods for the parameters of existing 4G/LTE networks provide an accelerated transition to the 5G ecosystem.

7 Conclusion

I The features of traditional solutions used in 4G/LTE mobile communication networks, which hinder and limit the transition to 5G networks, are noted.
II The concept of the 5G open cloud ecosystem, its advantages, evolutionary paths to the transition to the 5G network are reviewed and analyzed.
III The elements, systems and architecture of the traditional and new RAN models, the main network support subsystems for the transition to the 5G ecosystem are considered.
IV Deployment models and stages of a sequential evolutionary transition from 4G/LTE networks to an open 5G cloud network are proposed.

Findings

i The proposed concept, models and analysis methods for the parameters of existing 3G and 4G/LTE networks provide an accelerated evolutionary transition to the 5G ecosystem.
ii The first step in the transition should be the introduction of 5G automation and improvement of 4G network performance parameters.
iii The second step should be the joint use of 4G and 5G networks with the gradual shutdown of 4G base stations.
iv The proposed methodology is a scenario for implementing the 5G concept using a flexible, open, software-controlled network that will provide a wider range of better services at lower cost.

And finally Finland, China, South Korea and Japan have already begun to develop 6G technologies [7].

As the 5G active deployment phase is underway, Finland's 6G program, for example, is not only 6G. But above all, projects, tests and demonstrations aimed at improving 5G technology. At the same time, work is underway on the main technological innovations and solutions that will be required in 6G.

By 2030, it is expected that society will be driven by data and unlimited, virtually instantaneous, wireless mobile communications. But this is the topic of another material.

References

1. Cisco Annual Internet Report - Cisco Annual Internet Report (2018–2023) White Paper. https://www.cisco.com/c/en/us/solutions/collateral/executive-perspectives/annual-internet-report/white-paper-c11-741490.html. Accessed July 2020
2. Telecom Infra Project. (2020). https://telecominfraproject.com/. Accessed July 2020
3. 3GPP Technical specification 23.402: 3GPP System Architecture Evolution. www.3gpp.org
4. The 5G concept does not suit Russian vendors: 5G networks will be built by foreigners. (2020). https://www.tadviser.ru. Accessed July 2020
5. Virtual Evolved Packet Core (vEPC) Solution from Affirmed. (2019). https://www.affirmednetworks.com/products-solutions/virtual-evolved-packet-core/. Accessed July 2020
6. Poikselka, M., Niemi, A., Khartabil, H., Mayer, G.: The IMS: IP Multimedia Concepts and Services. John Wiley & Sons, New Jersey (2006)
7. Tomás, J.: '6G to satisfy the expectations not met with 5G': 6G Flagship Program (2020). https://www.rcrwireless.com/20200407/5g/6g-satisfy-expectations-not-met-5g-6g-flagship-program. Accessed July 2020

Analytical Modeling of Distributed Systems

The Simulation of Finite-Source Retrial Queueing Systems with Two-Way Communications to the Orbit and Blocking

János Sztrik, Ádám Tóth(✉), Ákos Pintér, and Zoltán Bács

University of Debrecen, Debrecen 4032, Hungary
{toth.adam,sztrik.janos}@inf.unideb.hu,
bacs.zoltan@econ.unideb.hu, apinter@science.unideb.hu

Abstract. A two-way communication, retrial queueing system is considered with a single server which from time to time is subject to random breakdowns. The investigated model is a M/M/1//N type of system where the number of sources is finite. After the service unit becomes idle it is able to call in customers residing in the orbit (outgoing call or secondary customers). Distribution of the service time of primary and secondary customers is exponential with rates μ_1 and μ_2, respectively. Every used random variable is assumed to be totally independent of each other in the model. Each time the server becoming in faulty state the operation of the system is blocked resulting that throughout this period customers can not enter the system. The novelty of this analysis is to study the effect of blocking in such system on the main performance measures using different distributions of failure time. Results are illustrated graphically with the help of a simulation program developed by the authors.

Keywords: Simulation · Blocking · Sensitivity analysis · Finite-source queueing system · Unreliable server · Retrial queue.

1 Introduction

Because of the increasing number of users and devices mainly due to the rapid development of technology it is not an easy task to cope with the question of designing communication systems or redesigning an existing pattern or scheme. Nowadays, every company possesses some kind of network infrastructure so it is unavoidable that the exchange of information would not take place therefore developing mathematical and simulation models and algorithms play quite an

The research was financed by the Higher Education Institutional Excellence Programme of the Ministry of Human Capacities in Hungary, within the framework of the NKFIH-1150-6/2019 thematic programme of the University of Debrecen.

© Springer Nature Switzerland AG 2020
V. M. Vishnevskiy et al. (Eds.): DCCN 2020, CCIS 1337, pp. 171–182, 2020.
https://doi.org/10.1007/978-3-030-66242-4_14

important role to deal with traffic growth. Applying retrial queues in such scenarios are useful and powerful tools to describe real-life problems emerging from main telecommunication systems like telephone switching systems, call centers, computer networks, and computer systems. Many researchers are dedicated to investigating this topic, some examples are mentioned which study retrial queueing systems with repeated calls like in [4,5]. The applicability of these models is utilized in many areas of science like improving the efficiency of systems for example in the case of local-area networks with random access protocols and with multiple access protocols [1,10].

The characteristics of two-way communication have a beneficial effect on most of the systems consequently its popularity is quite well-founded in recent years. This can be explainable by the fact that the operation of certain real-life systems can be matchable with models based on a two-way communication scheme. In terms of call-centers, this is especially appropriate considering that the service unit (or agent) apart from handling incoming calls may carry out other activities including selling, promoting, and advertising products. In this paper whenever the server gets to idle state after some random time it is capable of calling customers residing in the orbit. In such scenes, the utilization of the service unit (or workload of agents) is crucial and extensively examined by many papers like [3,12].

Scrutinizing the available literature on the internet relatively quite a high number of papers are found where the service facilities are presumed to be available all the time. Reliable operation is quite optimistic and an unrealistic approach because deterioration, power supply failure, or unforeseen circumstances can happen anytime modifying moderately the system characteristics. Regarding wireless communication, several components affect the transmission rate resulting in interruptions that can arise at any time throughout transmitting the packets. It is always a key question of how the property of unreliable operation alters the performance measures and the characteristics of the system. Recently published works about retrial queuing systems with a non-reliable server can be found for example in [8,9,11].

The main aim of this work is to explore the mechanism of blocking of the investigated system and to compare various distributions of failure time on main performance measures like the mean waiting time of an arbitrary customer or the total utilization of the server. The present paper is a natural continuation of [16] and we want to compare the achieved results with each other. Our self-developed simulation program is used to obtain every important performance measure using SimPack [6], which contains C/C++ libraries and executable programs for computer simulation. In this class, numerous algorithms can be found in connection with discrete-event, continuous, and combined (multi-model) simulation. Because of using other distributions apart from exponential and the fact that providing exact formulas is almost impossible we selected stochastic simulation to approximate the desired performance measures and to freely integrate any distribution in our code. The novelty of this paper is to present a sensitivity analysis of failure time on the main measures besides blocking using various distributions. Graphical illustrations are provided depicting an interesting phenomenon of sensitivity problems and comparison with the non-blocking system. This paper is the extended version of [17].

2 Model Description and Notations

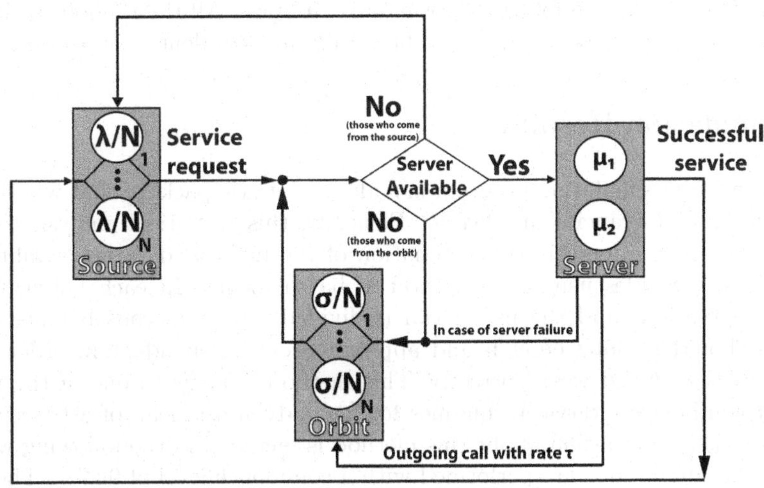

Fig. 1. System model

In Fig. 1 the considered finite-source queueing system is shown by the help of two-way communication with retrials which contains a non-reliable server. We considered a finite-source queueing system with the help of two-way communication with retrials which contains a non-reliable server. The source contains N customers and each of them produces requests (primary or ingoing customers) with rate λ/N resulting exponentially distributed inter-arrival time with parameter λ/N. Our model does not comprise queues thus in case of an idle server the service of an incoming customer starts immediately. The distribution of the service time of these customers is exponentially distributed with parameter μ_1. After being successfully served the customers return to the source. Alternatively, arriving customers from the orbit or source finding the server in a busy state are forwarded instantly to the orbit. Waiting an exponentially distributed time with parameter γ/N in this virtual waiting room customers launches another attempt to occupy the service unit. From time to time failure of the server may arise according to gamma, hypo-exponentially, hyper-exponentially, Pareto, and lognormal distribution with different parameters but with the same mean value. During this period customers can not enter the system because they are rejected in that instant, this is the so-called blocking. The recovery process begins instantaneously upon the failure of the server, which is also an exponentially distributed random variable with parameter γ_2. If the service unit breaks down during the service of a customer then that customer is transferred to the orbit immediately. Whenever the server becomes idle it may perform an outgoing call (secondary customers) towards the customers located in the orbit after an exponentially distributed random time with rate ν. The service of these customers is executed according to an exponential distribution with a rate of μ_2. Rates λ/N and σ/N are used because in [14,15] very similar systems are evaluated by an asymptotic method where N tends to infinity and was proved that the number of customers in the system follows a normal distribution. All the random variables in the model creation are assumed to be totally independent of each other.

3 Simulation Results

Our self-written simulation program includes a statistic package that was developed by Andrea Francini in 1994 [7]. Basically, this statistical analysis tool is suitable to make a quantitative estimation of the mean and variance values of the desired variables using the method of batch means. In each batch, there are n observations and the useful run is divided into numerous batches. The batches should be long enough and approximately independent in order that the estimation would work correctly. This method belongs to one of the most popular confidence interval techniques for a steady-state mean of a process. In more detailed information about this method is included in the following works [2,13]. The simulations are performed with a confidence level of 99.9%. The relative half-width of the confidence interval required to stop the simulation run is 0.00001.

3.1 Squared Coefficient of Variation is Greater Than One

To realize the sensitivity analysis four different distributions of failure time are selected to compare the performance measures with each other. The parameters are chosen in such a way that the mean value and variance would be equal, so we applied a fitting process that is necessary to be done. [18] contains a detailed description of the whole process characterizing every used distribution. We differentiated two main scenarios from each other. In the first one, the squared coefficient of variation is greater than one so I utilized hyper-exponential, gamma, Pareto, and lognormal distributions. Table 2 quantifies all the used input parameters of the various distributions of failure time while Table 1 shows the values of other parameters.

Table 1. Used numerical values of model parameters

N	λ/N	γ_2	σ/N	μ	μ_2	ν
100	0.01	1	0.01	1	1.2	0.02

The steady-state distributions are represented on Fig. 2 when λ/N is $= 0.01$ comparing the effect of all four applied distributions of failure time. It shows the probability that exactly i customers are located in the system. Averagely the same number of customers resides in the system, slight differences can be perceivable especially in the case of Pareto. Taking a closer look at the graphs all the curves correspond to normal distribution despite the characteristics of the various distribution.

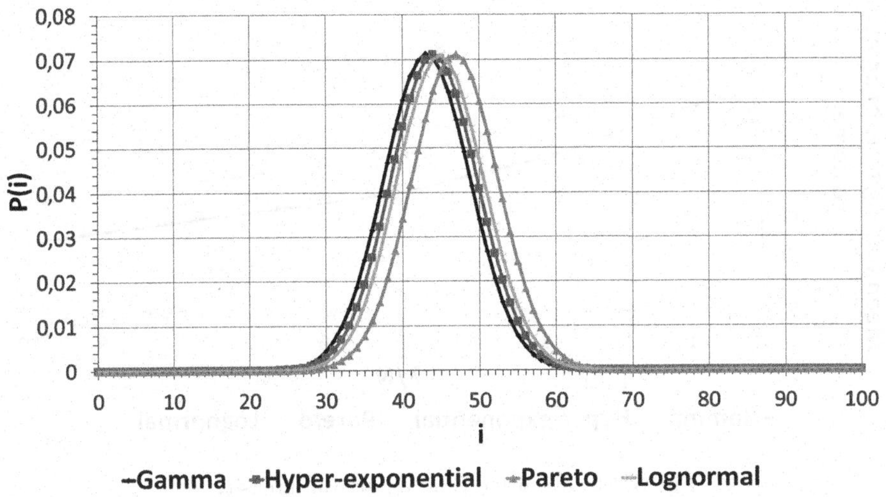

Fig. 2. Comparison of steady-state distributions when $\lambda/N = 0.01$

Table 2. Parameters of failure time

Distribution	Gamma	Hyper-exponential	Pareto	Lognormal
Parameters	$\alpha = 0.6$ $\beta = 0.5$	$p = 0.25$ $\lambda_1 = 0.41667$ $\lambda_2 = 1.25$	$\alpha = 2.2649$ $k = 0.67018$	$m = -0.3081$ $\sigma = 0.99037$
Mean	1.2			
Variance	2.4			
Squared coefficient of variation	1.6666666667			

In Fig. 3 the mean arbitrary response time is demonstrated as the request generation increases. Results clearly illustrate the effect of various distributions which is quite significant even though the first two moments are equal. The highest values are experienced at Pareto distribution while the lowest values at gamma distribution. The maximum property characteristic of a finite-source retrial queueing system arises which under suitable parameter setting occurs in spite of increasing arrival intensity.

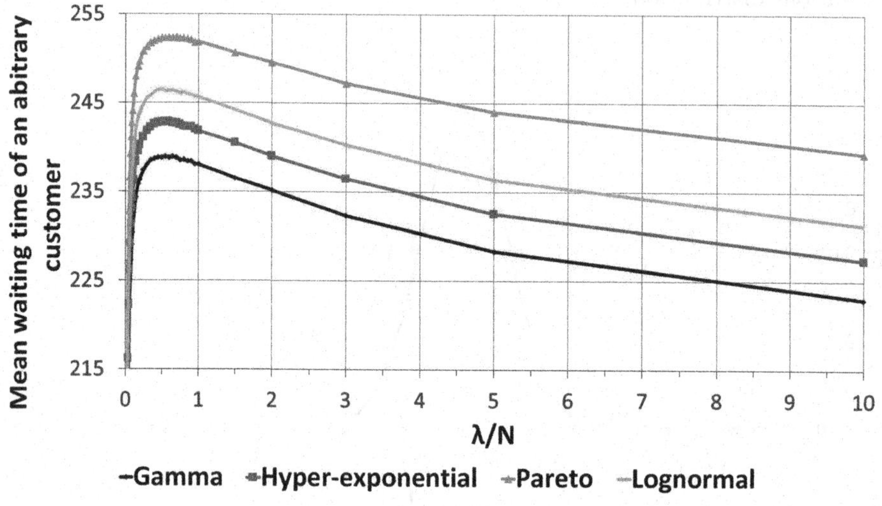

Fig. 3. Mean waiting time vs. arrival intensity

Figure 4 shows how the total utilization of the server escalates applying intensifying arrival intensity. Under total utilization, we mean every single service including the service of primary, secondary customers, and the interrupted ones, too. By examining closely the figure the received values are almost identical but the tendency is counteractive as we have seen in Fig. 3. As more and more customers enter the system the total utilization of the service unit increases.

Figure 5 emphasizes the effect of blocking on the mean waiting time versus arrival intensity. It is observable that in case of blocking the customers spend less time on average because during server failure the incoming customers go back to the source instead of waiting in the orbit. Besides the higher failure rate, the difference is more significant as well. At Fig. 5 the distribution of service time of the incoming customer is gamma, but the same tendency can be found in the case of the other distributions, too.

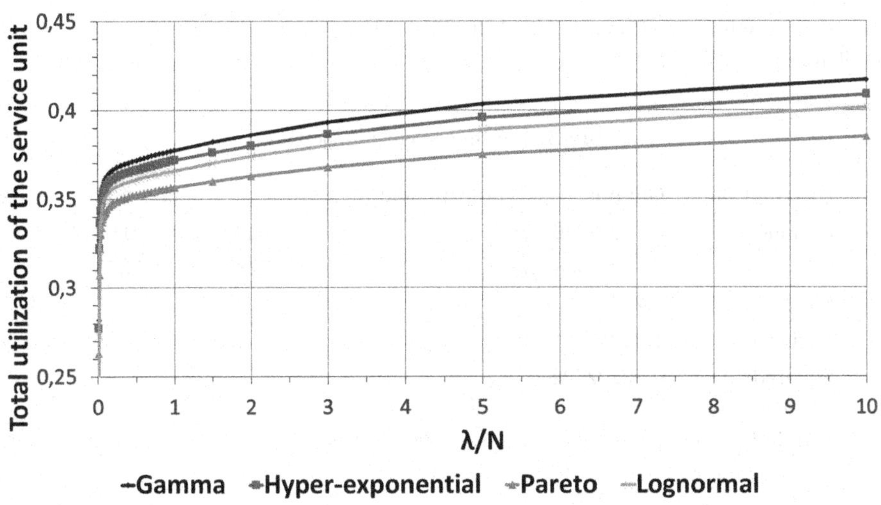

Fig. 4. Total utilization of the server vs. arrival intensity

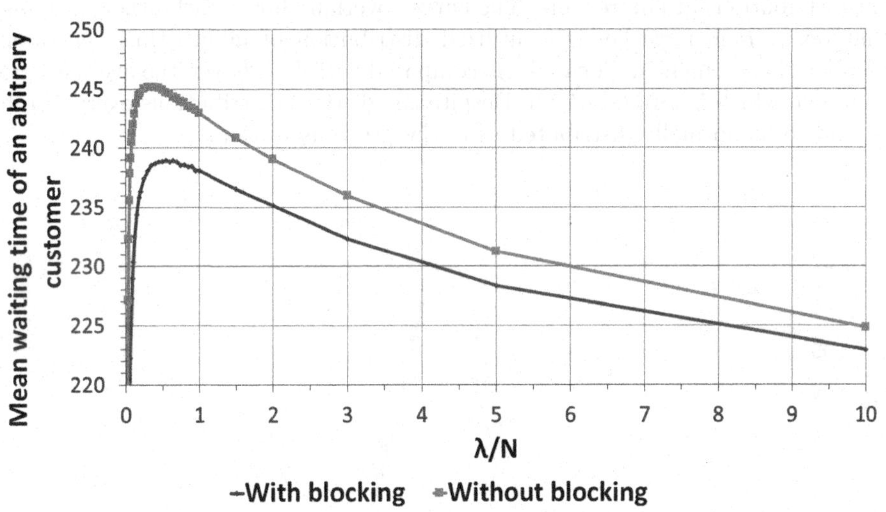

Fig. 5. The effect of blocking on the mean waiting time

3.2 Squared Coefficient of Variation Is Less Than One

Analyzing the results of the previous section we were curiously interested in how the modified parameters of the failure time alter the performance measures. In that case the parameters were chosen so that the squared coefficient of variation should be less than one. Instead of hyper-exponential we utilize hypo-exponential distribution because in the case of hypo-exponential distribution the squared coefficient of variation is always less than one. The same performance measures will be presented graphically as above but with using the new parameters of failure time which is shown in Table 3. The other parameters remain unchanged see Table 1.

Table 3. Parameters of failure time

Distribution	Gamma	Hypo-exponential	Pareto	Lognormal
Parameters	$\alpha = 1.3846$	$\mu_1 = 1$	$\alpha = 2.5442$	$m = -0.08948$
	$\beta = 1.1538$	$\mu_2 = 5$	$k = 0.7283$	$\sigma = 0.7373$
Mean	1.2			
Variance	1.04			
Squared coefficient of variation	0.72222222			

Figure 6 presents the comparison of the steady-state distribution using different distributions of failure time. The curves overlap almost each other meaning that on average regardless the utilized distribution of failure time the same number of customers are located. As compared with Fig. 2 even the averages are identical, which is around 65–66. Despite the distinct distributions every graph is tend to be normally distributed as in the previous section.

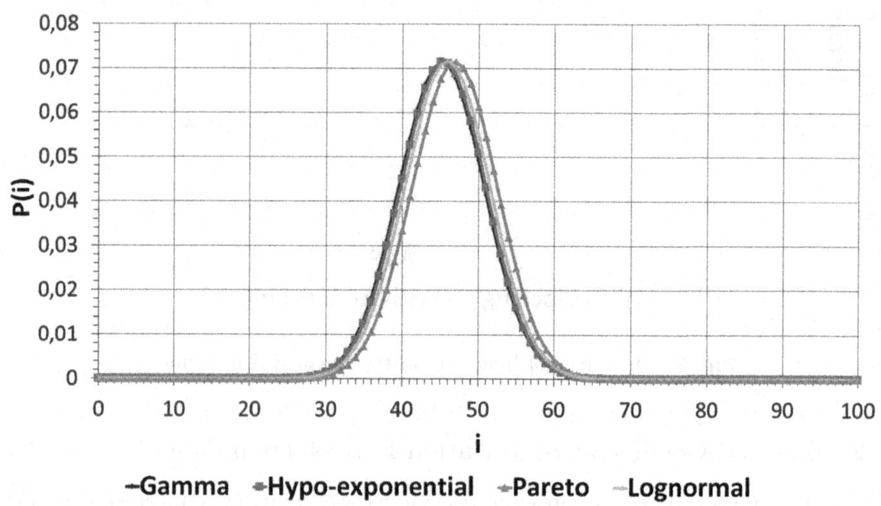

Fig. 6. Comparison of steady-state distributions

In Fig. 7 it can be seen how the mean waiting of a customer develops along with increasing arrival intensity. By examining closer the received graphs it is observable that they are much closer to each other compared to Fig. 3 although they can be differentiated from each other moderately. As in the case of Fig. 3 the highest values are experienced at Pareto distribution. When the squared coefficient of variation is less than one in terms of every distribution the values of mean waiting is higher as in the previous section.

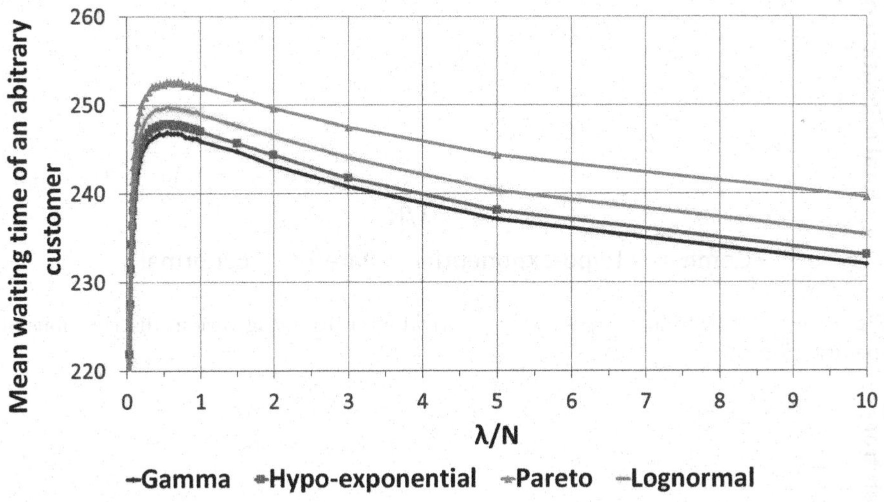

Fig. 7. Mean waiting time vs. arrival intensity using various distributions of operation time

Figure 8 demonstrates the utilization of the service unit in function of arrival intensity. No huge distinction can be discovered in terms of total utilization of the service unit among the various distributions observing the results. Basically we got back almost the same values except Pareto where the utilization is a little bit smaller. The same tendency can be noticeable comparing to Fig. 4 so at every distribution possess approximately same value of utilization. It is not surprising to see that the utilization increases with the increment of the arrival intensity.

Figure 9 represents the effect of blocking on the mean waiting time in the function of arrival intensity. The analyzed system with blocking results in lower average waiting time of an arbitrary customer due to that a customer can not enter the system when the service unit is faulty. Regardless of the used input parameters of the operation time the difference is similar to what we have observed in Fig. 5. The distribution of the operation time follows gamma distribution but it has to be noted that the same tendency can be noticed in the case of the other distributions, too.

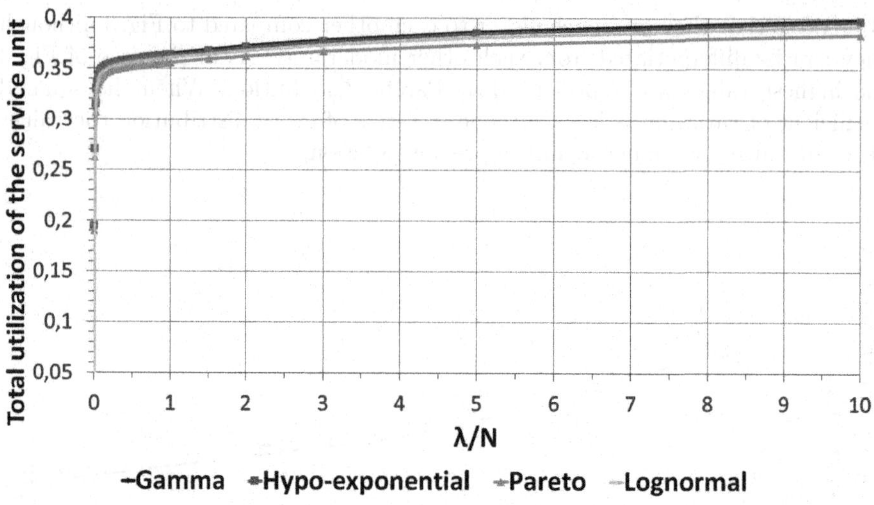

Fig. 8. Total utilization of the server vs. arrival intensity using various distributions of operation time

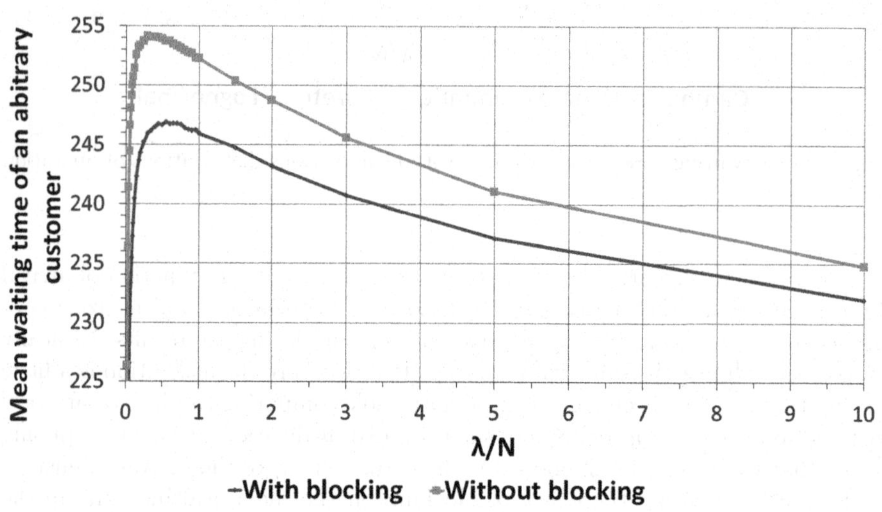

Fig. 9. The effect of blocking on the mean waiting time

4 Conclusion

A finite-source retrial queueing system with the help of two-way communication is introduced with applying blocking and an unreliable server which can make outgoing calls towards the customers of the orbit. The effect of the used distributions and blocking is illustrated by several figures on the mean arbitrary waiting time and the total utilization of the server. With the aid of stochastic simulation, the obtained results clearly revealed that in case the squared coefficient of variation is greater than one the disparity among the values of displayed performance measures is significant having the same mean and variance. In the future we would like to complete this system with other features like experimenting with more distributions, introducing some kind of impatience of the customers, or including more capacity of service.

References

1. Artalejo, J., Corral, A.G.: Retrial Queueing Systems: A Computational Approach. Springer, Cham (2008)
2. Chen, E.J., Kelton, W.D.: A procedure for generating batch-means confidence intervals for simulation: Checking independence and normality. SIMULATION **83**(10), 683–694 (2007)
3. Dragieva, V., Phung-Duc, T.: Two-way communication M/M/1//N retrial queue. In: Thomas, N., Forshaw, M. (eds.) ASMTA 2017. LNCS, vol. 10378, pp. 81–94. Springer, Cham (2017). https://doi.org/10.1007/978-3-319-61428-1_6
4. Falin, G., Artalejo, J.: A finite source retrial queue. Eur. J. Oper. Res. **108**, 409–424 (1998)
5. Fiems, D., Phung-Duc, T.: Light-traffic analysis of random access systems without collisions. Ann. Oper. Res. **277**, 311–3277 (2019). https://doi.org/10.1007/s10479-017-2636-7
6. Fishwick, P.A.: SimPack: Getting started with simulation programming in C and C++. In: In 1992 Winter Simulation Conference, pp. 154–162 (1992)
7. Francini, A., Neri, F.: A comparison of methodologies for the stationary analysis of data gathered in the simulation of telecommunication networks. In: Proceedings of MASCOTS 1996 - 4th International Workshop on Modeling, Analysis and Simulation of Computer and Telecommunication Systems, pp. 116–122 (February 1996)
8. Gharbi, N., Dutheillet, C.: An algorithmic approach for analysis of finite-source retrial systems with unreliable servers. Comput. Math. Appl. **62**(6), 2535–2546 (2011)
9. Gharbi, N., Nemmouchi, B., Mokdad, L., Ben-Othman, J.: The impact of breakdowns disciplines and repeated attempts on performances of small cell networks. J. Comput. Sci. **5**(4), 633–644 (2014)
10. Kim, J., Kim, B.: A survey of retrial queueing systems. Ann. Oper. Res. **247**(1), 3–36 (2015)
11. Krishnamoorthy, A., Pramod, P.K., Chakravarthy, S.R.: Queues with interruptions: a survey. TOP **22**(1), 290–320 (2012)
12. Kuki, A., Sztrik, J., Tóth, Á., Bérczes, T.: A contribution to modeling two-way communication with retrial queueing systems. In: Dudin, A., Nazarov, A., Moiseev, A. (eds.) ITMM/WRQ -2018. CCIS, vol. 912, pp. 236–247. Springer, Cham (2018). https://doi.org/10.1007/978-3-319-97595-5_19
13. Law, A.M., Kelton, W.D.: Simulation Modeling and Analysis. McGraw-Hill Education, New York (1991)
14. Nazarov, A., Sztrik, J., Kvach, A.: A survey of recent results in finite-source retrial queues with collisions. In: Dudin, A., Nazarov, A., Moiseev, A. (eds.) ITMM/WRQ -2018. CCIS, vol. 912, pp. 1–15. Springer, Cham (2018). https://doi.org/10.1007/978-3-319-97595-5_1

15. Nazarov, A., Sztrik, J., Kvach, A., Bérczes, T.: Asymptotic analysis of finite-source M/M/1 retrial queueing system with collisions and server subject to breakdowns and repairs. Ann. Oper. Res. **277**(2), 213–229 (2018)
16. Sztrik, J., Tóth, Á., Pintér, Á., Bács, Z.: Simulation of finite-source retrial queues with two-way communications to the orbit. In: Dudin, A., Nazarov, A., Moiseev, A. (eds.) ITMM 2019. CCIS, vol. 1109, pp. 270–284. Springer, Cham (2019). https://doi.org/10.1007/978-3-030-33388-1_22
17. Sztrik, J., Tóth, Á., Pintér, Á., Bács, Z.: Reliability analysis of finite-source retrial queueing systems with two-way communications to the orbit and blocking using simulation. In: Proceedings of International Conference on Distributed Computer and Communication Networks, DCCN 2020 (2020)
18. Toth, A., Sztrik, J., Kuki, A., Berczes, T., Efrosinin, D.: Reliability analysis of finite-source retrial queues with outgoing calls using simulation. In: 2019 International Conference on Information and Digital Technologies (IDT), pp. 504–511 (June 2019)

Token Based Parallel Processing Retrial Queueing System with a Probabilistic Joining Strategy for Priority Customers

Dhanya Babu(✉), V. C. Joshua, and A. Krishnamoorthy

Department of Mathematics, CMS College Kottayam, Kerala, India
{dhanyababu,vcjoshua,krishnamoorthy}@cmscollege.ac.in

Abstract. We consider a single server retrial queueing system with parallel queues of ordinary and priority customers. Priority customers join the queues according to a probabilistic joining strategy until the ordinary queue is full. We also assume that there is a waiting room for the blocked ordinary customers and they make retrials from there, to enter into the ordinary queue. Priority customers on arrival, seeing both ordinary and priority queue as full loss the system forever. Customers receive service on the basis of a token system. Ordinary customers, upon arrival finding the ordinary queue as full, enter into the orbit. Ordinary customers make their retrials from the orbit in an exponential duration of time intervals to enter into the ordinary queue. Ordinary customer from the orbit make their retrials in an exponential duration of time intervals. Whenever the queue size of ordinary queue is less than N then the ordinary customer in the orbit successfully enter into the ordinary queue. Two types of customers arrive according to Marked Markovian Arrival Process (MMAP). Service times are assumed to follow phase type distributions. Steady-state analysis of the model is done. Some system characteristics are evaluated.

Keywords: Retrial queues · Joining strategy · Marked Markovian Arrival Process

1 Introduction

The bibliographical information about retrial queues are given in [1,2]. Kulkarni analyzed a retrial queues with two types of calls for different retrial and service times in [11]. Falin extended the results to more than two types of calls in [6]. An $M/M/1$ queue with retrials and two streams of customers was studied in [16]. Choi gave a detailed survey of single server retrial queues with two types of calls for different variants such as different service times, geometric loss, feedback, vacations, recurrent calls, retrial rate control policy, threshold in retrial group and discrete time retrial queueing system in [4]. Multiple queues with simultaneous services are analysed in [3]. MMAP model is described in [5,13]. Steady-state probabilities are computed using Matrix Geometric methods

© Springer Nature Switzerland AG 2020
V. M. Vishnevskiy et al. (Eds.): DCCN 2020, CCIS 1337, pp. 183–194, 2020.
https://doi.org/10.1007/978-3-030-66242-4_15

in [14]. The rate matrix is computed using Ramaswami's Logarithmic Reduction Algorithm by [12]. In [7], Deepak T.G. et al. considered queues with postponed work. Analysis of various priority queues are described in [15].

The present work is an extension of [10] a single server queueing model with two independent Poisson arrival processes and phase type service distribution and an optimum joining strategy. This paper discusses the retrial version of the parallel queues of ordinary and priority customers where priority customers join the queues according to a probabilistic joining strategy. We consider the case when both ordinary queue and priority queue are of finite size. Priority customers join the queues according to a probabilistic joining strategy till the ordinary queue is full. We also assume that there is a waiting room for the blocked ordinary customers and they make retrials from there, to enter into the ordinary queue. Priority customers on arrival, seeing both ordinary and priority queue as full loss the system forever.

2 Model Description

We consider a single server parallel processing retrial queueing system with probabilistic joining strategy for priority customers. Here we assume that both ordinary and priority queues are of finite size. Ordinary customers (OC) form a queue say PC queue of size N and priority customers (PC) form a queue say OC queue of size M. Customers receive service on the basis of the token system. A token is circulating from 1, 2, ..., $K-1$, K if any of the queues are nonempty and whenever both the queues are empty, token number turns to 0. Even if the priority customer have the opportunity to get into service immediately as an ordinary customer than to stand in priority queue by joining strategy, we assume that some of priority customers take probabilistic decision whether to give up their additional benefit (reward) or not with a probability. Let p, $0 \leq p \leq 1$ be that probability of joining in PC queue by a priority customer even if according to a joining strategy he should join in ordinary queue and with probability $1-p$ he join in OC queue. This joining strategy occurs until ordinary queue attains a capacity of $N-1$. Ordinary customers, upon arrival finding the ordinary queue as full, enter into the orbit. Ordinary customers make their retrials from the orbit in an exponential duration of time intervals to enter into the OC queue. Ordinary customer from the orbit make their retrials in an exponential duration of time intervals with parameter μ. Whenever the queue size of ordinary queue is less than N then the ordinary customer in the orbit successfully enter into the OC queue. The arrival process is governed by a continuous time Markov chain $\{J_2(t), t \geq 0\}$ with state space $\{1, 2, \ldots n\}$. The sojourn time in the state i^1 is exponentially distributed with a positive λ_i^1, when the sojourn time in the state i^1 expires, the process jumps to the state j^1 without generation of a customer with a probability $\lambda_i^1 p_{i^1 j^1}(0) = d_{i^1 j^1}(0)$ where i^1, j^1 $=\{1, 2, \ldots n\}$. The process $J_2(t)$ jumps from state i^1 to j^1 with the generation of ordinary or priority customer with the probability $d_{i^1 j^1}(1)$ and $d_{i^1 j^1}(2)$ respectively $\lambda_i^1 p_{i^1 j^1}(1)$ $= d_{i^1 j^1}(1)$ and $\lambda_i^1 p_{i^1 j^1}(2) = d_{i^1 j^1}(2)$.

Let $D_0 = (d_{i^1j^1}(0))$, $D_1 = (d_{i^1j^1}(1))$ and $D_2 = (d_{i^1j^1}(2))$. The matrix $D = D_0 + D_1 + D_2$ represents the generator of the process $\{J(t), t \geq 0\}$. The average total arrival intensity λ is defined by $\lambda = \boldsymbol{\theta}(D_1 + D_2)\mathbf{e}$, where $\boldsymbol{\theta}$ is an invariant vector of the stationary distribution of the Markov chain $\{A(t), t \geq 0\}$. The vector $\boldsymbol{\theta}$ is the unique solution to the system $\boldsymbol{\theta}D = \mathbf{0}$ and $\boldsymbol{\theta}\mathbf{e} = 1$ where $\mathbf{e}$ is a column vector consisting of ones and $\mathbf{0}$ is a zero row vector. The average arrival intensities of ordinary and priority customers are given by $\lambda^{(1)} = \boldsymbol{\theta}D_1\mathbf{e}$ and $\lambda^{(2)} = \boldsymbol{\theta}D_2\mathbf{e}$. The two types of customer's arrive according to a Marked Markovian Arrival Process (MMAP) with representation (D_0, D_1, D_2) with order n. Service time of an ordinary customer follows phase type distribution with representation $PH(\alpha, T)$ of order s_1 and that of priority customer follows phase type distribution with representation $PH(\beta, S)$ of order s_2.

3 Mathematical Formulation of the Model

Let

- $N_1(t)$ be the number of customers at time t in orbit
- $N_2(t)$ be the number of customers at time t in ordinary queue
- $N_3(t)$ be the number of customers at time t in priority queue
- $R(t)$ be the token numbers from 1, 2, ..., $K-1$, K
- $J_1(t)$ be the phase of the service process
- $J_2(t)$ be the phase of the arrival process

The customer in service is counted either in first queue or in second queue depending on the status of the token.

Let $\{(N_1(t), N_2(t), N_3(t), R(t), J_1(t), J_2(t)); t \geq 0\}$ be the Markov Process on the state space
$\Omega = l^* \cup (\cup_{i=1}^{\infty} l(i))$ where
$l^* = \{(0, 0, 0, 0, 0, u) : 1 \leq u \leq m\}$ and for $i \geq 1$,
$l(i) = \{(i, j, k, r, u, v), i \geq 1, 0 \leq j \leq M, 1 \leq k \leq N, 1 \leq r \leq K, 1 \leq u \leq m, 1 \leq v \leq n\}$
where $m = \delta_{[r:1<K]}s_1 + \delta_{[r=K]}s_2$

$$\delta_{[condition]} = \begin{cases} 1, & \text{if condition is true} \\ 0, & \text{otherwise} \end{cases}$$

The Joining Strategy for priority customer is defined as follows:

Let $S_1 = \lfloor \frac{j+r-(K+1)}{K-1} \rfloor (K-1)$, $S_2 = (K-1)k$ and $S_1^* = \lfloor \frac{j}{K-1} \rfloor (K-1)$.
where

- j is the number of ordinary customers in the queue
- k is the number of priority customers in the queue
- r is the present token number

In effect with K-policy for taking for service, the PC customer arriving at the station decide to join PC or OC queue according as $S_1 > S_2$ or $S_1 \leq S_2$ with a OC customer in service; if a PC is in service is at his arrival epoch then he should join PC or OC according as $S_1^* > S_2$ or $S_1^* \leq S_2$.

Even though the priority customer get service immediately by joining in ordinary queue by this joining strategy, we assume that some of the priority customers join in priority queue with a probability p where $0 \leq p \leq 1$ by assuming some additional benefits.

4 Steady-State Analysis

The infinitesimal generator of the Markov chain is

$$Q = \begin{pmatrix} B & A_0 & & & & \\ A_2 & A_1 & A_0 & & & \\ & A_2 & A_1 & A_0 & & \\ & & A_2 & A_1 & A_0 & \\ & & & \cdots & \cdots & \cdots \\ & & & & \cdots & \cdots \cdots \end{pmatrix}$$

This model is a Level Independent Quasi-Birth-Death Process (LIQBD). We use matrix analytic method for the analysis of the system.

The transition matrices corresponding to the transitions (see Table 1) in the boundary level matrices are given as follows. The corresponding macro states are of the form (i, j, k, r) where i denotes the number of customers in the orbit j, k denotes the number of customers in the ordinary and priority queues respectively and r denotes the current status of the token.

The matrices A_0, A_1 and A_2 are as follows:

$$A_0 = \begin{pmatrix} \ddots & & \\ & \ddots & \\ & & U \end{pmatrix}$$

$$U = \begin{pmatrix} I_{(M+1)Ks_1} \otimes D_1 & & & \\ & I_{(M+1)Ks_1} \otimes D_1 & & \\ & & \ddots & \\ & & & I_{(M+1)Ks_2} \otimes D_1 \end{pmatrix}$$

Table 1. Intensities of transitions.

From	To	Condition	Transition rate
$(0,0,0)$	$(0,0,0,r)$	for $r=\{1,2,\ldots,K-1\}$ for $r=K$	$\alpha\otimes D_1$ $\beta\otimes D_2$
$(0,0)$	$(0,0)$		D_0
$(0,0,0,r)$	$(0,0)$	if $r=\{1,2,\ldots,K-1\}$	$T_0\otimes I_n$
$(0,0,0,K)$	$(0,0)$		$S_0\otimes I_n$
$(0,j,k,r)$	$(0,j,k,r)$	for $r=\{1,2,\ldots,K-1\}$ for $r=K$ $1\le j\le M$	$T\oplus D_0$ $S\oplus D_0$
$(0,0,k,r)$	$(0,0,k-1,r+1)$	for $r=\{1,2,\ldots,K-2\}$ $i\ge 0, 1\le j\le N, 1\le k\le M$	$(T_0\otimes\beta)\otimes I_n$
$(0,0,k,K-1)$	$(0,0,k-1,K)$	$i\ge 0, 1\le j\le N, 1\le j\le M$	$(S_0\otimes\beta)\otimes I_n$
$(0,j,0,r)$	$(0,j-1,0,r+1)$	for $r=\{1,2,\ldots,$ $K-2\}, 1\le j\le N,$	$(T_0\otimes\alpha)\otimes I_n$
$(0,j,0,K-1)$	$(0,j-1,0,1)$	$i\ge 0, 1\le j\le N, 1\le k\le M$	$(T_0\otimes\beta)\otimes I_n$
$(0,j,0,K)$	$(0,j-1,0,1)$	$i\ge 0, 1\le j\le N, 1\le k\le M$	$(S_0\otimes\beta)\otimes I_n$
$(i,0,0,0)$	$(i-1,0,0,1)$	$i\ge 0$	$\mu\alpha$
(i,j,k,r)	$(i-1,j+1,k,r)$	$i\ge 0, 0\le j\le N-1,$ $1\le k\le M-1, r=\{1,2,\ldots,K\}$	μ
$(i,0,0,r)$	$(i-1,0,0,r+1)$	$i\ge 0, r\le K-2$	$T_0\otimes\mu\alpha$
(i,j,k,r)	$(i,j,k+1,r)$	if $S_1\le S_2$ and $r=\{1,2,\ldots,K-1\}$ if $S_1^*>S_2$ and $r=K$ $i\ge 0,\ 0\le j\le N-1,$ $1\le k\le M-1$	$pD_2\otimes I_{s_1}$ $D_2\otimes I_{s_2}$
(i,j,k,r)	$(i,j+1,k,r)$	if $S_1\le S_2$ and $r=\{1,2,\ldots,K-1\}$ if $S_1>S_2$ and $r=\{1,2,\ldots,K-1\}$ if $S_1^*>S_2$ and $r=K$ for $i\ge 0, 0\le j\le N-1, 1\le k\le$ $M-1$	$(D_1+(1-p)D_2)\otimes I_{s_1}$ $D_1\otimes I_{s_1}$ $D_1\otimes I_{s_2}$
(i,j,M,r)	$(i,j+1,M,r)$	$r=\{1,2,\ldots,K-1\}, i\ge 0$, $0\le j\le N-1$ for $r=K,\ i\ge 0$	$(D_1+D_2)\otimes I_{s_1}$ $(D_1+D_2)\otimes I_{s_2}$
(i,j,k,r)	$(i,j-1,k,r+1)$	if $r=\{1,2,\ldots,K-2\}$	$(T_0\otimes\alpha)\otimes I_n$
(i,j,k,K)	$(i,j-1,k,1)$	$i\ge 0, 1\le j\le M$	$(S_0\otimes\alpha)\otimes I_n$

$$A_2=\begin{pmatrix} V_1 & V_0 & & & \\ & & V_0 & & \\ & & & V_0 & \\ & & & & \\ & & & V_1 & V_0 \end{pmatrix}$$

where

$$V_0 = \begin{pmatrix} I_{(M+1)K} \otimes \mu I_{s_1 n} & & & \\ & I_{(M+1)K} \otimes \mu I_{s_1 n} & & \\ & & \ddots & \\ & & & I_{(M+1)K} \otimes \mu I_{s_2 n} \end{pmatrix}$$

$$V_1 = \begin{pmatrix} V^* \ldots\ldots\ldots \\ \ldots\ldots\ldots\ldots \\ \ldots\ldots\ldots\ldots \\ \ldots\ldots\ldots\ldots \end{pmatrix}$$

$$V^* = \begin{pmatrix} \ldots & \mu\alpha \otimes I_n & & & \\ \ldots & & T^0 \otimes \mu\alpha \otimes I_n & & \\ \ldots & & & \ddots & \\ \ldots & & & & T^0 \otimes \mu\alpha \otimes I_n \\ \ldots & T^0 \otimes \mu\alpha \otimes I_n & & & \\ \ldots & S^0 \otimes \mu\alpha \otimes I_n & & & \end{pmatrix}$$

$$A_1 = \begin{pmatrix} A_{10} & A_{00} & & & & \\ A_2^* & A_{11} & A_{01} & & & \\ & A_2^* & A_{12} & A_{02} & & \\ & & \ldots & \ldots & \ldots & \\ & & & \ldots & \ldots & \ldots \\ & & & A_2^* & A_{1N-1} & A_{0N-1} \\ & & & & A_2^* & A_{1N} \end{pmatrix}$$

where

$$A_{10} = \begin{pmatrix} E_{10} & E_{00} & & & \\ E_2 & E_1 & E_0 & & \\ & E_2 & E_1 & E_0 & \\ & & \ldots & \ldots & \ldots \\ & & & \ldots & \ldots \\ & & & E_2 & E_1 \end{pmatrix}$$

$$E_{10} = \begin{pmatrix} D_0 & \alpha \otimes D_1 & & & \beta \otimes D_2 \\ T^0 \otimes I_n & T \oplus (D_0 - \mu I_n) & & & \\ \vdots & & \ddots & & \\ & & & \ddots & \\ T^0 \otimes I_n & & & T \oplus (D_0 - \mu I_n) & \\ T^0 \otimes I_n & & & & S \oplus (D_0 - \mu I_n) \end{pmatrix}$$

$$E_{00} = \begin{pmatrix} I_m \otimes D_2 & & & & \\ & I_m \otimes pD_2 & & & \\ & & \ddots & & \\ & & & I_m \otimes D_2 & \\ & & & & I_m \otimes pD_2 \end{pmatrix}$$

$$E_0 = \begin{pmatrix} I_m \otimes pD_2 & & & & \\ & I_m \otimes pD_2 & & & \\ & & \ddots & & \\ & & & I_m \otimes pD_2 & \\ & & & & I_m \otimes pD_2 \end{pmatrix}$$

$$E_2 = \begin{pmatrix} \dots\dots\dots\dots & T^0 \otimes \beta \otimes I_n \\ \dots\dots\dots\dots & T^0 \otimes \beta \otimes I_n \\ \dots\dots\dots\dots & \vdots \\ \dots\dots\dots\dots & \dots \\ \dots\dots\dots\dots & T^0 \otimes \beta \otimes I_n \\ \dots\dots\dots\dots & S^0 \otimes \beta \otimes I_n \end{pmatrix}$$

$$E_1 = \begin{pmatrix} T \oplus (D_0 - \mu I_n) & & & & \\ & T \oplus (D_0 - \mu I_n) & & & \\ & & \ddots & & \\ & & & T \oplus (D_0 - \mu I_n) & \\ & & & & S \oplus (D_0 - \mu I_n) \end{pmatrix}$$

for $1 \leq i \leq N-1$

$$A_{1i} = \begin{pmatrix} E_1^1 & E_0^1 & & & \\ E_2^1 & E_1^1 & E_0^1 & & \\ & E_2^1 & E_1^1 & E_0^1 & \\ & & \dots & \dots & \dots \\ & & & \dots & \dots & \dots \\ & & & & E_2^1 & E_1^1 \end{pmatrix}$$

where

$$E_1^1 = \begin{pmatrix} T \oplus (D_0 - \mu I_n) & & & & \\ & T \oplus (D_0 - \mu I_n) & & & \\ & & \ddots & & \\ & & & T \oplus (D_0 - \mu I_n) & \\ & & & & S \oplus (D_0 - \mu I_n) \end{pmatrix}$$

The transitions of E_0^1 depends on the strategy.

$$E_2^1 = \begin{pmatrix} \vdots & \cdots\cdots\cdots \\ \vdots & \cdots\cdots\cdots \\ \vdots & \cdots\cdots\cdots \\ \vdots & \cdots\cdots\cdots \\ \vdots & \cdots\cdots\cdots \\ T^0 \otimes \beta \otimes I_n & \cdots\cdots\cdots \\ \cdots & \cdots\cdots\cdots \end{pmatrix}$$

$$A_{1N} = \begin{pmatrix} E_1^1 & E_0^{11} & & & & \\ E_2^1 & E_1^1 & E_0^{11} & & & \\ & E_2^1 & E_1^1 & E_0^{11} & & \\ & & \cdots & \cdots & \cdots & \\ & & & \cdots & \cdots & \cdots \\ & & & & E_2^1 & E_1^1 \end{pmatrix}$$

$$E_0^{11} = \begin{pmatrix} I_m \otimes D_2 & & & & \\ & I_m \otimes D_2 & & & \\ & & \ddots & & \\ & & & I_m \otimes D_2 & \\ & & & & I_m \otimes D_2 \end{pmatrix}$$

$$A_2^* = \begin{pmatrix} F & & & & \\ & F_1 & & & \\ & & \ddots & & \\ & & & F_1 & \\ & & & & F_1 \end{pmatrix}$$

where

$$F = \begin{pmatrix} T_0 \otimes \alpha \otimes I_n & & & \\ & \ddots & & \\ & & T_0 \otimes \alpha \otimes I_n & \\ & & & T_0 \otimes \alpha \otimes I_n \\ T_0 \otimes \alpha \otimes I_n & & & \\ S_0 \otimes \alpha \otimes I_n & & & \end{pmatrix}$$

$$F_1 = \begin{pmatrix} T_0 \otimes \alpha \otimes I_n & & \\ & \ddots & \\ & & T_0 \otimes \alpha \otimes I_n \\ & & \\ S_0 \otimes \alpha \otimes I_n & & \end{pmatrix}$$

The transitions of A_{0i} for $0 \leq i \leq N-1$ depends on the strategy. The matrix $A = A_0 + A_1 + A_2$ can be written as

$$A = \begin{pmatrix} C_{10} & C_{00} & & & & & \\ C_{21} & C_{11} & C_{01} & & & & \\ & C_{22} & C_{12} & C_{02} & & & \\ & & \cdots & \cdots & \cdots & & \\ & & & \cdots & \cdots & \cdots & \\ & & & & C_{2(N-1)} & C_{1(N-1)} & C_{0(N-1)} \\ & & & & & C_{2N} & C_{1N} \end{pmatrix}$$

4.1 Stability Condition

Let π be the steady-state probability vector of A such that $\pi A = 0$ and $\pi \mathbf{e} = 1$

$$\pi_i = \pi_M \prod_{j=0}^{M-1-i} H_{M-1-j} \tag{1}$$

for $i = 0, 1, \ldots M-1$.

The sequence of matrices H_i are defined as
$H_i = -C_{2i}[H_{i-1}C_{0i} + C_{1i}]^{-1}$ for $i = 1, 2, \ldots, N-1$ and
$H_0 = -C_{20}[C_{10}]^{-1}$.

The vector π_M is obtained from the equation $\pi \mathbf{e} = 1$ where

$$\pi_N \left[\sum_{i=0}^{N-1} \prod_{j=0}^{N-1-i} H_{N-1-j} + I \right] \mathbf{e} = 1 \tag{2}$$

The stability condition is given by

$$\pi A_0 \mathbf{e} < \pi A_2 \mathbf{e}$$

where

$$\pi A_0 \mathbf{e} = \pi_N [I_{(M+1)(K+1)} \otimes D_1] \mathbf{e}$$

$$\pi A_2 \mathbf{e} = [\pi_0 + \pi_1 + \pi_2 + \ldots \pi_{N-1}] V_1 \mathbf{e} + [\pi_0 + \pi_1 + \pi_2 + \ldots \pi_{N-2}] I_{(M+1)(K+1)} \otimes \mu I_{mn} \mathbf{e}$$

4.2 Steady-State Probability Vector

The stationary distribution of the Markov process under consideration is obtained by solving the set of equations $\mathbf{x}Q = 0$, $\mathbf{x}\mathbf{e} = 1$. Let $\mathbf{x}$ be the steady-state probability vector of Q.

Partition this vector as: $\mathbf{x} = (\mathbf{x}_0, \mathbf{x}_1, \mathbf{x}_2, \ldots)$, where $\mathbf{x}_i = (\mathbf{x}_{i0}, \mathbf{x}_{i1}, \ldots \mathbf{x}_{iN})$ $\mathbf{x}_{ij} = (\mathbf{x}_{ij0}, \mathbf{x}_{ij1}, \mathbf{x}_{ij2}, \mathbf{x}_{ij3}, \ldots \mathbf{x}_{ijM})$ for $j = 0, 1, 2, \ldots, N$ whereas for $k = 0, 1, 2 \ldots M$, the vectors $\mathbf{x}_{ijk} = (\mathbf{x}_{ijk1}, \mathbf{x}_{ijk2}, \ldots, \mathbf{x}_{ijkK})$, where $\mathbf{x}_{ijkr}$ is the probability of being in state (i, j, k, r) for $r = 1, 2, \ldots K$, $i \geq 0$, $j = 0, 1, 2 \ldots N$, $k =$

$0, 1, 2 \ldots M$. For $r = 1, 2 \ldots K$, the vectors $\mathbf{x}_{ijkr} = (\mathbf{x}_{ijkr1}, \mathbf{x}_{ijkr2}, \ldots, \mathbf{x}_{ijkrm})$, where $\mathbf{x}_{ijkru}$ is the probability of being in state (i, j, k, r, u) for $r = 1, 2, \ldots K$, $i \geq 0$, $j = 0, 1, 2 \ldots N$, $k = 0, 1, 2 \ldots M$, $u = 1, 2, \ldots m$. For $u = 1, 2, \ldots, m$, the vectors

$\mathbf{x}_{ijkru} = (\mathbf{x}_{ijkru1}, \mathbf{x}_{ijkru2}, \ldots, \mathbf{x}_{ijkrun})$, where $\mathbf{x}_{ijkruv}$ is the probability of being in state (i, j, k, r, u, v) for $r = 1, 2, \ldots K$, $i \geq 0$, $j = 0, 1, 2 \ldots N$, $k = 0, 1, 2 \ldots M$, $u = 1, 2, \ldots m, v = 1, 2, \ldots n$ and $\mathbf{x}_{i0}$ is the probability of being in state $(i, 0, r, u, v)$.

Under the stability condition the steady-state probability vector is obtained as

$$\mathbf{x}_i = \mathbf{x}_{i-1} R, i \geq 1$$

$$\mathbf{x}_i = \mathbf{x}_0 R^i, i \geq 1$$

where the matrix R is the minimal non negative solution to the matrix quadratic equation

$$R^2 A_2 + R A_1 + A_0 = 0 \tag{3}$$

and R can be obtained by successive substitution procedure $R_0 = 0$ and $R_{k+1} = -V - R_k^2 W$ where $V = A_2 A_1^{-1}$, $W = A_0 A_1^{-1}$ by *Logarithmic Reduction Algorithm* developed by Latouche and Ramaswamy in [12].

The vectors $\mathbf{x}_0$ and $\mathbf{x}_1$ are obtained by solving

$$\mathbf{x}_0 B + \mathbf{x}_1 A_2 = 0$$

$$\mathbf{x}_{i-1} A_0 + \mathbf{x}_i A_1 + \mathbf{x}_{i+1} A_2 = 0, i \geq 1 \tag{4}$$

subject to the normalizing condition

$$\mathbf{x}_0 (I - R)^{-1} \mathbf{e} = 1.$$

4.3 Performance Measures

1. Expected Number of customers in the orbit

$$E[N_O] = \sum_{i=0}^{\infty} i \mathbf{x}_i \mathbf{e}$$

2. Expected Number of ordinary customers in the system

$$E[N_1] = \sum_{i=0}^{\infty} \sum_{j=0}^{N} j \mathbf{x}_i \mathbf{e}$$

3. Expected Number of priority customers in the system

$$E[N_2] = \sum_{i=0}^{\infty} \sum_{j=0}^{N} \sum_{k=0}^{M} k \mathbf{x}_{ijk} \mathbf{e}$$

4. Expected Number of customers in the system

$$E[N] = E[N_O] + E[N_1] + E[N_2]$$

5. Probability that the server is idle

$$t_0 = \sum_{i=0}^{\infty} x_{i00}$$

6. Probability that the server is busy with an ordinary customer

$$t_1 = \sum_{i=0}^{\infty}\sum_{j=0}^{N}\sum_{k=0}^{M}\sum_{r=1}^{K-1} \mathbf{x}_{ijkr}\mathbf{e}$$

7. Probability that the server is busy with a priority customer

$$t_2 = \sum_{i=0}^{\infty}\sum_{j=0}^{N}\sum_{k=0}^{M} \mathbf{x}_{ijkK}\mathbf{e}$$

8. The probability that a priority customer is blocked from entering the system upon arrival as priority customer

$$P_{b1} = \sum_{i=0}^{\infty}\sum_{j=0}^{N}\sum_{k=0}^{M}\sum_{r=1}^{K} \mathbf{x}_{ijMr}\mathbf{e}$$

9. The probability that a priority customer is blocked from entering the system upon arrival

$$P_{b2} = \sum_{i=0}^{\infty}\sum_{r=1}^{K} \mathbf{x}_{iNMr}\mathbf{e}$$

10. The overall rate of retrials at which the orbiting customers request service

$$RR = \mu^* = \sum_{i=0}^{\infty} \mu \sum_{j=0}^{N}\sum_{k=0}^{M}\sum_{r=0}^{K} \mathbf{x}_{ijkr}\mathbf{e}$$

11. The successful rate of retrials

$$SR = \mu^{**} = \sum_{i=0}^{\infty} \mu \sum_{j=0}^{(N-1)}\sum_{k=0}^{M}\sum_{r=0}^{K} \mathbf{x}_{ijkr}\mathbf{e}$$

12. The fraction of successful rate of retrials

$$FSR = \mu_{fr} = \frac{\mu^{**}}{\mu^*}$$

5 Conclusion

In this paper, we considered the retrial version of a parallel processing queueing system with a probabilistic joining strategy for priority customers. Here, joining strategy for priority customers is only up to the maximum capacity of ordinary queue. Blocked priority customers leave the system forever but the blocked ordinary customers have the chance to enter into the ordinary queue after successful retrial. We intend to find the optimum values for p, N, K and M to minimise the loss of priority customers.

References

1. Artalejo, J.R., Gomez-Corral, A.: Retrial Queueing Systems: A Computational Approach. Springer, Berlin (2008). https://doi.org/10.1007/978-3-540-78725-9
2. Artalejo, J.R.: Accessible bibliography of research on retrial queues. Math. Comput. Modell. **30**(3–4), 1–6 (1999)
3. Chakravarthy, S., Thiagarajan, S.: Two parallel queues with simultaneous services and Markovian arrivals. J. Appl. Math. Stochast. Anal. **10**(4), 383–405 (1997)
4. Choi, B.D., Chang, Y.: MAP1, MAP2/M/c retrial queue with the retrial group of finite capacity and geometric loss. Math. Comput. Modell. **30**, 99–113 (1999)
5. Dudin, S., Kim, C., Dudina, O.: MMAP/M/N queueing system with impatient heterogeneous customers as a model of a contact center. Comput. Oper. Res. **40**(7), 1790–1803 (2013)
6. Falin, G.I.: On a multiclass batch arrival retrial queue. Adv. Appl. Probab. **20**, 483–487 (1988)
7. Krishnamoorthy, A., Deepak, T.G., Joshua, V.C.: Queues with postponed work. Top **12**, 375–398 (2004). https://doi.org/10.1007/BF02578967
8. Krishnamoorthy, A., Manjunath, A.S.: On priority queues generated through customer induced service interruption. Neural Parallel Sci. Comput. **23**, 459–486 (2015)
9. Krishnamoorthy, A., Manjunath, A.S.: On queues with priority determined by feedback. Calcutta Stat. Assoc. Bull. **70**, 33–56 (2018)
10. Krishnamoorthy, A., Joshua, V.C., Babu, D.: A token based parallel processing queueing system with priority. In: Vishnevskiy, V.M., Samouylov, K.E., Kozyrev, D.V. (eds.) DCCN 2017. CCIS, vol. 700, pp. 231–239. Springer, Cham (2017). https://doi.org/10.1007/978-3-319-66836-9_19
11. Kulkarni, V.G.: Expected waiting time in a multi-class batch arrival retrial queue. J. Appl. Probab. **23**, 144–154 (1986)
12. Latouche, G., Ramaswami, V.: Introduction to Matrix analytic Methods in Stochastic Modelling. Siam (1999)
13. Mathew, A.P., Krishnamoorthy, A., Joshua, V.C.: A retrial queueing system with orbital search of customers lost from an offer zone. In: Dudin, A., Nazarov, A., Moiseev, A. (eds.) ITMM/WRQ -2018. CCIS, vol. 912, pp. 39–54. Springer, Cham (2018). https://doi.org/10.1007/978-3-319-97595-5_4
14. Neuts, M.F.: Matrix-Geometric Solutions in Stochastic Models - An Algorithmic Approach. The Johns Hopkins University Press, Baltimore (1981)
15. Takagi H.: Queueing Analysis. Volume 1: Vacations and Priority Systems. North-Holland, Amsterdam (1991)
16. Wang, J., Zhang, F.: Strategic joining in M/M/1 retrial queues. Eur. J. Oper. Res. **230**(1), 76–87 (2013)

A Disease Outbreak Managing Queueing System with Self-generation of Status and Random Clock for Quarantine Time

T. S. Sinu Lal(✉), V. C. Joshua, and A. Krishnamoorthy

Department of Mathematics, CMS College, Kottayam 686001, Kerala, India
{sinulal,vcjoshua,krishnamoorthy}@cmscollege.ac.in
http://www.cmscollege.ac.in

Abstract. We propose an efficient model of an epidemic management system with the notion of "self generation of status" and random clock for the quarantine period of a 'suspected' person (we designate them as 'customers'). Customers arrive for check up according to a Markovian arrival process. They enter into a multi server station; assume that the chance of a person being tested as infected is p, independent of others; if he/she falls in this category that person is directed to a quarantine pool; else (with probability 1-p), the customer leaves the system for ever. For each customer in the quarantine pool, two random clocks start ticking upon his entry. One of the random clocks has Erlang distributed life time; this is the quarantine duration, on realization of which, the customer can leave the system. On the other hand, a customer in quarantine generates as infected in a time period, measured from the time of his/her entry into the pool; this time duration has a Coaxian distribution. This is the second clock. The customer leaves the system upon the realization of the Erlang clock. The decision based on the clocks are: the customer can leave the system upon the realization of the Erlang clock, provided the Coxian clock does not realize until then. If the customer has generated into the infected status before the realization of the Erlang clock (thatis, Coxian clock realizes before the Erlang clock), then he is transferred to the specially designed care unit of finite capacity. This system is modeled as a continuous time Markov chain and is analyzed using matrix analytic method. The main concern is on finding the optimal capacity of the care unit so that maximum number of infected are admitted, taking into consideration several risk factors.

Keywords: Self generation of status · Matrix analytic methods · Quarantine pool · Random clock

1 Introduction

Probabilistic models of epidemic management systems have been a relevant area of research for the past several years. In this work, we develop an effective mathematical model for the management of epidemic outbreaks using queueing theoretical tools. We use the notion of self generation of status (of being infected or

© Springer Nature Switzerland AG 2020
V. M. Vishnevskiy et al. (Eds.): DCCN 2020, CCIS 1337, pp. 195–205, 2020.
https://doi.org/10.1007/978-3-030-66242-4_16

not) from a pool of suspected persons. This is achieved using two independent random clocks, for each person in the pool, one for the self generation of status and the other for the termination of quarantine period.

In [1] Trapman and Bootsma combines epidemiology and queueing theory and they establish a relation between the spread of infectious disease and dynamics of $M|G|1$ queues with processor sharing. Lopez-Garcia [2] analyses an SIR epidemic model for the disease spread throughout a small heterogeneous population of individuals and in this work two different orders for the state space of the underlying Markov chain yield a scalar and a matrix formalism. [3] Rania et al. gives a comprehensive and systematic review of mathematical models of disease transmission in healthcare settings and assess the application of contact and patient transfer network data over time and their impact on our understanding of transmission dynamics of infections. [4] also gives a useful model of disease infections.

Gómez-Corral, A. and López-García [5], present a more detailed description of the underlying piecewise-deterministic Markov process (PDMP), from which, they analyze the population transmission number and the infection probability of a certain susceptible individual. In [6] Amador and Gómez-Corral extreme values in an SIQS (susceptible infectious quarantined susceptible) model with two different states for quarantine and showed that the quarantine occupancy rate depends largely on the contact processes and the capacity of the quarantine compartment.

Queueing models with priority generations can be seen in [7,8]. A queueing inventory system as an organ transplantation model has been studied in [9] and idea of random clock is used for modelling search and perishing times. [10] describes a multi server tandem queue with phase type services and is analyzed using matrix analytic methods. For more details of matrix analytic methods see [12,13]. Description of Markovian arrival process can be found in [14].

A phase type distribution is defined as the distribution of the time until absorption in a Markov process with a finite state space and a single absorption state defined over nonnegative real line. A phase type distribution with transient states $\{1, 2, \ldots, m\}$ and an absorbing state $m + 1$ is represented by a two tuple of the form (α, T), where α is the probability vector of length m according to which the process selects the initial state from $\{1, 2, \ldots, m\}$ and T is an $m \times m$ matrix such that $\begin{pmatrix} \mathbf{T} & \mathbf{T^0} \\ \mathbf{0} & \mathbf{0} \end{pmatrix}$ generates the process, given the column vector $\mathbf{T^0}$ satisfies the condition $\mathbf{Te} + \mathbf{T^0} = \mathbf{0}$. $(\alpha, \mathbf{T})$ is called the representation of the phase type distribution. The distribution F of time until the chain gets absorbed into the state $m + 1$ is given by $F(x) = 1 - \alpha e^{\mathbf{T}x}\mathbf{e}$, $x \geq 0$.

In the coming sections, the mathematical model, stationary vector of the system process, performance characteristics and departure process of customers from the pool are included.

2 Mathematical Model

In this model customers arrive to the first station according to a MAP of order a. In the first station there are c identical servers whose service time follows an exponential distribution with parameter μ. If all the servers in the first station are busy then the arriving customers queue up in-front of the first station. At the first station a person is suspected to be infectious with a probability p and is sent to a quarantine pool of capacity N, other wise he can quit the system with the complimentary probability $1-p$. When a person reaches the quarantine pool, two random clocks starts ticking. One of the clocks gives the time for generation of status (positive or negative) of disease for a particular person and the other upon realization gives the total quarantine time.

The time for status generation are independent and identically distributed according to a Coaxian distribution of order h having phase type representation (α, C), where $\alpha = (1, 0, \ldots, 0)$ is the initial probability distribution and C is matrix giving the transition rates among the m phases. C takes the form

$$C = \begin{bmatrix} -\rho(1) & q_1\rho(1) & & & \\ & -\rho(2) & q_2\rho(2) & & \\ & & \ddots & \ddots & \\ & & & \rho(m-1) & q_{m-1}\rho(r-1) \\ & & & & -\rho(m) \end{bmatrix}.$$

where $\rho(i), 1 \leq i \leq r-1$ is the rate of transition from the phase i to $i+1$ and the process moves from the phase i to $i+1$ with probability q_i or enters the absorption state with the complimentary probability $1 - q_i$ provided the initial phase is 1.

The quarantine time follows an Erlang distribution of order g with density function.

$$f(t) = \frac{g\mu(g\mu t)^{r-1}e^{-g\mu t}}{(g-1)!}, t \geq 0, \mu > 0.$$

If a person is found positive then, he is sent to a special care unit with M identical servers, each with phase type distributed service times.

The process governing the system can be modeled as a continuous time Markov chain $\{\mathcal{X}(t), t \geq 0\}$

$$\mathcal{X}(t) = (N(t), m(t), e(t), c(t), s(t), a(t), t \geq 0)$$

The notations in the above definition are described below.

- $N(t)$ – Number of customers in the first stage. This includes the total count of customers in the queue along with those who are being served at the first station.
- $m(t)$ – Number of customers in the quarantine pool, $0 \leq m(t) \leq N$.
- $e(t) = (e_1(t), e_2(t), ..., e_g(t))$, where $e_i(t)$ is the number of Erlang clocks staying in the ith phase.

- $c(t) = (c_1(t), c_2(t), ..., c_h(t))$, where $c_k(t)$ is the number of Coxian clocks staying in the kth phase.
- $s(t) = (s_1(t), s_2(t), ..., s_r(t))$, where $s_i(t)$ is the number of servers at the second stage staying in the ith phase of service.
- a(t) – Phase of MAP.

The state space $\mathcal{S}$ of the system is described below.

$$\mathcal{S} = \cup_{i=0}^{\infty} \mathcal{L}(i)$$

Where $\mathcal{L}(i) = \mathcal{L}_1(i) \cup \mathcal{L}_2(i) \cup \mathcal{L}_3(i)$

$$\mathcal{L}_1(i) = \{(i, 0, j), 1 \leq j \leq a\}$$

$$\mathcal{L}_2(i) = \{(i, j, (e_1, e_2, ..., e_g), (c_1, c_2, ..., c_h), k), 0 \leq j \leq N, 1 \leq c_i, e_i \leq N, 1 \leq k \leq a\}$$

$\mathcal{L}_3(i) = \{(i, j, (e_1, e_2, ..., e_g), (c_1, c_2, ..., c_h), (s_1, s_2, ..., s_r), k), 0 \leq j \leq N, 1 \leq c_i, e_i \leq N, 1 \leq s_i \leq M, 1 \leq k \leq a\}$.

The infinitesimal generator takes the following form.

$$\mathcal{Q} = \begin{bmatrix} \mathcal{A}_{00} & \mathcal{A}_0 & & & & & \\ \mathcal{A}_{10} & \mathcal{A}_{11} & \mathcal{A}_0 & & & & \\ & \ddots & \ddots & \ddots & & & \\ & & \mathcal{A}_{c-10} & \mathcal{A}_{c-11} & \mathcal{A}_0 & & \\ & & & \mathcal{A}_0 & \mathcal{A}_1 & \mathcal{A}_0 & \\ & & & & \ddots & \ddots & \ddots \end{bmatrix}.$$

The block submatrices are described below.

For $1 \leq i \leq c-1$, $\mathcal{A}_{i-11} = i(1-p)\mu I + \mathcal{A}_i^*$, $\mathcal{A}_i^* = \begin{bmatrix} 0 & ip\mu I \\ 0 & 0 \end{bmatrix}$. For $1 \leq i \leq c-1$,

$$\mathcal{A}_0 = I \otimes D_0$$

For the sake of convenience we again devide the states in $\mathcal{L}_2(i)$ and $\mathcal{L}_3(j)$ further $\mathcal{L}_2(i, l)$ and $\mathcal{L}_3(j, k)$, where $0 \leq i, j \leq N$. $\mathcal{L}_h(i, l)$ is the group of states in the level i, for which l customers are in the pool for $h = 1, 2$.

$$\mathcal{A}_{ii} = \begin{array}{c|cccccccc} & \mathcal{L}_1(0) & \mathcal{L}_2(i,0) & \dots & \mathcal{L}_2(i,N) & \mathcal{L}_3(i,0) & \mathcal{L}_3(i,1) & \dots & \mathcal{L}_3(i,N) \\ \hline \mathcal{L}_1(0) & \Psi_0 & & & & & & & \\ \mathcal{L}_2(i,0) & \Phi_1 & \Psi_1 & & & \Theta_1 & & & \\ \vdots & & & \ddots & & & \ddots & & \\ \mathcal{L}_2(i,N) & & & \Phi_N & \Psi_N & & & & \Theta_N \\ \mathcal{L}_3(i,0) & \Lambda_* & & & & & & & \\ \mathcal{L}_3(i,1) & & \Lambda & & & & & & \\ \mathcal{L}_3(i,2) & & & & & \Lambda_1 & & & \\ \vdots & & & & & & \ddots & \ddots & \\ \mathcal{L}_3(i,N) & & & & & & & \Lambda_d & \Psi_b \end{array}$$

In this model customers arrive to the first station according to a MAP of order a. In the first station there are c identical servers whose service time follows an exponential distribution with parameter μ. If all the servers in the first station are busy then the arriving customers queue up in-front of the first station. At the first station a person is suspected to be infectious with a probability p and is sent to a quarantine pool of capacity N, other wise he can quit the system with the complimentary probability $1-p$. When a person reaches the quarantine pool, two random clocks starts ticking. One of the clocks gives the time for generation of status (positive or negative) of disease for a particular person and the other upon realization gives the total quarantine time.

The time for status generation are independent and identically distributed according to a Coaxian distribution of order h having phase type representation (α, C), where $\alpha = (1, 0, \dots, 0)$ is the initial probability distribution and C is matrix giving the transition rates among the m phases. C takes the form

$$C = \begin{bmatrix} -\rho(1) & q_1\rho(1) & & & \\ & -\rho(2) & q_2\rho(2) & & \\ & & \ddots & \ddots & \\ & & & \rho(m-1) & q_{m-1}\rho(r-1) \\ & & & & -\rho(m) \end{bmatrix}.$$

where $\rho(i), 1 \leq i \leq r-1$ is the rate of transition from the phase i to $i+1$ and the process moves from the phase i to $i+1$ with probability q_i or enters the absorption state with the complimentary probability $1 - q_i$ provided the initial phase is 1.

The quarantine time follows an Erlang distribution of order g with density function.

$$f(t) = \frac{g\mu(g\mu t)^{r-1} e^{-g\mu t}}{(g-1)!}, t \geq 0, \mu > 0.$$

If a person is found positive then, he is sent to a special care unit with M identical servers, each with phase type distributed service times.

The process governing the system can be modeled as a continuous time Markov chain $\{\mathcal{X}(t), t \geq 0\}$

$$\mathcal{X}(t) = (N(t), m(t), e(t), c(t), s(t), a(t), t \geq 0)$$

The notations in the above definition are described below.

- $N(t)$ – Number of customers in the first stage. This includes the total count of customers in the queue along with those who are being served at the first station.
- $m(t)$ – Number of customers in the quarantine pool, $0 \leq m(t) \leq N$.
- $e(t) = (e_1(t), e_2(t), ..., e_g(t))$, where $e_i(t)$ is the number of Erlang clocks staying in the ith phase.
- $c(t) = (c_1(t), c_2(t), ..., c_h(t))$, where $c_k(t)$ is the number of Coxian clocks staying in the kth phase.
- $s(t) = (s_1(t), s_2(t), ..., s_r(t))$, where $s_i(t)$ is the number of servers at the second stage staying in the ith phase of service.
- a(t) – Phase of MAP.

The state space $\mathcal{S}$ of the system is described below.

$$\mathcal{S} = \cup_{i=0}^{\infty} \mathcal{L}(i)$$

Where $\mathcal{L}(i) = \mathcal{L}_1(i) \cup \mathcal{L}_2(i) \cup \mathcal{L}_3(i)$

$$\mathcal{L}_1(i) = \{(i, 0, j), 1 \leq j \leq a\}$$

$$\mathcal{L}_2(i) = \{(i, j, (e_1, e_2, ..., e_g), (c_1, c_2, ..., c_h), k), 0 \leq j \leq N, 1 \leq c_i, e_i \leq N, 1 \leq k \leq a\}$$

$\mathcal{L}_3(i) = \{(i, j, (e_1, e_2, ..., e_g), (c_1, c_2, ..., c_h), (s_1, s_2, ..., s_r), k), 0 \leq j \leq N, 1 \leq c_i, e_i \leq N, 1 \leq s_i \leq M, 1 \leq k \leq a\}$.

The infinitesimal generator takes the following form.

$$\mathcal{Q} = \begin{bmatrix} \mathcal{A}_{00} & \mathcal{A}_0 & & & & \\ \mathcal{A}_{10} & \mathcal{A}_{11} & \mathcal{A}_0 & & & \\ & \ddots & \ddots & \ddots & & \\ & & \mathcal{A}_{c-10} & \mathcal{A}_{c-11} & \mathcal{A}_0 & \\ & & & \mathcal{A}_0 & \mathcal{A}_1 & \mathcal{A}_0 \\ & & & & \ddots & \ddots & \ddots \end{bmatrix}.$$

The block submatrices are described below.

For $1 \leq i \leq c-1$, $\mathcal{A}_{i-11} = i(1-p)\mu I + \mathcal{A}_i^*$, $\mathcal{A}_i^* = \begin{bmatrix} 0 & ip\mu I \\ 0 & 0 \end{bmatrix}$. For $1 \leq i \leq c-1$,

$$\mathcal{A}_0 = I \otimes D_0$$

For the sake of convenience we again devide the states in $\mathcal{L}_2(i)$ and $\mathcal{L}_3(j)$ further $\mathcal{L}_2(i,l)$ and $\mathcal{L}_3(j,k)$, where $0 \leq i,j \leq N$. $\mathcal{L}_h(i,l)$ is the group of states in the level i, for which l customers are in the pool for $h = 1,2$.

$$
\mathcal{A}_{ii} = \begin{array}{c|cccccccc}
 & \mathcal{L}_1(0) & \mathcal{L}_2(i,0) & \cdots & \mathcal{L}_2(i,N) & \mathcal{L}_3(i,0) & \mathcal{L}_3(i,1) & \cdots & \mathcal{L}_3(i,N) \\
\hline
\mathcal{L}_1(0) & \Psi_0 & & & & & & & \\
\mathcal{L}_2(i,0) & \Phi_1 & \Psi_1 & & & \Theta_1 & & & \\
\vdots & & & \ddots & & & & \ddots & \\
\mathcal{L}_2(i,N) & & & \Phi_N & \Psi_N & & & & \Theta_N \\
\mathcal{L}_3(i,0) & \Lambda_* & & & & & & & \\
\mathcal{L}_3(i,1) & & \Lambda & & & & & & \\
\mathcal{L}_3(i,2) & & & & & \Lambda_1 & & & \\
\vdots & & & & & & \ddots & \ddots & \\
\mathcal{L}_3(i,N) & & & & & & & \Lambda_d & \Psi_b
\end{array}
$$

$\mathcal{L}_1(1)$ corresponds to the states in the first level and has no customers in the pool or in the second station. Hence the transitions $\mathcal{L}_1(1) \to \mathcal{L}_1(1)$ are due to transitions in MAP without generating arrivals.

$$\Psi_0 = D_0$$

$$\Psi_i = \begin{bmatrix} \varphi & \vartheta & & \\ & \varphi & & \\ & & \ddots & \vartheta \\ & & & \varphi \end{bmatrix}, 1 \leq i \leq N.$$

where $\varphi = iC \oplus D_0, \vartheta = i\mu I$.

$$\Theta_i = i \begin{bmatrix} (1-q)\theta_1 I \\ (1-q)\theta_2 I \\ \vdots \\ \theta_m I \\ I \end{bmatrix}, 1 \leq i \leq N.$$

$$\Phi_i = e \otimes I, 1 \leq i \leq N.$$

For $N \leq i \leq b$,

$$\Psi_i = \begin{bmatrix} \varphi* & \vartheta* & & \\ & \varphi* & & \\ & & \ddots & \vartheta* \\ & & & \varphi* \end{bmatrix}, 1 \leq i \leq N.$$

$\varphi* = (iC \oplus S) \oplus D_0, \vartheta* = i\mu I.$
$\Lambda* = e \otimes S^0 \otimes I.$

$$\Lambda = \begin{bmatrix} U_{00} & U_{01} & & U_{0w} \\ \vdots & & \ddots & \\ U_{w0} & U_{w1} & & U_{ww} \end{bmatrix}, 1 \leq i \leq N.$$

Theorem 1. *The Markov chain described above is stable if and only if*

$$\sum_{i=0}^{\infty}\sum_{i=0}^{c} \pi_i I_r \otimes D_1 < c\mu$$

where $\pi = (\pi_0, \pi_1, ..., \pi_N)$ *is the stationary vector of the generator* $\mathcal{A} = \mathcal{A}_1 + \mathcal{A}_2 + \mathcal{A}_0$.

3 Stationary Vector

With the assumption of the stability condition, the steady state probability distribution exists. Let $x = (x_0, x_1, x_2, ...)$ be the steady state probability vector of the Markov Chain $\mathcal{X}(t)$. Then x is the unique solution to the system of equations $xQ = 0$ and $xe = 1$.

From $xQ = 0$ and $xe = 1$, we get the system of equations as

$$\begin{gathered} x_0\mathcal{A}_{00} + x_1\mathcal{A}_{10} = 0 \\ x_0\mathcal{A}_0 + x_1\mathcal{A}_{11} + x_2\mathcal{A}_{20} = 0 \\ \vdots \\ x_{c-1}\mathcal{A}_0 + x_c\mathcal{A}_1 + x_{c+1}\mathcal{A}_2 = 0 \\ x_i\mathcal{A}_0 + x_{i+1}\mathcal{A}_1 + x_{i+2}\mathcal{A}_2 = 0, i \geq c \end{gathered}$$

Now from Matrix analytic methods, $x_{c+i} = x_c R^i$, $i = 0, 1, 2 \ldots$, where R is the minimal nonnegative solution the matrix quadratic equation $R^2\mathcal{A}_2 + R\mathcal{A}_1 + \mathcal{A}_0 = 0$. R is computed algorithmically, using the logarithmic reduction algorithm [11].

For $i = 1, 2, \ldots, c$, $x_i = x_{i-1}\mathcal{A}_0(\mathcal{A}_{i1} + H_{i+1}\mathcal{A}_{i+10})$. Where $H_c = -\mathcal{A}_0(\mathcal{A}_1 + R\mathcal{A}_2)^{-1}$, for $i = 1, 2, \ldots, c-1$, $H_i = -\mathcal{A}_0(\mathcal{A}_{i1} + H_{i+1}\mathcal{A}_{i+10})$. Finally we reach at $x_0(\mathcal{A}_{00} + H_1\mathcal{A}_{10}) = 0$. Hence x_0 is obtained as the steady state distribution of a Markov chain on a finite state space, having the infinitesimal generator $\mathcal{A}_{00} + H_1\mathcal{A}_{10}$. x is calculated by dividing each x_i with the normalizing constant $\sum_{i=0}^{\infty} x_i e$.

4 Performance Characterestics

- Probability that the system is empty.

$$P_{em} = x_{00}e$$

- Probability that pool is full.

$$P_{pf} = \sum_{i=0}^{\infty} x_{iN}e$$

- Probability that there are k customers in pool and l customers are being served.

$$P_{kl} = \sum_{i=0}^{\infty} x_{kl}e$$

- Expected number of customers in the queue.

$$E_q = \sum_{i=0}^{\infty} ix_i e$$

- Expected number of busy servers in the first stage.

$$E_s^1 = \sum_{i=0}^{K} ix_i e + K\sum_{i=0}^{\infty} x_i e.$$

5 Departure Process of Customers from Pool

We define $\{\xi_m, m \geq 0\}$ as the departure process of the customers from the pool. ξ_m is a time at which a customer leaves the pool due to the realization of the Erlang clock or due to self generation of status and $\xi_0 = 0$. To analyse the process ξ_m, we consider the inter departure times $\xi_m^* = \xi_m - \xi_{m-1}$. The positive recurrence of $\mathcal{X}$ ensures that the random variables $\xi_1, \xi_2, ...$ are independent and identically distributed. Hence if is enough to find the distributions of ξ_0 and ξ_1.

Suppose that the distribution of ξ_1^* is F given by

$$F(t) = P(\xi_1^* \leq t)$$

Let the Laplace Steilges transform of ξ_1^* be

$$\Gamma(s) = E(e^{-s\xi_1^*}), Re(s) \geq 0.$$

Depending on the state of the process at ξ_0, the distribution F may be written as

$$F(t) = \sum_{i=0}^{\infty} x(i) F_i(t)$$

$$\Gamma(s) = \sum_{i=0}^{\infty} x(i) \Gamma_i(s)$$

where F_i is the conditional distribution of ξ_1 and $\Gamma_i(s)$ is the conditional Laplace Steilges transform conditioning on the fact that the process is at the level i at the epoch ξ_0.

Since the process $\mathcal{X}$ is level independent, the distributions $F_0, F_1, F_2 ...$ and $\Gamma_0, \Gamma_1, \Gamma_2, ...$ are identical. Clearly the departure process is independent of the arrivals at first station. Hence any time $\xi_i, i \geq 1$ can be thought of as the time until absorption takes place in the continuous time Markov chain $\mathcal{Z}^*$on the finite state space $\mathcal{S}^* = \{\#\} \cup \{(i, j)\}$. The corresponding infinitesimal generator takes the form

$$\mathcal{Q}^{\#} = \begin{bmatrix} \mathbf{0} & \mathbf{0} \\ \mathcal{N} & \mathcal{M} \end{bmatrix}.$$

where

$$\mathcal{M} = diag(M_1, M_2, ..., M_m), \mathcal{N} = [z, n]^T$$

The transitions in Z^* are either due to the realization of Erlang clock or due to the self generation of priority in the pool. The arrivals in $\mathcal{Z}$ has no effect but a sojourn in the current state of $\mathcal{Z}^*$. So Hence each ξ_i gas a phase type representation $(\alpha^*, \mathcal{H}m)$ The kth entry of F_i denoted by F_{ik} may be obtained as.

$$F_{ik}(t) = 1 - \alpha^*(e^{\mathcal{M}t})e$$

so

$$F_i(t) = e - (\mathbf{exp(\mathcal{M}t)})\mathbf{e}$$

Also the Laplace transform Γ_i is given by

$$\Gamma_i(s) = \alpha^*(sI - \mathcal{M})^{-1}\mathcal{N}e.$$

6 Conclusion

This paper proposes an effective Mathematical model for a disease outbreak management system in terms of queueing theoretic tools. The system is equipped with a quarantine pool and two random clocks giving the durations for status generation and quarantine time. Stability condition is derived and stationary distribution is obtained. Performance characteristics are also given. The departure process of customers from the pool is also studied.

Acknowledgment. Sinu Lal T S thanks University Grants Commission (UGC) of India for UGC-Junior Research Fellowship (Roll no.-432693, June 2017).

References

1. Trapman, P., Bootsma, M.C.J.: A useful relationship between epidemiology and queueing theory: the distribution of the number of infectives at the moment of the first detection. Math. Biosci. **219**(1), 15–22 (2009)
2. Lopez-Garcia, M.: Stochastic descriptors in an SIR epidemic model for heterogeneous individuals in small networks. Math. Biosci. **271**, 42–61 (2016)
3. Assab, R., et al.: Mathematical models of infection transmission in healthcare settings: recent advances from the use of network structured data. Curr. Opin. Infect. Dis. **30**(4), 410–418 (2017)
4. Lopez Garcia, M., Aruru, M., Pyne, S.: Health analytics and disease modeling for better understanding of healthcare-associated infections. BLDE Univ. J. Health Sci. **3**(2), 69–74 (2018)
5. Gómez-Corral, A., López-García, M.: On SIR epidemic models with generally distributed infectious periods: number of secondary cases and probability of infection. Int. J. Biomathematics **10**(02), 1750024 (2017)
6. Amador, J., Gómez-Corral, A.: A stochastic epidemic model with two quarantine states and limited carrying capacity for quarantine. Phys. A Stat. Mech. Appl. **544**, 121899 (2020)
7. Krishnamoorthy, A., Narayanan, V.C., Chakravarthy, S.R.: The impact of priority generations in a multi-priority queueing system-a simulation approach. In: Proceedings of the 2009 Winter Simulation Conference (WSC), pp. 1622–1633. IEEE (2009)
8. Gómez-Corral, A., Krishnamoorthy, A., Narayanan, V.C.: The impact of self-generation of priorities on multi-server queues with finite capacity. Stoch. Models **21**(2–3), 427–447 (2005)
9. Sinu Lal, T.S., Krishnamoorthy, A., Joshua, V.C.: A queueing inventory system with search and match - an organ transplantation model. In: Vishnevskiy, V.M., Samouylov, K.E., Kozyrev, D.V. (eds.) DCCN 2019. CCIS, vol. 1141, pp. 273–287. Springer, Cham (2019). https://doi.org/10.1007/978-3-030-36625-4_22
10. Sinu Lal, T.S., Krishnamoorthy, A., Joshua, V.C.: A multiserver tandem queue with a specialist server operating with a vacation strategy. Matematicheskaya Teoriya Igr i Ee Prilozheniya **11**(3), 31–52 (2019)
11. Latouche, G., Ramaswami, V.: Introduction to matrix analytic methods in stochastic modeling, vol. 5. SIAM (1999)
12. Neuts, M.F.: Matrix-Geometric Solutions in Stochastic Models: An Algorithmic Approach. Courier Corporation (1994)
13. Neuts, M.F.: Probability distributions of phase type. Liber Amicorum Prof. Emeritus H, Florin (1975)
14. Neuts, M.F.: A versatile Markovian point process. J. Appl. Probab. **16**(4), 764–779 (1979)

Stochastic Optimization of Local Purchase Quantities in a Geo/Geo/1 Production Inventory System

K.P. Jose[1(✉)] and M.P. Anilkumar[2]

[1] Department of Mathematics, St. Peter's College, Kolenchery 682 311, Kerala, India
kpjspc@gmail.com

[2] Department of Mathematics, T.M. Govt. College, Tirur 676 502, Kerala, India
anilkumarmp77@gmail.com

Abstract. We consider an (s, S) production inventory system in which demand occurs according to a Bernoulli process and service time follows a geometric distribution. The maximum inventory that can be accommodated in the system is S. When the on-hand inventory is reduced to a preassigned level of s due to service completion (and consequent purchase of exactly one item by each customer), production is started. The production time for each item (inter-production time) follows a geometric distribution. When the inventory level becomes zero, an instantaneous local purchase of one/s/S units is made to meet the demand. These three types of local purchases are discussed as three separate models. Using the closed-form solution obtained for the steady-state probability vector and by constructing an appropriate cost function, we compare these models with the help of a few numerical work.

Keywords: Discrete-time production inventory · Bernoulli process · Geometric distribution · Matrix-Analytic Method

1 Introduction

Queueing inventory with positive service time was introduced by Sigman and Simchi-Levi [23] in which demand arrives under a Poisson process and service time follows an arbitrary distribution. Melikov and Molchanov [16] also considered positive service time on calculating optimal stock reorder policies for transportation/storage systems in which exact as well as approximate solution methods are analysed. Later, queueing systems with inventory have been studied by several researchers. Krishnamoorthy and Jose [7] compared three retrial queueing inventory systems and analysed them using Matrix-Analytic Method. A study on developments in queueing inventory was done by Krishnamoorthy et al. [8]. Recent developments in queueing inventory with positive service time are mentioned in Krishnamoorthy et al. [10].

Berman et al. [4] introduced the processing time of inventory with deterministic service. This resulted in the situation for the analysis of queueing inventory

© Springer Nature Switzerland AG 2020
V. M. Vishnevskiy et al. (Eds.): DCCN 2020, CCIS 1337, pp. 206–220, 2020.
https://doi.org/10.1007/978-3-030-66242-4_17

system in which demands occur when inventory is stock out. In the production inventory system with positive service time and no restriction for customers to enter in the system when the inventory level zero is analysed by Krishnamoorthy and Narayanan [9]. The authors assumed that the server goes on vacation when either there is no inventory or no customers present.

The First reported work on product form solution in queueing inventory is of Schwarz et al. [21] in which the authors assumed different policies. The product form solution to this model is obtained only with the assumption that no customers are allowed to enter the system when the inventory level is zero. Later, there are few papers with product form solutions in queueing inventory system with positive service time and zero lead time. These works are mentioned in the survey paper by Krishnamoorthy et al. [8]. Schwarz and Daduna [20] developed approximation for performance measures in the M/M/1 queueing system in which the issue of inventory is considered as service with the assumption that customers can join in the system even when the inventory level is zero. To overcome the loss of customers, due to the lack of inventory in the product form solutions, Schwarz et al. [22] considered a queueing inventory model in which product form solution is obtained with the assumption that the demand that occurs during the stock out period is re-routed to other service stations. Instantaneous replenishment during the stock out the period with high replenishment cost is considered by Saffari and Haji [19] to obtain the product form solution. M/M/1 queueing inventory system under (r, Q) policy analyzed by Saffari et al. [18] in which, demand during the stock out period was assumed to be lost. An explicit expression for long-run performance measures was obtained and carried out cost optimization. The investigation of stochastic decomposition of production (s, S) inventory system in continuous time received much attention from researchers since Krishnamoorthy and Viswanath [12]. To get the explicit expression for the steady-state distribution, the authors restricted the entry of customers according to the inventory level. They obtained an explicit expression for the production cycle and optimized the cost function associated with the model with respect to maximum storage S. Deepthi [6] extended this model to discrete-time. Krishnamoorthy et al. [11] optimized N in (s, Q) inventory system in continuous time so that local purchase of $N + Q$ items is done when the on-hand inventory level is $s - N$. Recently, Krishnamoorthy et al. [13] also considered a continuous-time (s, S) production inventory system with positive service time and obtained stochastic decomposition of the system by introducing one unit of local purchase during the stock out period. This paper analyses the quantity of instantaneous replenishment during the stock out period. All the mentioned works on product form solutions are studied on the continuous-time setup.

Notable work on the discrete-time queue is done by Meisling [15]. Dafermos and Neuts [5] approximated a continuous-time model by a discrete-time single server queueing model as a limiting case. Lian et al. [14] introduced inventory in a discrete-time inventory system having a common life. Recently, Anilkumar and Jose [3] Analysed a discrete-time production (s, S) inventory system with the interruption during service. We use the discrete version of the Matrix-Analytic Method (MAM), explained in Alfa [1,2], to analyse the model. For elementary details of MAM, one can refer to Neuts [17]. The present paper is an attempt to avoid the loss of customers in the paper by Krishnamoorthy and Viswanath [12]. This is achieved by introducing local purchases in a discrete-time setup and obtained a closed-form solution. During the stock out period, one unit, s units or S unit is locally purchased with the high cost and these are studied in three different models. In the first two cases, the replenishment order is not canceled whereas in the third case it is canceled. These three models are compared based on a suitable cost function and determined the best model.

The rest of the paper is organized as follows. Sections 2, 3 and 4 provide mathematical modeling analysis, Stability condition is derived in section Steady-state probability vector and it's an explicit expression of models I, II and III. The probability of the number of times the local purchase is done during a specified time is discussed in Sect. 5. The relevant performance measures and their explicit forms are included in Sect. 6. Finally, Sect. 7 contains numerical experiments.

2 Model 1

In model I, we consider a single server (s, S) production inventory system in which the arrival of customers at each slot could only be singly with probability $p > 0$ (Bernoulli arrival process) so that the inter-arrival time follows geometric distribution with parameter p. The service time duration for each customer follows geometric distribution with parameter $q > 0$. Each customer receives one inventory after completing the service. When the inventory level depletes s due to demands, production starts. The production time of the individual item in the inventory follows a geometric distribution with parameter r. The production is stopped when the inventory is reached to the maximum level of S. We assume that arrival and service completion occurs at the beginning of the slot boundary and production of the individual item takes place at the end of the slot boundary. In any epoch, if the inventory level becomes zero due to service and production lag, an instantaneous local purchase of one unit is made with high purchasing cost.

Notations

$N(n)$: Number of customers in queue at an epoch n.

$I(n)$: Inventory level at the epoch n.

$$C(n) : \text{The production status, which is} \begin{cases} 0, \text{when production is off} \\ 1, \text{when the production is on} \end{cases}$$

$\bar{x} : 1 - x$, for $0 \leq x \leq 1$.

Then $\{(N(n), I(n)), c(n); n = 0, 1, 2, 3, ..\}$ is a Quasi Birth Death process with state space

$$\{(i, j); 1 \leq j \leq s\} \cup \{(i, j, k); s + 1 \leq j \leq S - 1, k = 0, 1\} \cup \{(i, S)\}, \text{for } i \geq 0$$

We consider order the state space as the dictionary order. Now, the transition probability matrix of the process is given by,

$$P_1 = \begin{bmatrix} D_1 & D_0 & & & \\ A_2 & A_1 & A_0 & & \\ & A_2 & A_1 & A_0 & \\ & & \ddots & \ddots & \ddots \end{bmatrix},$$

where, the blocks D_0, D_1, A_0, A_1 and A_2 square matrix of order $2S - s - 1$ and are given by

$$D_0 = \begin{bmatrix} p\bar{r} & pr & & & & & \\ & \ddots & \ddots & & & & \\ & & p\bar{r} & D_0^{0t} & & & \\ & & & D_0^1 & D_0^0 & & \\ & & & & \ddots & \ddots & \\ & & & & & D_0^1 & D_0^{0b} \\ & & & & & & p \end{bmatrix} \text{ with } D_0^{0t} = \begin{bmatrix} 0 & pr \end{bmatrix}, \quad D_0^1 = \begin{bmatrix} p & 0 \\ 0 & p\bar{r} \end{bmatrix}, \quad D_0^0 = \begin{bmatrix} 0 & 0 \\ 0 & pr \end{bmatrix}, D_0^{0b} = \begin{bmatrix} 0 \\ pr \end{bmatrix}$$

$$D_1 = \frac{\bar{p}}{p} D_0, \; A_0 = \bar{q} D_0$$

$$A_1 = \begin{bmatrix} t' & \bar{p}\bar{q}r & & & & & & \\ pq\bar{r} & t & \bar{p}\bar{q}r & & & & & \\ & \ddots & \ddots & \ddots & & & & \\ & & pq\bar{r} & t & A_1^{0t} & & & \\ & & & A_1^{2t} & A_1^1 & A_1^0 & & \\ & & & & A_1^2 & A_1^1 & A_1^0 & \\ & & & & & \ddots & \ddots & \ddots \\ & & & & & A_1^2 & A_1^1 & A_1^{0b} \\ & & & & & & A_1^{2b} & \bar{p}\bar{q} \end{bmatrix}$$

with $t' = pq + \bar{p}\bar{q}\bar{r}$, $t = \bar{p}\bar{q}\bar{r} + pqr$

$$A_1^{0b} = \begin{bmatrix} 0 \\ \bar{p}\bar{q}r \end{bmatrix}, \; A_1^0 = \begin{bmatrix} 0 & 0 \\ 0 & \bar{p}\bar{q}r \end{bmatrix}$$

$$A_1^2 = \begin{bmatrix} pq & 0 \\ 0 & pq\bar{r} \end{bmatrix}, \; A_1^{2b} = \begin{bmatrix} pq & 0 \end{bmatrix}$$

$$A_1^{0t} = \begin{bmatrix} 0 & \bar{p}\bar{q}r \end{bmatrix}$$

$$A_1^1 = \begin{bmatrix} \bar{p}\bar{q} & 0 \\ 0 & \bar{p}\bar{q}\bar{r} + pqr \end{bmatrix}$$

$$A_2 = \begin{bmatrix} \bar{p}q & & & & & & \\ \bar{p}q\bar{r} & \bar{p}qr & & & & & \\ & \ddots & \ddots & & & & \\ & & \bar{p}q\bar{r} & \bar{p}qr & & & \\ & & & A_2^{2t} & A_2^1 & & \\ & & & & A_2^2 & A_2^1 & \\ & & & & & A_2^2 & A_2^1 \\ & & & & & & A_2^{2b} \; 0 \end{bmatrix} \text{ with } A_2^{2t} = \begin{bmatrix} \bar{p}q \\ 0 \end{bmatrix}, \quad A_2^1 = \begin{bmatrix} 0 & 0 \\ 0 & \bar{p}qr \end{bmatrix}, \quad A_2^2 = \begin{bmatrix} \bar{p}q & 0 \\ 0 & 0 \end{bmatrix}, \quad A_2^{2b} = \begin{bmatrix} \bar{p}q & 0 \end{bmatrix}$$

2.1 Stability Condition

Theorem 1. *The above system is stable, if and only if* $p < q$.

Proof. Consider the matrix $A = A_0 + A_1 + A_2$. Then, we have

$$A = \begin{bmatrix} \bar{q}\bar{r}+q & \bar{q}r & & & & & & & \\ q\bar{r} & u^* & q\bar{r} & & & & & & \\ & \ddots & \ddots & \ddots & & & & & \\ & & q\bar{r} & u^* & A^{0t} & & & & \\ & & & A^{2t} & A^1 & A^0 & & & \\ & & & & A^2 & A^1 & A^0 & & \\ & & & & & \ddots & \ddots & \ddots & \\ & & & & & & A^2 & A^1 & A^{0b} \\ & & & & & & & A^{2b} & \bar{q} \end{bmatrix}$$

with $u^* = \bar{q}\bar{r} + qr$,
$A^{0b} = \begin{bmatrix} 0 \\ \bar{q}r \end{bmatrix}$, $A^0 = \begin{bmatrix} 0 & 0 \\ 0 & \bar{q}r \end{bmatrix}$,
$A^2 = \begin{bmatrix} q & 0 \\ 0 & q\bar{r} \end{bmatrix}$, $A^{2b} = \begin{bmatrix} q & 0 \end{bmatrix}$
$A^{0t} = \begin{bmatrix} 0 & \bar{q}r \end{bmatrix}$, $A^{2t} = \begin{bmatrix} q \\ q\bar{r} \end{bmatrix}$,
$A^1 = \begin{bmatrix} \bar{q} & 0 \\ 0 & u^* \end{bmatrix}$

Let $\widehat{\Psi} = (\psi_1, \ldots, \psi_s, \psi_{s+1,0}, \psi_{s+1,1}, \ldots, \psi_{S-1,0}, \psi_{S-1,1}, \psi_S)$ be the steady-state probability vector of A. Then, Ψ is obtained by solving $\Psi A = \Psi$ and $\Psi \boldsymbol{e} = 1$, which leads to

$$\psi_j = \frac{q}{(r-q)}(1 - v^{S-s})v^{s-j}\psi_S \text{ for } 1 \le j \le s,$$

$$\psi_S = \frac{(1-v)(r-q)}{q(v^S - v^s) + r(1-v)(S-s)},$$

$$\psi_{j,1} = \frac{q}{r-q}(1 - v^{S-j})\psi_S \text{ for } s+1 \le j \le S-1,$$

$$\psi_{s+1,0} = \psi_{s+2,0} = \cdots = \psi_{S-1,0} = \psi_S, \text{ where } v = \frac{q\bar{r}}{\bar{q}r}.$$

The Markov chain considered above is stable if and only if the mean left drift exceeds the mean right drift (see Neuts [17]).

That is, $\Psi A_0 \boldsymbol{e} < \Psi A_2 \boldsymbol{e}$.

On simplification, we get $p\bar{q} < \bar{p}q$, which leads to $p < q$. □

2.2 Steady-State Analysis

To find the steady-state probability vector of P, consider the production inventory system with negligible service time. Then, the corresponding Markov chain $(j(n), c(n))$ having the finite state space given by

$$\cup_{j=1}^{s}\{j\} \cup \cup_{i=s+1}^{S-1}\{(j,0),(j,1)\} \cup S$$

The transition probability matrix $\widehat{p}_1$ is given by

$$\widehat{p}_1 = \begin{bmatrix} \bar{p}\bar{r}+p & p\bar{r} & & & & & & & \\ p\bar{r} & t^* & p\bar{r} & & & & & & \\ & \ddots & \ddots & \ddots & & & & & \\ & & p\bar{r} & t^* & E^{0t} & & & & \\ & & & E^{2t} & E^1 & E^0 & & & \\ & & & & E^2 & E^1 & E^0 & & \\ & & & & & \ddots & \ddots & \ddots & \\ & & & & & & E^2 & E^1 & E^{0b} \\ & & & & & & & E^{2b} & \bar{p} \end{bmatrix} \quad \begin{array}{l} \text{with } t^* = \bar{p}\bar{r}+pr \\ E^{0b} = \begin{bmatrix} 0 \\ \bar{p}r \end{bmatrix}, E^0 = \begin{bmatrix} 0 & 0 \\ 0 & \bar{p}r \end{bmatrix} \\ E^2 = \begin{bmatrix} p & 0 \\ 0 & p\bar{r} \end{bmatrix}, E^{2b} = \begin{bmatrix} p & 0 \end{bmatrix} \\ E^{0t} = \begin{bmatrix} 0 & \bar{p}r \end{bmatrix}, E^{2t} = \begin{bmatrix} p \\ p\bar{r} \end{bmatrix} \\ E^1 = \begin{bmatrix} \bar{p} & 0 \\ 0 & \bar{p}\bar{r}+pr \end{bmatrix} \end{array}$$

Let $\widehat{\Pi}^{(1)} = (\widehat{\pi}_1^{(1)}, \ldots, \widehat{\pi}_s^{(1)}, \widehat{\pi}_{s+1,0}^{(1)}, \widehat{\pi}_{s+1,1}^{(1)}, \ldots, \widehat{\pi}_{S-1,0}^{(1)}, \widehat{\pi}_{S-1,1}^{(1)}, \pi_S^{(1)})$ be the steady-state probability vector of $\widehat{P}_1$.

Simplifying the expressions $\widehat{\Pi}^{(1)}\widehat{P}_1 = \widehat{\Pi}^{(1)}$ and $\widehat{\Pi}^{(1)}\boldsymbol{e} = 1$ leads to

$$\begin{aligned} \widehat{\pi}_j^{(1)} &= \frac{p}{(r-p)}(1-k^{S-s})k^{s-j}\widehat{\pi}_S^{(1)} \text{for } 1 \le j \le s, \\ \widehat{\pi}_S^{(1)} &= \frac{(1-k)(r-p)}{p(i-k-k^{s+1}+k^{S+1})+r(1-k)(S-s)}, \\ \widehat{\pi}_{j,1}^{(1)} &= \frac{p}{r-p}(1-k^{S-j})\widehat{\pi}_S^{(1)} \text{ for } s+1 \le j \le S-1, \\ \widehat{\pi}_{s+1,0}^{(1)} &= \widehat{\pi}_{s+2,0}^{(1)} = \cdots = \widehat{\pi}_{S-1,0}^{(1)} = \widehat{\pi}_S^{(1)}, \text{ where } k = \frac{p\bar{r}}{\bar{p}r}. \end{aligned}$$

Theorem 2. *The steady-state probability vector $\Pi^{(1)} = (\pi_0, \pi_1, \pi_2, \ldots)$ of P is given by*

$$\pi_i^{(1)} = \begin{cases} (\frac{q-p}{q})\widehat{\Pi}^{(1)} \text{ for } i = 0 \\ (\frac{q-p}{q})\frac{p}{\bar{p}q}\rho^{i-1}4\widehat{\Pi}^{(1)} \text{ for } i \ge 1 \end{cases} \tag{1}$$

where $\rho = \dfrac{p\bar{q}}{\bar{p}q}$.

Proof. From the structure of the transition probability matrix P and $\widehat{P}$, we have

$$D_1 + \frac{p}{\bar{p}q}A_2 = \widehat{P}_1$$

$$\frac{\bar{p}q}{p}D_0 + A_1 + \frac{p\bar{q}}{\bar{p}q}A_2 = \widehat{P}_1$$

$$\frac{\bar{p}q}{p\bar{q}}A_0 + A_1 + \frac{p\bar{q}}{\bar{p}q}A_2 = \widehat{P}_1$$

Now

$$\begin{aligned}
\pi_0^{(1)} D_1 + \pi_1^{(1)} A_2 &= (\frac{q-p}{q})\widehat{\Pi}^{(1)}(D_1 + \frac{p}{\bar{p}q} A_2) \\
&= (\frac{q-p}{q})\widehat{\Pi}^{(1)}\widehat{P}_1 = (\frac{q-p}{q})\widehat{\Pi}^{(1)} = \pi_0^{(1)} \\
\pi_0^{(1)} D_0 + \pi_1^{(1)} A_1 + \pi_2^{(1)} A_2 &= (\frac{q-p}{q})\frac{p}{\bar{p}q}\widehat{\Pi}^{(1)}(\frac{\bar{p}q}{p} D_0 + A_1 + \frac{p\bar{q}}{\bar{p}q} A_2) \\
&= (\frac{q-p}{q})\frac{p}{\bar{p}q}\widehat{\Pi}^{(1)}\widehat{P}_1 = \pi_1^{(1)}
\end{aligned}$$

and for $i \geq 2$,

$$\begin{aligned}
\pi_{i-1}^{(1)} A_0 + \pi_i^{(1)} A_1 + \pi_{i+1}^{(1)} A_2 &= (\frac{q-p}{q})\frac{p}{\bar{p}q}(\frac{p\bar{q}}{\bar{p}q})^{i-1}\widehat{\Pi}^{(1)}(\frac{\bar{p}q}{p\bar{q}} A_0 + A_1 + \frac{p\bar{q}}{\bar{p}q} A_2) \\
&= (\frac{q-p}{q})\frac{p}{\bar{p}q}(\frac{p\bar{q}}{\bar{p}q})^{i-1}\widehat{\Pi}^{(1)}\widehat{P}_1 = \pi_i
\end{aligned}$$

Hence we have $\Pi^{(1)} P = \Pi^{(1)}$, and on summing, we have $\Pi^{(1)} \boldsymbol{e} = 1$. □

3 Model II

In Model II, we consider all the assumptions in the previous model except that, in any epoch, if the inventory level becomes zero due to service and production lag, an instantaneous local purchase of s unit is made with high purchasing cost without canceling the production order.

Then, as before, $\{(N(n), I(n)), c(n); n = 0, 1, 2, 3, ..\}$ is a Quasi Birth Death process with state-space

$$\{(i,j); 1 \leq j \leq s\} \cup \{(i,j,k); s+1 \leq j \leq S-1, k = 0,1\} \cup \{(i,S)\}, \text{for } i \geq 0$$

We order the state-space as the dictionary order. The transition probability matrix of the process is given by,

$$P_2 = \begin{bmatrix} D_1 & D_0 & & & \\ B_2 & B_1 & A_0 & & \\ & B_2 & B_1 & A_0 & \\ & & \ddots & \ddots & \ddots \end{bmatrix},$$

The blocks B_1 and B_2 square matrix of order $2S - s - 1$ and are given by

$$[B_1]_{ij} = \begin{cases} \bar{p}\bar{q}\bar{r} + pqr & \text{for } i = 1, j = 1 \\ pq\bar{r} & \text{for } i = 1, j = s \\ [A_1]_{ij} & \text{otherwise} \end{cases}$$

$$[B_2]_{ij} = \begin{cases} \bar{p}qr & \text{for } i = 1, j = 1 \\ \bar{p}q\bar{r} & \text{for } i = 1, j = s \\ [A_2]_{ij} & \text{otherwise} \end{cases}$$

The blocks D_0, D_1 and D_2 are as in Model I. There is no change in the stability condition. That is, the system is stable, if and only if $p < q$.

3.1 Steady State Analysis

Assuming the service time is negligible, the corresponding Markov chain has state space $(j(n), c(n))$ having the finite state space given by

$$\cup_{j=1}^{s}\{j\} \cup \cup_{i=s+1}^{S-1}\{(j,0),(j,1)\} \cup S$$

Corresponding transition probability matrix $\widehat{p}_2$ is given by

$$[\widehat{p}_2]_{ij} = \begin{cases} \bar{p}\bar{r} + pr & \text{for } i = 1, j = 1 \\ p\bar{r} & \text{for } i = 1, j = s \\ [\widehat{p}_1]_{ij} \end{cases} \tag{2}$$

Let $\widehat{\Pi}^{(2)} = (\widehat{\pi}_1^{(2)}, \ldots, \widehat{\pi}_s^{(2)}, \widehat{\pi}_{s+1,0}^{(2)}, \widehat{\pi}_{s+1,1}^{(2)}, \ldots, \widehat{\pi}_{S-1,0}^{(2)}, \widehat{\pi}_{S-1,1}^{(2)}, \pi_S^{(2)})$ be the steady-state probability vector of $\widehat{P}_2$.

Simplifying the expressions $\widehat{\Pi}\widehat{P} = \widehat{\Pi}$ and $\widehat{\Pi}^{(2)}e = 1$ leads to

$$\widehat{\pi}_i^{(2)} = \frac{p(k^s - k^S)(1-k^i)}{(r-p)(1-k^s)k^s}\widehat{\pi}_S^{(2)} \text{ for } 1 \le j \le s,$$

$$\widehat{\pi}_S^{(2)} = \frac{(1-k^s)(r-p)}{r(1-k^s)(S-s) - ps(k^S - k^s)},$$

$$\widehat{\pi}_{j,1}^{(2)} = \frac{p}{r-p}(1-k^{S-j})\widehat{\pi}_S^{(2)} \text{ for } s+1 \le j \le S,$$

$$\widehat{\pi}_{s+1,0}^{(2)} = \widehat{\pi}_{s+2,0}^{(2)} = \cdots = \widehat{\pi}_{S-1,0}^{(2)} = \widehat{\pi}_S^{(2)}, \text{ where } k = \frac{p\bar{r}}{\bar{p}r}.$$

Theorem 3. *The steady-state probability vector* $\Pi^{(2)} = (\pi_0^{(2)}, \pi_1^{(2)}, \pi_2^{(2)}, \ldots)$ *of* P_2 *is given by*

$$\pi_i^{(2)} = \begin{cases} (\frac{q-p}{q})\widehat{\pi}^{(2)} \text{ for } i = 0 \\ (\frac{q-p}{q})\frac{p}{\bar{p}q}\rho^{i-1}4\widehat{\pi} \text{ for } i \ge 1 \end{cases} \tag{3}$$

where $\rho = \frac{p\bar{q}}{\bar{p}q}$.

4 Model III

In this case, we conser all the assumptions in the Model I except that, in any epoch if the inventory level becomes zero due to service and production lag, an instantaneous local purchase of S unit is made with high purchasing cost by canceling the production order.

Then, $\{(N(n), I(n)), c(n); n = 0, 1, 2, 3, ..\}$ is a Quasi Birth-Death process with state space mentioned above. Now, the transition probability matrix of the process is given by,

$$P_3 = \begin{bmatrix} D_1 & D_0 & & & \\ C_2 & C_1 & A_0 & & \\ & C_2 & C_1 & A_0 & \\ & & \ddots & \ddots & \ddots \end{bmatrix},$$

where, the blocks C_1 and C_2 square matrix of order $2S - s - 1$ and are given by

$$[C_1]_{ij} = \begin{cases} pq\bar{r} & \text{for } i = 1, j = 2S - s - 1 \\ 0 & \text{for } i = 1, j = s \\ [B_1]_{ij} & \text{otherwise} \end{cases}$$

$$[C_2]_{ij} = \begin{cases} \bar{p}q\bar{r} & \text{for } i = 1, j = 2S - s - 1 \\ 0 & \text{for } i = 1, j = s \\ [B_2]_{ij} & \text{otherwise} \end{cases}$$

As in the above case, the system is stable if and only if $p < q$.

4.1 Steady State Analysis

Assuming the service time is negligible, then the corresponding Markov chain has state space $(j(n), c(n))$ having the finite state space given by

$$\cup_{j=1}^{s}\{j\} \cup \cup_{i=s+1}^{S-1}\{(j,0),(j,1)\} \cup S$$

Corresponding transition probability matrix $\widehat{p}_3$ is given by

$$[\widehat{p}_3]_{ij} = \begin{cases} p\bar{r} & \text{for } i = 1, j = s \\ [\widehat{p}_2]_{ij} & \text{otherwise} \end{cases} \tag{4}$$

Let $\widehat{\Pi}^{(3)} = (\widehat{\pi}_1^{(3)}, \ldots, \widehat{\pi}_s^{(3)}, \widehat{\pi}_{s+1,0}^{(3)}, \widehat{\pi}_{s+1,1}^{(3)}, \ldots, \widehat{\pi}_{S-1,0}^{(3)}, \widehat{\pi}_{S-1,1}^{(3)}, \pi_S^{(3)})$ be the steady-state probability vector of $\widehat{P}_3$.

Simplifying the expressions $\widehat{\Pi}^{(3)}\widehat{P}_3 = \widehat{\Pi}^{(3)}$ and $\widehat{\Pi}^{(3)}\boldsymbol{e} = 1$ leads to

$$\widehat{\pi}_i^{(3)} = \frac{pk^{(s-i)}(1-k^{(S-s)})(1-k^i)}{(r-p)(1-k^S)}\widehat{\pi}_S^{(3)} \text{ for } 1 \le j \le s,$$

$$\widehat{\pi}_S^{(3)} = \frac{(1-k^S)(r-p)}{r(1-k^s)(S-s) - ps(k^s - k^S)},$$

$$\widehat{\pi}_{j,1}^{(3)} = \frac{p(1-k^s)}{(r-p)(1-k^S)}(1-k^{S-j})\widehat{\pi}_S^{(2)} \text{ for } s+1 \le j \le S-1,$$

$$\widehat{\pi}_{s+1,0}^{(3)} = \widehat{\pi}_{s+2,0}^{(2)} = \cdots = \widehat{\pi}_{S-1,0}^{(2)} = \widehat{\pi}_S^{(2)}, \text{ where } k = \frac{p\bar{r}}{\bar{p}r}$$

Theorem 4. *The steady-state probability vector* $\Pi^{(3)} = (\pi_0^{(3)}, \pi_1^{(3)}, \pi_2^{(3)}, \ldots)$ *of* P_3 *is given by*

$$\pi_i^{(3)} = \begin{cases} (\frac{q-p}{q})\widehat{\pi}^{(3)} \text{ for } i = 0 \\ (\frac{q-p}{q})\frac{p}{\bar{p}q}\rho^{i-1}\widehat{\pi}^{(3)} \text{ for } i \geq 1 \end{cases} \tag{5}$$

where $\rho = \frac{p\bar{q}}{\bar{p}q}$.

5 Performance Measures

Let $\Pi^{(k)} = (\pi_0^{(k)}, \pi_1^{(k)}, \pi_2^{(k)}, \ldots)$ for the model $k = 1, 2, 3$. Then the corresponding important system performance measures considered are give below.

i) Expected queue length, $EQ = \sum_{i=0}^{\infty}\sum_{j=1}^{S} i\pi_{ij}^{(k)} = \frac{p\bar{p}}{(q-p)}$.
ii) Expected inventory level, $EIL = \sum_{i=0}^{\infty}\sum_{j=1}^{S} j\pi_{ij}^{(k)}$.
iii) Expected rate of production, EPR, is $EPR = r\sum_{j=1}^{s}\widehat{\pi}_i^{(k)} + r\sum_{i=s+1}^{S}\widehat{\pi}_{i,1}^{(k)} = (1-(S-s)r\widehat{\pi}_S^{(k)})$.
iv) Expected local purchase rate, $ELP = q(1-r)\sum_{i=1}^{\infty}\pi_i^{(k)} = \widehat{\pi}_1^{(k)}p(1-r)$.
v) Expected production switching on rate, $E_{ON} = q\sum_{i=1}^{\infty}(\frac{q-p}{q})\frac{p}{\bar{p}q}\rho^{i-1}\widehat{\pi}_{s+1,0}^{(k)}$ $= p\widehat{\pi}_S^{(k)}$.

6 Distribution of Number of Local Purchase in Specified Time

In order to calculate the probability duration of the number of customers being served in given time duration, first we truncate the size of the queue. For this, choose $\epsilon > 0$ and N large enough so that

$$\sum_{i=N+1}^{\infty} \rho^{i-1} < \frac{\epsilon}{(\frac{q-p}{q})\frac{p}{\bar{p}q}}$$

On simplification, it reduces to

$$\rho^N < \frac{\epsilon}{q}$$

Let $L(n)$ denote the number of local purchases during the time $[o, n]$. N(n), I(n) and c(n) respectively denote the number of customers, inventory level and server-status at an epoch n. Consider the Markov chain $\{(L(n), N(n), I(n), c(n)); n \geq 0\}$ with state space $\{\Delta\} \cup \{0, 1, 2 \ldots\} \times \{0, 1 \ldots, N\} \times \{1, 2, \ldots, s, (s+1, 0), (s+1, 1), \ldots, (S-1, 0), (S-1, 1), S\}$, where Δ represents the absorbing state on the realization of the random clock which is geometrically distributed with parameter

δ and N is the truncation level mentioned above. The transition probability matrix of the process P_L, is given by

$$P_L = \begin{bmatrix} 1 & \mathbf{0} \\ \boldsymbol{\delta} & \boldsymbol{U} \end{bmatrix}$$

in which $\boldsymbol{U} = \begin{bmatrix} N_1 & N_0 & & \\ & N_1 & N_0 & \\ & & \ddots & \ddots \end{bmatrix}$, and $\boldsymbol{\delta} = \begin{bmatrix} \delta e \\ \delta e \\ \vdots \end{bmatrix}$.

where the entry matrices are

$$N_1 = \bar{\delta} \begin{bmatrix} B_1 & B_0 & & \\ N_1^2 & N_1^1 & A_0 & \\ & \ddots & \ddots & \ddots \end{bmatrix}, \quad N_0 = \bar{\delta} \begin{bmatrix} \mathbf{0} & \mathbf{0} & \\ N_0^1 & N_0^0 & \\ & \ddots & \ddots \end{bmatrix}$$

$$[N_1^1]_{ij} = \begin{cases} \bar{p}\bar{q}\bar{r} & \text{for } i = 1, j = 1 \\ [A_1]_{ij} & \text{otherwise} \end{cases} \qquad [N_1^2]_{ij} = \begin{cases} 0 & \text{for } i = 1, j = 1 \\ [A_2]_{ij} & \text{otherwise} \end{cases}$$

N_0^1 and N_0^2 depends on the number of items locally purchased.

For Model I,

$$[N_0^2]_{ij} = \begin{cases} \bar{p}q\bar{r} & \text{for } i = 1, j = 1 \\ 0 & \text{otherwise} \end{cases}, \quad [N_0^1]_{ij} = \begin{cases} pq\bar{r} & \text{for } i = 1, j = 1 \\ 0 & \text{otherwise} \end{cases}$$

For Model II,

$$[N_0^2]_{ij} = \begin{cases} \bar{p}q\bar{r} & \text{for } i = 1, j = s \\ 0 & \text{otherwise} \end{cases}, \quad [N_0^1]_{ij} = \begin{cases} pq\bar{r} & \text{for } i = 1, j = s \\ 0 & \text{otherwise} \end{cases}$$

For Model III,

$$[N_0^2]_{ij} = \begin{cases} \bar{p}q\bar{r} & \text{for } i = 1, j = S \\ 0 & \text{otherwise} \end{cases}, \quad [N_0^1]_{ij} = \begin{cases} pq\bar{r} & \text{for } i = 1, j = S \\ 0 & \text{otherwise} \end{cases}$$

Let x_k denote the probability that k local purchase is done before the realization of the random clock. In order to calculate x_k, we consider the top left sub matrix $\boldsymbol{U}^*$of $\boldsymbol{U}$ having order $(k+1)(2S-s)$.

Then the probability that absorption will take place at k^{th} level is $\boldsymbol{\beta} N' \delta \boldsymbol{e}$, where N' is the $(k+1)^{th}$ block of the first row of $(\boldsymbol{I} - \boldsymbol{U}^*)^{-1}$ and $\boldsymbol{\beta}$ is the initial probability vector. Hence,

$$x_0 = -\delta \boldsymbol{\beta} (I - N_1)^{-1} \boldsymbol{e},$$
$$x_k = (-1)^{k-1} \delta \boldsymbol{\beta} ((I - N_1)^{-1} N_0)^k (I - N_1)^{-1} \boldsymbol{e} \text{ for } k > 0.$$

The initial probability vector $\boldsymbol{\beta} = (\beta_0, \beta_1, \beta_2, \ldots, \beta_N)$, where β_i's are given by $\beta_0 = \dfrac{1 - \frac{p}{q}}{1 - \frac{p}{q}\rho^N} \widehat{\Pi}$ and for $i \geq 1$, $\beta_i = \dfrac{(1 - \frac{p}{q}) \frac{p}{\bar{p}q} \rho^{i-1}}{1 - \frac{p}{q}\rho^N} \widehat{\Pi}$, where $\widehat{\Pi} = \widehat{\Pi}^{(i)}$ for $i = 1, 2, 3$ depending on the model.

7 Numerical Experiments

7.1 Cost Function

We define a suitable cost function on the basis of system performance measures. For this, we define individual cost c_0, c_1, c_2, c_3 and c_4 as

c_0 : switching cost for the production
c_1 : production per unit inventory per unit time
c_2 : holding of inventory per unit per unit time
c_3 : cost due to local purchase unit items
c_4 :cost per local purchase
c_5 :cancellation cost of a production process due to local purchase ($c_5 = 0$ for model I and II)

Define expected total cost (ETC) per unit time as

$$ETC = c_0 E_{ON} + c_1 EPR + c_2 EIL + (c_3 Q + c_4 + c_5) ELP$$

where Q is the number of items locally purchased.

7.2 Graphical Illustrations

To determine the profitable model, the comparison of three models on the basis of the expected total cost is made. For this, first, fix all parameters and individual costs associated with the model except one. Then compare the cost associated with these models from the graph.

Figure 1 illustrates the variation ETC with p corresponding to the other parameters in the figure. From this figure, when $p = 0.34$, ETC for all the three models are approximately the same. When p increases from there, ETC first decreases for these models. For $0.34 < p < 0.5$, the Model 1 is the most profitable and Model 2 is profitable than Model 3. As p increases further, Model 2 is the most profitable and model 1 is more profitable than model 2. For $p > 0.74$ the model 3 is profitable than model 1. The greater ETC for Model 3 in the interval $0.34 \leq p \leq 0.74$ is due to the cancellation cost c_5 of the production associated with that model. If we increase c_5 further, the Model 3 will be the least profitable model for all the values of p.

Figure 2 illustrates the variation of ETC with r. For lower values of r, that is for $r \leq 0.5$, model 2 is the best model and for $r > 0.5$, model 1 will be the suitable model. This indicates that when the replenishment rate is high, the minimum unit of local purchase makes the firm more profitable.

Since all performance measures considered for cost function are independent of q, variations in q will not affect ETC. Figure 3 shows that for the other parameters mentioned in the figure, model II is the most profitable one and model I is profitable than model III. From all the figures, it is clear that the Model III is less acceptable compared to the model I and model II.

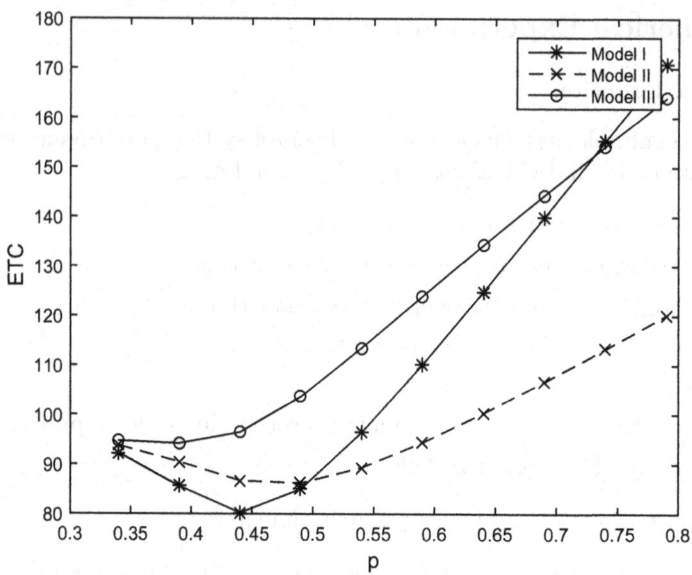

Fig. 1. ETC vs. p. $c_0 = 250; c_1 = 100; c_2 = 5; c_3 = 120; c_4 = 200 : c_5 = 600 : q = 0.9; r = 0.4; s = 8; S = 20$

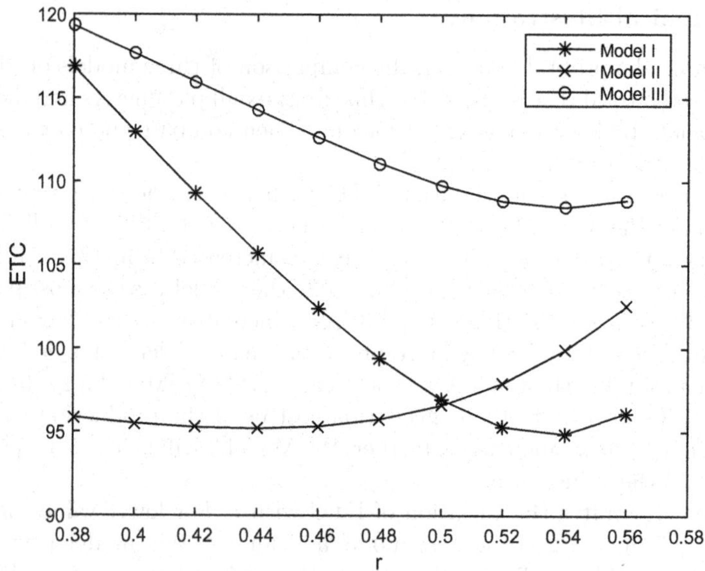

Fig. 2. ETC vs. r. $c_0 = 250; c_1 = 100; c_2 = 5; c_3 = 120; c_4 = 200 : c_5 = 100 : q = 0.8; p = 0.6; s = 8; S = 20$

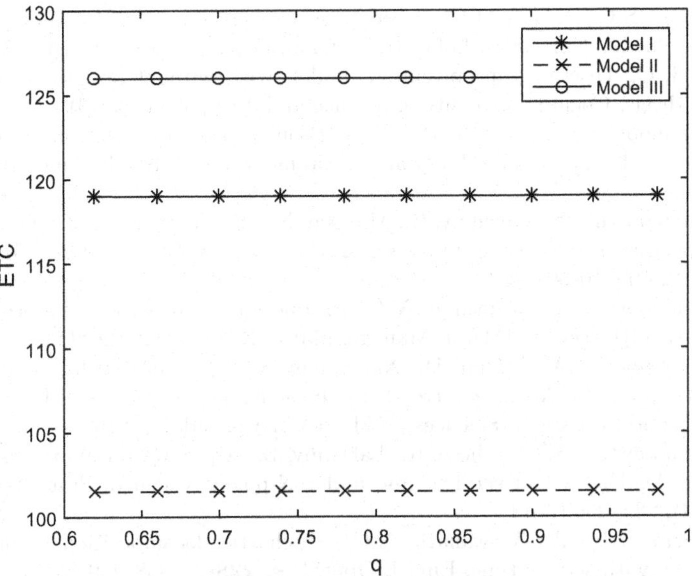

Fig. 3. ETC vs. q. $c_0 = 250; c_1 = 100; c_2 = 5; c_3 = 120; c_4 = 200 : r = 0.4; p = 0.6; s = 8; S = 20$

Concluding Remarks

This article analyzes a $Geo/Geo/1$ production inventory system with a local purchase during the stock-out period. We analyzed three models based on the number of items locally purchased. A closed-form solution is obtained for all these three models. Based on suitable cost function we compared the models by varying the parameters. The distribution of the number of times locally purchased during a specified period is also calculated. A simple extension of this model can be done by taking the variations in the local purchase quantity instead of these three cases. Further extensions are also possible by considering a discrete-time MAP for the arrival process or discrete-time phase-type distributions for service time and lead-time or both.

References

1. Alfa, A.S.: Applied Discrete-Time Queues. Springer, New York (2016). https://doi.org/10.1007/978-1-4939-3420-1
2. Alfa, A.S.: Discrete time queues and matrix-analytic methods. Top **10**(2), 147–185 (2002)
3. Anilkumar, M.P., Jose, K.P.: A $Geo/Geo/1$ inventory priority queue with self induced interruption. Int. J. Appl. Comput. Math. **6**(4), 1–14 (2020). https://doi.org/10.1007/s40819-020-00857-8
4. Berman, O., Kaplan, E.H., Shevishak, D.G.: Deterministic approximations for inventory management at service facilities. IIE Trans. **25**(5), 98–104 (1993)

5. Dafermos, S.C., Neuts, M.F.: A single server queue in discrete time. Technical report, Purdue University, Lafayette Indiana, Department of Statistics (1969)
6. Deepthi, C.: Discrete time inventory models with/without positive service time. Ph.D. thesis, Cochin University of Science and Technology (2013)
7. Krishnamoorthy, A., Jose, K.P.: Comparison of inventory systems with service, positive lead-time, loss, and retrial of customers. Int. J. Stoch. Anal. **2007**, 1–23 (2008)
8. Krishnamoorthy, A., Lakshmy, B., Manikandan, R.: A survey on inventory models with positive service time. Opsearch **48**(2), 153–169 (2011). https://doi.org/10.1007/s12597-010-0032-z
9. Krishnamoorthy, A., Narayanan, V.C.: Production inventory with service time and vacation to the server. IMA J. Manage. Math. **22**(1), 33–45 (2011)
10. Krishnamoorthy, A., Shajin, D., Narayanan, V.: Inventory with positive service time: a survey. In: Advanced Trends in Queueing Theory: Series of Books "Mathematics and Statistics", Sciences. ISTE & Wiley, London (2019)
11. Krishnamoorthy, A., Varghese, R., Lakshmy, B.: An (s; Q) inventory system with positive lead time and service time under N-policy. Calcutta Stat. Assoc. Bull. **66**(3–4), 241–260 (2014)
12. Krishnamoorthy, A., Viswanath, N.C.: Stochastic decomposition in production inventory with service time. Eur. J. Oper. Res. **228**(2), 358–366 (2013)
13. Krishnamoorthy, A., Varghese, R., Lakshmy, B.: Production inventory system with positive service time under local purchase. In: Dudin, A., Nazarov, A., Moiseev, A. (eds.) ITMM 2019. CCIS, vol. 1109, pp. 243–256. Springer, Cham (2019). https://doi.org/10.1007/978-3-030-33388-1_20
14. Lian, Z., Liu, L., Neuts, M.F.: A discrete-time model for common life time inventory systems. Math. Oper. Res. **30**(3), 718–732 (2005)
15. Meisling, T.: Discrete-time queuing theory. Oper. Res. **6**(1), 96–105 (1958)
16. Melikov, A., Molchanov, A.: Stock optimization in transportation/storage systems. Cybern. Syst. Anal. **28**(3), 484–487 (1992)
17. Neuts, M.F.: Matrix-geometric solutions in stochastic models: an algorithmic approach. Courier Corporation (1994)
18. Saffari, M., Asmussen, S., Haji, R.: The M/M/1 queue with inventory, lost sale, and general lead times. Queueing Syst. **75**(1), 65–77 (2013)
19. Saffari, M., Haji, R.: Queueing system with inventory for two-echelon supply chain. In: Proceedings of the 2009 International Conference on Computers and Industrial Engineering, pp. 835–838. IEEE (2009)
20. Schwarz, M., Daduna, H.: Queueing systems with inventory management with random lead times and with backordering. Math. Methods Oper. Res. **64**(3), 383–414 (2006)
21. Schwarz, M., Sauer, C., Daduna, H., Kulik, R., Szekli, R.: M/M/1 Queueing systems with inventory. Queueing Syst. **54**(1), 55–78 (2006)
22. Schwarz, M., Wichelhaus, C., Daduna, H.: Product form models for queueing networks with an inventory. Stoch. Models **23**(4), 627–663 (2007)
23. Sigman, K., Simchi-Levi, D.: Light traffic heuristic for an M/G/1 queue with limited inventory. Ann. Oper. Res. **40**(1), 371–380 (1992)

Mathematical Model of Horizontal VNF Scalability in virtualized Evolved Packet Core

Alexey Tsarev(✉) and Pavel Abaev

Peoples' Friendship University of Russia (RUDN University),
6 Miklukho-Maklaya Street, Moscow 117198, Russian Federation
atsarev@sci.pfu.edu.ru, abaev_po@rudn.university
http://www.rudn.ru

Abstract. Next generation networks such as 5G provide a new approach for building highly scalable network infrastructure owing to NFV/VNF and SDN technologies. Infrastructure scalability allows a network operator to offer fine-graded on-demand services deployment. We consider virtualized Evolved Packet Core which functions are represented in form of several Virtual Network Functions. Computational capacity of every function could be extended via scaling mechanism with non-instantaneous activation/deactivation. In this paper we build a mathematical model of horizontal scaling for vEPC function in terms of queuing model with finite buffer size, several group of servers and queue thresholds. As a result, analytical formulas were derived and QoS metrics were proposed. Analytical evaluation shows that correct thresholds selection allows to significantly improve system performance.

Keywords: Virtualized EPC · Evolved packet core · Queuing model · Horizontal scalability · Non-instantaneous activation/deactivation · VNF · SDN · 5G

1 Introduction

Rapid development and evolution of modern computer networks is a network operator's response to increasing data consumption [1]. Cisco analytics predicted that two-thirds of humanity will have access to the Internet by 2023, and, consequently, generate data traffic. Further, the number of mobile users will reach 5.7 billion and the number of mobile connections will reach 13.1 billion [2]. Beyond that, one should take into the account increased user mobility and network traffic heterogeneity that implies different QoS/QoE requirements.

In order to cope with the mentioned challenges, telecommunication operators need properly update network infrastructure. One of the emerging trends in next

The publication has been prepared with the support of "RUDN University Program 5-100" program.

© Springer Nature Switzerland AG 2020
V. M. Vishnevskiy et al. (Eds.): DCCN 2020, CCIS 1337, pp. 221–234, 2020.
https://doi.org/10.1007/978-3-030-66242-4_18

generation networks became 5G technology standardized by ITU. According to forecasts, more than 50 operators from 38 countries plan to launch 5G between 2018 and 2023 [3].

Since 5G includes concepts such as Software Defined Networking and Network Functions Virtualization [4], telecommunication operators could build more flexible and scalable infrastructure. The transition of Evolved Packet Core to the virtual one with help of Virtual Network Functions allows to increase the use of network resources and reduce OPEX [5]. At the same time one of key features of virtualization is possibility of resource scaling. What is more, the network load by user data usually has peaks during the day [6]. Therefore, one of the optimal strategies for mobile operators could be resource scaling in order to cope the bottlenecks.

One of approaches for modelling scale in/out process and overcoming either overloads or resource allocation issues alongside with the hysteresis technique successfully used in [7–10,17–19] could be non-instantaneous activation and deactivation of additional resources studied in this work.

The rest of the paper is organized as follows. Section 2 provides a brief overview of scalability researches for virtualized EPC and SDN. Section 3 describes a system model and approach for scale in/out process. Section 4 deals with a mathematical model in terms of queueing model with finite buffer, finite servers number and queue thresholds. Section 5 provides results of analytical evaluation. Section 6 contains conclusions.

2 Background

Every new standard of next generation networks forces the network owner to update mobile core network, especially Evolved Packet Core, due to the tightening requirements for Quality of Service and Quality of Experience. The emergence of some modern technologies, such as cloud computing, network virtualization and software-defined networking changed traditional approaches for building and updating network. In particular, network core virtualization allows to replace expensive proprietary equipment with similar not so expensive ones with an open architecture. Moreover, in such case Evolved Packet Core as well as network control plan could be moved to the cloud. This fact allows to use benefits of scalability. In the literature researchers focus on different aspects of virtual EPC scalability.

Abaev et al. study in [9] case of hybrid evolved packet core with assumption there are legacy and virtualized parts of EPC. The system has two thresholds: one for scaling in and another for scaling out. Once scaling process initiated, VNF orchestrator deploys new VNF consecutively one by one. Proposed model is analyzed in a custom simulator.

Tobar et al. develop in [10] a scaling mechanism for evolved packet core based on network functions virtualization taking into the account both horizontal and vertical scalability. Proposed framework allows to find optimal workload between horizontal and vertical scaling and avoid waste of resources. Also, extensive scalability evaluation was performed in AWS on the base of vEPC from Indian Institute of Technology Bombay.

Liebsch et al. propose in [11] extension for traditional VNF scaling. The main idea of developed approach consists in scaling VNF from vEPC Data Plane that allows to avoid failure of overloaded Data Plane functions. In order to validate the proposal, the authors perform testing in real environment. Evaluated results demonstrate efficiency of runtime offload.

Kempf et al. consider in [12] the possibility to move EPC to the cloud. The main contribution of this work consists in describing how SDN could be applied Evolved Packet Core. The authors describe enhancements for Open Flow protocol and Open Flow switches necessary for EPC virtualization. Neither analytical nor numerical evaluation carried out.

Arteaga et al. investigate in [13] adaptive scaling mechanism in application to NFV-based EPC. The authors propose adaptive scaling algorithm based on Q-Learning and Gaussian Processes. Simulation shows better accuracy then algorithms with static thresholds.

Khakimov et al. focuse in [14] on IoT topic and propose Fog based structure of the network in SDN enabled environment. The authors describe IoT-Fog based structure and introduce data offloading algorithm that consists in scaling of so-called Fog machines in case of resources allocation. Experimental evaluation of the proposed framework shows efficiency of dynamic Fog nodes deployment.

Muthanna et al. study in [15] multi-controller SDN networks. The authors consider SDN controllers hierarchy consisting of two levels: single main controller and a set of distributed child controllers. According to considered framework master controller is responsible for distributed controllers clustering. Load balancing algorithm with dynamic clustering of child controllers is proposed and evaluated. Dynamic clustering using described algorithm achieves higher efficiency then existing static ones.

Muthanna et al. provide in [16] a framework that integrates at the same time such concepts as IoT, Fog computing, SDN and blockchain. The first layer of the proposed architecture consists of IoT devices and sensors, the second layer consists of dynamically deployed Fog nodes and the last layer represents remote cloud. SDN is used for control and management of Fog nodes. In order to provide high-security level to the proposed Iot network blockchain is used for SDN and Fog nodes connection. The authors propose data offloading algorithm and analytical traffic model.

Overall, all the mentioned papers have no analytical results. Alternatively, if we ignore SDN/VNF topic we can find interesting analytical works regarding either search of optimal hysteresis strategy in multi-server system [17] and analysis of multi-server threshold systems with hysteresis in [18,19].

Kim et al. consider in [17] the problem of building optimal strategy for hysteresis control in multi-server queuing system. The authors show a complexity of dynamic resources optimization and highlight differences between hysteresis strategy and threshold strategy. As a main result the authors provide a framework for performance evaluation of MMAP/PN/N queuing system with hysteresis control. Numerical evaluation realized by the authors demonstrates efficiency of hysteresis strategy.

Lui and Golubchik consider in [18] three variants of multi-server queuing systems with thresholds. The first considered system has homogeneous servers with Poisson input flow, the other system has heterogeneous servers with Poisson input flow and the last one has homogeneous servers with bulk Poisson input flow. The authors obtain closed-form analytical solution for homogeneous case under assumption there are no restrictions on the servers number or requests waiting time. Moreover, the algorithmic solution for variations problem was proposed for the system with heterogeneous servers and bulk arrivals under the same assumptions. In order to validate derived formulas and check system performance the authors realize analytical evaluation using the systems response time as a performance indicator.

Chou et al. study in [19] multi-server system with limited resources with heterogeneous workload classes and non-instantaneous servers activation. The achieve clarify that it is important not only achieve the best performance, but also appropriate cost/performance ratio. The authors provide iterative approximation method for solving considered set of models. As a conclusion, the authors provides numerical validations in order to prove the accuracy of the proposed methods, demonstrate the impact of the servers activation rate and highlight the benefits of the resources sharing among heterogeneous classes of the workload.

In this paper we study VNF deployment of virtual EPC function that supports horizontal resources scaling and we propose analytical model for mentioned VNF deployment in form of multi-server queuing model with thresholds and non-instantaneous activation and deactivation.

3 System Model

Evolved Packet Core is an important element of any modern network. Furthermore, EPC is responsible for authentication, authorization, channel bandwidth calculation, user mobility processing, etc. The main functions of the EPC include Access and Mobility Management Function (AMF), Session Management Function (SMF), User Plane Function (UPF), Unified Data Management (UDM) and Policy Control Function (PCF). SDN support in 5G networks allows network operators to virtualize mentioned functions via software and hardware decoupling.

In the literature can be found different approaches of Evolved Packet Core virtualization. For instance, in [20] are described three main approaches of Evolved Packet Core architecture virtualization, all the others ways can be considered as a mix of the presented ones.

The first EPC architecture is based only on the virtualization with NFV concept. In such implementation all the functions of the EPC are implemented as software running in some virtual machine/cloud container. Multitenancy concept could be considered for this implementation since multiple vEPC could be deployed simultaneously. However, there are challenges at adding new VNF to vEPC since they should be correctly configured as well as deployed in a coherent way. Also, scaling could be not so efficient due to different resource requirements for data and control plane at the gateway.

The other virtual EPC implementation includes a branch of architectures based on SDN and NFV concepts. Here SDN-controller is used as a bridge between control and data planes. Since control plane is decoupled from data plane, thus they be scaled independently. This virtualization approach also is a key driver for mobile edge computing since virtualized EPC functions could be deployed over the network in the closest way to the users. Despite the benefits of mentioned approach, SDN-controller scalability issue remains one of the main challenges.

The last one approach of EPC virtualization consists in using of only SDN concept. In this case all the network elements are replaced with simple data forwarding elements and network functions of EPC are replaced with special software installed at SDN-controller. The main advantage of clear SDN EPC architecture are high flexibility and full programmability as well as flat network structure. As a drawback the following issues could be considered: control plan complexity that affects QoS and the fact that described approach is not compatible with present working EPC.

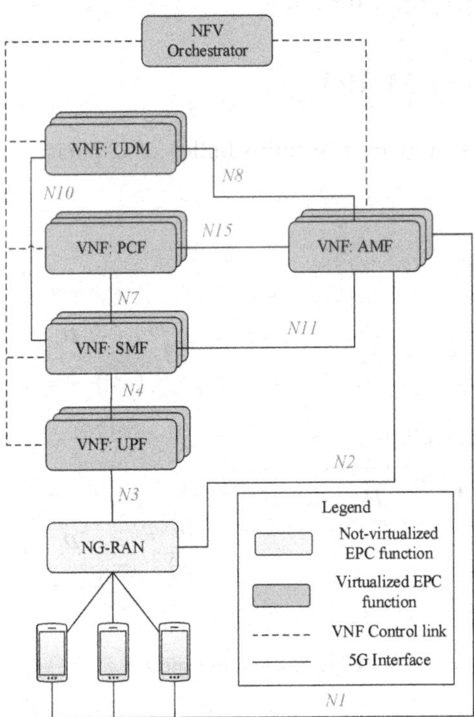

Fig. 1. Virtualized EPC architecture

In scope of this work we consider virtual EPC implemented according to the second approach using SDN and NFV due to the following reasons: a) this approach is applicable to the majority of the existing core networks and b) unlike the purelly NFV approach there is no problems with VNF scalability witin deploed virtual EPC.

EPC virtualization involves the transformation of EPC functions into Virtual Network Function (VNF), whose lifecycle is managed using VNF Manager. The decision whether to deploy a new function or disable the old one is realized by NFV Orchestrator. Thus, if a bottleneck occurs at the level of any of the EPC functions, network functions management and orchestration will allow to react to the situation promptly and provide additional computing power to the problem node. An example of a virtualized EPC architecture is shown in the Fig. 1.

We consider a VNF function of a certain type. The minimum number of available rsources is enough for the normal functioning of the network, we denote this group as c_0. When the bottleneck occurs, it is feasible to allocate additional computational resources to the VNF in order to provide required QoS level, particularly for delay critical applications. Since new resources deployment is not instant and requires some time, it is worthwhile to use batch deploy approach. We consider the same assumption for undeployment procedure. We consider reaching the threshold H_i in the queue of waiting requests as a trigger for the resources group $c_i, 1 \leq i < k$ deploy.

4 Mathematical Model

The model is represented by the finite buffer of capacity r and C servers, as shown in the Fig. 2.

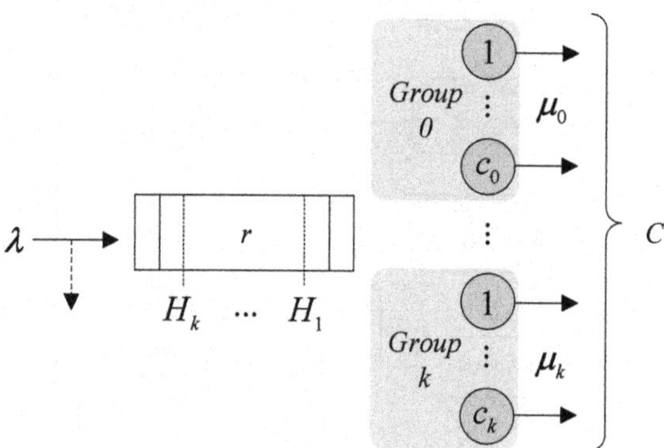

Fig. 2. Queuing model

The Poisson requests flow reaches the system with rate $\lambda, 0 < \lambda < \infty$. We consider FCFS behavior for all incoming requests. Any server accepts one request at a time. All the servers are divided into $k+1$ groups with exponentially distributed service rate $\mu_i, 0 < \mu_i < \infty, 0 \leq i \leq k+1$. The system's buffer has a set of thresholds $H = H_i, 0 < H_i < r, 1 \leq r \leq k$.

Threshold H_i correlates to server group i. Initially, only servers from group 0 handle incoming requests. As soon as the buffer size reaches the threshold H_i, all the servers from group i become available for requests handling after scale in procedure that follows exponential distribution with rate $\theta, 0 < \theta < \infty$; and vice versa in case buffer size drops below the threshold H_i, all the servers from group i become unavailable after scale out procedure that follows exponential distribution with rate $\gamma, 0 < \gamma < \infty$.

Let us use the following assumption: during activation or deactivation of a servers group the number of requests in the buffer cannot change. Another assumption is that the server, that provides service to a request, can not be changed if requests servicing is in process.

Assume $X(t) = (n(t), m(t))$ is a two-dimensional Poisson process over the set of states $\chi(t) = (n, m) : (0 \leq n \leq r, 0 \leq m \leq k)$ where n defines the number of requests in the system and m defines the number of working server groups. It is evident $X(t)$ is ergodic and its stationary distribution exists. Let us denote $p(i,j) = \lim_{t\to\infty} P\{n(t) = i, m(t) = j\}$ as stationary distribution. Transition rate graph for described system is presented in the Fig. 3.

Assume $g(m)$ as a set of states for group m taking into the account the following assumption: $H_0 = 0$ and $H_{k+1} = r$:

$$g(m) = \{(H_m, m), ..., (H_{m+1}, m)\}, 0 \leq m \leq k \tag{1}$$

Note, that:

$$\sum_{m=0}^{k} \sum_{(i,j)\in g(m)} p(i,j) = 1 \tag{2}$$

Consider for brevity $\mu(n, m)$ as serving rate for n requests in the system and m deployed server groups:

$$\begin{aligned} \mu(n,0) &= min(n, c_0) \cdot \mu_0 \\ \mu(n,m) &= \textstyle\sum_{j=0}^{m-1} c_j \cdot \mu_j + min(n - \sum_{u=0}^{m-1} c_u, c_k) \cdot \mu_k, m > 0 \end{aligned} \tag{3}$$

For case $m = 0$ stationary distribution could be expressed as:

$$p(n,0) = \frac{\lambda^n}{\prod_{i=1}^{n} \mu(i,0)} \cdot p(0,0), 1 \leq n \leq H_1 \tag{4}$$

Note, that partial balance equations exist for neighboring groups:

$$p(n,m) = \frac{\theta}{\gamma} \cdot p(n, m-1), 1 \leq m \leq k, n = H_m \tag{5}$$

Taking into the consideration (5), for case $m = 1$ we obtain:

$$p(n,1) = \frac{\theta}{\gamma} \cdot \lambda^{n-H_1} \prod_{i=1}^{n-H_1} \mu(i,1) \cdot \frac{\lambda^{H_1}}{\prod_{i=1}^{H_1} \mu(i,0)} \cdot p(0,0), H_1 \leq n < H_2 \tag{6}$$

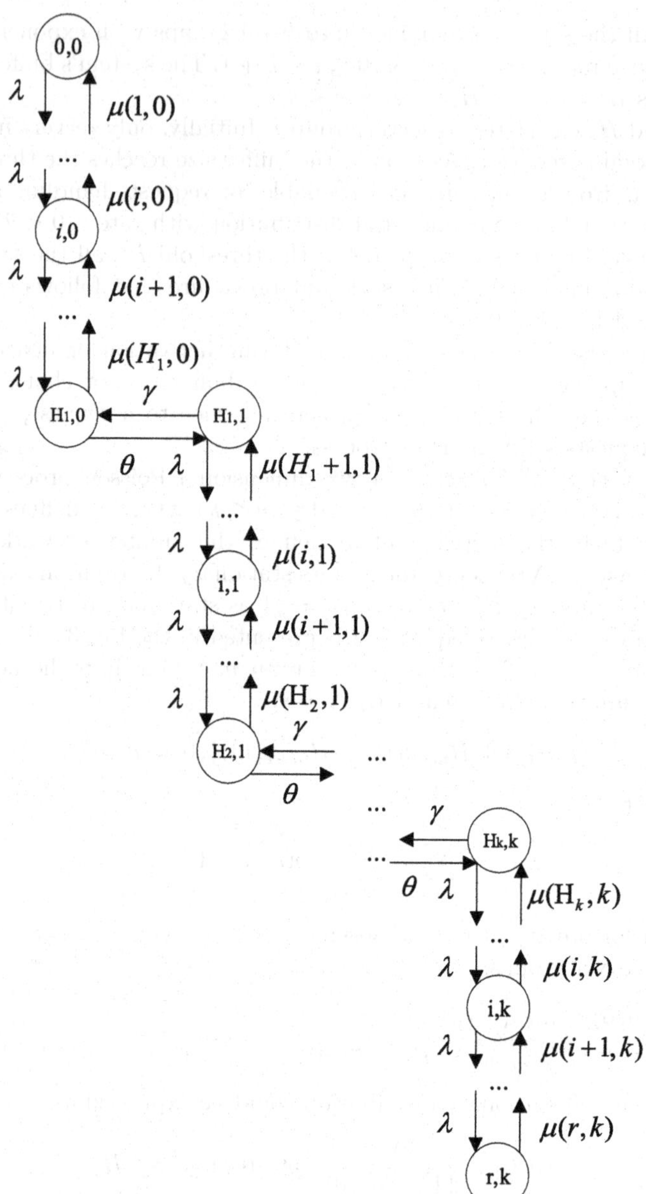

Fig. 3. Transition rate graph

Correspondingly, in general case we have:

$$p(n,m) = \lambda^n \cdot \left(\frac{\theta}{\gamma}\right)^m \cdot \prod_{v=0}^{m} \prod_{(i,j)\in g(v)} \frac{1}{\mu(i,j)} \cdot p(0,0) \tag{7}$$

Since all $p(n, m)$ are expressed through $p(0, 0)$ we have:

$$p(n, m) = q(n, m) \cdot p(0, 0) \tag{8}$$

Hence, from (2), (7) and (8) we obtain $p(0, 0)$:

$$p(0, 0) = \frac{1}{\sum_{v=0}^{m} \sum_{(n,m)\in g(v)} q(n, m)} \tag{9}$$

Thus, stationary distribution could be calculated analytically using (8) and (9). Main performance metrics could be calculated by the formulas (10), (11), (12), (13) and (14).

Let B denote the blocking probability:

$$B = p(r, k) \tag{10}$$

Let Q denote mean queue size:

$$Q = \sum_{v=0}^{k} \sum_{(n,m)\in g(v)} \left(n - min(n, \sum_{j=0}^{m} c_j) \right) \cdot p(n, m) \tag{11}$$

Let N denote mean number of requests in the system:

$$N = \sum_{v=0}^{k} \sum_{(n,m)\in g(v)} n \cdot p(n, m) \tag{12}$$

According to Little's Law, let W_Q denote mean time that a request spends in the queue and W_N denote mean time that a request spends in the system:

$$W_Q = \frac{Q}{\lambda \cdot (1 - B)} \tag{13}$$

$$W_N = \frac{N}{\lambda \cdot (1 - B)} \tag{14}$$

5 Analytical Evaluation

We provide analytical evaluation applied to four difference scenarios. Input values for evaluation are provided in Table 1. We consider that μ^{-1} equals 1 as a time unit. Server groups number equals 3 and includes zero group, hence we have only 2 thresholds.

Table 1. Considered input for evaluation

Parameter	Comment	Value
λ	Incoming requests rate	1–20
Q	Queue length	50
θ	Servers group deployment rate	0.005
γ	Servers group undeployment rate	0.00005
c_0	Group 0 servers count	10
c_1	Group 1–2 servers count	5
μ_0	Service rate in group 0–2	1
H_1	Threshold for group 1	12
H_2	Threshold for group 2	25

5.1 Varying Queuing Size

In this scenario we investigate the influence of buffer size on the system performance. Evaluation demonstrates that buffer capacity growth leads to packet

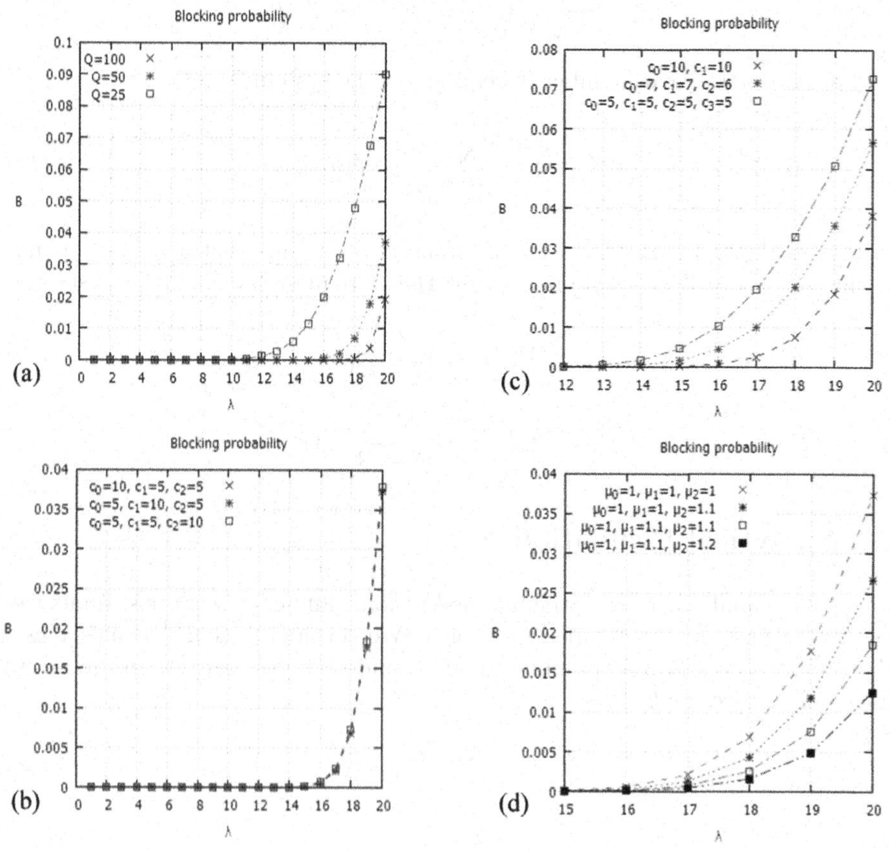

Fig. 4. Probability to block incoming request

loss reduction illustrated on the Fig. 4(a) and mean queue size rise illustrated on the Fig. 5(a). Although, there is no effect of added thresholds control on the blocking probability, anyway we can see serious benefit in terms of mean waiting time and slight impact on queue size.

5.2 Varying the Biggest Group Number

In this scenario we analyze resource allocation among fixed number of server groups. Figure 4(b) shows that there is no difference in terms of blocking probability for different group setups. However, it is worth noting that allocating more resources to the last servers group improves QoS: mean queue size drastically falls on the Fig. 5(b) as well as mean waiting time on the Fig. 6(b).

5.3 Varying Groups Count

In this scenario we study optimal number of groups for case of equal resource allocation. We consider 2 groups of 10 servers, 3 groups of 6 and 7 servers and,

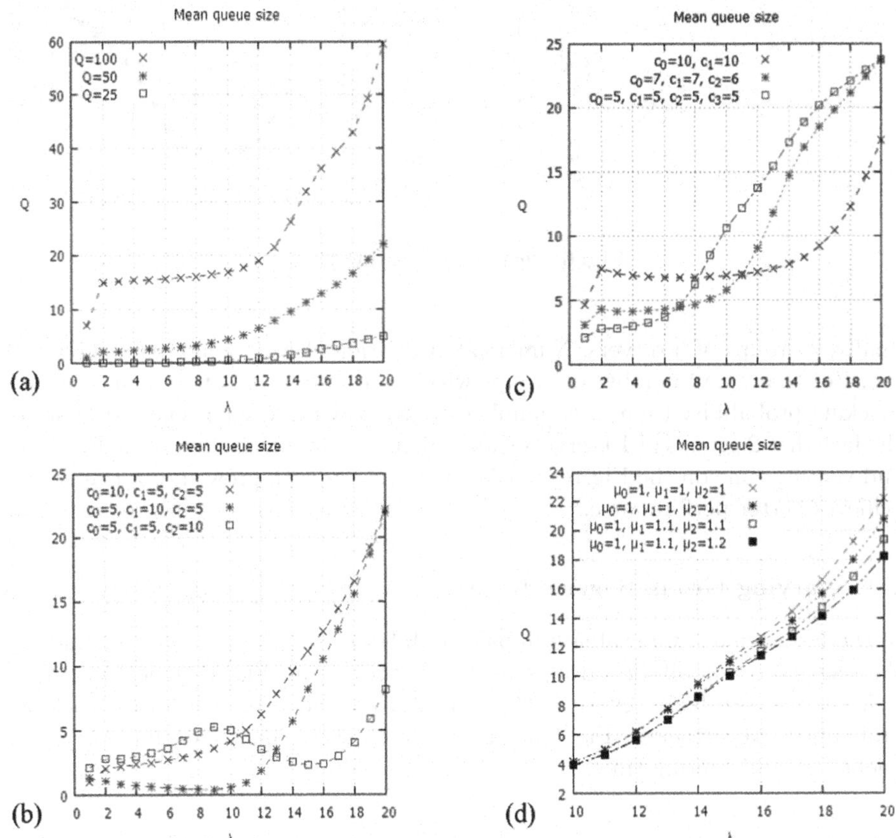

Fig. 5. Mean length of system's queue

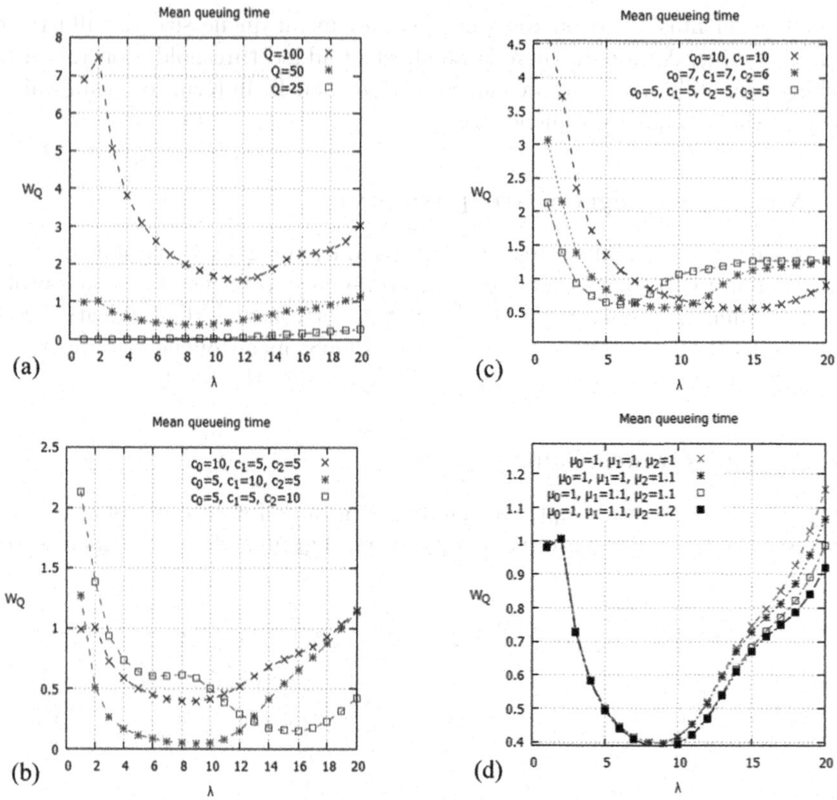

Fig. 6. Mean serve waiting time

finally, 4 groups of 5 servers. Numerical evaluation clearly demonstrates on the Fig. 4(c) that small number of groups with more resources is better in terms of blocking probability than large number of groups with fewer resources. Despite the fact that 2 groups of 10 servers shows higher mean queue size on the Fig. 5(c) and waiting time on the Fig. 6(c) at low values of λ, with growth of λ this setup achieves better performance.

5.4 Varying Group Service Rate

In the last scenario we evaluate impact of different service rates on the system. We use $\mu = 1$ for all the groups as a reference result.Fig. 4(d) demonstrates that service rate boost by 10%, 20% and 30% considerable decreases blocking probability. Nevertheless, Fig. 5(d) and Fig. 6(d) show slightly decline for mean queue size and waiting time.

6 Conclusions

In this paper we give an overview of virtualized Evolved Packet Core which throughput could be extended by allocation of additional computational resources in result of scaling in procedure. The thresholds control approach was applied in order to handle horizontal scaling of VNF. Mathematical model in terms of queuing theory was proposed and analytically evaluated. It was shown that horizontal resource scaling could significantly improve the system performance.

The following topics could be covered during further researches. First, in order to find optimal server groups partition and thresholds position optimization problem should be formulated and solved. Second, computation issues could appear for high values of r parameter. Third, more realistic approximation of input flow and activation/deactivation mechanism could be used that implies mathematical model sophistication and, correspondingly, results refinement.

7 Appendix

This paper is an extension of work originally presented in 23rd International Conference on Distributed Computer and Communication Networks: Control, Computation, Communications (DCCN 2020).

References

1. Soliman, S., Song, B.: Fifth generation (5G) cellular and the network for tomorrow: cognitive and cooperative approach for energy savings. J. Netw. Comput. Appl. **85**, 84–93 (2017)
2. Cisco annual internet report (2018–2023). White Paper (2020)
3. Global progress to 5G - trials, deployments and launches. GSA Report (2018)
4. Agiwal, M., Roy, A., Saxena, N.: Next generation 5G wireless networks: a comprehensive survey. IEEE Commun. Surv. Tutorials **18**(3), 1617–1655 (2016)
5. Economic benefits of virtualized evolved packet core. White Paper, IDC (2016)
6. Abaev, P., Razumchik, R., Uglov, U.: Statistical analysis and modeling of SIP traffic for parameter estimation of server hysteretic overload control. J. Telecommun. Inf. Technol. **4**, 22–31 (2013)
7. Abaev, P., Gaidamaka, Y., Samouylov, K., Pechinkin, A., Razumchik, R., Shorgin, S.: Hysteretic control technique for overload problem solution in network of SIP servers. Comput. Inf. **33**, 218–236 (2014)
8. Abaev, P., Gaidamaka, Y., Samouylov, K.E.: Modeling of hysteretic signaling load control in next generation networks. In: Andreev, S., Balandin, S., Koucheryavy, Y. (eds.) NEW2AN/ruSMART -2012. LNCS, vol. 7469, pp. 440–452. Springer, Heidelberg (2012). https://doi.org/10.1007/978-3-642-32686-8_41
9. Abaev, P., Tsarev, A.: Hysteretic mechanism for 5G hybrid evolved packet core resource management. In: Proceedings of the 10th International Congress on Ultra-Modern Telecommunications and Control Systems and Workshops (ICUMT), pp. 1–6 (2018). https://doi.org/10.1109/ICUMT.2018.8631209

10. Tobar Arteaga, C.H., Anacona, F.B., Tobar Ortega, K.T., Caicedo Rendon, O.M.: A scaling mechanism for an evolved packet core based on network functions virtualization. IEEE Trans. Netw. Serv. Manage. **17**, 1–14 (2019). https://doi.org/10.1109/TNSM.2019.2961988
11. Liebsch, M., Faqir, F.: Virtualized EPC - runtime offload for fast data-plane scaling. In: Proceedings of the IEEE International Symposium on Personal, Indoor and Mobile Radio Communications (PIMRC), pp. 1–6 (2016). https://doi.org/10.1109/PIMRC.2016.7794939
12. Kempf, J., Johansson, B., Petterson, S., Luening H., Nilsson, T.: Moving the mobile evolved packet core to the cloud. In: Proceedings of the 8th International Conference on Wireless and Mobile Computing, Networking and Communications, pp. 784–791 (2012). https://doi.org/10.1109/WiMOB.2012.6379165
13. Tobar, C., Risso, F., Caicedo, O.: An adaptive scaling mechanism for managing performance variations in network functions virtualization: a case study in an NFV-based EPC. In: Proceedings of the International Conference on Network and Service Management (CNSM), pp. 1–7 (2017). https://doi.org/10.23919/CNSM.2017.8255982
14. Khakimov, A., Gudkova, I., Ateya, A.A., Markova, E., Muthanna, A., Koucheryavy, A.: IoT-Fog based system structure with SDN enabled. In: ACM International Conference Proceeding Series, Article No. 62 (2018). https://doi.org/10.1145/3231053.3231129
15. Muthanna, A., et al.: SDN multi-controller networks with load balanced. In: ACM International Conference Proceeding Series, Article No. 57 (2018). https://doi.org/10.1145/3231053.3231124
16. Muthanna, A., et al.: Secure and reliable IoT networks using fog computing with software-defined networking and blockchain. J. Sens. Actuator Netw. **8**(1), 15 (2019). https://doi.org/10.3390/jsan8010015. Article No. 8010015
17. Kim, C.S., Dudin, A., Dudin, S., Dudina, O.: Hysteresis control by the number of active servers in queueing system with priority service. Perform. Eval. **101**, 20–33 (2016). https://doi.org/10.1016/j.peva.2016.04.002
18. Lui, J., Golubchik, L.: Stochastic complement analysis of multi-server threshold queues with hysteresis. Perform. Eval. **35**(1–2), 19–48 (1999). https://doi.org/10.1016/S0166-5316(98)00043-1
19. Chou, C., Golubchik, L., Lui, J.: Multiclass multiserver threshold-based systems: a study of noninstantaneous server activation. IEEE Trans. Parallel Distrib. Syst. **18**(1), 96–110 (2007). https://doi.org/10.1109/TPDS.2007.253284
20. Nguyen, V., Brunstrom, A., Grinnemo, K., Taheri, D.: SDN/NFV-based mobile packet core network architectures: a survey. IEEE Commun. Surv. Tutorials **19**(3), 1567–1602 (2017). https://doi.org/10.1109/comst.2017.2690823IEEE

The Method for User Localization in the Local Wireless Network in an Emergency

A. I. Paramonov[1,2](✉), T. M. Tatarnikova[3], and R. V. Shamilova[4]

[1] The Bonch-Bruevich Saint-Petersburg State University of Telecommunications, Bolshevikov, 22, Saint-Petersburg, Russia
alex-in-spb@yandex.ru

[2] RUDN University, Miklukho-Maklaya str.6, Moscow, Russia

[3] Russian State Hydrometeorological University, Voronezhskaya ulitsa, 79, Saint-Petersburg, Russia

[4] ITMO University, Kronverkskiy pr., 49, Saint-Petersburg, Russia

Abstract. The paper provides an analysis of the use of a number of capabilities of local wireless broadband access networks in the management of rescue people in emergency situations. The article discusses such basic features as localization of users and increasing the availability of information for people who find themselves in dangerous conditions and for the rescue control center. Possible localization methods are considered and an original method is proposed that allows localization to be performed elementwise, i.e. in rooms or in certain areas of the service area.

Keywords: Wireless local area network · Broadband access network · Localization · Coordinate determination · Local area network

1 Introduction

Wireless Broadband Access Networks (WLANs) are now widely distributed around the world. In particular, for example, technologies based on the IEEE 802.11 family of standards [1] are used to build networks of various sizes: private networks (PAN - Private Area Network) within private residential premises, device-to-device connections (D2D - Device-to-Device) between mobile terminals or other devices, corporate networks (WLANs - wireless local area networks) of various organizations, public WLANs, both indoors and outdoors.

To date, the number of networks built using these technologies is so large that almost anywhere in the city territory can receive signals from several access points of various networks [2–6] WLAN. In promising mobile networks, WLANs also play an important role, according to which they become an integral part of these networks [7].

WLAN networks are just a tool that potentially allows you to receive information about people (coordinates, dialogue), the state of the environment, the

Supported by RUDN University.

© Springer Nature Switzerland AG 2020
V. M. Vishnevskiy et al. (Eds.): DCCN 2020, CCIS 1337, pp. 235–246, 2020.
https://doi.org/10.1007/978-3-030-66242-4_19

transfer of information to coordinate actions in case of emergencies. In this paper, we consider only the possibility of using WLAN in the tasks of localizing users of devices connected to the network.

We will also assume that a person in the emergency zone may have a subscriber terminal that supports the protocols of this group of standards. We believe that if such a terminal exists, then it is with the user. The user localization function can be implemented by means of a WLAN (controller) or external application software installed on an external server relative to the WLAN equipment.

In principle, the localization task is similar to the positioning task. There are many works devoted to positioning problems. As a rule, the positioning problem is considered as the problem of determining the coordinates of an object in space. This problem can be viewed both in 3D space and on a plane. This work also deals with determining the position of an object, but its purpose is not to obtain numerical estimates of coordinates. The purpose of localization is to determine whether the coordinate of an object belongs to a particular area in the network service area. Despite the fundamental similarity of these tasks, the solution to the localization problem may differ from the positioning problem.

In real wireless access networks, various ways of implementing this function are possible, they depend on the specific network and the manufacturer of the WLAN equipment, configuration, software and controller settings. In this paper, we propose a localization method that can be implemented both as part of the controller software and on equipment external to the network, provided that there is access to the necessary information.

In many WLAN implementations, such functions are not provided at all, however, the potential capabilities of user terminals (smartphones) make it possible to implement it in the form of application software installed on terminals and some server equipment.

In this paper, the purpose of localization will mean determining the location of clients, i.e. people in the room in question. The result of solving such a problem is information about the element of the room where people are. In what follows, we will understand a room element as a certain part of it (room, corridor, hall, etc.) or the conditional border of such an element, introduced when forming the floor plan. Thus, the result of the functioning of such a system is information about the elements of the room in which people are. Of course, to solve this problem, the functioning of the WLAN or part of it, as well as the client equipment, is necessary. Let us consider further the options for implementing the localization system, depending on the technical capabilities of the WLAN used.

2 Implementation Options

As noted in the introduction, the functions under consideration can be implemented in two ways: by means of WLAN, i.e. using the means of the WLAN controller, and using additional means, i.e. software developed for user terminals and server hardware. Let's consider the first way. The ability to implement localization functions by means of WLAN assumes the presence of a WLAN controller, the corresponding software installed on the controller or on an additional server that interacts with the controller, as shown in Fig. 1.

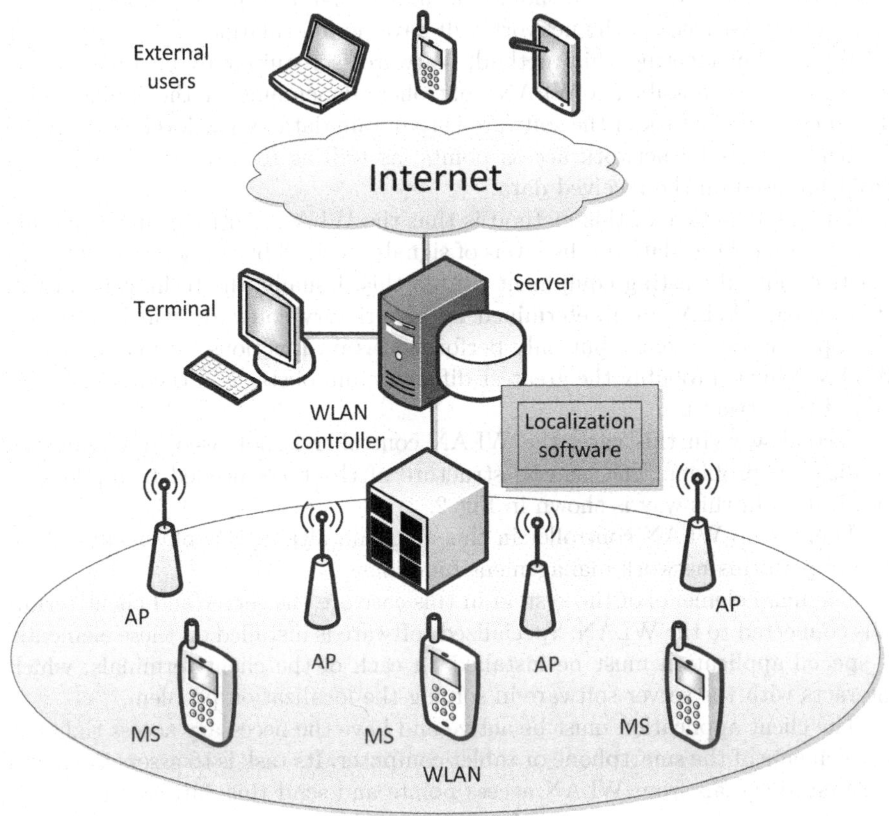

Fig. 1. Service organization structure by means of WLAN

In this case, the WLAN controller allows either to implement the localization functions in whole or in part, and also allows access to the data necessary for solving the localization problem from the server side, on which specialized software is installed. The software is controlled by the administrator of the service that solves the problem under consideration.

In general, the server can communicate with the controller via a communication network, i.e. be at a considerable distance from the WLAN. It can also be part of an overall system deployed to serve multiple WLANs. Possible implementation options depend on specific applications.

The server, possibly in conjunction with the WLAN controller, provides basic localization functions, and also provides access to localization results from external clients, which can be various services, for example, building administrators, security services, rescue teams or other services.

The advantages of this structure are that clients for which the localization task (MS) is performed can be any clients that have devices turned on and connected to the WLAN. It should be noted that a WLAN connection is a prerequisite. Otherwise, the network will have no information.

When implementing this method, software is required to be installed on the server and, possibly, a WLAN controller, depending on the implementation specifics. The task of the software is to obtain data on the level of reception of client signals by network access points, as well as to solve the localization problem based on the received data.

The disadvantage of this method is that the WLAN controller must provide the necessary data (data on the levels of signals received by the access points). In practice, not all existing equipment can do this. Usually, the technical features of building a WLAN are determined by proprietary software, which does not give open access to data, but only performs certain functions for managing the WLAN. This is probably the greatest difficulty and obstacle in the implementation of this structure.

Second way. In this case, the WLAN controller is not used in solving the localization problem. The general structure of the tools needed to implement localization in this way is shown in Fig. 2.

There is no WLAN controller in this diagram, although it physically exists, it only performs network management functions.

The main elements of the system in this case are the server and client terminals connected to the WLAN. Specialized software is installed on these elements. A special application must be installed on each of the client terminals, which interacts with the server software in solving the localization problem.

The client application must be active and have the necessary access rights to the elements of the smartphone or tablet computer. Its task is to assess the levels of received signals from WLAN access points and send this data to the server. Server software based on received from clients solves the problem of localization. Of course, the client application may have other functions that are of interest both in this task and in other tasks, which may differ in each specific case.

The advantage of this method is independence from the specific technical implementation of the WLAN. To implement it, you only need a server, appropriate software and client software. From the point of view of technical implementation, this method may be the simplest.

The disadvantage of this method is that the client software needs to be installed and properly functioning. This is a significant drawback that can become an obstacle to the implementation of such a system in public places.

The administrator of this system must be able to influence the behavior of clients in such a way that they independently install client applications. It is probably possible to implement this in corporate networks, where customers are employees, visitors or other persons with a certain interest in following the rules.

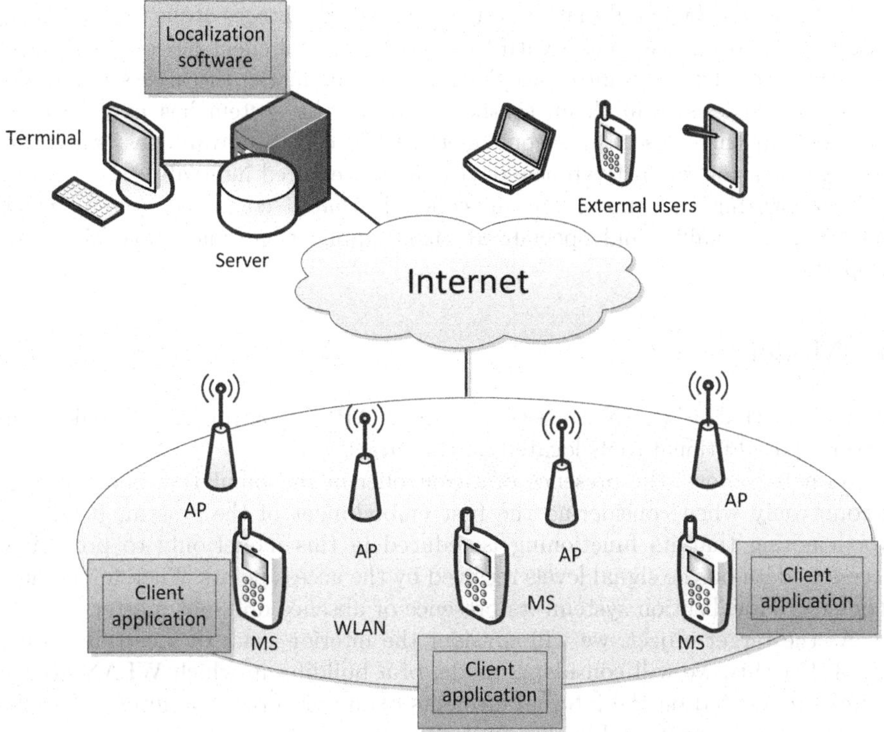

Fig. 2. Service organization structure by additional means

As can be seen from the description, these options are similar to the localization method, namely, the measurement of the received signal power level is used to determine the position.

In the first method, measurements are taken by the access points. The levels of the signal received from client devices are measured. The measurement results are transmitted to the server, processed and used to solve the localization problem. In the second method, measurements are taken by client devices running a special client application. This evaluates the signal power levels of the access points. The results are sent by each client terminal to the localization server. The server processes the received data and solves the localization problem.

The results of solving the localization problem are available to users, who can be people, automatic devices or control systems that have access to them.

Probably, if there is a technical feasibility, the first option is preferable because it is possible to work with any clients without the need to install client software. However, in the absence of technical feasibility, the second implementation option in a number of cases makes it possible to increase the efficiency of solving localization problems.

It should also be noted that, despite the noted advantages of the first method, an error may occur associated with the passive state of client devices, which, in order to save energy, go into "sleep" mode. In this mode, the access points do not receive signals from them, so the data that the system has may become outdated. In this sense, the second method has certain advantages, since the client application can activate the device at the required intervals.

Summarizing the above, we can conclude that a combined system using WLAN functionality and specialized client applications may have the best properties.

3 Model

We assume that a WLAN consists of many n APs, a controller, an application server, and additional tools located on the target.

As noted above, the presence of a controller in the simulation is taken into account only when considering the first embodiment of the system. It is also worth noting that its functioning is reduced in this model only to providing access to data on the signal levels received by the access points. Therefore, when modeling a localization system, its presence or absence does not matter.

As the target object, we will consider the interior space of a certain room, Fig. 3. For this, we will consider a model of a building in which WLAN access points are located on the internal elements of the premises, the internal rooms are separated by walls and floors, similarly, auxiliary elements of the system can be placed. MS network users can be anywhere in the building.

We use the approach implemented by classification methods [8]. We divide the room in question into some units of area or volume (slices) and number them $q = 1, ..., S$. The unit may be, for example, a room, square or cube of a certain size. The choice of units depends on the characteristics of the room and the convenience of their localization. Due to the fact that the WLAN located in the room has fairly stable parameters, there is reason to believe that when the terminal is placed at the same point multiple times, the signal level received by different access points will be quite close at each attempt. This problem can be considered as the task of assigning an unknown element to one of the known classes - slices.

In the classification problem, each of the slices represents a class, and the measured signal level by the access point is a discriminant variable. Based on the values of the discriminant variables, a decision is made on whether the observed object belongs to one or another class.

Suppose that, as a result of preliminary experiments on measuring the signal level (training) generated by the terminal at a set of m slice points q, the level value is obtained, $j = 1, ..., n$; $r = 1, ..., m$. Each of m points is characterized by a vector of n elements $u^{(r)} = \{u_1, u_2, ..., u_n\}, r = 1, ..., m$ that correspond to signal levels $u^{(r)}$ received by n access points, i.e. has n coordinates. For each slice i, the center of mass is calculated, which is a point in n - dimensional space whose coordinates are defined as (1).

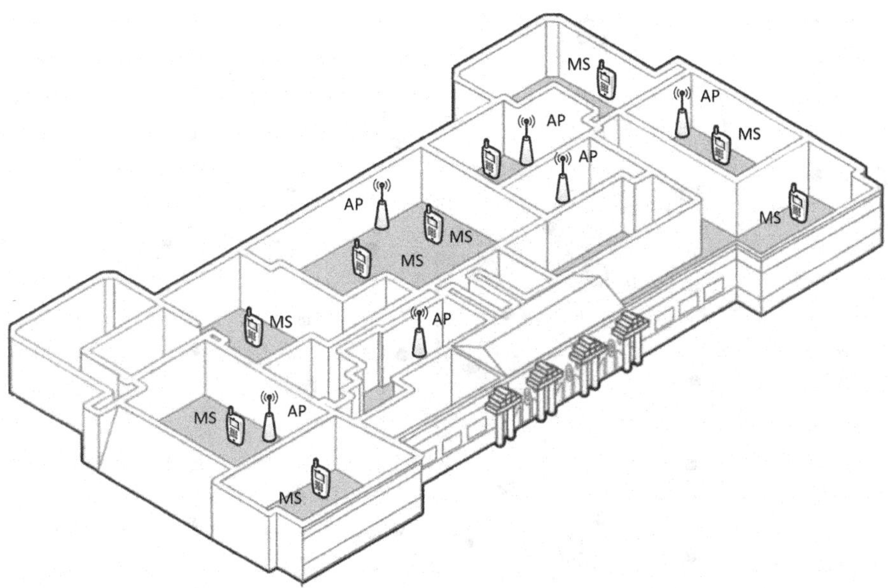

Fig. 3. Placement of access points, auxiliary equipment and users in the premises

$$\mu_q = \{\mu_1, \mu_2, ..., \mu_n\}, q = 1, ...S, where\ \mu_j = \frac{1}{m}\sum_{r=1}^{m} u_j^{(r)}\ j = 1, ..., n \quad (1)$$

Figure 4 is an example of conditional division of the served area into slices.

To localize the subscriber terminal x, it is necessary to determine its proximity to the centers of mass of each of the slices defined at the stage of system training. We estimate the degree of proximity by calculating the Mahalanobis distance [9] given by (2).

$$M_q(x) = \sqrt{(x - \mu_q)^T S^{-1}(x - \mu_q)} \quad (2)$$

where S^{-1} is the covariance matrix x and μ.

Calculating (2) for each slice, we obtain the distance vector $\mathbf{M}$ of size S. The most probable localization is determined by the minimum value search operation.

$$q_{min} = arg\ qmin\{\mathbf{M}\} \quad (3)$$

The result of (3) is a single slice number, however, in a real situation it may turn out that several estimates (3) have close values. Therefore, it is advisable to introduce some empirical quantity η, taking into account the localization error. Then the solution to the problem will be a set of slice numbers for which the Mahalonobis distance between the center of mass and point x does not exceed $qmin\{\mathbf{M} + \eta\}$.

$$Q = \{q|(M_q \leq qmin[\mathbf{M} + \eta],\ q = 1, ..., S)\} \quad (4)$$

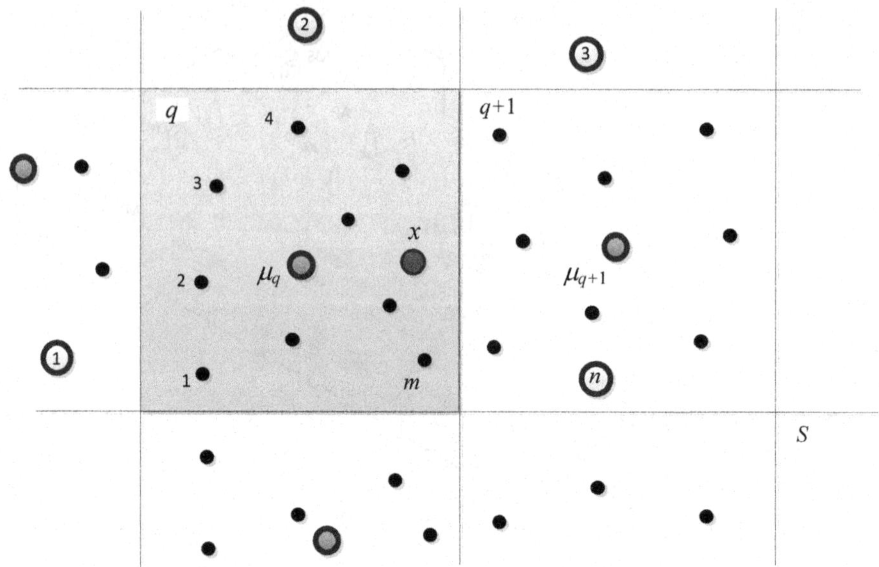

Fig. 4. An example of representing a service area as slices

The last expression is of great practical importance, because it allows you to define a set of slices in the practical situation of rooms where people are most likely to be (subscriber terminal). Along with expression (5), one can use the approach using the assumption of the distribution of M, which, in the case of the normal distribution of x, has a Hotelling distribution T^2 [9].

$$F(M) = I_{\frac{M(m-n+1)}{m\left(\frac{M(m-n+1)}{m}+m-n+1\right)}}\left(\frac{n}{2}, \frac{1}{2}(m-n+1)\right) \tag{5}$$

where I is the regularized incomplete beta function, n is the dimension, m is the scale factor. Then the expression for the solution will have the form

$$Q_H = \{q|(M_q \leq arg[F(M_0) \leq P_0])\} \tag{6}$$

where P_0 is the specified probability that the Mahalonobis distance does not exceed M_0.

It should be noted the obvious sensitivity of this method to the failure of access points. The previously expressed expression (2) is calculated based on the data of all access points n, therefore, when solving the localization problem (4), (5) it is also required to have data from all n access points. In case of failure of some access points, i.e. the absence of part of the initial data, the solution may have a significant error.

In order to reduce this error, it is advisable to store the source data, on the basis of which expressions (2) are calculated, and if some access points fail, recount (2) taking into account the data of only functioning equipment. The advantages of this method include the fact that it is invariant to units of measurement (power level) and does not require calibration of access points, which greatly simplifies its implementation.

The localization method can be represented by the following algorithm, Fig. 5.

The training stage, a lot of measurements of the signal level are made at various points of various slices. For these measurements, the average values $\mu_q = \{\mu_1, \mu_2, ..., \mu_n\}, q = 1, ...S$ are calculated according to (1). The measurement cycle continues until the probability of a localization determination error exceeds a preset value of Po or until its value ceases to decrease with an increase in the number of measurements. Upon completion of this step, the average values will be determined with the required accuracy.

In practice, measurements during the training phase are experiments that involve placing client devices at different points in the selected slices. Device placement positions can be randomly selected within slices. This stage is similar for both the first and second variants of the system implementation, except that in the second variant the client devices are equipped with a specialized client application.

In practice, other elements can be used as elements for dividing the area or volume of the considered room "slices", and not necessarily of equal size. For example, rooms, halls, corridors and other room elements that are easy enough to localize can be selected as such elements. It is likely that elements of complex configuration, such as long winding corridors, will be difficult to describe unambiguously. In such cases, it makes sense to introduce a conditional division of these elements into some fragments. In any case, as a result of dividing the premises into elements to be localized, a plan must be built, according to which further "training" of the system is carried out.

At the operation stage, measurements of signal levels from subscriber devices are carried out and for each of the measurements, the distance of Mahalanobis (2) to the center of each of the slices is calculated. After that, each of the devices is associated with that slice, the distance to which is minimal (3).

This correspondence is the solution to the localization problem. Further, the results of localization are displayed, if necessary, and the work cycle is repeated. The repetition period of the localization cycle depends on the degree of user mobility and in practical tasks can be from seconds to several minutes.

The repetition period of the localization cycle depends on the degree of user mobility and in practical tasks can range from seconds to several minutes. It is also worth noting here the feature of subscriber terminals associated with the transition to the low consumption mode "sleep mode". As noted above, in this state, it becomes impossible to track the position of the terminal; therefore, the measurement frequency should be selected taking into account the peculiarities of the terminals and their software.

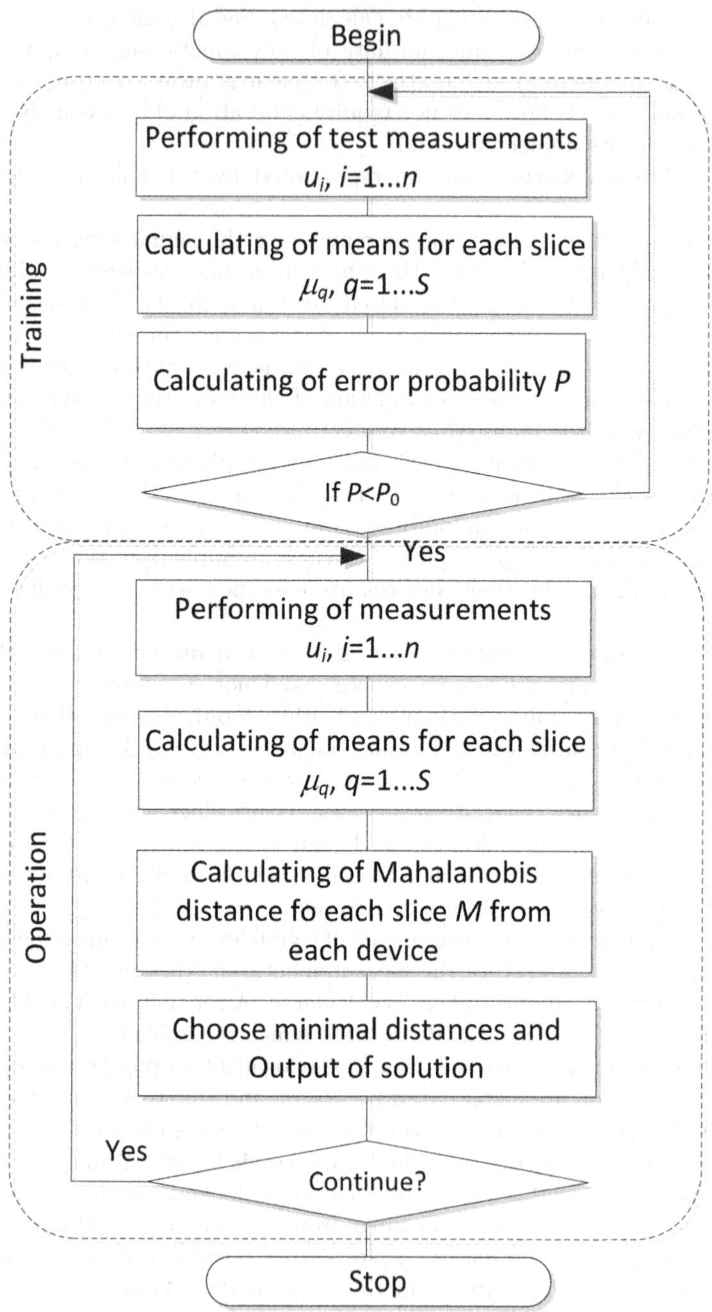

Fig. 5. The algorithm of the positioning system

When using client applications, this task is simplified due to the fact that they can potentially have access to the sensors of subscriber terminals and react to human behavior. In this case, the client terminal can initiate the update of the position data.

4 Conclusion

The analysis showed that the use of WLAN for localization tasks in the event of an emergency can significantly increase the availability of information both about people in the danger zone and information coming to them from the control center.

The ability to localize users via the web is the most important factor in improving the efficiency of search and rescue operations.

Therefore, when building modern WLANs and integrating them into companies' information systems, attention should be paid to the aspect of the applicability of these systems for localization tasks in the event of an emergency. This concerns both the information systems themselves from the point of view of software implementation at the level of solving applied problems, and the development of specialized client software.

The ability to localize users using web-based tools or additional tools is a critical factor in improving the efficiency of search and rescue operations.

The possibility of organizing a stable ad hoc network based on direct connections between subscriber terminals makes it possible to promptly deliver information to users about action plans in a given situation, as well as deliver information to the control center about the current state.

Summarizing the analysis, it should also be noted that the greatest efficiency of the system is achieved with the integrated use of all available tools to increase the availability of information. Therefore, when building modern WLANs and integrating them into information systems, a company should pay attention to the aspect of the applicability of these localization systems in the event of an emergency. This applies to both information systems in terms of software implementation at the level of solving applied problems, and the development of specialized client software.

Acknowledgment. The publication was prepared with the support of the "RUDN University Program 5–100".

References

1. IEEE Std 802.11 - 2016. IEEE Standard for Information technology – Telecommunications and information exchange between systems. Local and metropolitan area networks – Specific requirements. Part 11: Wireless LAN Medium Access Control (MAC) and Physical Layer (PHY) Specifications, 3534 p. IEEE, New York (2016)

2. Muthanna, A., et al.: Analytical evaluation of D2D connectivity potential in 5G wireless systems. In: Galinina, O., Balandin, S., Koucheryavy, Y. (eds.) NEW2AN/ruSMART -2016. LNCS, vol. 9870, pp. 395–403. Springer, Cham (2016). https://doi.org/10.1007/978-3-319-46301-8_33
3. Makolkina, M., Koucheryavy, A., Paramonov, A.: Investigation of Traffic Pattern for the Augmented Reality Applications. In: Koucheryavy, Y., Mamatas, L., Matta, I., Ometov, A., Papadimitriou, P. (eds.) WWIC 2017. LNCS, vol. 10372, pp. 233–246. Springer, Cham (2017). https://doi.org/10.1007/978-3-319-61382-6_19
4. Abdellah, A.R.: IoT traffic prediction using multi-step ahead prediction with neural network. In: Abdellah, A.R., Mahmood, O.A.K., Paramonov, A., Koucheryavy, A (eds.) 11th International Congress on Ultra Modern Telecommunications and Control Systems and Workshops (ICUMT), p. 8970675 (2019)
5. Paramonov, A., Vikulov, A., Scherbakov, S.: Practical results of WLAN traffic analysis. In: Galinina, O., Andreev, S., Balandin, S., Koucheryavy, Y. (eds.) NEW2AN/ruSMART/NsCC -2017. LNCS, vol. 10531, pp. 721–733. Springer, Cham (2017). https://doi.org/10.1007/978-3-319-67380-6_68
6. Paramonov, A., Hussain, O., Samouylov, K., Koucheryavy, A., Kirichek, R., Koucheryavy, Y.: Clustering optimization for out-of-band D2D communications. Wirel. Commun. Mobile Comput. **2017**, 1–11 (2017). Article ID: 6747052
7. Andreev, S., Galinina, A., Pyattaev, S.: 5G Multi-RAT LTE-WiFi ultra-dense small cells: performance dynamics, architecture, and trends 1.0. IEEE J. Sel. Areas Commun. **33**(6), 1224–1240 (2015)
8. Kim, J.-O., Mueller, C.W.: Factor Analysis: Statistical Methods and Practical Issues. Sage, Beverly Hills (1986). Eleventh Printing, 215 p
9. Ahrens, H., Lauter, J.: Mehrdimensionale Varianzanalyse. Academie-Verlag, Berlin (1981). 226 p
10. Vladyko, A., Paramonov, A., Kirichek, R., Koucheryavy, A.: Using the IEEE 802.11 family of standards for communication between robotic systems. In: Advances in Intelligent Systems Research. Vol. 133, pp. 153–157 (2016)

Analysis of the Using of D2D Communications for the Ad Hoc Network Based on Subscriber Terminals

A. Paramonov[1,2](✉), T. Tatarnikova[3], and A. Marochkina[3]

[1] The Bonch-Bruevich Saint-Petersburg State University of Telecommunications, Bolshevikov, 22, Saint-Petersburg, Russia
alex-in-spb@yandex.ru

[2] RUDN University, Miklukho-Maklaya str.6, Moscow, Russia

[3] Russian State Hydrometeorological University, Voronezhskaya ulitsa, 79, Saint-Petersburg, Russia

Abstract. The article presents the results of an analysis of the potential capabilities of a network built using D2D communications as an ad hoc network. It is shown that with a sufficient density of devices, it can provide a high probability of connectivity. It is also shown that for a network with moving nodes, the probability of connectivity depends on time. Such a network can operate as a network with acceptable delays. Analytical expressions are obtained that relate the probability of connectivity for both a static network and a network with mobile nodes with its main parameters.

Keywords: Device to device · Ad hoc · Connectivity probability · Network with mobile nodes · Delivery time · Network with acceptable delays

1 Introduction

Mobile network subscriber terminals are currently high-tech devices that combine the capabilities of a computer and a high-speed wireless access modem. Moreover, these devices are equipped with transceivers that support many standards for wireless data transmission. Almost all subscriber terminals have support for standards of wireless access networks of the IEEE 802.11 family. These are extremely popular technologies that enable high-speed data transmission over relatively short distances. Currently, this functionality has been supplemented by technologies based on IEEE 802.11 standards, which allow creating direct device-to-device D2D connections [1]. This feature allows you to simplify the connection of the smartphone to various devices (TV, printer, monitor) as well as to other smartphones for data exchange. This technology is called WiFi-direct. There are many client applications that use this technology to organize communication

Supported by RUDN University.

© Springer Nature Switzerland AG 2020

V. M. Vishnevskiy et al. (Eds.): DCCN 2020, CCIS 1337, pp. 247–258, 2020.
https://doi.org/10.1007/978-3-030-66242-4_20

between external devices and between users: chats, voice communication, file transfer, screen sharing, etc. The possibility of establishing communication and data transmission to neighboring devices located in the communication zone opens up the potential for the implementation of transit transmission (from device to device) over distances exceeding the communication range between two neighboring nodes. In fact, this means the possibility of building an ad hoc network [2,3] using subscriber terminals. Of course, the capabilities of such a network are determined by the location and behavior of its nodes. For example, if the nodes are mobile, that is, there is a non-zero probability that the node that received the message, but does not have the ability to broadcast it immediately after receiving, after some time move and will be in the communication zone of the destination node and deliver this message. If the requirements for delivery time are not too strict, but allow a certain delay, then such a network will allow it to be implemented even in the absence of coverage of the mobile communication network [4]. Probably, such an opportunity can be very useful in emergency situations or in cases of low user density and the absence of other means of communication. On the other hand, in the case of a high density of subscriber units, which takes place in modern communication networks, there is a potential for delivering messages over considerable distances [2]. This capability allows delivering a portion of the traffic using only the capabilities of the terminals, i.e. bypassing the operator's network infrastructure. This makes it possible to somewhat relieve the operator's communication network and, in some cases, improve the quality of traffic service.

2 Model

We will consider a flat surface, assuming that users are distributed on it randomly. The user area is not limited. User density ρ number of users per unit area $1/m^2$.

It should be noted that in the general case, one should consider the network in some three-dimensional space, since in practice, it can be located in a multistorey building or structure. However, in reality, the model described above does not reduce the generality of further reasoning. The transition from a flat model to a three-dimensional model consists only in the introduction of one more coordinate and a corresponding transition from the description of flat geometric figures to three-dimensional, which is not very difficult. However, this can complicate the display and reduce the visibility of the model. In the proposed model, we will assume that the surrounding space is isotropic; the receiving-transmitting antennas of the network nodes are also isotropic. Under such assumptions, the communication area of a node can be described by a circle with a certain radius R. Most of the short-range wireless communication technologies, including the IEEE 802.11 standards, assume some dependence of the communication quality (channel) on the distance between the nodes: the receiver and transmitter, sources of interference, i.e. on the signal-to-noise ratio, depending on the attenuation of the signal in the environment. The channel quality is characterized by the data transfer rate. In this problem, we will assume that the admissible

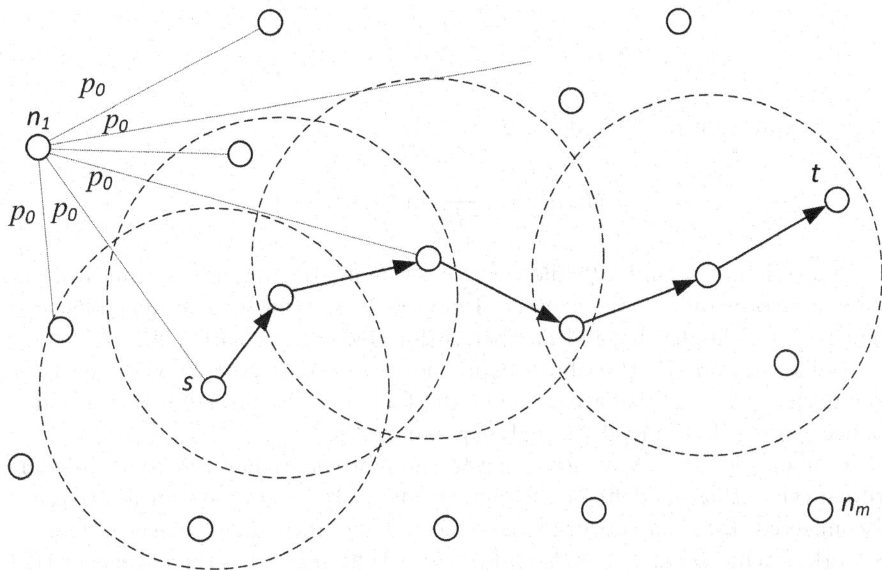

Fig. 1. Network model

delivery time is much longer than the data transmission time over the communication channel, even in the case of its low quality. Therefore, this dependence practically does not affect the functioning of the network. Based on this, the node's communication zone model only describes the communication possibility, i.e. the presence or absence of the possibility of data transmission (Fig. 1).

Let us describe the network of users in a random graph. Then the probability of the existence of a route between any two terminals (taking into account transit) for a sufficiently large number of subscribers can be estimated according to one of the theorems of Erdös-Renyi [5–7] or Bollobash-Riordan [7], depending on the density of nodes, as shown in [8]. At a high density of network nodes, for example, in a city, the connectivity is determined by the communication distance of the node (communication radius) and the number of nodes, according to the Erdés-Renyi theorem, with a "critical" value of the connection probability of the node

$$p_0 = \frac{\ln n}{n} \tag{1}$$

where n is the number of nodes in the network. For a flat model, the probability of falling into the communication zone of a node of at least one neighboring node, with their uniform distribution will be determined as

$$\hat{p}_0 = 1 - p(0) = 1 - e^{-\pi R^2 \rho} \tag{2}$$

where $p(0)$ is the probability that there will be no nodes in the communication zone. Equating (1) and (2) we obtain the expression for R

$$R_0(\rho, n) = \sqrt{-\frac{1}{\pi\rho} \ln\left(1 - \frac{\ln n}{n}\right)} \tag{3}$$

and expression for node density

$$\rho_0(R, n) = -\frac{1}{\pi R^2} \ln\left(1 - \frac{\ln n}{n}\right) \tag{4}$$

Expressions (1) and (2) allow us to estimate the minimum values of the communication radius and density of devices, respectively. The dependence of the radius on the density and number of devices is shown in Fig. 2.

Unlike [5] and [7], the above expressions do not require knowledge of the boundaries of the service area, due to the fact that the probability of a node's connectivity is determined through expression (2).

It should be noted that according to the aforementioned theorem, these are critical values that are defined as an intermediate state, a phase transition, from a disconnected state of a network to a connected one. Providing connectivity in the network is achieved at $R > kR_0$ and $\rho > k\rho_0$ Where $k \geq 1$ is the coefficient that ensures the network goes into a connected state. There is no exact expression for estimating this coefficient, however, from the results of simulation it was found that for $k \approx 2$ the probability of network connectivity tends to unity.

From the above dependence it is seen that the required radius of the communication zone of the node decreases with increasing density of nodes and their number. This is the expected result, since with the growth of these parameters the likelihood of a neighboring node getting into the communication zone also increases.

It should be noted here that in this model, we do not take into account the interference created by neighboring nodes, the total power of which also increases with increasing node density. This negatively affects the quality of communication, which physically manifests itself in a decrease in the radius of the actually possible communication. However, this effect manifests itself in the case of activity of neighboring nodes, i.e. in case of relatively high network load. We consider a relatively low network load with traffic, while we assume that the probability of activity of neighboring nodes at the time of message transmission is negligible.

With a low density of nodes, according to (1), the likelihood of connectivity is low; therefore, a static model with fixed nodes shows that this network is practically unsuitable for communications.

However, With moving nodes, the network properties change, because routes in such a network are not static. If the movement of the node is accidental and in no way limited, then for a sufficiently long time, with a probability close to unity, it will be in the communication zone of the destination node. Therefore, in such a network, one should not consider a static characteristic of connectivity, but the route length and delivery time.

Let the node move with an average speed v, then in time τ the communication zone of the node will cover an area equal to

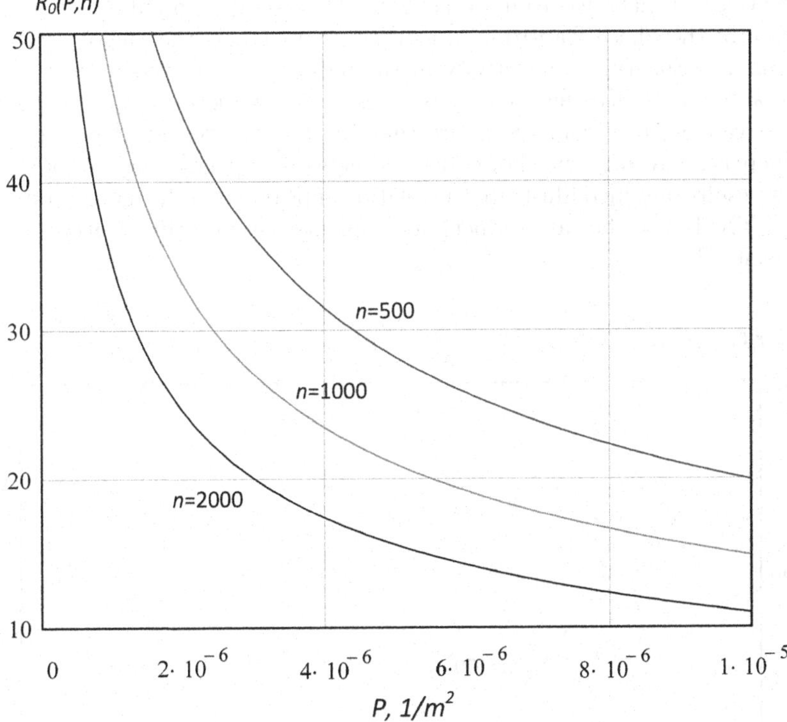

Fig. 2. Dependence of the minimum communication radius on the density and number of network nodes.

$$S(\tau) = 2Rv\tau \tag{5}$$

then the probability of at least one node falling into this area

$$p_{>0}(\tau) = 1 - e^{-S(\tau)\rho} = 1 - e^{-2Rv\rho\tau} \tag{6}$$

Equating expressions (5) and (1) we obtain the expression for τ

$$\tau_0(v, n) = -\frac{1}{2R\rho v} \ln\left(1 - \frac{\ln n}{n}\right) \tag{7}$$

The dependence of the time during which the connectivity of the mobile network is provided on the number of nodes and the average speed is shown in Fig. 3.

The figure shows that the average delivery time decreases with an increase in the average speed of nodes and network capacity (number of nodes). The results obtained should be an illustration of the relationship between the main parameters of the DTN. In specific cases of network implementation, the peculiarities of the movement should be taken into account.

The time estimate according to (7) gives only the lower boundary τ at which the state of the phase transition is reached, if we follow the analogy with the static model. Ensuring connectivity in the network is achieved when $\tau > k\tau_0$. Where $k \geq 1$ is the coefficient that ensures the network goes into a connected state. It can be seen from the figure that the average delivery time decreases with increasing average speed of nodes and network capacity (number of nodes).

The results obtained illustrate the relationships between the main parameters of the DTN. In specific cases of network implementation, traffic features should be considered.

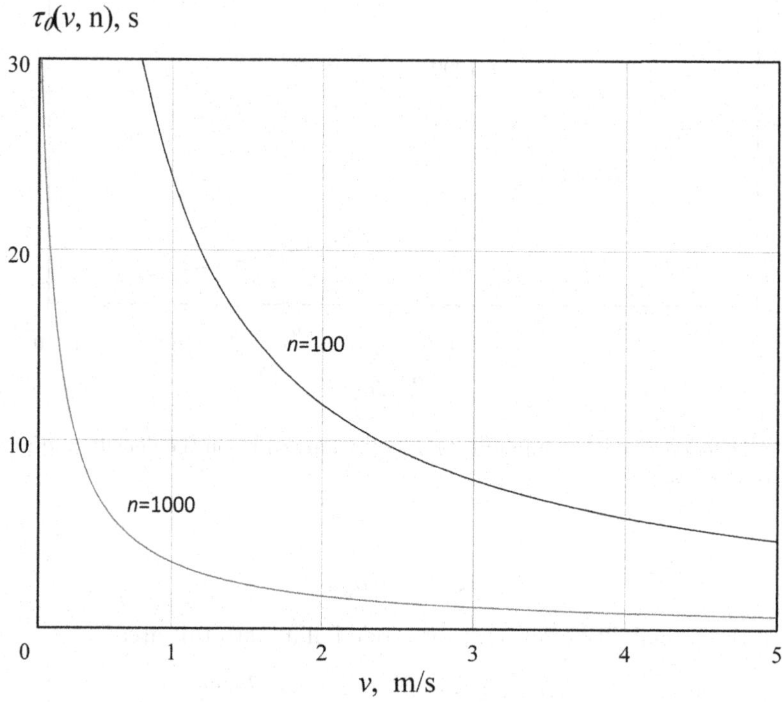

Fig. 3. The dependence of the connection time on the average speed of nodes and their number.

According to the Erdés-Renyi model, the average path length is of the order of $g(n) \propto \ln n$, while for the Bollobash-Riordan model this value is $g(n) \propto \frac{\ln n}{\ln \ln n}$.

$$g(n) \propto \frac{\ln n}{\ln \ln n} \tag{8}$$

An illustration of the dependence of the average path length on the number of nodes in the network for the Erdos-Renyi (I) and Bollobash-Riordan (II) models is shown in Fig. 4. It can be seen from the above graph that in the case of a network with a low density of nodes (sparse graph), the expected path length (in the number of transits) is significantly less than for a network with a relatively high density.

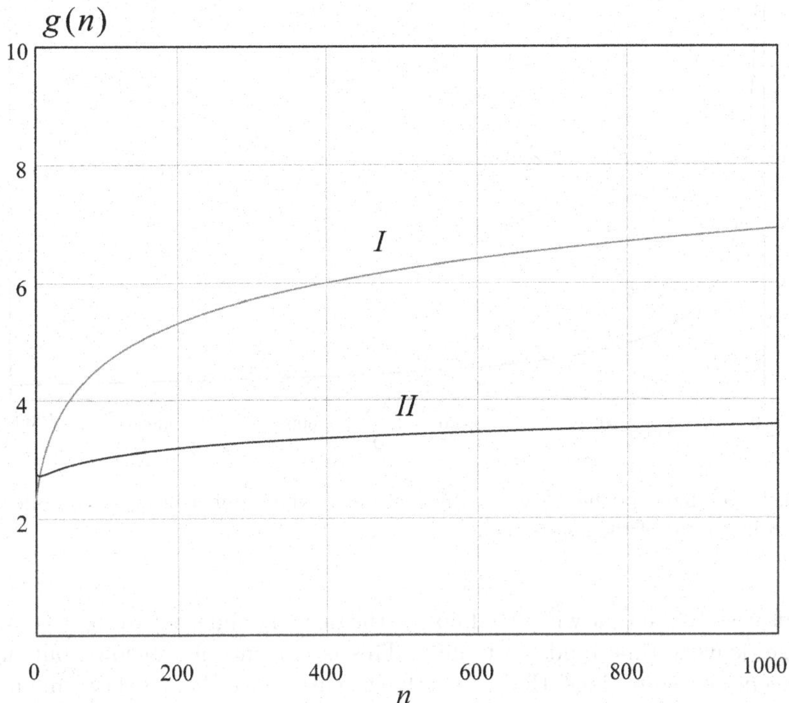

Fig. 4. Dependence of the average path length in the network for models I and II on the number of nodes.

We will assume that the delivery processes on each of the route sections are independent, then for the full route the delivery time will be determined as

$$g(n) = \tau(v, n)g(n) \tag{9}$$

The dependence of the delivery time on the number of network nodes and the average speed of the nodes movement is shown in Fig. 5.

The graph shows that the delivery time decreases with an increase in the number of network nodes and the average speed of their movement. This is the expected result since an increase in the number of nodes entails an increase in the probability of connectivity according to (1), and an increase in the speed of movement of nodes leads to its increase according to (6).

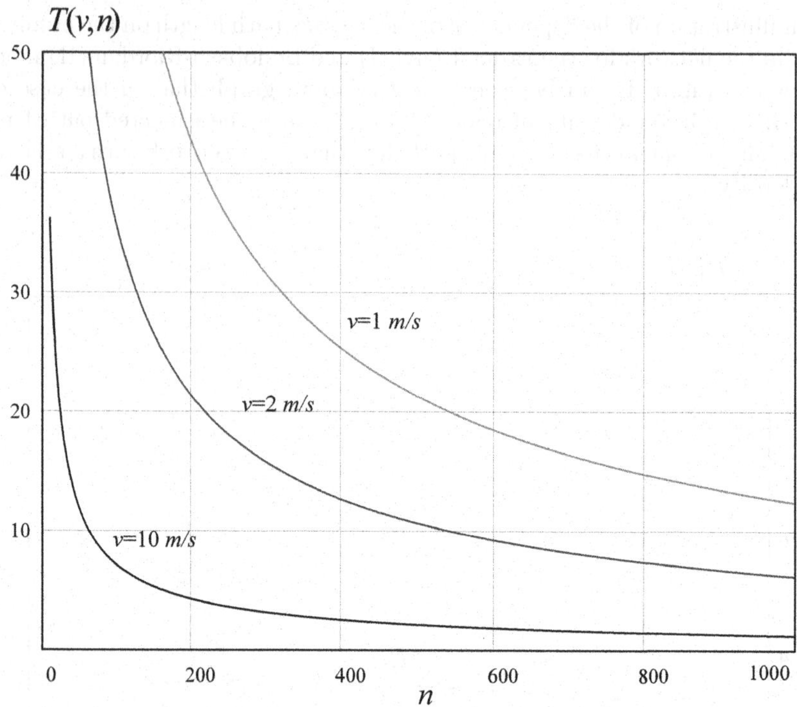

Fig. 5. Dependence of the average delivery time on the number of network nodes and the average speed of their movement.

It is easy to see that with fixed nodes, the delivery time, according to (7), the average delivery time tends to infinity. This is not entirely obvious, but under the accepted assumptions, this is exactly the case, since the delivery time in the considered network is determined only by the time of "waiting" for the connected state. At a low density of nodes, when the distances between them significantly exceed the communication range, the probability of connectivity tends to zero, i.e. communication is only achieved at a non-zero speed.

The important factors that determine the functioning of the network under consideration are the probability of message delivery and the number of transmitted messages. The mobility of nodes is a random process, therefore, the probability is nonzero that for a certain time interval τ_0 a node will not be able to send a message due to the fact that there will be no connection with other nodes. According to the chosen model, it is equal to

$$p_0(\tau_0) = e^{-2Rv\rho\tau_0} \tag{10}$$

Then, for a route of k sections, it will be

$$p_L(\tau_0, k) = 1 - (1 - e^{-2Rv\rho\tau_0})^k \tag{11}$$

where τ_0 is the message lifetime (information relevance). Then the probability of message delivery

$$p_D(\tau_0, k) = (1 - e^{-2Rv\rho\tau_0})^k \quad (12)$$

Expression (12) gives the probability of delivery of a message sent k times on each of the route sections if its lifetime (information relevance) is τ_0. This dependence is shown in Fig. 6, where $v = 6\,\text{m/s}$ and $\rho = 1 * 10^{-6}$ users/m^2.

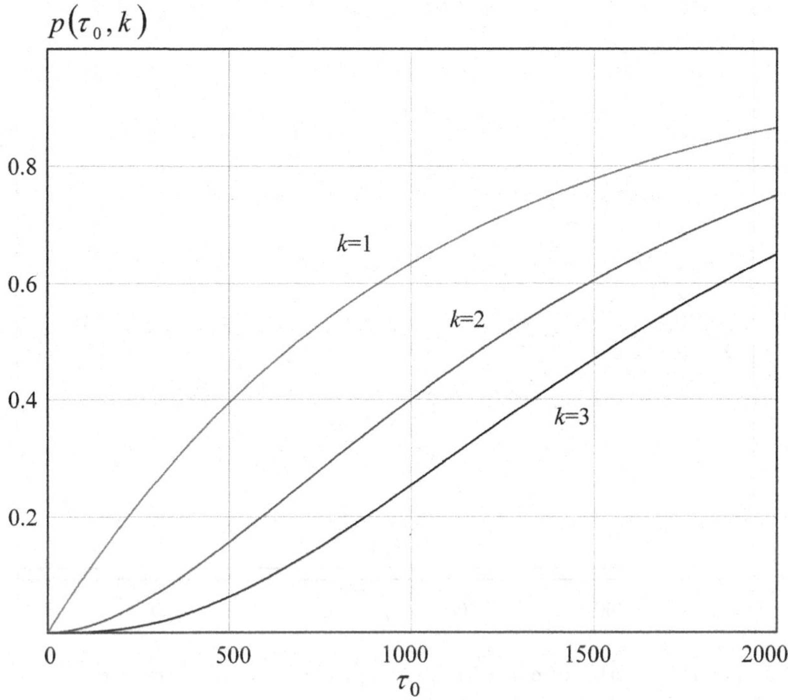

Fig. 6. Dependence of the probability of message delivery on its lifetime and the number of repetitions.

It can be seen from the given dependence that the probability of delivery increases with an increase in the message lifetime and the number of its repeats. An increase in the number of repetitions of a message leads to a significant increase in traffic in the considered network. If we assume that on each of the route sections the transit node performs an equal number of repetitions equal to k, then the total number of messages (message copies) in the network generated by the delivery of a single message will be

$$\eta(n, k) = k^d = k^{g(n)} \quad (13)$$

Where k is the number of repetitions, d is the average length of the route, determined according to (8).

Figure 7 shows the dependence of the number of messages produced in the network on the number of repetitions and the length of the route, expressed in terms of the number of nodes in the network, according to (8).

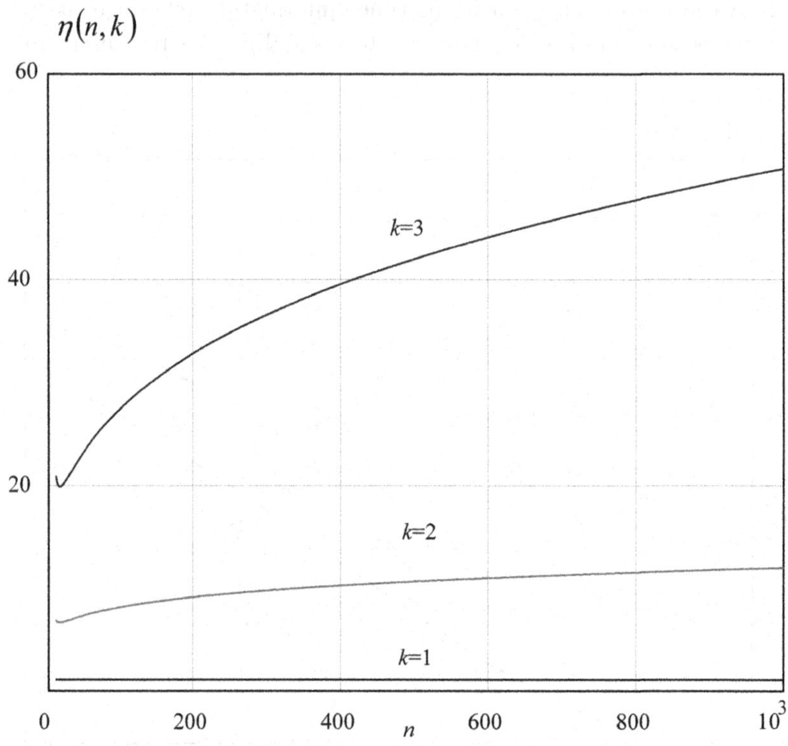

Fig. 7. Dependence of the number of messages produced in the network on the number of repetitions and the number of network nodes.

It can be seen from the above illustration that as the number of repetitions of a message increases, the number of messages generated in the network increases rapidly. This is the expected result. However, it should be noted that this growth rate, in the case of a network with a low density of nodes, is "softened" by the dependence of the route length on the number of network nodes, according to the chosen model (8). Due to this, the number of messages produced in the network, per one outgoing message, grows according to the logarithmic law with the increase in the size of the network, which makes it possible to obtain values that are within the actually permissible limits (tens of messages).

Under the above assumptions about low node density and low traffic intensity, such values are quite acceptable, and the time indicators are within reasonable limits.

Thus, modeling has shown that the time and probability of message delivery in a network with a low density of nodes with admissible delays makes it possible to obtain an information transfer service that is quite suitable for practical applications.

3 Conclusion

The above reasoning and models describe the potential capabilities of a communication network built using D2D communications. Believing that it can be organized as a peer-to-peer ad hoc network. The properties of such a network significantly depend on the density of nodes, which determines the probability of connectivity in a static mode, i.e. with stationary network nodes.

Of particular interest is a network with mobile nodes, which potentially allows the delivery of messages with a low density of terminals, such that in a static mode the probability of network connectivity is close to zero.

With moving nodes and low density, the probability of connectivity is determined by time. The constructed models allow us to estimate the average time required to ensure connectivity, i.e. delivery of the message. This time depends on both the average speed of movement of the nodes and the average length of the route. In theory, all messages on such a network will be delivered if their lifetime (life) is not limited. In practice, given the limited lifetime of a message, there is a nonzero probability of loss.

To increase the probability of delivery, it is necessary to increase the number of sent messages (attempts to send). This leads to an increase in network traffic, due to the fact that the number of message copies increases at each segment of the route. However, with a relatively low density of network nodes, this number is quite acceptable.

In general, the analysis showed that the potential capabilities of modern subscriber terminals make it possible to organize an ad hoc network with a fairly high probability of connectivity. This possibility exists both in static networks with fixed nodes and in networks with a relatively low density of mobile nodes, as a network with acceptable delays.

Network parameters such as connectivity probability, required communication range, node density, and message delivery time can be estimated using theorems from random graph theory. The practical implementation of such a network requires the development of appropriate protocols and client applications that can be used as an additional means of communication in cases of low availability of other networks.

Acknowledgment. The publication was prepared with the support of the "RUDN University Program 5–100".

References

1. Paramonov, A., Hussain, O., Samouylov, K., Koucheryavy, A., Kirichek, R., Koucheryavy, Y.: Clustering optimization for out-of-band D2D communications. Wirel. Commun. Mobile Comput. **2017**, 1–11 (2017). Article Id: 6747052

2. Mahmood, O., Paramonov, A.: Peculiarity of the internet of thing traffic routing, selection of the gateway location. In: Workshops ICUMT 2018 (2018)
3. Mahmood O.A., Paramonov, A.: Optimization of routes in the Internet of Things. In: 18th International Conference on Next Generation Wired/Wireless Networking (NEW2AN), pp. 584–593 (2018)
4. Muthanna, A., et al.: Analytical evaluation of D2D connectivity potential in 5G wireless systems. In: Galinina, O., Balandin, S., Koucheryavy, Y. (eds.) NEW2AN/ruSMART -2016. LNCS, vol. 9870, pp. 395–403. Springer, Cham (2016). https://doi.org/10.1007/978-3-319-46301-8_33
5. Paramonov, A., Nurilloev, I., Koucheryavy, A.: Provision of connectivity for (heterogeneous) self-organizing network using UAVs. In: Galinina, O., Andreev, S., Balandin, S., Koucheryavy, Y. (eds.) NEW2AN/ruSMART/NsCC -2017. LNCS, vol. 10531, pp. 569–576. Springer, Cham (2017). https://doi.org/10.1007/978-3-319-67380-6_53
6. Nurilloev, I., Paramonov, A., Koucheryavy, A.: Connectivity estimation in wireless sensor networks. In: Galinina, O., Balandin, S., Koucheryavy, Y. (eds.) NEW2AN/ruSMART -2016. LNCS, vol. 9870, pp. 269–277. Springer, Cham (2016). https://doi.org/10.1007/978-3-319-46301-8_22
7. Abdulkareem Mahmood, O., Muthanna, A., Paramonov, A.: Data delivery algorithm for latency sensitive IoT application. In: Galinina, O., Andreev, S., Balandin, S., Koucheryavy, Y. (eds.) NEW2AN/ruSMART -2019. LNCS, vol. 11660, pp. 467–480. Springer, Cham (2019). https://doi.org/10.1007/978-3-030-30859-9_40

Channel Switching Threshold Strategies for Multichannel Controllable Queuing Systems

A. S. Mandel[1(✉)] and V. A. Laptin[2]

[1] V.A. Trapeznikov Institute of Control Sciences RAS, Profsoyuznaya 65, Moscow, Russia
almandel@yandex.ru

[2] M.V. Lomonosov Moscow State University, Lenin hills 1, Moscow, Russia
straqker@bk.ru

Abstract. This paper deals with a controllable queuing system in which the number of switching service channels monitor and modify at control time points spaced apart by a fixed time step. At transition from step to step, the intensity of the simplest incoming flow changes in accordance with a Markov's chain. The system is in a stationary mode between the steps. A cost function is the minimization of the total average cost of the system over a multi-step planning period. The problem is to find a channel switching strategy. The parametric structure of an optimal strategy significantly simplifies its construction.

Keywords: Controllable queuing systems · Markovian incoming flow · Switching strategies · Strategy parameterization

1 Introduction

In this paper, we consider one of the problems related to the scope of the theory of controlled queuing systems, which began with the works of V. Rykov and summed up by him in his fundamental survey article [1], see also [2–4]. The peculiarity of the proposed work is that the process of channel switching control in a multichannel queuing system is considered using the criterion for obtaining the maximum average profit from the operation of the system on a multistep time interval. The solution is reduced to the analysis of stochastic dynamic programming algorithms and develops the results obtained in the monograph [5], taking into account the specifics of the problem under study. In accordance with the conclusions of the general theory [1], control algorithms for channel switching processes turn out to be threshold.

So, let us consider a problem of the theory of controllable queuing systems (QS) which develops and generalizes problem statements, that have been discussed in works presented at conferences DCCN15 [6], DCCN17 [7] and DCCN18 [8]. In the investigated QS the number of service channels can change at the

© Springer Nature Switzerland AG 2020
V. M. Vishnevskiy et al. (Eds.): DCCN 2020, CCIS 1337, pp. 259–270, 2020.
https://doi.org/10.1007/978-3-030-66242-4_21

moments of control, standing apart from each other by the value of fixed time step. In this case, as in [6], it is believed that the QS receives the simplest incoming flow, the intensity of which at the moments of control undergoes sudden changes, taking a finite number of values λ_i, $i \in \overline{1,k}$, from a discrete set Λ. We use the assumption that the duration of the control step (which is what we choose for the unit of time) is sufficient to establish in the QS a stationary, in the probabilistic sense, mode of operation.

The aim is to form a strategy for switching service channels (disabling redundant service channels or introducing backup channels) in order to minimize the average cost of QS in a given N-step planning period.

2 Problem Statement

We discuss the QS, in which the number of service channels is a controllable value and can be changed at control moments periodically located on the time axis (with step 1) of control of the QS state. A matrix of transition probabilities of the corresponding homogeneous Markov's chain $P = \|p_{ij}\|$, is given, where p_{ij} is the probability of transition (at the moment of control) from the intensity λ_i, $i \in \overline{1,k}$, at the previous step to the intensity λ_j, $j \in \overline{1,k}$ at the next step.

We use the assumption that the duration of the control step is sufficient to establish a stationary, in the probabilistic sense, operation mode in the QS in question at this step. If the intensity of the incoming flow at this step is equal to λ_i, while the intensity of service in one service channel is μ, then it is obvious [9,10], that the number of service channels in the QS should be chosen to satisfy the following inequality:

$$u \geq u_{\text{critical}}(\lambda_i) = \underline{u_i} = \left[\frac{\lambda_i}{\mu}\right] + 1. \tag{1}$$

Let us introduce a control quality function which for given: (a) number of steps n, remaining till the end of the planning period ($n \leq N$), (b) intensity of the incoming flow λ_i, $i \in \overline{1,k}$, (c) actual number of service channels m and (d) decision on the switched number of service channels u is described by the value of average total cost $C_n(\lambda_i, m, u)$. The aim is to minimize $C_n(\lambda_i, m, u)$ by choosing the channel switching strategy. This cost is mathematical expectation of a sum of one-step costs at the remaining n steps, along the trajectory of the incoming flow whose intensity changes in the Markovian jump-wise manner.

Let us write an equation for one-step costs in the first step $C^{(1)}(\lambda_i, m, u)$:

$$C^{(1)}(\lambda_i, m, u) = C_{\text{oper}} + C_{\text{queue}} + C_{\text{switch}}. \tag{2}$$

Here, the operating costs of the active systems in the first step $C_{\text{oper}} = c_1 u$, the cost of queuing in the stationary mode is $C_{\text{queue}} = d\bar{l}_{\text{queue}}$, where $\bar{l}_{\text{queue}}$ is the average queue length, while the switching cost C_{switch} can be presented as:

$$C_{\text{switch}} = \begin{cases} A_1, & \text{if } u > m; \\ 0, & \text{if } u = m; \\ A_2 + c_2(m - u), & \text{if } u < m. \end{cases} \tag{3}$$

We will now use classical results [9,10] to write down:

$$\bar{l}_{\text{queue}} = \left[\sum_{k=1}^{u-1} \frac{(u\rho_i)^k}{k!} + \frac{(u\rho_i)^u}{u!(1-\rho_i)}\right]^{-1} \frac{(u\rho_i)^u \rho_i}{u!(1-\rho_i)^2}, \tag{4}$$

where $\rho_i = \frac{\lambda_i}{u\mu}$.

Now, if under the assumptions made we denote through $C_n^*(\lambda_i, m)$ the minimum possible value of the average total cost in the last n steps of the control process, it is logical to write the following system of discrete dynamic programming equations:

$$C_1^*(\lambda_i, m) = \min_{u \geq \underline{u_i}} C^{(1)}(\lambda_i, m, u), \tag{5}$$

$$C_n^*(\lambda_i, m) = \min_{u \geq \underline{u_i}} (C^{(1)}(\lambda_i, m, u) + \alpha \sum_{j=1}^{k} p_{ij} C_{n-1}^*(\lambda_j, u)), \tag{6}$$

where $i \in \overline{1,k}$, α is the discount factor, $0 \leq \alpha \leq 1$, while $n \in \overline{2,N}$.

3 *A*-convexity in Discrete Problems of Optimization

We consider function $g(i)$ of the variable i, taking values from the finite segment of the natural series: $i = 1, 2, \ldots, k$.

Definition 1. *The function $g(i)$ is called convex (downward) if for all natural i and j:*

$$g(i+j) - g(i) - \Delta^{(1)} g(i) \times j \geq 0, \tag{7}$$

where $\Delta^{(1)} g(i)$ is the first difference of $g(i)$ function in point i: $\Delta^{(1)} g(i) = g(i) - g(i-1)$.

The necessary and sufficient condition for the convexity (downward) of the function is fulfilling the inequality $\Delta^{(2)} g(i) \geq 0$, where $\Delta^{(2)} g(i)$ is the second difference of $g(i)$ function in point i. In fact, let j from definition 1 be 1, then formula (7) is written as $\Delta^{(1)} g(i+1) \geq \Delta^{(1)} g(i)$, i.e. the first difference of $g(i)$ is rejected by the monotonically increasing function, i.e. $\Delta^{(2)} g(i) \geq 0$. Prove of the sufficiency is similar.

Definition 2. *Function*[1] *$g(i)$ is called A-convex ($A \geq 0$), if for all natural i and j:*

$$A + g(i+j) - g(i) - \Delta^{(1)} g(i) \times j \geq 0. \tag{8}$$

[1] The concept of A-convexity was proposed by Herbert Scarf [12] to analyze the properties of optimal decision making (control) strategies in inventory control problems in the presence of fixed supply costs that do not depend on the size of supply, adding to them the amounts determined by the size of the supply batch.

For functions of discrete variables, all the properties of A-convex functions noted by Herbert Scarf [12] are retained. Namely:

Property 1. If $g(i)$ is A-convex, then for any natural j the function $g(i+j)$ is also A-convex.

Property 2. If function $g_1(i)$ is A_1-convex, and function $g_2(i)$ is A_2-convex, then for any $\theta_1, \theta_2 > 0$, function $g(i) = \theta_1 g_1(i) + \theta_2 g_2(i)$ is also $(\theta_1 A_1 + \theta_2 A_2)$-convex.

Property 3. If function $g(i)$ is A_1-convex, then it is also A_2-convex for any $A_2 > A_1$.

Property 4. Let i be a random value with distribution $p_i{}_1^k, p_i > 0, \sum_{i=1}^k p_i = 1$, and let function $g(i)$ be A-convex. Then, function $\alpha \sum_{i=1}^k p_i g(i)$, with $\alpha \geq 0$, is also A-convex.

4 Properties of Optimal Channel Switching Strategies

Let us revisit the problem of investigating properties of solutions to the system of dynamic programming equations (5)–(6). Some words about the plan of this section to present the following results:

- review of the qualitative features of "myopic" inventory control strategies;
- broadening of the concept of A-convexity for the problems of the examined type;
- transfer of results obtained for "myopic" strategies to multistep (dynamic) control problems.

4.1 "Myopic" Strategies of Control

This subsection briefly (with minor corrections) retells the content of the work [8] by the same authors.

In order to build an optimal "myopic" switching strategy, one has to find:

$$\min_{u \geq \underline{u_i}} C^{(1)}(\lambda_i, m, u) = \min_{u \geq \underline{u_i}} \{c_1 u + d\bar{l}_{\text{queue}} + \min \begin{cases} A_1 \mathbb{1}(u-m), \\ 0, \\ A_2 \mathbb{1}(m-u) + c_2(m-u), \end{cases} \}, \quad (9)$$

where $\mathbb{1}(u)$ denotes the function of a unit jump (Heaviside's function) which is equal to 1, if $u > 0$, or 0 in all other cases.

The main results obtained in [8] are that for each state $i = 1, 2, ..., k$, there are five critical parameters that fully determine the "myopic" inventory control strategy: $\underline{u_i}$ (defined by formula (1)), $r_{1,i}^{(1)}$, $R_{1,i}^{(1)}$, $r_{1,i}^{(2)}$ and $R_{1,i}^{(2)}$. Moreover, the last four parameters are arranged as follows: $r^{(1)}_{1,i} < R^{(1)}_{1,i} \le R^{(2)}_{1,i} < r^{(2)}_{1,i}$, while a "myopic" channel switching strategy obeys the formula[2]:

$$u = \begin{cases} R^{(1)}_{1,i}, & \text{if } m \le r^{(1)}_{1,i} \text{ (switching on)}, \\ m, & \text{if } r^{(1)}_{1,i} < m \le R^{(1)}_{1,i}, \\ m, & \text{if } R^{(2)}_{1,i} \le m < r^{(2)}_{1,i}, \\ R^{(2)}_{1,i}, & \text{if } m \ge r^{(2)}_{1,i} \text{ (switching off)}. \end{cases} \tag{10}$$

It should be noted that in [8], as already noted, the discussion was conducted at a purely qualitative level and did not take into account, in particular, the discreteness of one of the QS state variables. Using the technique briefly outlined in Sect. 3 of this paper, it is possible to reformulate and strictly prove all the statements made in [8]. In contrast to [8], in this paper, the last four characteristic parameters of an optimal "myopic" strategy have one more sub-index, in this case equal to 1. In one of the subsequent paragraphs, this unit will transform into n the number of steps left until the end of the planning period for the multistep problem).

If this has not been the case, we would have to take care of "sliding" modes, when at first we would have to switch off part of the channels (bringing the number of active channels to $R^{(2)}_{1,i}$), but immediately after that, under certain circumstances, we would have to switch on part of the channels, returning the QS to another parameter – $R^{(1)}_{1,i}$), etc. It does not seem to be much of a trouble, but there is one more and probably more difficult task.

The above fact necessitates one more decomposition of $C_n(\lambda_i, m, u)$ criterion of the initial problem (5)–(6). In other words, two alternative managerial decision options are considered separately: (a) to enable additional channels or, conversely, (b) to disable some active channels. At the same time, the fee for enabling is described by the function $B_{\text{switch on}}(u) = c_1 u + d\bar{l}_{\text{queue}}$, while the fee for disabling is described by the function $B_{\text{switch off}}(u) = (c_1 - c_2)u + c_2 m + d\bar{l}_{\text{queue}}$, where $\bar{l}_{\text{queue}}$ is obtained from the formula (4) and all designations coincide with those introduced above. It is the type of functions that gives rise to the inequality $R^{(1)}_{1,i} \le R^{(2)}_{1,i}$. It is also important to note that $R^{(1)}_{1,i}$ is the point of the absolute minimum of the one-step cost function when channels are enabled, and $R^{(2)}_{1,i}$ is the point of absolute minimum of one-step cost function when channels are disabled.

4.2 Broadening of the Notion of *A*-convexity

The proof of the facts listed in paragraph 4.1 based on the analogy between the problems of inventory control theory and those of queuing systems theory, which was important feature discussed in the present and previous works of the first

[2] The application of formula (9) is associated with certain caveats, the essence of which is that in some cases listed in [8], the parameter $R^{(1)}_{1,i}$ is assigned equal to $\underline{u_i}$. There are other nuances as well.

author. An example of the original analysis of such analogies is the paper [13]. The problem of inventory control discussed in [13] the authors called "fantastic"[3]. Nonetheless, we cannot but admit that for the class of problems under consideration (both "fantastic" from the theory of inventory control and the problem of channel switching of QS theory), some unpreparedness and unsuitability of the existing theory of optimization and the theory of convexity for the solution of these problems was evident. The point is that in the mathematical description of the class of problems in question, not one but two different concepts of optimality (optimality criteria) are used. Speaking a language adequate to this class of problems, it is the *optimality of enabling channels* (when placing orders in inventory control theory), when we move along the axis of an integer variable u *left-to-right*, and the *optimality of disabling channels* (returning goods in inventory control theory), when we move along the axis of the variable *right-to-left.* We shall wrap this remark into new definitions.

Definition 3. *Function $g(i)$ is referred to as A-convex from the left ($A \geq 0$), if for all natural i and j, which are to the left of the point of its absolute minimum,*

$$A + g(i + j) - g(i) - \Delta^{(1)} g(i) \times j \geq 0.$$

Definition 4. *Function $g(i)$ is referred to as A-convex from the right ($A \geq 0$), if for all natural i and j, which are to the right of the point of its absolute minimum,*

$$A + g(i + j) - g(i) - \Delta^{(1)} g(i) \times j \geq 0.$$

We now can state that the enabling function in the one-step ("myopic") problem is A_1-convex from the left, whereas the disabling function for the same problem is A_2-convex from the right.

4.3 Multi-step Dynamic Problems of Channel Switching

Let us return to the solution of the multistep problem of channel switching control, which is described in Eqs. (5)–(6). We repeat the notation of goal function in Eq. (6):

$$C_n^*(\lambda_i, m) = \min_{u \geq \underline{u_i}} (C^{(1)}(\lambda_i, m, u) + \alpha \sum_{j=1}^{k} p_{ij} C_{n-1}^*(\lambda_j, u)). \quad (11)$$

The expression in braces in the right hand side (r.h.s.) of this equation, regardless of which alternative is estimated (on or off), contains a non-modifying term $\alpha \sum_{j=1}^{l} p_{ij} C_{n-1}^*(\lambda_j, u)$. The one-step add-on $C^{(1)}(\lambda_i, m, u)$, by contrast, varies when the alternative is changed. From this we can conclude that there is every reason to believe that the expression in braces in the r.h.s. of formula (6)

[3] Fantastic in [13] was the assumption that a warehouse could not only submit replenishment orders, but also return the goods to their suppliers.

(also two-variant) is simultaneously A_1-convex from the left and A_2-convex from the right.

Assume mathematical induction that this statement is true for any n. As it follows from [8], for the case $n = 1$ it is so. Suppose now that this statement is true for the case $n-1$. Further proof of the truth of this statement for the case n is quite tedious (in terms of mathematical computation), but, in fact, is carried out according to the classical scheme of proof of optimality, but for two-level strategies [11,12].

As noted above, an important feature that helps avoiding the risk of "pitfalls", like sliding modes, is the ordering of the local minimum points of alternative functions in the form of inequalities $R_{n,i}^{(1)} \leq R_{n,i}^{(2)}$, $n \in \overline{1,N}, i \in \overline{1,k}$. It should be admitted here that the authors managed to prove this statement for strictly $n = 1$ case only. As for the arbitrary value of n, we can only argue that in the entire range of computational experiments performed there was not a single case of failure to fulfill these inequalities. The authors hope that in the near future they will manage to find a strong solution to this problem.

5 Simulation Example

The system of equations for discrete dynamic programming (11) has one interesting feature in the case when the intensities of the incoming flow for all steps of the process under study are known before planning. Of course, in this case, it is not difficult to calculate the minimum total costs, however, with a large number of steps and a large number of service channels, the calculations can require a lot of computer time (for the exhaustive search method). This system of equations allows not only to do this in a linear (on the number of steps) time, but also to find out the correct order of switching the service channels at each step of the process under study (route).

To begin with, let us show that the threshold channel switching strategy, which is valid for the one-step case with a fixed initial number of service channels m, turns out to be non-optimal already for the multistep case, when the intensity of the incoming flow for each step is known in advance (before planning). If we are guided by this policy and at each step choose the value of the number of service channels that implements the minimum costs specifically at this step, then the minimum costs for the entire process cannot be found. However, the threshold strategies themselves take place, since they help to calculate all the available options for switching service channels.

To do this, consider an example of a simple 3-steps process. Moreover, it will be enough to find the optimal strategy simply by going through all the available options (this is possible due to the small size). Then find a step that does not correspond to the optimal threshold strategy for the one-step case with a given initial number of service channels. You can even give an example of a thought experiment when the number of channels is the same in the previous and next steps. In order not to spend money on switching, you can incur large costs at the current step, but win in the whole process.

Let's move on to an example of a 3-steps multistep process. Let's choose the following values for this example: $c_1 = 1, c_2 = 0, A_1 = A_2 = 1$, the set $\Lambda = \{28, 18, 28\}, d = 1, u_{max} = 10$.

Based on these data, it is possible to compile tables of the dependence of the total cost functional on the initial number of service channels and the number of channels after making a control decision for each step separately, shown in Figs. 1, 2 and 3. In these tables, the rows correspond to the value of the number of service channels at the beginning of the step, and the columns represent the number of channels after the control decision is made. That is, if at the 1st step we had the initial number of service channels equal to 6 (corresponds to the 6th row in the table in Fig. 1), then the minimum total costs will be realized if we connect two more service devices (see Fig. 1 – 6th row, 8th column).

Cells with a *NaN* value correspond to less than the critical number of channels. The critical values are 6, 4 and 6, respectively. It is clearly seen at the 2nd and 3rd steps (see Fig. 2 and 3) not only the number of available rows and columns is not equal to each other, but also at the previous and current steps the critical values of the service channels are different.

	1	2	3	4	5	6	7	8	9	10
1	NaN	NaN	NaN	NaN	NaN	NaN	NaN	NaN	NaN	NaN
2	NaN	NaN	NaN	NaN	NaN	NaN	NaN	NaN	NaN	NaN
3	NaN	NaN	NaN	NaN	NaN	NaN	NaN	NaN	NaN	NaN
4	NaN	NaN	NaN	NaN	NaN	NaN	NaN	NaN	NaN	NaN
5	NaN	NaN	NaN	NaN	NaN	NaN	NaN	NaN	NaN	NaN
6	NaN	NaN	NaN	NaN	NaN	17.533288	9.949279	9.633551	10.23406	11.088745
7	NaN	NaN	NaN	NaN	NaN	18.533288	8.949279	9.633551	10.23406	11.088745
8	NaN	NaN	NaN	NaN	NaN	18.533288	9.949279	8.633551	10.23406	11.088745
9	NaN	NaN	NaN	NaN	NaN	18.533288	9.949279	9.633551	9.23406	11.088745
10	NaN	NaN	NaN	NaN	NaN	18.533288	9.949279	9.633551	10.23406	10.088745

Fig. 1. Dependence of total costs on the initial and selected number of service channels for the 1st step.

After going through all the available options, we find that the minimum total costs for the 3-steps process are about 25 units, while the optimal route is described by the set $\{7, 7, 7\}$. That is, having chosen the initial number of channels equal to 7 at the first step, it is most advantageous not to switch at all anymore. Although the minimum value for each step separately is not implemented in any of the cases of this route.

Let us now try to compose another route, choosing at each step the minimum value of the total costs. For the 1st step, the minimum will be delivered by the initial choice of the number of channels equal to 8, it is also the final one for the

	1	2	3	4	5	6	7	8	9	10
1	NaN	NaN	NaN	NaN	NaN	NaN	NaN	NaN	NaN	NaN
2	NaN	NaN	NaN	NaN	NaN	NaN	NaN	NaN	NaN	NaN
3	NaN	NaN	NaN	NaN	NaN	NaN	NaN	NaN	NaN	NaN
4	NaN	NaN	NaN	NaN	NaN	NaN	NaN	NaN	NaN	NaN
5	NaN	NaN	NaN	NaN	NaN	NaN	NaN	NaN	NaN	NaN
6	NaN	NaN	NaN	12.170492	7.079926	6.302722	8.093845	9.02912	10.008726	11.002487
7	NaN	NaN	NaN	12.170492	7.079926	7.302722	7.093845	9.02912	10.008726	11.002487
8	NaN	NaN	NaN	12.170492	7.079926	7.302722	8.093845	8.02912	10.008726	11.002487
9	NaN	NaN	NaN	12.170492	7.079926	7.302722	8.093845	9.02912	9.008726	11.002487
10	NaN	NaN	NaN	12.170492	7.079926	7.302722	8.093845	9.02912	10.008726	10.002487

Fig. 2. Same as in Fig. 1, but for the 2nd step.

	1	2	3	4	5	6	7	8	9	10
1	NaN	NaN	NaN	NaN	NaN	NaN	NaN	NaN	NaN	NaN
2	NaN	NaN	NaN	NaN	NaN	NaN	NaN	NaN	NaN	NaN
3	NaN	NaN	NaN	NaN	NaN	NaN	NaN	NaN	NaN	NaN
4	NaN	NaN	NaN	NaN	NaN	18.533288	9.949279	9.633551	10.23406	11.088745
5	NaN	NaN	NaN	NaN	NaN	18.533288	9.949279	9.633551	10.23406	11.088745
6	NaN	NaN	NaN	NaN	NaN	17.533288	9.949279	9.633551	10.23406	11.088745
7	NaN	NaN	NaN	NaN	NaN	18.533288	8.949279	9.633551	10.23406	11.088745
8	NaN	NaN	NaN	NaN	NaN	18.533288	9.949279	8.633551	10.23406	11.088745
9	NaN	NaN	NaN	NaN	NaN	18.533288	9.949279	9.633551	9.23406	11.088745
10	NaN	NaN	NaN	NaN	NaN	18.533288	9.949279	9.633551	10.23406	10.088745

Fig. 3. Same as in Fig. 1, but for the 3rd step

1st step, the corresponding value of the minimum cost at the step is 8.63. Then, at the 2nd step, the initial number of channels will be 8, and the minimum will be realized when 5 channels are turned on and equal to 7.08. At the 3rd step, the initial number of channels will be 5, and the minimum will be at 8 and equal to 9.63. Then the total costs will be approximately 25.35 and more than 25, obtained by the exhaustive search method. Moreover, it turns out that the optimal values of the service channels for each step lie between the values of the number of channels that realize the minimum cost for each step taken separately. This is all true only in the case when the intensity of the incoming flow for each step is known in advance, even before the start of planning.

From the results obtained above, the problem arises of finding the optimal route that minimizes the functional of total costs for the entire process. You can represent this in a more visual form by writing everything down in a table (see Fig. 4), where the rows correspond to the step number (written in direct time), and the columns correspond to the initial number of channels. There is a feeling that the optimal channel switching strategy in this case can be "lost" in this dynamic programming equation. However, after running in the opposite direction, you can restore all the values of the number of service devices that should be selected.

	1	2	3	4	5	6	7	8	9	10
1	NaN	NaN	NaN	NaN	NaN	25.992402	24.992402	25.296222	25.947537	25.992402
2	NaN	NaN	NaN	NaN	NaN	15.936273	16.043123	16.662671	16.713477	16.713477
3	NaN	NaN	NaN	9.633551	9.633551	9.633551	8.949279	8.633551	9.234060	9.633551

Fig. 4. Total costs for a 3-steps process

To restore the "route" obtained using the dynamic programming equation, it will be necessary to create a table similar to the table in Fig. 4 (that is, the rows also correspond to the numbers of steps, and the columns correspond to the number of initial service channels) with the values of the number of service channels after making the control decision (see Fig. 5). Next, we find the minimum value of the total costs $C_n(\xi)$ according to the table in Fig. 4 and the corresponding initial number of channels m. For this initial number m, we find the corresponding column in the table in Fig. 5 (1st row, 7th column) is already the number of service channels after making the control decision u. Next, go to the 2nd row of the table in Fig. 4 and select a column equal to the number of channels u, with which step 1 was completed. After going through all the rows sequentially, you can get the entire route; for this example it will be $\{7, 7, 7\}$, which corresponds to the result obtained by enumerating all the options. You can take a different initial number of channels as an example. Let, for example, $m = 9$. Then the total costs (see Fig. 4) will be approximately equal to 25.95, and the optimal route itself $\{9, 5, 8\}$.

	1	2	3	4	5	6	7	8	9	10
1	NaN	NaN	NaN	NaN	NaN	7	7	8	9	7
2	NaN	NaN	NaN	NaN	NaN	6	7	8	5	5
3	NaN	NaN	NaN	8.0	8.0	8	7	8	9	8

Fig. 5. The number of service channels depending on the initial number

Hence, we can conclude that the dynamic programming equation (11) helps in a multistep process, with the intensity of the incoming flow for all steps known before planning, to choose the optimal solution that implements the minimum total costs. Moreover, the algorithm linearly depends on the number of steps, and its complexity is $O(m_{max}^2 n)$.

6 Conclusion

The paper concerns with channel switching strategies in a multi-line queuing system with a Markov description of the input flow intensity change process. The objective function in strategy development is to minimize the total average cost in a multistep planning period. Thus it is assumed, that the procedure of channel switching (enabling or disabling) leads to the expenses which consist of the fixed payments and costs which size depends on the number of enabled or disabled channels. As a result, the optimal channel switching strategies turn out to be parametric and the optimal solution at each step depends only on four characteristic parameters. The proof process required introducing some generalizations of the A-convexity concept, classic for the operations research.

References

1. Rykov, V.: Controllable queueing systems: from the very beginning up to nowadays. Reliab. Theory Appl. **12**, 39–61 (2017)
2. Rykov, V.V., Efrosinin, D.V.: To the slow server problem. Autom. Rem. Control **12**, 81–91 (2009)
3. Rykov, V.: On a slow server problem. In: Li, H., Li, X. (eds.) Stochastic Orders in Reliability and Risk. Lecture Notes in Statistics, vol. 208, pp. 351–361. Springer, New York (2013). https://doi.org/10.1007/978-1-4614-6892-9_18
4. Rykov, V.V., Kozyrev, D.V.: Optimal control for scheduling with correction. In: Proceedings of the XII All-Russian Workshop on Control Problems, Moscow, 6–19 June 2014, pp. 8850–8854. ITP Ras (2014)
5. Sennott, L.I.: Stochastic Dynamic Programming and the Control of Queueing Systems. John Wiley & Sons, New York (1999). 358 p.
6. Mandel, A.: Econometric models of controllable multiple queuing systems. In: Vishnevsky, V., Kozyrev, D. (eds.) DCCN 2015. CCIS, vol. 601, pp. 296–304. Springer, Cham (2016). https://doi.org/10.1007/978-3-319-30843-2_31
7. Mandel, A., Bakulin, K.: Models of controllable multiple queuing systems for channel switching myopic strategies. In: Proceedings of the 20th International Conference, Distributed Computer and Communication Networks, DCCN 2017, Moscow, Russia, Tecnhosphera, pp. 534–542 (2017). (in Russian)
8. Mandel, A., Laptin, V.: Myopic channel switching strategies for stationary mode: threshold calculation algorithms. In: Vishnevskiy, V.M., Kozyrev, D.V. (eds.) DCCN 2018. CCIS, vol. 919, pp. 410–420. Springer, Cham (2018). https://doi.org/10.1007/978-3-319-99447-5_35
9. Gnedenko, B., Kovalenko, I.: Introduction to The Queuing Systems Theory. Nauka, Moscow (1966). (in Russian)
10. Vishnevskiy, V.: Theoretical Principles of Computer Networks Design. Tecnhosphera, Bengaluru (2003). (in Russian)

11. Hadley, G., Whitin, T.M.: Analysis of Inventory Control Systems. Prentice Hall Inc., Englwood Cliffs (1967)
12. Arrow, K., Karlin, S., Scarf, H.: Studies in the Mathematical Theory of Inventory and Production. Stanford University Press, Stanford (1958)
13. Mandel, A., Granin, S.: Investigation of analogies between the problems of inventory control and the problems of the controllable queuing systems. In: Proceedings of 2018 Eleventh International Conference Management of Large-scale System Development (MLSD 2018), pp. 1–4. IEEE (2018). https://doi.org/10.1109/MLSD.2018.8551852

Modeling and Simulation of Reliability Function of a k-out-of-n:F System

Nika Ivanova[1,2(✉)]

[1] Peoples' Friendship University of Russia (RUDN University), 6 Miklukho-Maklaya St., 117198 Moscow, Russian Federation
nm_ivanova@bk.ru

[2] V.A.Trapeznikov Institute of Control Sciences of Russian Academy of Sciences, 65 Profsoyuznaya street, 117997 Moscow, Russia

Abstract. A hot standby repairable k-out-of-n system is studied. In one of the previous papers, reliability function for the case of $k = 2$ and $k = 3$ with exponential life and general repair time distribution have been found. Moreover, sensitivity analysis of their reliability characteristics to the shape of the repair time distribution of the system's elements has been performed. In this paper, the problem of the asymptotic insensitivity of such a system when both life and repair time have general distribution is considered with a simulation approach. The results are presented for a special case of a 3-out-of-6:F system.

Keywords: k-out-of-n system · Reliability function · Mathematical modeling and simulation · Markovization method · Sensitivity analysis · Python

1 Introduction

There are many examples of systems of k-out-of-n – type in various fields of human activity. They are used in areas such as telecommunication, transmission, transportation, manufacturing, and service applications. Therefore, the reliability study of k-out-of-n systems is useful not only from a theoretical point of view but also from a practical one. Unfortunately, the calculation of some reliability characteristics by analytical methods becomes difficult even when considering relatively simple systems. This task becomes more complicated in the case when the repair time of the system's elements is supposed non-exponential. For this problem solution, the simulation methods can be applied for the necessary characteristics' calculation.

A k-out-of-n system is a repairable system that consists of n elements, and its behavior depends on how we determine the parameter "k". If the system remains operational when k out of n elements fail, it defines as a k-out-of-n:G system.

The publication has been prepared with the support of "RUDN University Program 5-100" program.

© Springer Nature Switzerland AG 2020
V. M. Vishnevskiy et al. (Eds.): DCCN 2020, CCIS 1337, pp. 271–285, 2020.
https://doi.org/10.1007/978-3-030-66242-4_22

In the other case, if the system fails, when k of its element fails, the system is denoted as a k-out-of-n:F system. In this paper, we consider the k-out-of-n:F system [1,2].

The importance of k-out-of-n systems' application is confirmed by many studies as well as by the presence of various applied problems that are currently being solved using such models. For example, one of the crucial applications in this area is the reliability study of high-altitude unmanned rotor-craft platforms [3–5]. The multi-rotor architecture of such platforms allows a platform with n rotary-wing engines to stay operational even after $k-1$ engines fail.

Due to the continuous development and complication of such systems, their study is also becoming increasingly difficult and requires the usage of new methods. Therefore, researchers resort to new calculation methods, among which there is a simulation. One of the earliest results of these studies belongs to Barlow and Heidtmann. In 1984, they created a linear-time algorithm and its short computer program in BASIC for k-out-of-n:G system reliability computation [6]. In [7] the study was continued, and an improved version of the method for reliability calculation of the k-out-of-n:G system was presented. In the paper [8], the study of the reliability function of a homogeneous hot double redundant repairable system is extended with the help of a discrete-event simulation model. In [9], a simulation method is applied to calculate the steady-state probabilities of a heterogeneous double redundant hot standby repairable system.

Further, some papers with a more complex k-out-of-n:G system structure appear. For example, one of the widely used configurations is a weighted k-out-of-n:G(F) system, where the weight associated with each component can be considered as a load/capacity of that component. A weighted k-out-of-n:G systems are widely applied in various scenarios, such as multiple line (power/oil transmission line) transmission systems, where the accuracy of complex calculations is so significant, and simulation is also applied [10,11]. Note that the simple k-out-of-n:G system is a unique case of the weighted k-out-of-n:G system where the weight of each component is 1. For a weighted k-out-of-n system, the accuracy of the data obtained is essential since any error in the application of the model can have adverse consequences in practice. Thus, on the one hand, the simulation model will probably have a more complex structure, but on the other hand, it can help improve the results of analytical calculations.

Simulation tools have many applications in different types of analysis and calculations. Among them, the calculation of stationary and non-stationary characteristics, sensitivity, and reliability analysis is of the key importance [12,13]. Nowadays, there are many various situations under which the above tasks are solved, and simulation modeling is applied. For example, consecutive k-out-of-n system [14]; a multi-state consecutive k-out-of-n:G [15]; three-unit series system under warm standby [16]; repairable and non-repairable systems with two types of independent failures [17]. Also, there are many works devoted to the study of the systems' reliability as well as the sensitivity of its characteristics to the shapes of life and repair time distributions (see [18–22]).

This paper continues these studies related to reliability and sensitivity and considers a hot standby repairable system using simulation with Python programming language. The article is organized as follows. In the next section, the problem setting and some notations will be introduced. In Sect. 3, a k-out-of-n:F system will present in a specific case of 3-out-of-6:F system, and analytical results of its reliability function and the meantime to failure will study. These results are performed in case of Poisson failure flow and general repair time distribution. In the 4-th Section, we present simulation results of the considered system with general distributions for both life and repair time in comparison with the analytical ones. The paper ends with a conclusion and some problem's description.

2 Problem Setting and Notation

Consider a hot redundant k-out-of-n:F ($k < n$) reparable system. Such a system can be viewed as a system in parallel elements' connection, which loses its work capability if any of its k elements fail (Fig. 1). The dashed line means that the system requires at least k of n elements to operate. The allocation of these elements is conditional since they are independent items and any k of n components may fail.

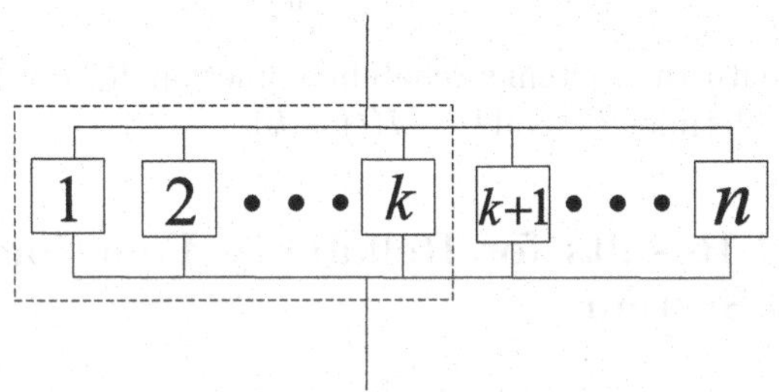

Fig. 1. A k-out-of-n : F system.

Suppose that

- the system works till it first enters state k, which is the state of the system failure;
- there is only one unit to repair a failed element;
- the elements fail according to a Poisson flow with intensity α (for analytical results section only);
- the random repair time of elements are independent and their common cumulative distribution function (c.d.f.) $B(t)$ is absolutely continuous with probability density function (p.d.f.) $b(t) = B'(t)$.

The system state space can be represented as $\mathbf{E} = \{0, 1, ...k\}$, where:

- 0 means all n elements operate;
- j means that j elements out of n ($j = \overline{1, k-1}$) have failed, one of them is being repaired, and others $(n-k)$ operate;

- k means that k elements have failed that is the system failure and its restoration, "DOWN" state.

Using the so-called *markovization method* [23] introduce as a supplementary variable $X(t)$ — the elapsed repair time of the element under repair, and consider a two-dimensional stochastic process Z

$$Z = \{Z(t) = (J(t), X(t)),\ t \geq 0\},$$

where the value $J(t)$ represents the number of failed elements at time t. Due to the supplementary variable the process Z is a Markov one.

Denote its micro-state p.d.f.'s concerning the supplementary variable in domain $0 \leq x \leq t < \infty$ by

$$\pi_j(t; x)dx = \mathbf{P}\{J(t) = j,\ x < X(t) \leq x + dx\}\ \ (j = \overline{1, k})$$

and corresponding macro-state probabilities for $t \geq 0$ by

$$\pi_j(t) = \mathbf{P}\{J(t) = j\} = \int_0^t \pi_j(t; x)dx.$$

The paper deals with the system's reliability function $R(t) = \mathbf{P}\{T > t\}$, where T is the system's lifetime, $T = \inf\{t:\ J(t) = k\}$.

3 Analytical Results for Reliability Function for 3-out-of-6 System

In order to avoid cumbersome calculations and being focused on applying the system to the multicopter model (for example, see [4,5]), this paper deals with the special case of a k-out-of-n: F system, when $k = 3$ and $n = 6$. To construct the appropriate Markov process we introduce the following notations and present the transition graph of the 3-out-of-6 system as a two-dimensional stochastic process Z (see Fig. 2).

- j — the number of elements in the "DOWN" state,
- $\lambda_j = (n - j)\alpha$ — the system failure intensity in its j-th state,
- $\beta(x) = \frac{B'(x)}{1-B(x)}$ — conditional repair density of elements, given elapsed repair time is x,
- $\tilde{b}(s)$ — Laplace transform (LT) of the p.d.f. $b(t)$,
- $b = \int_0^\infty (1 - B(x))dx$ — mean repair time of a failed element.

According to the Fig. 2, the system of Kolmogorov forward partial differential equations for process Z with absorbing state 3 in the scope $0 < x < t < \infty$ has the following form

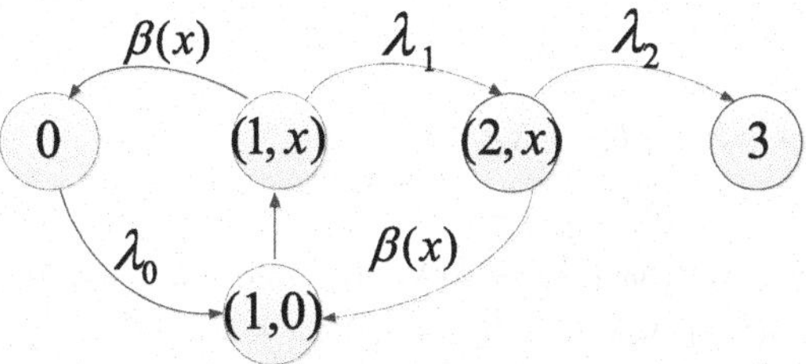

Fig. 2. Transition graph of the 3-out-of-6 system with absorption

$$
\begin{aligned}
\frac{d}{dt}\pi_0(t) &= -\lambda_0\pi_0(t) + \int_0^t \pi_1(t,x)\beta(x)dx,\\
\left(\frac{\partial}{\partial t} + \frac{\partial}{\partial x}\right)\pi_1(t;x) &= -(\lambda_1 + \beta(x))\pi_1(t;x),\\
\left(\frac{\partial}{\partial t} + \frac{\partial}{\partial x}\right)\pi_2(t;x) &= -(\lambda_2 + \beta(x))\pi_2(t;x) + \lambda_1\pi_1(t;x)\\
\frac{d}{dt}\pi_3(t) &= \lambda_2\int_0^t \pi_2(t;x)dx.
\end{aligned}
\tag{1}
$$

jointly with the initial

$$\pi_0(0) = 1, \tag{2}$$

and the boundary conditions

$$
\begin{aligned}
&\pi_1(t;0) = \lambda_0\pi_0(t) + \int_0^t \pi_2(t;x)\beta(x)dx,\\
&\pi_2(t;0) = 0.
\end{aligned}
\tag{3}
$$

Remark 1. Note that the second boundary condition follows from the fact that the process never occurs into the state 2 with the elapsed time x equal to zero since the process enters this state only as a result of a failure of another element and the transition from the state $(1,x)$ with the same elapsed repair time.

The proof of the following Theorem 1 is presented in [22] using the method of characteristics [24], so it will not be considered in this work. The objectives of the study are to build a simulation model and to compare its results with the analytical ones that were obtained earlier.

Theorem 1. *The Laplace Transforms (LT) $\tilde{R}(s)$ of the reliability function $R(t)$ of the 3-out-of-6:F system with Poisson failure flow and general repait time distribution is*

$$\tilde{R}(s) = \frac{C_3(s)\cdot s^2 + C_2(s)\cdot s + C_1(s) + C_0}{\Delta}, \tag{4}$$

where

$$
\begin{aligned}
C_3(s) &= \lambda_1\left(1+\tilde{b}(s+\lambda_1)-\tilde{b}(s+\lambda_2)\right)-\lambda_2,\\
C_2(s) &= \lambda_1(1-\tilde{b}(s+\lambda_2))(\lambda_0+\lambda_1)-\\
&\quad -\lambda_2(1-\tilde{b}(s+\lambda_1))(\lambda_0+\lambda_2)-\lambda_1\lambda_2(\tilde{b}(s+\lambda_2)-\tilde{b}(s+\lambda_1)),\\
C_1(s) &= \lambda_2\tilde{b}(s+\lambda_1)(\lambda_1^2+\lambda_0\lambda_2)-\lambda_1^2\tilde{b}(s+\lambda_2)(\lambda_0+\lambda_2),\\
C_0 &= (\lambda_1-\lambda_2)\left(\lambda_0(\lambda_1+\lambda_2)+\lambda_1\lambda_2\right),\\
\Delta &= (s+\lambda_1)(s+\lambda_2)\left((s+\lambda_0)(\lambda_1(1-\tilde{b}(s+\lambda_2))-\lambda_2)+\tilde{b}(s+\lambda_1)(s\lambda_1+\lambda_0\lambda_2)\right)
\end{aligned}
$$

Corollary 1. *The expectation of the system lifetime has the form*

$$
\begin{aligned}
\mathbf{E}[T] = \frac{1}{\lambda_2} &+ \frac{(\lambda_2-\lambda_1)(1-\tilde{b}(\lambda_1))}{\lambda_1\left[\lambda_2(1-\tilde{b}(\lambda_1))-\lambda_1(1-\tilde{b}(\lambda_2))\right]}+\\
&+\frac{\lambda_2-\lambda_1(1+\tilde{b}(\lambda_1)-\tilde{b}(\lambda_2))}{\lambda_0\left[\lambda_2(1-\tilde{b}(\lambda_1))-\lambda_1(1-\tilde{b}(\lambda_2))\right]}.
\end{aligned}
$$

Proof. The proof of this corollary can be done by substitution $s = 0$ in $\tilde{R}(s)$.

4 Simulation Results

In this section, we present the results of simulation of the 3-out-of-6: F system with one repair unit and general distributions of both life and repair time of its elements.

To build a simulation model of the studied system, a programming language Python was chosen. Simulation modeling is conducted with the help of the discrete-event modeling method. It means that the system is modeled as a process, i.e., a sequence of operations being performed across entities. The constructed simulation model is showed as a process flowchart in Fig. 3. As a result of the algorithm, we get the empirical reliability function $\hat{R}(t)$, the mean system lifetime $E[T]$.

The goals of constructing a simulation model are to calculate the empirical reliability function of the 3-out-of-6: F system and its average lifetime for the usage of non-exponential distributions of lifetimes and repairs as well as to show the insensitivity of the system to the shape of its components' repair time distributions.

All simulation experiments were conducted with the total simulation time $T = 10^4$. Moreover, for clarity, all the graphs are built on the scale of the meantime to failure of the system.

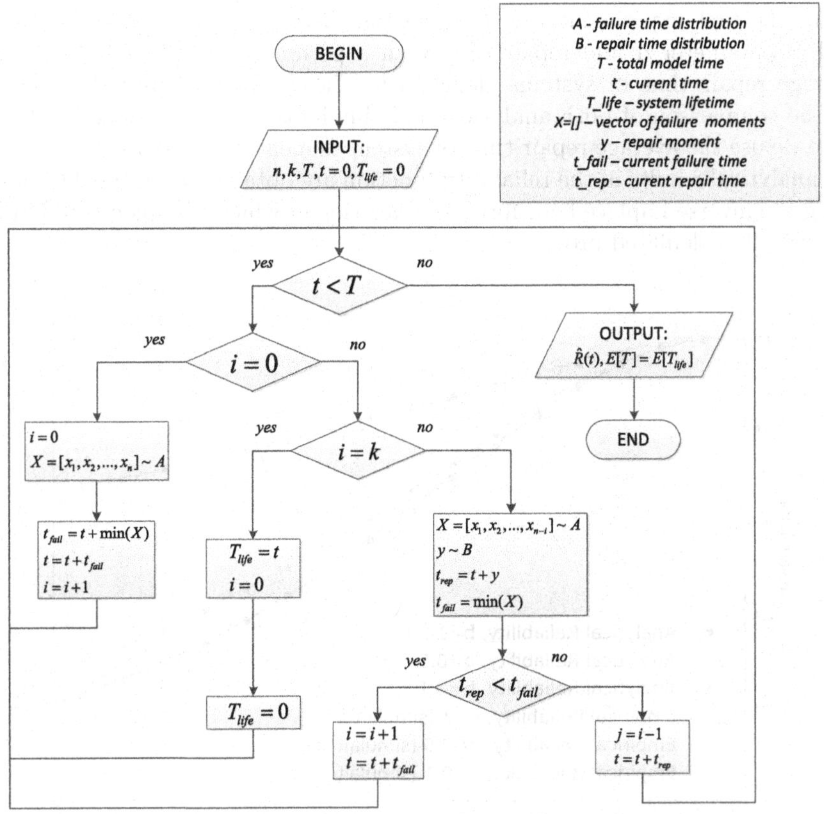

Fig. 3. Flowchart of the discrete-event simulation model

In our experiments the following distributions with its characteristics are used for the repair time:

– Gamma ($G = G(k,\theta)$)

$$b = k \cdot \theta, \quad c = 1/\sqrt{k};$$

– Gnedenko-Weibull ($GW = GW(k,\lambda)$)

$$b = \lambda \cdot \Gamma\left(1 + \frac{1}{k}\right), \quad c = \frac{\sqrt{\lambda^2 \cdot \Gamma(1 + 2/k) - b^2}}{b};$$

– Pareto ($P = P(k,x_m)$)

$$b = \frac{k \cdot x_m}{k-1}, \quad c = \frac{1}{k} \cdot \sqrt{\frac{k}{k-2}};$$

For the lifetime we used Exponential ($Exp = Exp(\alpha)$, $a = \alpha^{-1}$) and Gamma (with the same parameters as above) distributions.

In the first experiment, the Exponential distribution is used for both life with a parameter α and repair time with a parameter β with decreasing the average repair time of systems elements b. The average failure time a equals 1. The comparison of both analytical and simulation results is shown in Fig. 4. Here we use the average repair time of system elements $b = 1, 0.5, 0.1$. Note that the analytical results of the reliability function are obtained from the Theorem 1 using the inverse Laplace transform, the function of which is implemented in the numerically calculated program.

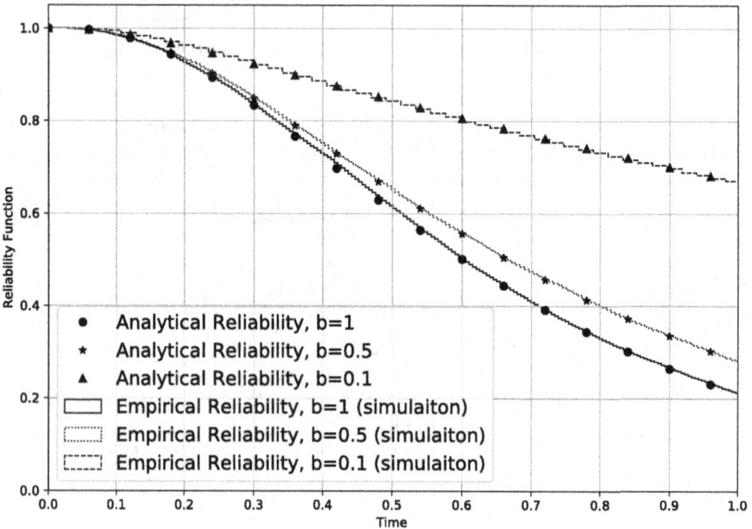

Fig. 4. Reliability function with exponential distribution for both life and repair time. Analytical and simulation results.

As can be seen in Fig. 4, curves of analytical calculation and simulation for reliability functions are the same for all cases over the entire period. With the decreasing value of b, the considered system has higher reliability and its mean lifetime among all the results of this example. The results of this experiment confirm the correctness of the simulation model.

The values of the average lifetime $\mathbf{E}[T]$ of the system for each case are shown in Table 1. These values of $\mathbf{E}[T]$ are confirmed by the behavior of the curves in the graph above.

In the second experiment, we show asymptotic insensitivity of system behavior to the repair time distribution. Here we use Exponential distribution for the elements' lifetime and Gamma, Gnedenko-Weibull, and Pareto distributions for the repair time of the system's elements (see Fig. 5). The parameters of all distributions are selected to fix the average repair time b and the coefficient of variation c. The average failure time $a = 1$. For the repair time distribution, we compare the following cases:

Table 1. The meantime to failure $\mathbf{E}[T]$ of the system with Exponential life and repair time distribution

	$Exp(1)$, $b = 1$	$Exp(2)$, $b = 0.5$	$Exp(10)$, $b = 0.1$
Analytical	0.708171	0.816581	2.284287
Simulation	0.708333	0.816667	2.283334

1. the average repair time of the elements is $b = 1$, while its coefficient of variation $c = 0.5$ (Fig. 5 (a)),
2. the average repair time of the elements is $b = 0.1$ with the same coefficient of variation (Fig. 5 (b)).

In this experiment, we also compare analytical and simulation results. Analytical curves were built with the same average time for life and repair but for this Exponential and Gamma distributions are used respectively. The graph shows that simulation curves are very close to the analytical ones. The differences in the two cases of the mean repair time are almost negligible (see Table 2). As we see, different cases of repair time distributions with the same parameters (average repair time of elements and coefficient of variation) show very close results for the empirical reliability function and mean system lifetime. This result confirms the asymptotic insensitivity to the form of the repair time distribution.

Table 2. The meantime to failure $\mathbf{E}[T]$ of the system with Exponential distribution for the lifetime and different distribution for the repair time

	$G(k, \theta)$	$GW(k, \lambda)$	$P(k, x_m)$	$G(k, \theta)$ - analytical
$b = 1$	0.652441	0.649719	0.648695	0.65221
$b = 0.1$	2.595789	2.619817	2.60269	2.5989

In the next two experiments, we use only the simulation approach due to the impossibility of analytical calculations. In the following examples, we consider the distribution above for the repair time, for the life we use Gamma distribution. Here we compare the same cases of the average repair time with the different values of the coefficient of variation.

In the next example for the repair time we use the distribution above with $b = 1$ and $c = 0.5$. For the lifetime we compare the different cases of the coefficient of variation with the same mean time. Let's $a = k \cdot \theta = 1$ and $c = 1/\sqrt{k} = 0.5; 2$ (Fig. 6).

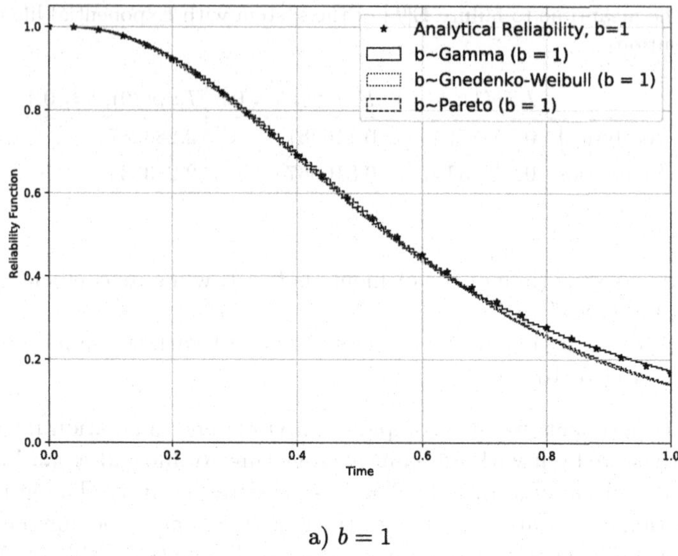

a) $b = 1$

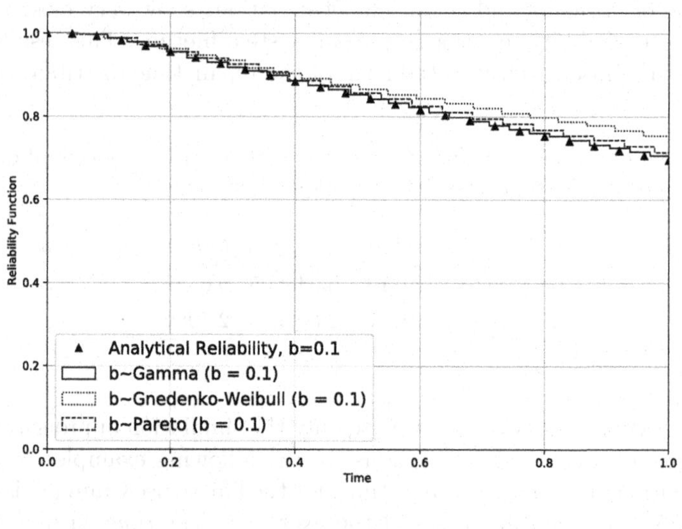

b) $b = 0.1$

Fig. 5. Reliability function with gamma distribution for lifetime and different distributions for repair time. Analytical and simulation results

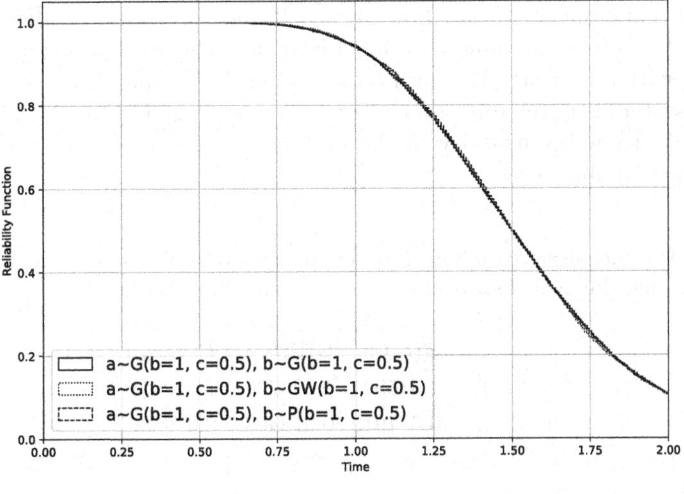

a) $c = 0.5$

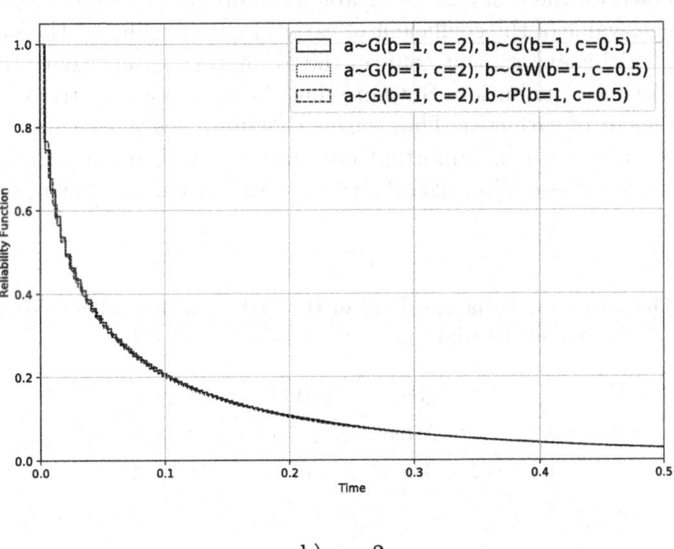

b) $c = 2$

Fig. 6. Reliability function with gamma distribution for lifetime and different distributions for repair time

In this experiment we see how the coefficient of variation affects system behavior. With a small value $c = 0.5$ (Fig. 6 (a)), the system has a fairly high reliability with an average $\mathbf{E}[T] = 1.5$ (see Table 3). At a high value $c = 2$ (Fig. 6 (b)), the system is unreliable and has a very low average $\mathbf{E}[T] = 0.07$. Moreover, both graphs show insensitivity of the system to the shape of its components' repair time distributions.

Table 3. The meantime to failure $\mathbf{E}[T]$ of the system with Gamma distribution for the lifetime and different distribution for the repair time ($b = 1$)

	$G(k, \theta)$	$GW(k, \lambda)$	$P(k, x_m)$
a) $c = 0.5$	1.51592	1.525796	1.512018
b) $c = 2$	0.077912	0.073408	0.076536

In the final example, we consider the same cases as above but the average repair time is equaled 0.5. We see that the average lifetime of the system has almost doubled for the first case (see Table 4) compared to the previous example. For the bigger value of the coefficient of variation, we see almost the same result. The results of this experiment confirm the asymptotic insensitivity to the form of the repair time distribution. But note that the system is sensitive to the values of the coefficient of variation. That means that from the practical point of view of solving some tasks, it is important not only the mathematical expectation of the incoming parameters but also the value of variance or coefficient of variation (Fig. 7) .

Table 4. The meantime to failure $\mathbf{E}[T]$ of the system with Gamma distribution for the lifetime and different distribution for the repair time ($b = 0.5$)

	$G(k, \theta)$	$GW(k, \lambda)$	$P(k, x_m)$
a) $c = 0.5$	3.406536	3.541664	4.419625
b) $c = 2$	0.068732	0.071757	0.070166

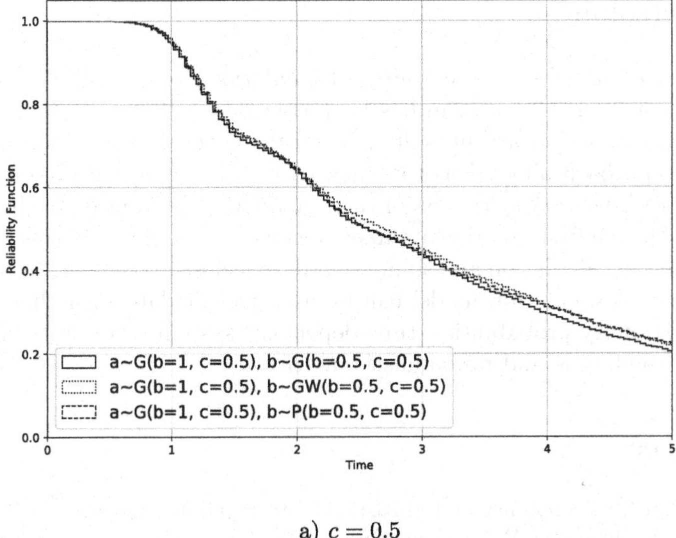

a) $c = 0.5$

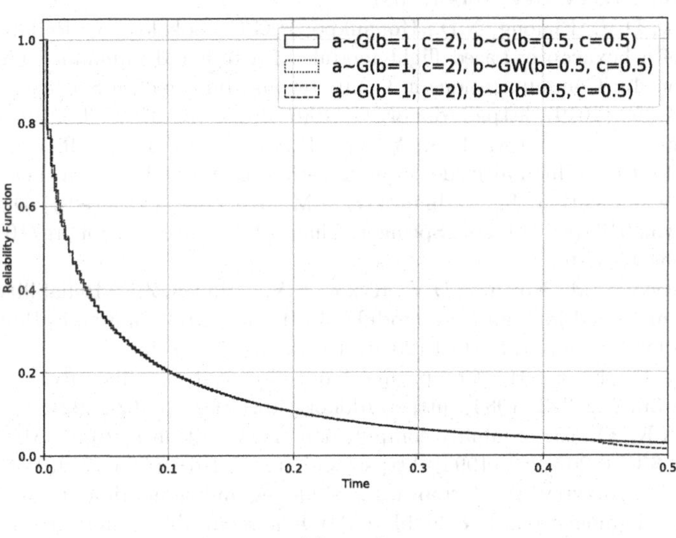

b) $c = 2$

Fig. 7. Reliability function with gamma distribution for lifetime and different distributions for repair time.

5 Conclusion

The problem of analytical calculation and simulation assessment of the reliability function for a k-out-of-n system has been considered. The analytical results of the reliability function are presented in terms of the Laplace transform. The simulation approach allowed us to demonstrate the asymptotic insensitivity of the considered system to the form of the repair time distribution of the system's elements. The analysis of the obtained results shows that the results of the exact analytical calculation and simulation have a close agreement. In the future, the constructed simulation model can be used to calculate such characteristics as quasi-stationary probabilities, time-dependent system state probabilities and other important time and probabilistic metrics.

References

1. Trivedi, K.S.: Probability and Statistics with Reliability. Queuing and Computer Science Applications. Wiley, New York (2002)
2. Shepherd, D.K.: k-out-of-n Systems. Encyclopedia of Statistics in Quality and Reliability. Wiley, New York (2008)
3. Kozyrev, D.V., Phuong, N.D., Houankpo, H.G.C., Sokolov, A.: Reliability evaluation of a hexacopter-based flight module of a tethered unmanned high-altitude platform. In: Communications in Computer and Information Science, 1141 CCIS, pp. 646–656 (2019). https://doi.org/10.1007/978-3-030-36625-4_52
4. Perelomov, V.N., Myrova, L.O., Aminev, D.A., Kozyrev, D.V.: Efficiency enhancement of tethered high altitude communication platforms based on their hardware-software unification. In: Vishnevskiy, V.M., Kozyrev, D.V. (eds.) DCCN 2018. CCIS, vol. 919, pp. 184–200. Springer, Cham (2018). https://doi.org/10.1007/978-3-319-99447-5_16
5. Vishnevsky, V.M., Kozyrev, D.V., Rykov, V.V., Nguyen, Z.F.: Reliability modeling of an unmanned high-altitude module of a tethered telecommunication platform. Inf. Technol. Comput. Syst. **4** (2020). (in Russian)(in print)
6. Barlow, R., Klaus, H.: Computing k-out-of-n system reliability. IEEE Trans. Reliab. **30**, 322–323 (1984). https://doi.org/10.1109/TR.1984.5221843
7. Sarje, A.K.: On the reliability computation of a k -out-of-n system. Microelectron. Reliab. **33**(2), 267–269 (1993). https://doi.org/10.1016/0026-2714(93)90487-j
8. Rykov, V., Kozyrev, D., Zaripova, E.: Modeling and simulation of reliability function of a homogeneous hot double redundant repairable system. In: Proceedings of the 31st European Conference on Modelling and Simulation, ECMS2017, pp. 701–705 (2017). https://doi.org/10.7148/2017-0701
9. Rykov, V., Zaripova, E., Ivanova, N., Shorgin, S.: On sensitivity analysis of steady state probabilities of double redundant renewable system with Marshall-Olkin failure model. In: Vishnevskiy, V.M., Kozyrev, D.V. (eds.) DCCN 2018. CCIS, vol. 919, pp. 234–245. Springer, Cham (2018). https://doi.org/10.1007/978-3-319-99447-5_20
10. Wu, J.S., Chen, R.J.: An algorithm for computing the reliability of a weighted k-out-of-n system. IEEE Trans. Reliab. **43**(2), 327–328 (1994). https://doi.org/10.1109/24.295016

11. Li, X., You, Y., Fang, R.: On weighted k-out-of-n systems with statistically dependent component lifetimes. Probab. Eng. Inform. Sci. **30**(4), 533–546 (2016). https://doi.org/10.1017/S0269964816000231
12. Harchol-Balter, M.: Performance Modeling and Design of Computer Systems: Queueing Theory in Action. Cambridge University Press, Cambridge (2013). https://doi.org/10.1017/CBO9781139226424
13. Cui, L., Frenkel, I., Lisnianski, A.: Stochastic Models in Reliability Engineering. CRC Press, Boca Raton (2020). https://doi.org/10.1201/9780429331527
14. Kuo, W., Zuo, M.J.: Consecutive k-out-of-n system. In: Optimal Reliability Modeling: Principles and Applications, pp. 328–383. Wiley, Hoboken, NJ, USA (2003)
15. Huang, J., Zuo, M., Wu, Y.: Generalized multi-state k-out-of-n: G systems. IEEE Trans. Reliab. **49**, 105–11 (2000). https://doi.org/10.1109/24.855543
16. Goyal, N., Ram, M., Amoli, S., Suyal, A.: Sensitivity analysis of a three-unit series system under k-out-of-n redundancy. Int. J. Q. Reliab. Manag. **34**(6), 770–784 (2017). https://doi.org/10.1108/IJQRM-07-2016-0106
17. El-Damcese, M., Shama, M.S.: Reliability analysis of a new k-out-of-n: G Model. World J. Model. Simul. **16**(1), 3–17 (2020)
18. Kozyrev, D.V.: Analysis of asymptotic behavior of reliability properties of redundant systems under the fast recovery. Math. Inf. Sci. Phys. **3**, 49–57 (2011). (in Russian). Bulletin of Peoples' Friendship University of Russia
19. Rykov, V., Ngia, T.A.: On sensitivity of systems reliability characteristics to the shape of their elements life and repair time distributions. Vestnik PFUR. Ser. Math. Inform. Phys. **3**, 65–77 (2014). (in Russian)
20. Efrosinin, D., Rykov, V., Vishnevskiy, V.: Sensitivity of reliability models to the shape of life and repair time distributions. In: 9th International Conference on Availability, Reliability and Security (ARES 2014), pp. 430–437. IEEE (2014). https://doi.org/10.1109/ARES.2014.65
21. Rykov, V.: On reliability of renewable systems. In: Vonta, I., Ram, M. (eds.) Reliability Engineering: Theory and Applications, pp. 173–196. CRC Press, Boca Raton (2018)
22. Rykov, V., Kozyrev, D., Filimonov, A., Ivanova, N.: On reliability function of a k-out-of-n system with general repair time distribution. Probab. Eng. Inform. Sci. 1–18 (2020). https://doi.org/10.1017/S0269964820000285
23. Cox, D.: The analysis of non-Markovian stochastic processes by the inclusion of supplementary variables. Math. Proc. Camb. Philos. Soc. **51**(3), 433–441 (1955). https://doi.org/10.1017/S0305004100030437
24. Petrovsky I.G.: On Cauchy problem for a system of partial differential equations. In: Selected Works, Nauka, pp. 34–97 (1986). (in Russian)

Two Ways of Group Polling Method Application for Sensors Detecting in Unsynchronized Structured WSNs

Ivan Tsitovich(✉)

Institute for Information Transmission Problems (Kharkevich Institute) RAS, Bolshoy Karetny per. 19, build.1, Moscow 127051, Russia
cito@iitp.ru

Abstract. It is investigated the problem of detecting alarming sensors in large monitoring networks when there are objects where sensors at the object can activate simultaneously. We propose a generalization of the method of a sensor signal coding for an alarm signalization when a sensor signal cannot be synchronized in time. This method bases on the method of group polling for alarming sensors identification for a synchronized network and has similar characteristics of complexity but apply in mainly for detecting objects with active sensors. Our methods has more complicated structure than previous ones. For a network with very large object, we propose to use a sub-network of sensors at the object based on the Wi-Fi HaLow technology. For a network with middle size of objects, it is proposed the method with partly synchronized in time sensors at every object.

Keywords: Wireless sensor network · Sensor for an alarm signalization · Group polling · Unsynchronized time · Heterogeneous sensors

1 Introduction

Studies of wireless sensors monitoring networks (WSNs) are widely conducted in different directions. One of the most interesting is the network organization which provides a quick identification of a sensor that has an urgent message for sending to the situation center (SC). The lack of common communication protocols that ensure the interests of all users suggests that networks are too diverse and can be fundamentally different in their characteristics. One such aspect covers in this paper, where it is suppose that probabilities of possible emergencies are very unlikely. The scope of the proposed method is low-power wide-area networks [1] oriented onto an environmental control. We examine the interaction of sensors with the SC via the WSN, where three nature sensor stages of activity are possible:

Supported by grant AAAA-A19-119022590088-5.

© Springer Nature Switzerland AG 2020
V. M. Vishnevskiy et al. (Eds.): DCCN 2020, CCIS 1337, pp. 286–298, 2020.
https://doi.org/10.1007/978-3-030-66242-4_23

I. the sensor transmits the information about its current state according to a given schedule sharing information with the SC;
II. the sensor has detected an emergency and sends the alarm signal into the communication radio channel which is common for all sensors;
III. the sensor transmits information about the emergency at the request of the SC through the dedicated communication channel.

It is evident that for sensors in the first stage of activity we have a stable information flow such as the timetable of sensors activity is known. Number of sensors in the third stage of activity is small and their information flow is determined by the protocols of an emergency information. Therefore, sensors at the first and third stages generate a predictable traffic and are not under investigation in this paper by this reason.

The main interest consists in constructing such sensors and organizing the WSN in such way that it is possible to ensure energy efficiency of sensors, reliability, and timeliness of communications in case of a large number of sensors connected to a common wireless channel when sensors are in the second stage of activity. The main feature is the following: most of the sensors does not send a signal of the second stage for the rest of their lives. Therefore, it is not energy efficient to maintain continuous communication with such sensors. If the SC detects a sensor as active then it switches the sensor into the third stage and the sensor becomes be inactive. Since the number of active sensors (sending alarm signal) at the same time is small, it is advisable to allocate a common radio channel for them and, therefore, it is the problem to identify such sensors based on their common signal in the channel such as sensor signals are mixed.

In the paper [2], it was proposed the method of group polling for detecting of alarming sensors in a WSN network and properties of this method were investigated under the assumption of independent activity of the alarming sensors. It is supposed that the WSN is very large and contains thousands of sensors but all sensors synchronize in time their alarm signals. The last demand is difficult for its practical realization. However, proposed in [2] method ensures the fulfilment of a short time of an alarming sensors detection, i.e. if t is a number of sensors in the WSN then the detection time is $O(\log t)$. In the papers [3] and [4], it is proposed a generalization of this method onto a case of unsynchronized in time alarming signal sending. It is showed that the group polling method for alarming sensors identification is applicable at this case but its computation complexity is such that it is difficult to use the method for online detection or it is very expensive. In [6] it is proposed a more complex profiles of sensors which give us possibility to detect alarming sensors in time similar to a WSN with synchronized in time alarming signal sending.

A principle scheme is on Fig. 1. The main property is that every emergency actives one sensor only.

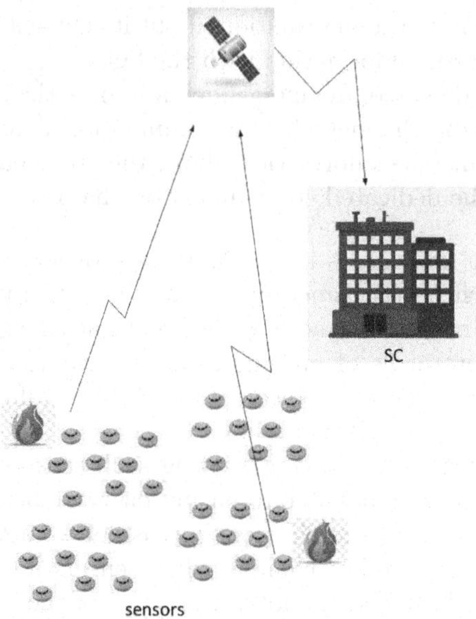

Fig. 1. WSN with homogeneous sensors as in [2]–[4],[6]. Two sensors detected emergencies and send the signals into the common channel that they have detected emergencies. The SC separates the obtained signals and identifies the sensors such as codes of the signals give this possibility.

However, it was supposed that the WSN consists with homogeneous sensors and only one of them is active in the case of an emergency at an object. Really we have WSNs where there are many objects with sensors for emergencies detection. If there is an emergency at the object then several sensors at this object begin to send the alarm signal for a real WSN. Such as a number of simultaneously sending sensors is a critical parameter for the stable detecting of active sensors (see [2]), it is necessary to investigate WSNs with heterogeneous sensors. Sense of sensor heterogeneity consists in their dependent activity for the sensors at the same object when the object has an emergency. This problem is investigated in [5] where it is supposed that all sensors at the object are active in the case of an emergency at this object. The motivation of such approach lies in the fact that the analysis of the data from all sensors makes the possibility of more accurately characterize of the emergency. If it is transferred the data from inactive sensors then it only leads to additional traffic in the communication channels and can be considered as an insignificant increase in the spending by the emergency detection. This assumption is valid only in the case when number of sensors at the object is small (this restriction was introduced in [5]). Now we suppose more real situation when a number of sensors at the object can be large: a few tens or even thousands.

The formulation of the problem is presented in Sect. 2. In Sect. 3 it is described the algorithm of WSN output signal modeling when alarming sensors begin their signals in random time moments and take into account digitization in time of the output signal. In the next section, we present two way of active sensors communications with the SC in a structured WSN.

2 Setting of the Problem

For the mathematical model description, we follow the notations from [5]. The WSN consists of B objects, such that n_j sensors are mounted at the jth object, $j = 1, \ldots, B$. We assume that the number B is relatively large and $n_j \ll B$ for all j. Thus, we have in the WSN

$$t = \sum_{j=1}^{B} n_j$$

sensors and develop the polling strategy aimed at the fastest identification of objects $j_1, \ldots, j_r$, whose sensors are ready for data transmission, and corresponding sensors (r is a number of simultaneous objects with dangers and is unknown).

Let T be the average time of active sensors detection and λ be the intensity of Poisson process of disasters appearing on objects. Then the average value of r is λBT, and since the T is small, λ is extremely small, the value r is just as small.

The objects with numbers $j_1, \ldots, j_r$ are named as active objects and, therefore, it is necessary to find the set $S^o = \{j_1, \ldots, j_r\}$.

It was supposed in [5] that the data from all sensors at the object are received in the case of emergency at this object. This assumption is valid only in the case when n_j is small (this restriction was introduced in [5]). Such as we suppose now that n_j can be large we need to find alarming sensors also. Such sensors we name active ones. Their set is denoted by $S^s = \{i_1, \ldots, i_s\}$. Therefore, we have a more complicate problem than in [5].

In this problem, we assume that the emergency probability at the jth object p_j is unknown but is relatively small:

$$p_j \leq p$$

where p is the predetermined probability, which is similar to the quantity $p = \frac{s}{t}$ from [6]. Now we suppose that $p = \frac{s}{B}$ and $s \leq 5$.

We assume that z_j sensors are simultaneously activated at the jth object in the presence of the emergency, so that z_j is a discrete random value distributed on the set $\{1, \ldots, n_j\}$. Therefore we have the set

$$S_j = \{i_1^j, \ldots, i_{z_j}^j\}$$

of active sensors at the jth object and the set

$$S^s = \cup_{j \in S^o} S_j$$

of all active sensors in the WSN.

The distribution of z_j depends of the object, the reasons of emergencies, the positions of sensors, etc. These parameters are either unknown or difficult to take into account. However, we assume that the probability of the activation more than one sensor at one object is relatively high. For example, if it is supposed that any sensor can be active with a conditional probability p_j^a independently of another sensors at the object when the object is active one then it is natural to suppose that

$$p_j^a \gg p. \tag{1}$$

In contrast with [5] when all sensors begin to send their signal simultaneously, now they start to send signals separately.

Let $\hat{j}_1, \ldots, \hat{j}_{\hat{s^o}}$ be detected emergency object numbers, $\hat{s^o}$ be the number of identified emergencies as in [5] and additionally $\hat{i}_1, \ldots, \hat{i}_{\hat{s^s}}$ be detected active sensor numbers and $\hat{s^o}$ be the number of detected sensors.

The quality of algorithm characterizes by the probability of correct identification of emergencies. Let $\hat{S^o} = \{\hat{j}_1, \ldots, \hat{j}_{\hat{s^o}}\}$ then P_1 is the probability of the emergency object missing, i.e. $P_1 = \mathbf{P}(S^o \not\subseteq \hat{S^o})$, and P_2 is the probability of the active sensor missing, i.e. $P_2 = \mathbf{P}(S^s \not\subseteq \hat{S^s})$. We calculate the probabilities under the assumption that the uniform distribution is determined in the admissible domain S.

3 Profiles for Signals of Active Sensors

Now profiles of sensors are a combination of profiles from [5] and [6].

Every sensor has the unique code in [2] as a sensor at the object $\mathbf{o} = (o^1, \ldots, o^{M_o})$, and for the jth object its code is denoted by $\mathbf{o}_j$. Coordinates of the jth code have the values 1 or 0. For creating the codes we construct the Boolean matrix $\mathbf{A} = (\mathbf{a}_i, i = 1, \ldots, B)$, where a_i^j are independent random numbers 0 or 1 with a proper probability p^0 for 1 in the matrix. The value p^0 will be specified by the formula (6). Also the sensor at the jth object has the unique code $\mathbf{s} = (s^1, \ldots, s^{M_j})$ where M_j depend on n_j. We propose several ways of such codes constructing in dependent of M_j and a type of sensors communication.

The total code of the jth object's sensor is $\mathbf{a} = (o^1, \ldots, o^{M_o}, s^1, \ldots, s^{M_j})$. This code gives us the possibility to determine the number of sensor's objects and its number at this object.

The code $\mathbf{a}_i$ generates a profile for a signal of the ith sensor when this sensor is active by the following way (see [6]). The profile consists of two part; every part has three portions.

Let us begin with the first part. Let us suppose that the i-th sensors begins to be active at time u_i. On the first portion the sensor is passive, i.e.

$$a_i(u) = 0, \text{ if } u < u_i. \tag{2}$$

On the second portion the sensor is sending a special signal

$$a_i(u) = 1, \text{ if } u_i \le u < u_i + \Delta L, \tag{3}$$

where L is one of parameters for the profile. This is a new element in the profile in contrast with [3].

On the third portion we followed in general to the profile from [3] by the following

$$a_i(u) = \begin{cases} 1, & \text{if } a_i^0 = 0 \text{ for } u_i + \Delta L \le u < u_i + \Delta(L+1) \text{ and} \\ 0 & \text{for } u_i + (L+1)\Delta \le u < u_i + (L+k+1)\Delta, \\ 1, & \text{if } a_i^0 = 1 \text{ for } u_i + \Delta L \le u < u_i + \Delta(L+k) \\ a_i^j, & \text{if } a_i^j = a_i^{j-1} \text{ for the next time interval } k\Delta, \\ a_i^{j-1}, & \text{if } a_i^j \ne a_i^{j-1} \text{ for the next time interval } \Delta \text{ and} \\ a_i^j & \text{for the next time interval } (k+1)\Delta, \\ & \text{and so on,} \end{cases} \tag{4}$$

where $k, k > 1$, is a number of repeating of code symbols a_i^j and is a parameter of the profile. It is proposed in [6] $k = 3$.

Therefore, the length of the third portion depends on the vector $\mathbf{a}_i$ and is denoted by $N_i\Delta$. Under the conditions of the vector $\mathbf{a}_i$ constructing N_i is a random value with the mean $k + 4p^0(1 - p^0)$.

On the second part the first portion of the profile is as for the first part but has a random length v with the exponential distribution with a parameter $\lambda > 0$. This parameter is a common parameter for all profiles. We add this element in the profile by the same reason as it is used in aloha systems for minimizing conflicts.

The resulting output continuous signal is dropped onto short time intervals with the length Δ; as a result, the continuous function $f(u)$ drops onto the group of observations $(f_1, f_2 \dots)$, that can be 0, 1, or *nil* where *nil* means that the output signal cannot be interpreted correctly. We have $f_j = 0$ if in the jth time interval of digitation $f(u) \equiv 0$, $f_j = 1$ if $f(u) \equiv 1$ in the interval. When the function $f(u)$ at the interval changes its value (from 0 to 1, or 1 to 0), i.e. $\exists u_1, u_2 : f(u_1) \ne f(u_2)$, we cannot interpret the received signal correctly. This situation is a conflict and we should mark such signals as *nil*. Thus, for the jth Δ-interval

$$f_j = \begin{cases} 1, & \text{if } f(u) \equiv 1, \\ 0, & \text{if } f(u) \equiv 0, \\ nil, & \text{if } \exists u_1, u_2 : \; f(u_1) \ne f(u_2). \end{cases} \tag{5}$$

By the same way we construct from (2)–(4) discrete values $\hat{a}_i(j)$ that can be also 0, 1 or *nil* (i is the number of the sensor and j is the number of the corresponding time interval).

Then, the result of an output signal for the jth interval

$$f_j = (\hat{a}_1(j) \wedge x_1) \vee, \dots, \vee(\hat{a}_t(j) \wedge x_t),$$

providing that $1 \vee nil \equiv 1$ and $0 \vee nil \equiv nil$.

We assume that data transmission errors are possible in the network. This means that the value f_j is known with a certain error. Therefore, results 0 or 1 of f_j can be transformed in accordance with the matrix

$$\mathbf{W} = \begin{pmatrix} 1-\beta_0, & \beta_0 \\ \beta_1, & 1-\beta_1 \end{pmatrix}$$

and we get the vector of observation $\mathbf{g} = (g_1, g_2, \dots)$. If $f_j = nil$ then g_j equals 0 or 1 with probability 0.5 and, therefore, the observations $f_j = nil$ cannot help for an alarming sensor detection. Therefore, we lose a part of information in contrast with the case from [2] and need to have longer sensors codes for an alarming sensor identification with the same quality.

For minimizing a part of uninformative observation we repeat signal of a sensor k times and send additionally signals by the way as in (4) when the sensor changes his signal. Therefore, in mean we have k informative signal per $4p^0(1-p^0)$ uninformative ones. By this reason, for optimization a capacity of the common channel we use p^0 such as

$$p^0 = \frac{2+k-4p_0-\sqrt{(2+k)^2 - -16(1+k)p_0(1-p_0)}}{4-8p_0} \tag{6}$$

for $p_0 < 0.5$ and $p^0 = 0.5$ for $p_0 = 0.5$, where

$$p_0 = 1 - \sqrt[s_0]{\frac{\frac{1}{2}-\beta_0}{1-\beta_0-\beta_1}}$$

and s_0 is supposing number of active sensors.

4 Two Possible Ways of Group Polling Method Application for Sensors Detecting in Unsynchronized Structured WSNs

It is followed from (1) that the proposed in [6] group polling method is not applicable for coding sensors at an object if the object is relatively large, for example $n_j > 5$. For these cases, we propose two different methods of communication sensors with the situation center.

If objects of the WSN are little then we can use the method from [6] applied for objects detection. All sensors on the detected object consider as active ones. This method is effective if $\mathbf{E}z_j = O(n_j)$.

Group Polling Method Application for WSNs with medium-sized objects (up to a hundred sensors) and large objects (a few hundred or thousands of sensors) are given further in this section.

4.1 Structured WSN with Medium-Sized Objects

If $n_j < 100$ we propose binary orthogonal code design with sharing access time for sensors at the jth object. It means the following. All sensors at the object are synchronized in time. This means that a sensor starts to send signal in fixed times and the sequence of possible start times is common for all sensors at the object. In contrast with a synchronized in time WSN [2,5] we have no large number of sensors and it is technically feasible requirement.

A principle scheme is represented on Fig. 2.

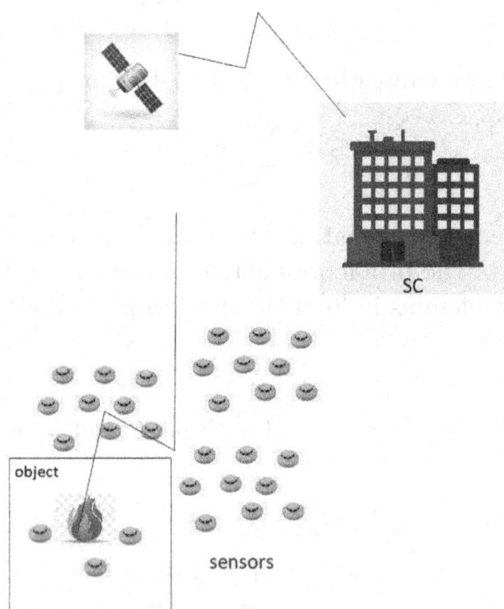

Fig. 2. WSN with middle size of objects. All sensors at the object send synchronized signals simultaneously

All sensors have the same part of the code that corresponds to the object and have different codes of the part that corresponds to the sensor. The length of the last part of the code is $3n_j$. Every sensor has its number in the objects $0, \ldots, n_j - 1$; let, for example, it is i. Then $\mathbf{s}_i = (s^1, \ldots, s^{3n_j})$ has three 1 in positions $3i$, $3i+1$, and $3i+2$ and 0 in different positions.

The transformation of codes $\mathbf{s}_i$ into the sensor signal differs from (4). This part of the signal is the following

$$a_i(u) = \begin{cases} 0 & \text{for } 0 \le u < 3i\Delta, \\ 1 & \text{for } 3i\Delta \le u < u_i + 3(i+1)\Delta, \\ 0 & \text{for } 3(i+1)\Delta \le u < 3n_j\Delta. \end{cases} \tag{7}$$

It is followed from (7) and (5) that is case of the unmistakable signal transmission we detect 2 times symbol 1 in positions that corresponds to the ith sensor. Signals of other active sensors do not mix. We detect as active all sensors at the detected objects that the output signal $\mathbf{g}$ has at list one symbol 1 in the positions that correspond to the symbols 1 in the second part of the sensor's signal $a(u)$ in (7). Therefore,

$$\mathbf{P}(S^s \not\subseteq \hat{S}^s | S^o \subseteq \hat{S}^o) \leq \beta_1^2 \sum_{j \in S^o} n_j$$

and the conditional mean number of detected sensors has the following inequality

$$\mathbf{E}(\hat{s^s} | S^o \subseteq \hat{S}^o) \leq 3\beta_0 \sum_{j \in \hat{S}^o} n_j.$$

Using these formulas, we can estimate P_2 if P_1 is known:

$$P_2 \leq \beta_1^2 \sum_{j \in S^o} n_j + P_1.$$

Therefore the problem of estimating the quality properties of active sensors detection is reduced to the problem of active objects detection that gives possibility to solve this problem as in [6] if the detection method applies to the active objects detection.

Table 1. Increasing of the alarm signal length on B and n_j when $\beta_0 = \beta_1 = 0.01$, $c = 7.5$, $s_0^o = 2$

B	n_j	N	L	δ
1000	50	126	3546	5.6
5000	50	156	4387	4.6
5000	100	156	4387	9.2
10000	100	168	1260	8.5
25000	100	186	1395	7.7
100000	100	216	1620	6.6
100000	150	216	1620	9.9

The proposed method, in addition to synchronizing sensors, increases the length of the alarm signal. The Table 1 shows the results of the alarm signal length calculation by numbers of objects B and sensors at object n_j when $c = 7.5$ (the parameter c in [6]), $s_0^o = 2$ (the number of active sensors), $L = 18$ (proposed values in [6]). Here N is the code of object length, $L = 18 + cN(3 + 4p^0(1 - p^0))$ is the mean value of the length of the first part of alarm signal generated by $\mathbf{o}$, δ is the additional length of alarm signal generated by $\mathbf{s}$ as a percentage of L. It is followed from this data that for $n_j \leq 100$ or $n_j \leq 100$ for very large WSN the additional signal length and the time of active sensors identification grow less than 10%.

4.2 Structured WSN with Large Objects

One of possible way of the problem solving consists in using the Wi-Fi HaLow technology [7], based on the IEEE 802.11ah standard. This way gives a possibility to organize a communication between sensors and the SC when an object has hundreds and even some thousands of sensors. The communication is set up via a Wi-Fi access point (AP). An AP uses for sending the alarm signal of any sensor at the object. If a sensor detects a danger then it sends to the AP a signal that it is ready to send an information to the SC. The AP send its profile signal in the common wireless radio channel to the SC. If the SC detects the AP then the AP sends an information from the active sensors at the object to the SC.

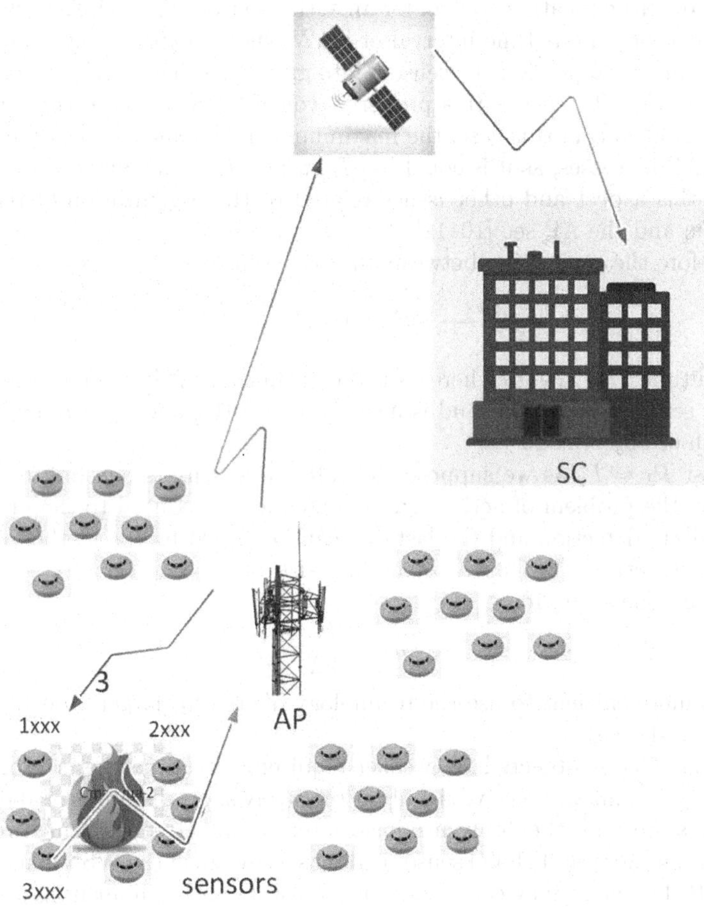

Fig. 3. WSN with large size of objects. All sensors near the fire are ready to send the signals but only the sensor 3xxx has such possibility. Its code the AP sends to the SC

For energy loss minimizing, a sensor set-up a link with the AP when the sensor detects an emergency only. It is followed from (1) that a number of such sensors is relatively large. The IEEE 802.11ah standard identifies many channel-level mechanisms that allow you to use the channel more efficiently in scenarios where a large number of sensors are connected to a single Wi-Fi hotspot. One of the most important mechanisms is the Restricted Access Window (RAW), which allows the AP to divide the sensors into groups and each group to set aside a separate channel time interval during which only the sensors of a given group can transmit data. Together with the use of standard methods of random access to the channel, this mechanism significantly improves the efficiency of the network (see, for example, [8,9]).

On Fig. 3 we illustrate the technology RAW for minimizing of collisions on the stage of authentication control for an active sensor. To minimize the number of active sensors in one time interval of RAW when an emergency occurs on an object, you need to portion the sensors into groups so that close sensors are into different groups. To ensure this property, you need to increase the number of groups g. However, in this case, the maximum time to connect the active sensor with the AP increases, as it is equal to gT_c, where T_c is the duration of one time slot. For this aspect and other issues related to the organization of the sensor connection and the AP see [10–12].

Therefore the connection between an active sensor and the SC has two stage

$$S \longleftrightarrow AP \text{ and } AP \longrightarrow SC$$

and has a time $T = T_1 + T_2$ where T_1 is the mean time of the connection between an active sensor and the AP and is investigated in [12] and $T_2 = L\Delta$ where L is the length of object code.

Almost $P_2 = P_1$ as we suppose that all active sensors can link to the AP. Therefore, the problem of active sensors detection is reduced to the problem of active objects detection and the last one can be solved by the method of [6].

For a structured WSN with a middle size of objects $T = T_3 + T_2$ where $T_3 = 3n_j\Delta$. Therefore, if

$$n_j \geq \frac{T_1}{3\Delta}$$

then it is more efficient to use the technology offered for larger objects than for medium-sized ones.

Sensors at large objects have weaker requirements than other sensors. First, they must transmit and receive signals at relatively short distances of about a few kilometeres. Second, they only use local internal object codes, which represent only their numbers, and global sensor codes as elements of the WSN can be stored on the AP. Therefore,they can use simpler software and equipment and this may offset the cost of the AP. At medium-sized objects, sensors are integrated into the global network and must maintain local time synchronization.

5 Conclusions

The main result consists in that for unsynchronized in time structured WSN it can be constructed a group polling with $O(\log t)$ time for alarming sensors detection. A time of sensors detection is similar to one for a WSN with independent activity of sensors as in [6].

It is proposed the method with partially synchronized sensors when objects in the WSN have less than 100 sensors and its effectiveness is investigated.

For a WSN with large objects it is proposed the method of sensors detection that is based on the Wi-Fi HaLow technology.

The computational complexity is growing no so much as in the case of [6] if a WSN structure is not taken into account.

The decoding procedure for unsynchronized in time structured WSN can be similar to one for a WSN with independent activity of sensors as in [6].

For a WSN with heterogeneous objects and sensors the emergency probability p_j can vary in a wide range. For this case preferably use codes with the signal profile length in dependence of p_j and this problem will be investigated later.

References

1. Xiong, X., Zheng, K., Xu, R., Xiang, W., Chatzimisios, P.: Low power wide area machine-to-machine networks: key techniques and prototype. IEEE Commun. Mag. **53**(9), 64–71 (2015)
2. Malikova, E.E., Tsitovich, I.I.: Group polling upon the independent activity of sensors in the monitoring networks. J. Commun. Technol. Electron. **56**, 1556–1563 (2011)
3. Tsitovich, I.I., Shtokhov, A.N.: Method of group polling upon the independent activity of sensors in non-synchronized monitoring networks. Inform. Process. **16**, 237–245 (2016)
4. Shtokhov, A., Tsitovich, I., Poryazov, S.: On the method of group polling upon the independent activity of sensors in unsynchronized wireless monitoring networks. Commun. Comput. Inform. Sci. **678**, 266–278 (2016)
5. Malikova, E.E., Tsitovich, I.I.: Analysis of the efficiency of group polling upon the dependent activity of sensors in the monitoring networks. J. Commun. Technol. Electron. **56**, 1552–1555 (2011)
6. Tsitovich, I.: Group polling method upon the independent activity of sensors in unsynchronized wireless monitoring networks. In: Vishnevskiy, V.M., Samouylov, K.E., Kozyrev, D.V. (eds.) DCCN 2019. CCIS, vol. 1141, pp. 436–448. Springer, Cham (2019). https://doi.org/10.1007/978-3-030-36625-4_35
7. Khorov, E., Lyakhov, A., Krotov, A., Guschin, A.: A survey on IEEE 802.11 ah: an enabling networking technology for smart cities. Comput. Commun. **58**, 53–69 (2015)
8. Tian, L., Famaey, J., Latré, S.: Evaluation of the IEEE 802.11ah restricted access window mechanism for dense IoT networks. In: 2016 IEEE 17th International Symposium on World of Wireless, Mobile and Multimedia Networks (WoWMoM), pp. 1–9. A/IEEE (2016)

9. Khorov, E., Lyakhov, A., Yusupov, R.: Two-Slot based model of the IEEE 802.11 ah restricted access window with enabled transmissions crossing slot boundaries. In: 2018 IEEE 19th International Symposium on A World of Wireless, Mobile and Multimedia Networks (WoWMoM), pp. 1–9. IEEE (2018)
10. Khorov, E., Krotov, A., Lyakhov, A.: Modelling machine type communication in IEEE 802.11ah networks. In: Proceedings of the 2015 IEEE International Conference on Communication Workshop (ICCW), London, UK, 8–12 June 2015, pp. 1149–1154 (2015)
11. Kureev, A., Bankov, D., Khorov, E., Lyakhov, A.: Improving efficiency of heterogeneous Wi-Fi networks with joint usage of TIM segmentation and restricted access window. In: Proceedings of the 2017 IEEE 28th Annual International Symposium on Personal, Indoor, and Mobile Radio Communications (PIMRC), Montreal, QC, Canada, 8–13 Oct 2017, pp. 1–5 (2017)
12. Bankov, D., Khorov, E., Lyakhov, A., et al.: What is the fastest way to connect stations to a Wi-Fi Ha Low Network? Sensors **18**(9), 1–23 (2018). http://www.mdpi.com/1424-8220/18/9/2744

Regeneration Analysis of Non-Markovian System with Simultaneous Service and Speed Scaling

Ruslana Nekrasova[1,2(✉)]

[1] Institite of Applied Mathematical Research KarRC RAS, Petrozavodsk 185000, Russia
ruslana.nekrasova@mail.ru

[2] Petrozavodsk State University, Petrozavodsk 185000, Russia

Abstract. We consider a multi-server queuing system under speed scaling policy, speed regimes are switched at arrival/departure instants according to the corresponding transition probability matrices. The model admit simultaneous service discipline: arriving customer occupies a random number of servers, which is simultaneously captured and then released after the service is finished. The paper concentrates on model with general distribution of service times. We present the regenerative structure of the model and rely on regenerative approach to construct confidence intervals for the steady-state system performance measures. As stability results were obtained for only for Markovian case, we rely on monotonicity properties to establish steady-state regime. Simulation results illustrate, that regenerative method provides more accurate estimation in comparison with bounds based on monotonicity properties. Obtained results are applicable in analysis of modern simultaneous service systems like wireless transmitters and laptops or parallel computing technologies.

Keywords: Non work-conserving model · Simultaneous service · Speed scaling · Regenerative approach · Regenerative estimation

1 Introduction

Most of modern computing systems like distributed computing or high performance clusters contain simultaneous service technique. We deal with the model, where arrival simultaneously demands a random number of servers and join the queue, if resources are not available. Such a model admits idle servers, while the queue is not empty. Thus, service discipline is non work-conserving. That makes

The study was carried out under state order to the Karelian Research Centre of the Russian Academy of Sciences (Institute of Applied Mathematical Research KRC RAS). The research is partly supported by Russian Foundation for Basic Research, projects 18-07-00147, 18-07-00156, 19-07-00303, 19-57-45022.

© Springer Nature Switzerland AG 2020
V. M. Vishnevskiy et al. (Eds.): DCCN 2020, CCIS 1337, pp. 299–310, 2020.
https://doi.org/10.1007/978-3-030-66242-4_24

the analysis much more complicated (see, at instance [4,6]). Stability criterion for the Markovian system with simultaneous service policy was obtained in [13], authors based on a matrix-analytic approach. In more general case we have to rely on assumptions or simulation.

Regenerative method is a powerful instrument in stochastic modelling and performance analysis. In recent work [12] regenerative confidence estimation was applied for mean queue size in $M/M/2$-type model with simultaneous service and speed scaling policy. Another application of regeneration theory for analysis of a simultaneous service model was presented in [3], where hypoexponential distribution of service time was considered.

In this paper we construct m-server model with a general distribution of service time and speed scaling policy: both servers simultaneously switches speed regimes in arrival/departure instants. Under stability condition for dominating single server queueing model, we apply regenerative confidence estimation for original system with simultaneous service and illustrate confidence intervals for mean number of customers in the system and mean queue size, considering particular cases of Pareto and Weibull distributions of service time.

The paper is organised as follows. In Sect. 2 we present a detailed description of a model under consideration. Section 3 is dedicated to regenerative approach for analysis of the system. We briefly discuss regenerative confidence estimation in discrete time and present the conditions of its applicability. Section 4 contains simulation results of regenerative confidence estimation for mean number of customers and mean queue size in simultaneous servers systems, considering Pareto or Weibull distribution of service time. Section 5 concludes the paper.

2 Description of a Model

We construct m-server queueing model with infinite buffer. Arrivals join the system at instants $\{t_n;\, n \geq 1\}$ according to renewal input with a rate λ. The n-th customer is characterized by a pair of random parameters: (S_n, C_n) which define the amount of work and a number of required servers, respectively.

Note, that current customer tries to capture $C_n \in \{1, \dots, m\}$ servers simultaneously. If the resources are not available at instant t_n, the customer joins a queue (organised according to FIFO discipline) until the service is possible. Then after completion of the required work amount S_n, all C_n servers are seized and released simultaneously. The sequences $\{S_n;\, n \geq 1\}$ is independent and identically distributed (iid) and independent of an iid sequence $\{C_n;\, n \geq 1\}$. Denote the corresponding generic elements of such sequences by S and C, respectively. We consider *general* distribution of S and discrete distribution of C such, that

$$P(C = k) := p_k,\; k = 1, \dots, m,$$

where p_k are given probabilities.

Other significant feature of the system under consideration is a *speed scaling* technique: the servers can process L distinct speeds with "rates" $\mu_1 < \cdots < \mu_L$, and speed switching simultaneously affects to both servers. Namely, if the servers work at rate μ_i (in the i-th mode) a work amount S is completed in S/μ_i time interval.

We consider L modes, and assume, that speeds is altered only at arrival or departure epochs according to (corresponding) Markov chains with transition matrices

$$||a_{ij}||, \quad ||d_{ij}||, \quad i, j = 1, \dots, L.$$

Namely, speed μ_i may be switched to speed μ_j at arrival or departure instant with probability a_{ij} or d_{ij}, respectively.

3 Regenerative Structure

Regenerative method is a strong instrument in stochastic analysis. In this section we present a detailed description of regeneration structure for a model under consideration. First, we define by $\nu(t) \in \{0, 1, \dots, m\}$ and $Q_n(t) \in \{0, 1, \dots\}$ the number of customers on service and the queue size at time instant $t \geq 0$, respectively. Note, that $\nu(t)$ is an actual number of customers on service, which may not coincide with the number of busy servers. In particular, if $C_n = k \leq m$ (the n-th arrival occupies k servers), while and other $(m-k)$ servers are stay idle at instant t_n, then $\nu(t_n) = 1$. The process, associated with total number of customers, is defined by

$$X(t) = \{\nu(t) + Q(t);\, t \geq 0\}. \tag{1}$$

We additionally assume, that system has zero initial state ($X(0) = 0$) and construct the following sequence:

$$T_k = \min_n\{t_n > T_{k-1} : X(T_n^-) = 0\}, \quad k \geq 1,\, T_0 = 0. \tag{2}$$

Namely, $\{T_k, k \geq 1\}$ indicates the instants, when arrival joins totally empty system: $\nu(T_k) = Q(T_k) \equiv 0$.

The system at instants T_k starts over in stochastic sense or *regenerates*. Random segments (cycles) $\{X(t);\, T_{k-1} \leq t < T_k\}, \quad k \geq 1$ of a process X are iid, and $\{X(t);, t \geq 0\}$ is called a *regenerative* process [1]. Note, that the sequence of regeneration cycles lengths is iid

$$T_k - T_{k-1} \;=_{st} T$$

and denote the generic length by T.

If the mean cycle length ET is finite, the process $\{X(t), t \geq 0\}$ is called *positive recurrent* [9,14]. In general, positive recurrence also includes the condition for the first cycle:

$$T_1 < \infty, \tag{3}$$

which automatically holds, since the process was considered without delay, $X(0) = 0$ (the first cycle is stochastically equal to a generic one).

If the condition $ET < \infty$ holds, then the regenerative method is applicable. Then we construct "time average" value of the process X at $[0, t]$ as

$$r(t) = \frac{1}{t}\int_0^t X(u)du, \tag{4}$$

The following limit (if exists)

$$\lim_{t\to\infty} r(t) = r, \tag{5}$$

is a steady-state performance measure (mean number of customers in the system). Conditions of existence for the limit 5 are rather simple. Define the iid sequence of accumulated values over regeneration cycles as

$$Z_n = \int_{T_{n-1}}^{T_n} X(u)du, \quad n \geq 1. \tag{6}$$

If process X is positive reccurent and $\int_0^T |X(u)|du < \infty$, then

$$\frac{1}{t}\int_0^t X(u)du \to \frac{EZ}{ET}. \tag{7}$$

More detailed description of regeneration approach is well-presented in [1,8–10].

3.1 Regenerative Estimation in Discrete Time

In this section we discuss regenerative method of confidence estimation, defining the basic processes in discrete time scale. Remind $\{t_n, n \geq 0\}$ the sequence on arrival instants in the system under consideration. Denote by

$$\tau_n = t_n - t_{n-1}, \quad n \geq 1$$

iid inter-arrival times with a generic element τ. Next we construct the discrete analogue of the process X as follows:

$$\{X_n := X(t_n^-), n \geq 0\}.$$

Note, that X_n is associated with the total number of customers in the system just before the n-th arrival. (Define by ν_n, Q_n the number of customers on service and queue length before the instant t_n, respectively.)

Next consider the sequence of regeneration points

$$\beta_k = \min_n\{n > \beta_{k-1} : X_n = 0\}, \quad k \geq 1, \beta_0 = 0. \tag{8}$$

Namely, $\{\beta_k, k \geq 1\}$ indicates actual numbers of customers, that arrive into totally empty system: $X_{\beta_k} = 0$. Note, that regeneration instants, defined in continuous and discrete time, are related as $T_n = t_{\beta_n}$.

Next we define the sequence of iid regeneration cycles length discrete time (with a generic length α) by

$$\alpha_n = \beta_n - \beta_{n-1}, \quad n \ge 1 \tag{9}$$

and iid accumulated numbers of customers over each regeneration cycle by

$$Y_n = \sum_{k=\beta_{(n-1)}}^{(\beta_n)-1} X_k, \quad n \ge 1.$$

Note, that mean cycle lengths in both time scales are connected via Wald's identity as

$$ET = E\alpha E\tau.$$

Thus, if $E\tau < \infty$, the conditions $ET < \infty$ and $E\alpha < \infty$ are equivalent and imply the positive recurrence of regenerative process $\{X_n\}$.

Hence, if

$$E\alpha < \infty \tag{10}$$

the following limit exists:

$$r_n := \frac{\sum_{k=1}^{n} Y_k}{\sum_{k=1}^{n} \alpha_k} \to \frac{EY}{E\alpha} =: r, \qquad n \to \infty, \tag{11}$$

where Y is a generic element of a sequence $\{Y_n, \, n \ge 1\}$.

Note, that r_n coincides with an average number of customers in a system within interval $[0, \, t_{\beta_n})$:

$$r_n = \frac{1}{\beta_n} \sum_{k=1}^{\beta_n} X_k. \tag{12}$$

Actually, the result (11) means, that with a growth of cycle number, time average value of regeneration process converges to the ratio of mean cumulative value over cycle to mean cycle length. Namely, in case of positive recurrence, the behavior of regenerative process could be described by its cycle characteristics.

By other significant result from regeneration theory [5], we obtain, that if α is aperiodic, the following weak convergence holds

$$X_i \Rightarrow X^{(S)}, \qquad i \to \infty, \tag{13}$$

where $X^{(S)}$ is a *stationary analogue* of regenerative process X. From (13), we easily obtain

$$\lim_{n \to \infty} r_n = EX^{(S)}. \tag{14}$$

By Proposition 4.1 from [2] the estimator r_n satisfies the following Central Limit Theorem

$$\sqrt{n}(r_n - r) \Rightarrow \mathbf{N}(0, \, \sigma^2), \qquad n \to \infty, \tag{15}$$

where

$$\sigma^2 = \frac{E[Y - r\alpha]^2}{(E\alpha)^2}.$$

Hence, if limits (11) and (13) exist, then weak convergence (15) holds and implies the following $100(1-\gamma)\%$ confident interval:

$$EX^{(S)} \in \Big[r_n - \Delta_n, r_n + \Delta_n\Big], \tag{16}$$

with accuracy

$$\Delta_n = \frac{z_\gamma \overline{\sigma}_n}{\sqrt{n}}.$$

Note, that γ is a given reliability and

$$\overline{\sigma}_n^2 = \frac{n^2}{n-1} \frac{\sum_{i=1}^n \left(Y_i - r_n \alpha_i\right)^2}{\left(\sum_{i=1}^n \alpha_i\right)^2}.$$

(The value z_γ defines $(1-\gamma/2)$-quantile of the standard normal law.)

3.2 Application for Speed Scaling System

A point estimator r_n coincides with mean average, which could be obtained without regenerative approach. Our goal is to apply regenerative method (RM) to build more informative interval estimator (16) for mean number of customers in considered system with speed scaling policy and simultaneous service. In this section we present the conditions for applicability of regeneration confidence estimation.

First, we construct an auxiliary infinite buffer queueing system $G/G/1$ as follows: the same renewal stream with a rate λ as in original system with speed scaling, and service times are iid and stochastically equivalent to S/μ_1 (time, demanded to serve a work amount S on the "slowest" regime in original system). By monotonicity properties [11] such a new system dominates the original system in a sense of load. Thus, its stability implies the stability of considered system with speed scaling.

Remind generic inter-arrival time τ defined by and assume, that the following conditions hold

$$\rho := \lambda ES/\mu_1 < 1, \tag{17}$$

$$P\Big(\tau > S/\mu_1\Big) > 0. \tag{18}$$

Note, that $E\tau = 1/\lambda$ and ρ defines a load coefficient in dominating queuing system. Remind considered zero initial state in original system: $X_0 = 0$. Under condition $\rho < 1$ the dominating system is stable. Hence, we obtain the stability of original model, which also yields, that the process $\{X_n\}$ is stochastically bounded. Then basing on results from regeneration theory and under (18) we derive the positive recurrence: $E\alpha < \infty$, where α is a generic regenerative cycle length, defined in (9) (see [9] for details).

The condition (18) also implies, that with a positive probability there exist a regeneration cycle based the only arrival ($P(\alpha = 1) > 0$), which means that α is aperiodic. Thus, (17)–(18) allow to apply RM for confident estimation of mean number of customers.

4 Simulation

In this section we present simulation results for the particular case of $M/G/2$-type system with simultaneous service and speed scaling for a few distributions of S under conditions (17)–(18).

We define speed scaling transition matrices, associated with arrival and departure instants by

$$||a_{ij}|| = \begin{bmatrix} 1-a & a \\ 0 & 1 \end{bmatrix}, \quad ||d_{ij}|| = \begin{bmatrix} 1 & 0 \\ d & 1-d \end{bmatrix}, \tag{19}$$

respectively.

Our goal is to build confident intervals for the mean number of customers in the system $EX^{(S)}$ and mean queue size $EQ^{(S)}$, where $Q^{(S)}$ is a stationary analogue of the queue process $\{Q_n\}$.

In Poisson input case the condition $P(\tau > S/\mu_1) > 0$ automatically holds and $EX^{(S)}$ is bounded above by the mean number of customers in dominating system $M/G/1$ (denoted by EL). The value EL is obtained by Pollaczek-Khinchine formula [7] as

$$EL = \rho + \frac{\rho^2 + \lambda^2 DS/\mu_1^2}{2(1-\rho)}. \tag{20}$$

while the mean queue (upper bound of $Q^{(S)}$) is obtained by

$$EN = \frac{\rho^2 + \lambda^2 DS/\mu_1^2}{2(1-\rho)}. \tag{21}$$

Note, that the process Q regenerates at the same instants as original process X. Thus, conditions (17)–(18) allow to construct the confident interval af follows:

$$EQ^{(S)} \in \left[q_n - \frac{z_\gamma \overline{\sigma}_n^{(Q)}}{\sqrt{n}}, \, q_n + \frac{z_\gamma \overline{\sigma}_n^{(Q)}}{\sqrt{n}} \right], \tag{22}$$

where

$$q_n = \frac{1}{\beta_n} \sum_{k=1}^{\beta_n} Q_k,$$

$$\overline{\sigma}_n^{(Q)} = \left[\frac{n^2}{n-1} \frac{\sum_{i=1}^n \left(Y_i^{(Q)} - q_n \alpha_i \right)^2}{\left(\sum_{i=1}^n \alpha_i \right)^2} \right]^{1/2},$$

$$Y_n^{(Q)} = \sum_{k=\beta_{(n-1)}}^{(\beta_n)-1} Q_k, \quad n \geq 1.$$

4.1 Pareto Service

First we consider Pareto distribution of S with a scale parameter fixed and equal to 1 and shape parameter denoted by $\mathcal{K}$. Thus,

$$P(S \leq x) = 1 - x^{-\mathcal{K}}, \qquad x \geq 1, \tag{23}$$

and corresponding characteristics are defined by

$$ES = \frac{\mathcal{K}}{\mathcal{K}-1}, \qquad DS = \frac{\mathcal{K}}{(\mathcal{K}-1)^2(\mathcal{K}-2)}. \tag{24}$$

Note, that in this case dominating queuing system $M/G/1$ has Pareto distribution of service time with a scale parameter $1/\mu_1$ and shape parameter $\mathcal{K}$, while load coefficient is defined by

$$\rho = \frac{\lambda\mathcal{K}}{\mu_1(\mathcal{K}-1)}. \tag{25}$$

We explored the system behavior, considering one of three combinations of speed scaling rates (μ_1, μ_2). For the "light" speed scaling we considered $(\mu_1, \mu_2) = (0.9, 1.1)$, for the "midium" speed scaling we fixed $(\mu_1, \mu_2) = (0.7, 1.3)$ and finally we simulated the system with a "hard" speed scaling, assuming $(\mu_1, \mu_2) = (0.5, 1.5)$. We also varied the value of ρ and illustrate the dynamics of total process X in case of "light", "medium" and "hard" load of dominating system with corresponding load coefficients $\rho = 0.45$, $\rho = 0.70$, $\rho = 0.92$. (Note, that in all considered cases we fixed $\mathcal{K} = 3$, which implies $ES = 1.5$, and varied λ to obtain the appropriate value of ρ.)

The results, related to confidence estimation of $EX^{(S)}$ in $M/Pareto/2$-type system with speed scaling and simultaneous service are presented in Table 1. Note, that all the experiments were done for fixed probabilities $a = 0.4$, $d = 0.6$, $p_1 = 0.5$, other results had shown that values of a, d, p_1 do not seriously affect to the accuracy of obtained intervals.

Table 1. Confidence estimation of mean number of customers in $M/Pareto/2$-type speed scaling system, $\mathcal{K} = 3$.

Test	ρ	λ	μ_1	μ_2	r_n	n	EL	Δ_n	$EL - r_n$
1	0.450	0.270	0.9	1.1	0.530	12497	0.695	0.015	0.166
2	0.450	0.210	0.7	1.3	0.413	13668	0.695	0.012	0.282
3	0.450	0.150	0.5	1.5	0.339	14526	0.695	0.010	0.357
4	0.700	0.420	0.9	1.1	0.989	8968	1.789	0.049	0.800
5	0.700	0.330	0.7	1.3	0.714	10520	1.789	0.026	1.075
6	0.700	0.233	0.5	1.5	0.542	11949	1.789	0.015	1.247
7	0.920	0.552	0.9	1.1	1.609	6182	7.973	0.115	6.364
8	0.920	0.429	0.7	1.3	1.053	8186	7.973	0.046	6.921
9	0.920	0.307	0.5	1.5	0.796	9661	7.973	0.029	7.178

The first three tests in Table 1 correspond to the system with "light" load (the load coefficient in dominating system $\rho = 0.45$). We based on 20 000 arrivals and obtained $n = 12497$ regenerations for the "light" speed scaling $\mu_1 = 0.9$, $\mu_2 = 1.1$. In this case the accuracy of regenerative estimation $\Delta_n \approx 0.015$, while the estimator, based on monotonicity properties is nearly in 10 times less accurate: $EL - r_n \approx 0.166$, where $EL \approx 0.695$. The comparison of obtained estimators (test 1) is illustrated on Fig. 1.

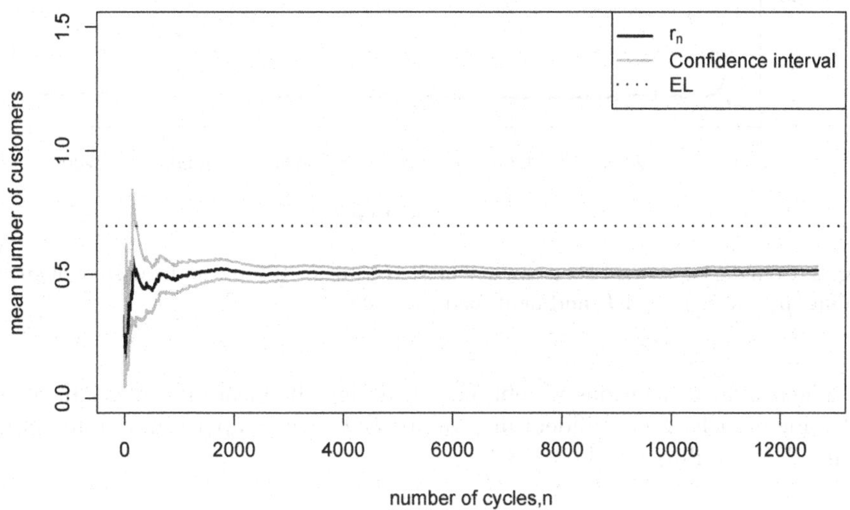

Fig. 1. Confidence interval for mean number of customers in $M/Pareto/2$-type model with light speed scaling, $\mu_1 = 0.9$, $\mu_2 = 1.1$, and light load $\rho = 0.45$.

With a growth of speed scaling rates the difference between dominating queueing system and original speed scaling system increases. Thus, for the case hard speed scaling (test 3) we obtained $EL - r_n = 0.357$, while regenerative estimation is much more accurate: $\Delta_n = 0.010$. Note, that the bigger values (μ_1, μ_2) for a fixed ρ implies the decreasing of λ. The system becomes less loaded, which yields more regenerations, thus the confident interval in case 3 became a little thinner in comparison with the case 1.

Tests 4–6 and 7–9 correspond to "medium" and "heavy" load, respectively. Smaller values of μ_1 force the difference in load between dominating and original systems, and estimation r_n by EL becomes less accurate. This fact is easily explained by (20). The increasing of such a difference is more notable in light load case. Meanwhile, varying of parameters μ_1, μ_2, ρ does not strongly affect to the accuracy of interval estimator, obtained by RM. In particular, for $\mu_1 = 0.5$, $\mu_2 = 1.5$, $\rho = 0.92$ (case 9), we obtained $\Delta_n \approx 0.029$, while $EL - r_n = 7.178$. Such a result illustrates the advantages of regeneration estimation (at least in comparison with bounds build basing on monotonicity properties).

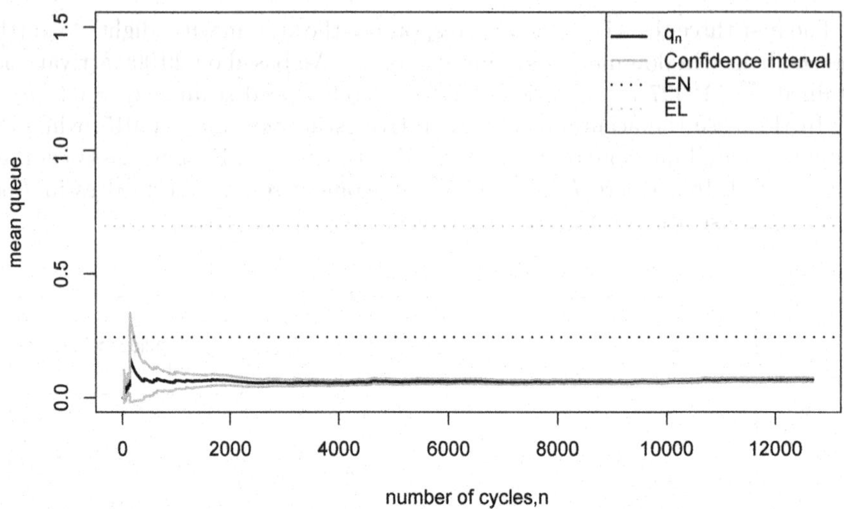

Fig. 2. Confidence interval for mean queue in $M/Pareto/2$-type model with light speed scaling, $\mu_1 = 0.9$, $\mu_2 = 1.1$, and light load $\rho = 0.45$.

The similar conclusions we obtain, estimating the mean queue size in corresponding models. The confident interval for $EQ^{(S)}$ in comparison with its upper bound EN are illustrated on Fig. 2.

4.2 Weibull Service

For Weibull distribution of S we define a scale parameter fixed and equal to 1 and shape parameter denoted by w. Thus,

$$P(S \le x) = 1 - e^{-x^w}, \qquad x \ge 0. \tag{26}$$

and corresponding parameters are defined via Gamma-function as

$$ES = \Gamma\Big(1 + \frac{1}{w}\Big), \qquad DS = \Gamma\Big(1 + \frac{2}{w}\Big) - (ES)^2. \tag{27}$$

The dominating queuing system $M/G/1$ has also Weibull distribution of service time with shape parameter w, but scale parameter is equal to $1/\mu_1$. Its load coefficient is defined by

$$\rho = \frac{\lambda}{\mu_1}\Gamma\Big(1 + \frac{1}{w}\Big). \tag{28}$$

Numerical results for regenerative estimation of mean number of customers in $M/Weibull/2$-type model (for fixed values $w = 2$, $a = 0.4$, $d = 0.6$, $p_1 = 0.5$) are presented in Table 2.

Table 2. Confidence estimation of mean number of customers in $M/Weibull/2$-type speed scaling system, $w = 2$.

Test	ρ	λ	μ_1	μ_2	r_n	n	EL	Δ_n	$EL - r_n$
1	0.45	0.457	0.9	1.1	0.513	12633	0.684	0.014	0.171
2	0.450	0.254	0.5	1.5	0.333	14552	0.684	0.009	0.352
3	0.920	0.934	0.9	1.1	1.580	6015	7.656	0.083	6.076
4	0.920	0.519	0.5	1.5	0.770	9671	7.656	0.022	6.885

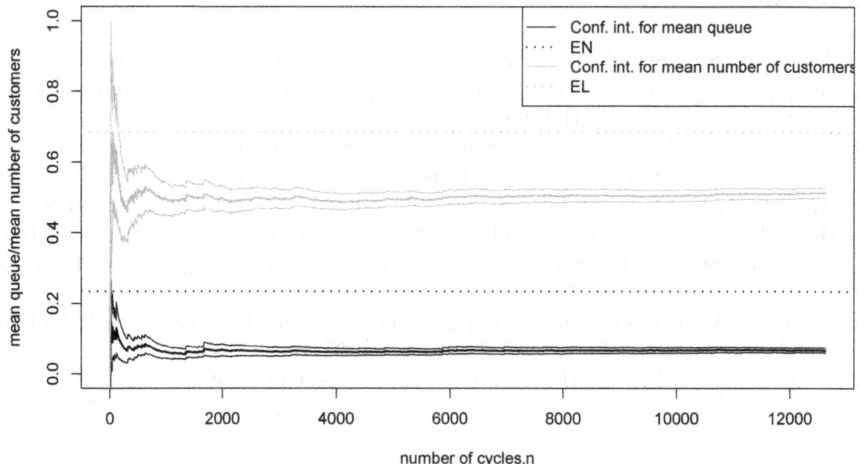

Fig. 3. Confidence intervals for mean number of customers and mean queue size in $M/Weibull/2$-type model with light speed scaling, $\mu_1 = 0.9$, $\mu_2 = 1.1$, and light load $\rho = 0.45$.

Similar to the case of Pareto service, we obtained, that regenerative confidence estimation provides more accurate results (in comparison with theoretical bound EL), specially with a growth of load coefficient.

Confident intervals of $EX^{(S)}$ and $EQ^{(S)}$ for the system under light speed scaling policy and with $\rho = 0.45$ are illustrated on Fig. 3. Note, that the original model in such a configuration is the most close to dominating system and bounds EN and EL are the most accurate (in comparison with other tests, see Table 2). Moreover, that interval for mean queue (black solid lines) is thinner, than for mean number of customers. Thus, we obtained the variance relation $\sigma_n^{(Q)} < \sigma_n$, is explained by simultaneous service.

5 Conclusion

We considered a general case of a system with simultaneous service under speed scaling policy. Such a model could be useful in simulation of modern parallel computing systems. As stability results were obtained only for models under exponential assumptions, we relied on stability condition for the dominating queuing system. Basing on regenerative approach, we applied confidence estimation for mean number of customers and mean queue size, considering 2-server model with Poisson input in stable regime. Numerical results, obtained for different configurations and particular cases of Pareto or Weibull distributions of service time, had shown the advantages of applied method in comparison with interval estimation, based on monotonicity properties.

References

1. Asmussen, S.: Applied Probability and Queues. Wiley, New York (2003)
2. Asmussen, S., Glynn, P.W.: Stochastic Simulation: Algorithms and Analysis. Springer-Verlag, New York (2007)
3. Afanaseva, L., Bashtova, E., Grishunina, S.: Stability analysis of a multi-server model with simultaneous service and a regenerative input flow. Methodol. Comput. Appl. Probab. **22**, 1–17 (2019). https://doi.org/10.1007/s11009-019-09721-9
4. Brill, P., Green, L.: Queues in which customers receive simultaneous service from a random number of servers: a system point approach. Manag. Sci. **30**(1), 51–68 (1984)
5. Crane, M.A., Lemoine, A.J.: An Introduction to the Regenerative Method for Simulation Analysis. Springer-Verlag, Berlin (1977)
6. Fletcher, G.Y., Perros, H., Stewart, W.: A queueing system where customers require a random number of servers simultaneously. Eur. J. Oper. Res. **23**, 331–342 (1986)
7. Haigh, J.: Probability Models. Springer, London (2002)
8. Morozov, E.: The tightness in the ergodic analysis of regenerative queueing processes. Queueing Syst. **27**, 179–203 (1997)
9. Morozov, E.: Weak regeneration in modeling of queueing processes. Queueing Syst. **46**, 295–315 (2004)
10. Morozov, E., Delgado, R.: Stability analysis of regenerative queues. Autom. Remote control **70**, 1977–1991 (2009)
11. Morozov, E., Rumyantsev, A., Peshkova, I.: Monotonicity and stochastic bounds for simultaneous service multi-server systems. In: 2016 8th International Congress on Ultra Modern Telecommunications and Control Systems and Workshops (ICUMT) on Proceedings, pp. 294–297. IEEE, Moscow (2016)
12. Nekrasova, R., Rumyantsev, A.: Regenerative estimation of a simultaneous service multi-server system with speed scaling. In: Proceedings of the 26th Conference of Open Innovations Association FRUCT, pp. 346–351. Helsinki, Finland (2020)
13. Rumyantsev, A., Morozov, E.: Stability criterion of a multi-server model with simultaneous service. Ann. Oper. Res. **252**(1), 29–39 (2015). https://doi.org/10.1007/s10479-015-1917-2
14. Sigman, K., Wolff, R.W.: A review of regenerative processes. SIAM Rev. **35**, 269–288 (1993)

Integrated Tolerant Distributed Computing Network

Oleg Brekhov(✉)

Moscow Aviation Institute (National Research University) (MAI), Volokolamskoye shosse, Moscow 125993, Russia
obrekhov@mail.ru

Abstract. Distributed computer network (DCN) consists of computing modules (CM). Two analytical performance models of tolerant computing networks are described in this paper. The first model is itself based on two models: a model for evaluating performance depending on the number of serviceable CM and a performance model depending on the method of ensuring the tolerance of the computer network. This model assumes that a computing module can't be restored. The second model assumes that during the life cycle of a tolerant DCN its CM can be in one of three possible states: non-functional, functional – working, and functional – controlled. It is also assumed that the CM can be restored.

Keywords: Model · Analytical · Performance · Computing network · Distributed · Tolerant · Restoration work

1 Performance Model of the DCN with Non-restorable Computing Modules

The performance of a distributed computing network (DCN) can be determined in relation to various levels of functioning (commands, programs, tasks, and jobs). Here, the performance of the DCN is related to the level of task completion. The task can be completed when a certain event occurs for it and there is a free resource. The impact of task readiness can be taken into account when using the model of the CM functioning as a system with a dynamic change in the number of tasks. In this case, each task executed by the processor brings a certain number of tasks to the ready state out of the maximum tasks to be completed. The availability of a free resource undoubtedly depends on the distribution of tasks across the CM, where they should be performed.

It is necessary to provide measures to ensure tolerance so that the degree of performance reduction is the lowest. We will limit ourselves to software tools. We will build an integrated analytical model of the performance of a tolerant DCN based on two models: a model for evaluating performance based on the number of serviceable CM, and a performance model based on the method of ensuring tolerance.

© Springer Nature Switzerland AG 2020
V. M. Vishnevskiy et al. (Eds.): DCCN 2020, CCIS 1337, pp. 311–326, 2020.
https://doi.org/10.1007/978-3-030-66242-4_25

1.1 Model of the Number of Virtual Computing Modules Depending on the Means of Ensuring Network Tolerance

During the life cycle of the DCN, CM may be in good or faulty condition. We will use a functional approach to control the state of CM using software tools. On this path, when using a diagnostic test, performing a single task requires one (d = 1) CM. In another implementation of module health monitoring, parallel execution of d, $d \geq 2$, copies of one task for each copy requires its own module, i.e. d computing modules. To ensure CM control, CM must be distracted from performing basic functional tasks. The time required to detect and locate a fault, as well as to logically exclude a faulty module, depends significantly on the tools and methods used in this case. As a rule, this time is longer if the number of CM running copies of tasks at the same time is less. If you use d, $d \geq 2$, copy tasks, program actions related to determining the health of modules are performed on another CM with minimal time spent on the controlled CM. When using a diagnostic test, control measures require time from both the controlled and controlling modules. In our model, without loss of generality, we will assume that the controlling activities spend the time resources of only one module. We will also assume that the activation of the functional control mechanism is not necessarily associated with the completion of each functional task. Denote $p_{i,j}(t)$ the probability that at time t with $i + j$ functioning CM i CMs are in the state of performing functional tasks and j CMs are performing control functions.

Next, our goal is to build a model of the DCN, which will allow us to determine the probabilities $p_{i,j}(t), i, j = \overline{0, n}$.

We assume that the DCN subsystem contains n serviceable CM. Let the operating time intervals and control time intervals be random variables with exponential distribution functions with intensities λ_i and μ_j, depending on i operating and j controlling CM, respectively. As a result of control measures with probability r, the fault of the computing module is fixed and the faulty module is excluded from the list of serviceable one with an additional probability $(1 - r)$, the number of serviceable CM remains the same.

Due to the exponential distribution functions of the duration of processes, the model of DCN operation in non-stationary mode is described by the following graph of states (see Fig. 1).

The graph contains C^2_{n-d+3} states, and there are no states (i, j) with an index $i < d - 1$, since one task is required in the CM to perform in copies. Thus, the state $(d - 1, j), j = \overline{0, n - d + 1}$, corresponds to a situation when the CM does not perform functional tasks, but only performs controlling functions.

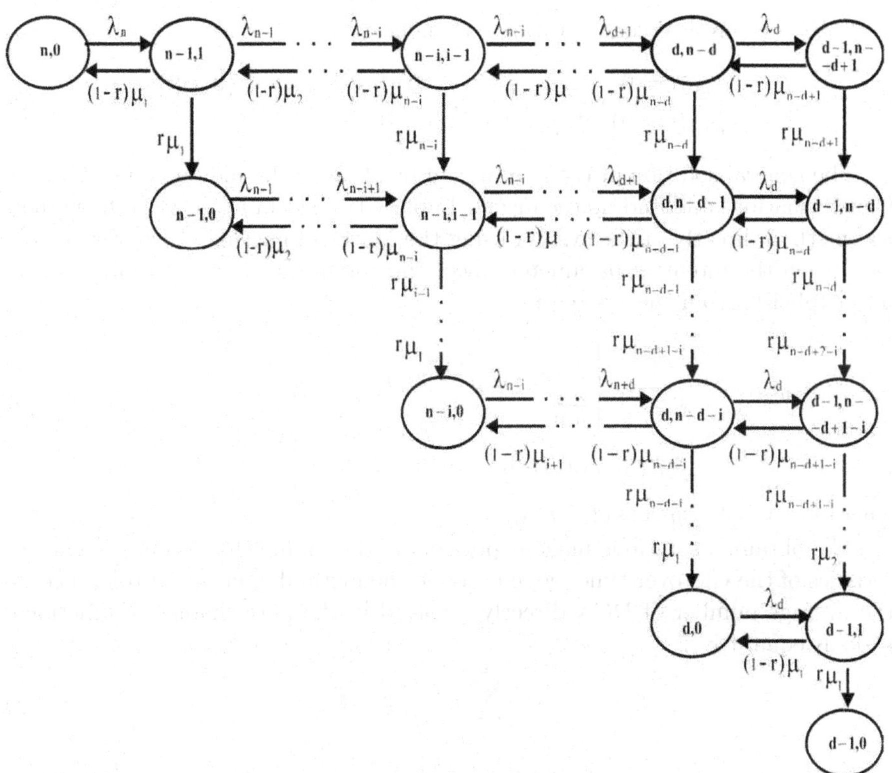

Fig. 1. .

In accordance with the graph of States we have a system (1) of equations:

$$
\begin{aligned}
p'_{n,0}(t) &= -\lambda_n p_{n,0}(t) + (1-r)\mu_1 p_{n-1,1}(t);\\
p'_{n-j,j}(t) &= -(\mu_j - \lambda_{n-j}) p_{n-j,j}(t) + \lambda_{n-j+1} p_{n-j+1,j-1}(t)\\
&\quad + (1-r)\mu_{j+1} p_{n-j-1,j+1}(t), j = \overline{1, n-d};\\
p'_{d-1,n-d+1}(t) &= -\mu_{n-d+1} p_{d-1,n-d+1}(t) + \lambda_d p_{d,n-d}(t);\\
p'_{n-i,0}(t) &= -\lambda_{n-i} p_{n-i,0}(t) + r\mu_1 p_{n-i,1}(t)\\
&\quad + (1-r)\mu_1 p_{n-i-1,1}(t), i = \overline{1, n-d};\\
p'_{n-i-j,j}(t) &= -(\mu_j + \lambda_{n-i-j}) p_{n-i-j,j}(t) + \lambda_{n-i-j+1} p_{n-i-j+1,j-1}(t)\\
&\quad + (1-r)\mu_{j+1} p_{n-i-j-1,j+1}(t) + r\mu_j p_{n-i-j,j+1}(t),\\
&\quad i = \overline{1, n-d-1}, j = \overline{1, n-i-d};\\
p'_{d-1,n-d-i+1}(t) &= -\mu_{n-d-i+1} p_{d-1,n-i-d+1}(t) + \lambda_d p_{d,n-i-d}(t)\\
&\quad + r\mu_{n-d-i+2} p_{d-1,n-i-d+2}(t), i = \overline{1, n-d};\\
p'_{d-1,0}(t) &= r\mu_1 p_{d-1,1}(t);
\end{aligned}
\tag{1}
$$

Based on physical considerations, we believe that:

$$\begin{aligned} p_{n,0}(t=0) &=1, p_{d-1,0}(t \to \infty) = 0 \text{ for } i,j \in \{\overline{0,n}\}; \\ p_{i,j}(t=0) &=0, p_{i,j}(t \to \infty) = 0 \end{aligned} \tag{2}$$

The general solution of the system can be obtained by moving on to Laplace transformations and finding the inverse Laplace transform $p_{\{i,j\}}(s)$, or by numerical method directly of the system using the Mathcad package. However, for the case when the intensity parameters meet the condition, $\lambda_i = \lambda$ and $\mu_i = \mu$, we find explicitly from the system (3):

$$\begin{aligned} p_{i,n-i}(s) =& \frac{1}{\lambda^{i-d+1}} \left[\sum_{j=0} (-1)^j a^{i-d-2j} b^j (c^{j-1}_{i-d-j} a + c^j_{i-d-j}(a-\lambda)) \right] \\ & \times p_{d-1,n-d+1}(s), i = \overline{d,n}; \end{aligned} \tag{3}$$

where $a = s + \lambda + \mu, b = (1-r)\lambda\mu$.

The obtained solutions make it possible to determine changes in the characteristics of the CM over time, depending on the method of ensuring tolerance, so the average number of DCN directly involved in the performance of functional tasks is equal to:

$$n_{cp} = \sum_{i=d}^{n} i \sum_{j=0}^{n-i} j p_{i,j}(t). \tag{4}$$

Since a single task requires d (d$\geq$1), CM, the number of CMs directly involved in performing functional tasks should be reduced by a factor of d, i.e. the number of virtual (actual) modules is equal to:

$$\nu = \frac{n_{cp}}{d} \tag{5}$$

Table 1 and graphs on Fig. 2 show the average number of VMS engaged in performing functional tasks for the following parameter values: the average time for performing functional tasks, the average control time is equal to: 1 - when using a diagnostic test (d = 1), 0.2 - when performing two-copy tasks (d = 2), and 0.05 - when performing three-copy tasks (d = 3), the probability of CM failure and, the initial number of CM is n = 32.

Figure 3 shows the graphs of changes in the number of virtual VMS depending on the life of the CM while maintaining the above parameters.

Based on the calculations performed, we can offer an approximating formula for calculating the average number of virtual DCN:

$$\widetilde{\nu} = \frac{n}{d^2(1+\frac{1}{\mu})} \left[1 - \frac{r^t}{1.1^d(1+\frac{1}{\mu})} \right] \tag{6}$$

which, if the condition is met, provides an accuracy of up to 5%. Note that at the time scale of sec. the value corresponds to the operating time of the aircraft for about 2.5 h.

Table 1. .

d	r	n_{cp}		
		t = 0	t = 4000	t = 8000
1	10^{-6}	16	15.96	15.92
1	10^{-5}	16	15.68	15.36
2	10^{-6}	26.67	26.58	26.49
2	10^{-5}	26.67	25.78	24.89
3	10^{-6}	30.48	30.36	30.24
3	10^{-5}	30.48	29.31	28.14

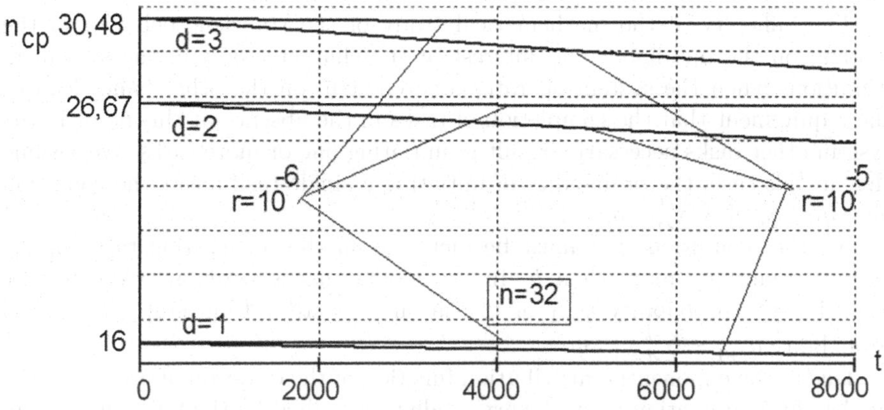

Fig. 2. .

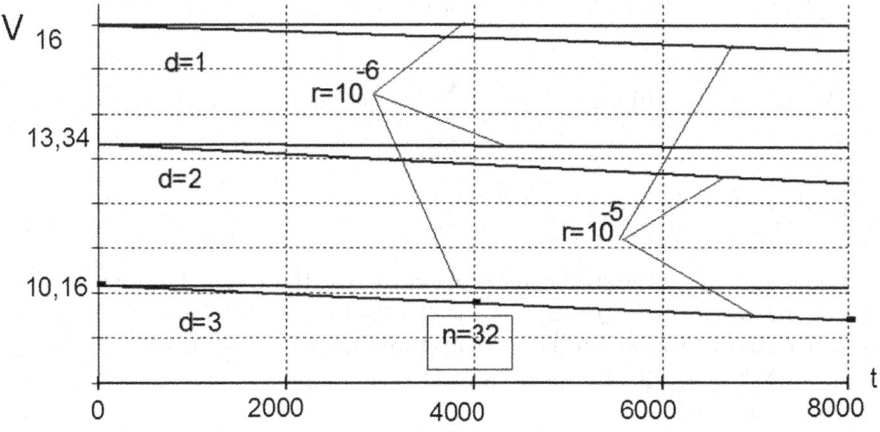

Fig. 3. .

1.2 DCN Performance Estimation Model

Let the DCN contain v of the same type of virtual CM. The functioning of the CM consists in performing the same type of tasks, and one task requires exactly one CM to complete it. Without generality restriction, we assume that the maximum number of tasks to be completed is h, where $h \geq v$. The completed task initiates i tasks with the probability $q_i, \sum_{j=2}^{k} q_i = 1, i = \overline{0,k}, 2 \leq k \leq h$, each of which starts running when there is a free VM. Let the runtime τ_B CM tasks be a random variable with an exponential distribution function $p(\tau_B < t) = 1 - e^{-\gamma t}$ with parameter γ.

To evaluate the performance of the DCN used a variety of performance indexes. We will get a solution for the DCN subsystem, based on which various indexes of DCN performance are determined, in particular, the average number of tasks and the average time to complete tasks.

The difficulty of the mathematical study of the transient behavior of the system, on the one hand, and interest in the characteristics of the system in conditions when the system operates long enough, on the other hand, lead to the requirement that the completion of tasks in the absence at this point in the system other tasks necessarily result in initiating one or more tasks. We assume that in this case, the probability of initiating a single task meets the condition $q_1' = q_1 + q_0, q_1' + \sum_{j=2}^{k} q_i = 1$.

Another condition that must be met: the number of existing tasks in the system cannot be more than h. In the model this is taken into account by replacing the probability q_{j+1} of transition to state h from state $(h - j)$ to probability $q_{j+1}' = \sum_{l=j+1}^{k} q_l$.

Due to the exponential distribution function of the duration of processes, the model of CM operation in stationary mode is described by the following system of equations:

$$\sum_{j=1}^{k-1} q_j(i+j-k)p_{i+j-k}b_i = ip_i, j = \overline{2,\nu}, p_{i+j-k} = 0 \text{ for } i+j-k \leq 0$$
$$\sum_{j=1}^{k-i-1} (\nu+i+j-k)p_{\nu+i+j-k}b_i + \nu \sum_{j=0}^{i-1} p_{v+j}b_{k-i-j} = \nu p_{n+i}, i = \overline{1,k-2} \quad (7)$$
$$\sum_{j=1}^{k-1} p_{\nu+i+j}b_i = p_{\nu+k+i}, i = \overline{-1,h-\nu-k}.$$

Where $p_i, i \leq v$ is the stationary probability that the system has i-active VMS (initiated I tasks); $p_{n+j}, j = \overline{1,h-v}$ is the stationary probability that n VMS are active, and j tasks are in the waiting state (initiated $v + j$ tasks), $Q_i = q_i/q_0$ and $b_j = \sum_{l=k-j+1}^{k} Q_l$. It can be proved that for probabilities $p_l, l = \overline{1,h}$ the relation (8) is valid

$$min(\nu,l)p_l = p_1 \sum_{i_1=0}^{\left[\frac{l-1}{k-1}\right]} b_1^{i_1} \times \sum_{i_2=0}^{\left[\frac{l-1-(k-1)i_1}{k-2}\right]} b_2^{i_2} \times \sum_{i_3=0}^{\left[\frac{l-1-\sum_{j=1}^{2}(k-1)i_j}{k-3}\right]} b_3^{i_3}$$
$$\times \sum_{i_{k-2}=0}^{\left[\frac{l-1-\sum_{j=1}^{k-3}(k-j)i_j}{2}\right]} b_k^{i_{k-2}} b_{k-1}^{l-1-\sum_{j=1}^{k-3}(k-j)i_j} P\left[i_1, i_2, \ldots, i_{k-2}, l-1-\sum_{j=1}^{k-2}(k-j)i_j\right] \tag{8}$$

The average number of active issues is:

$$h_{cp} = \sum_{l=1}^{h} min(\nu,l)p_l = p_1 (\sum_{i_1=0}^{\left[\frac{h-1}{k-1}\right]} b_1^{i_1} \times \sum_{i_2=0}^{\left[\frac{h-1-(k-1)i_1}{k-2}\right]} b_2^{i_2} \times \sum_{i_3=0}^{\left[\frac{h-1-\sum_{j=1}^{2}(k-1)i_j}{k-3}\right]} b_3^{i_3} \times \ldots$$
$$\times \sum_{i_{k-2}=0}^{\left[\frac{l-1-\sum_{j=1}^{k-3}(k-j)i_j}{2}\right]} b_k^{i_{k-2}} \sum_{i_{k-1}=0}^{\frac{h-1-\sum_{j=1}^{k-3}(k-j)i_j}{k-3}} (1 + max(\frac{i_{k-1}-\nu+1}{\nu}, 0)) b_{k-1}^{i_{k-1}}$$
$$\times P(i_1, i_2, \ldots, i_{k-1})), \tag{9}$$

Based on the obtained solution, we find the average time to solve problems:

$$T_{cp} = \frac{1}{\gamma}\frac{h_{cp}}{\nu} \tag{10}$$

In particular, if the number of initiated tasks is limited to no more than two for (K = 2), the average number of tasks activated in the DCN is equal to

$$h_{cp} = \frac{Q_2^h - 1 + \frac{Q_2}{\nu}\frac{(h-\nu)Q_2^{\nu+1}-(h-\nu+1)Q_2^{h-\nu}+1}{Q_2^{h-\nu}-1}}{(Q_2-1)\left[\frac{Q_2^{h-\nu}}{\nu}\frac{Q_2^{h-\nu}-1}{Q_2-1} + \sum_{i=1}^{\nu}\frac{Q_2^{i-1}}{i}\right]} \tag{11}$$

which when $Q_2 = q_2/q_0 = 1$ can be approximated as:

$$\widetilde{h_{cp}} = \frac{h + \frac{h-\nu(h-\nu+1)}{2\nu}}{\frac{h-\nu}{\nu} + 0.6 + ln(\nu)} \tag{12}$$

Figure 4 shows the values of the normalized average time γ for completing tasks with parameters: the maximum number of tasks is $h = 128$, the probability of initiation γ, the tasks correspond $Q_2 = 1$, and the number of virtual VMS varies from 8 to 16. We see that even when switching from 16 CM to 15, the task execution time increases by 8.57%.

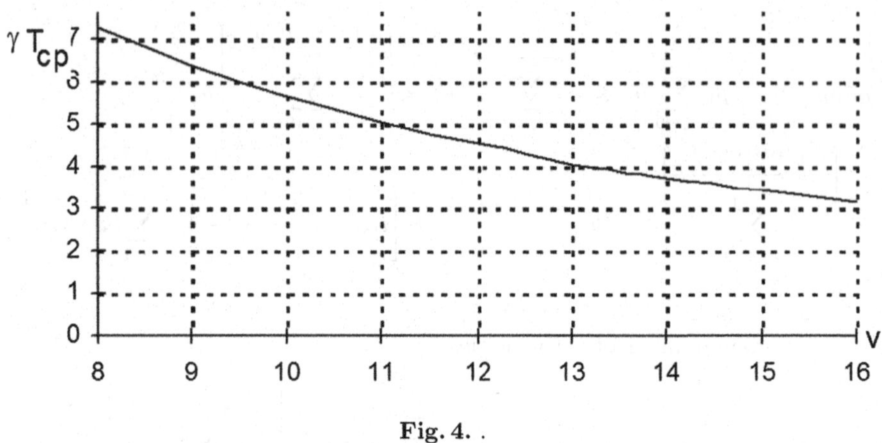

Fig. 4. .

1.3 An Integrated Model of the DCN

Combining the models developed above: the model of the number of virtual VMS depending on the means of ensuring tolerance and the model of the execution time of functions depending on the number of virtual VMS, we get an integrated model.

In accordance with the results and formulas, graphs of changes in the execution time of functional tasks depending on the life of the system are given at Fig. 5 for the values of parameters: H = 128, Q2 = 1, k = 2, n = 32, $\lambda = \gamma = 1$.

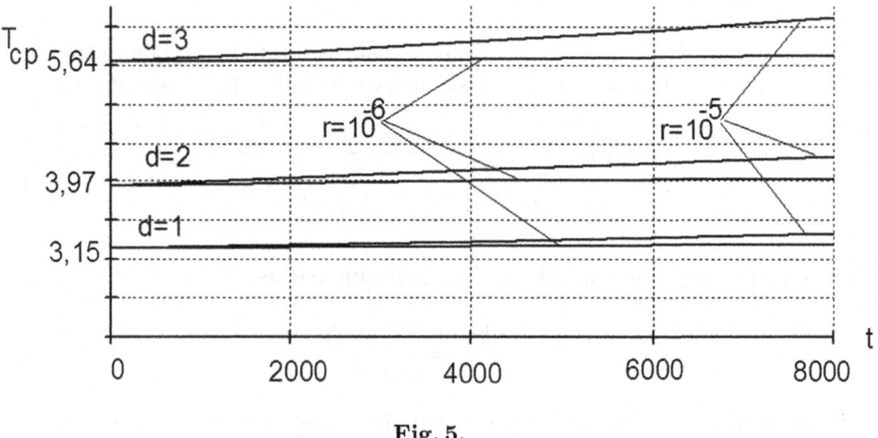

Fig. 5. .

We see that the execution time of functional tasks is the minimum, i.e. the best, when using the diagnostic test (d = 1). The ranking of methods for ensuring tolerance at d = 1, 2, 3 based on the execution time of functional tasks is not complete, of course.

2 Performance Model of the Tolerant DCN with Restorable Computing Modules

Here the main attention is paid to the definition of the functioning and development of models for the effectiveness of tolerant DCN (TDCN) with a variable number of TDCN and with a variable number of functional and service tasks. The variable number of TDCN is due to the fact that the CM can be in one of three possible states: not working, working-working, and working-controlled. It is also assumed that the CM can be restored. In turn, the CM can perform work and service tasks when it is in a working state. Performance of service tasks should be associated with ensuring the tolerance of computing systems. To ensure TDCN tolerance, a functional approach can be used, in which the state control of CM is implemented by software, using diagnostic testing or d-copy task execution.

When using a diagnostic test, performing a single task requires one (d = 1) module. When performing d-copies of tasks, d-copies of the same task are executed in parallel, when each copy requires its own CM. In each of these cases, to ensure control, the CM must be distracted from performing functional tasks. When using a diagnostic test, control measures require time from the controlled and controlling CM. If you use a d-copy, task execution, program activities related to determining the health of modules are performed on another CM with minimal time spent on the controlled CM. Further, it is accepted that monitoring activities spend time resources of only one module.

2.1 Statement of the Problem

Initially, the TDCN system contains n serviceable TDCN. The time intervals of CM operation and control are random variables with exponential distribution functions with intensities and, depending on the number of operating i and controlling j TDCN, respectively.

As a result of control measures with probability r, the CM is not working properly, and the non-working CM is excluded from the list of working, with an additional probability $(1 - r)$ the number of working TDCN remains the same. The time interval for restoring a CM is a random variable with an exponential distribution function, with an intensity that depends on the number i of non-functioning TDCN.

Due to the exponential distribution functions of the duration of processes, the model of TDCN is described by the following graph of States, see Fig. 6. The graph contains States, and there are no States (i, j) with an index, since d instances of a single task require d CM. Thus, the state $(d - 1, j)$, corresponds to a situation when TDCN does not perform functional tasks, but only performs controlling functions. In accordance with the state graph, the TDCN behavior can be studied for non-stationary (transient) and stationary cases.

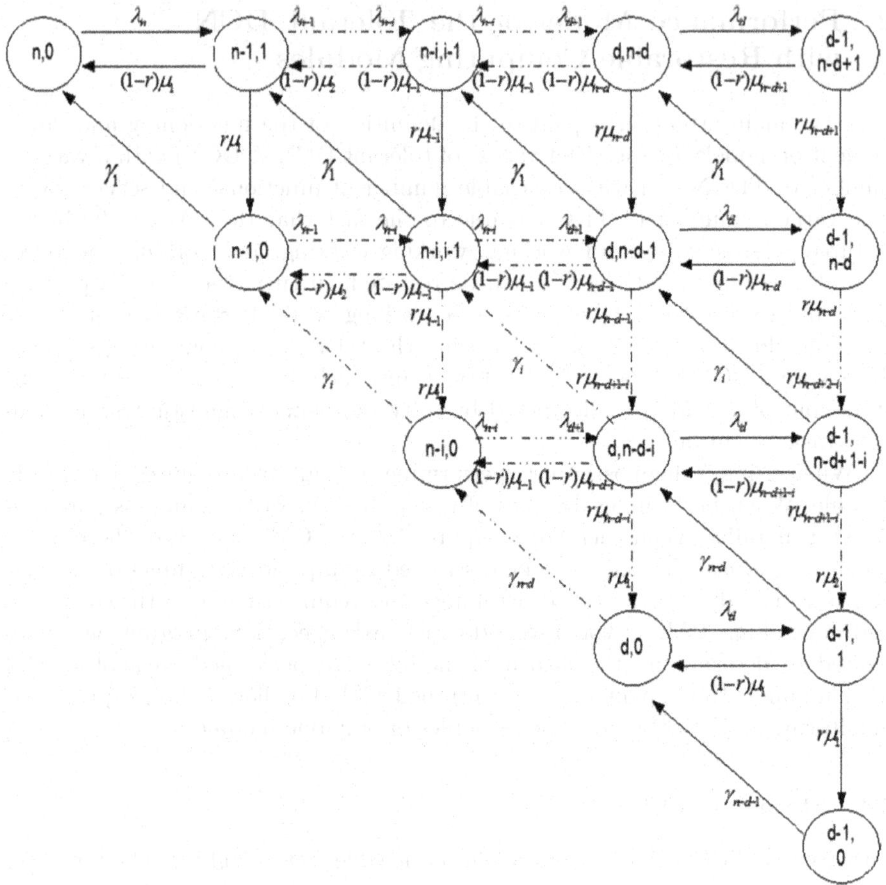

Fig. 6. .

2.2 System of Differential Equations for Studying the Transition Process of TDCN Behavior

Denote the probability that at time t with $i+j$ functioning TDCN, i TDCN are in the state of performing functional tasks and j TDCN are performing control functions.

The TDCN behavior is described by the system 13 of differential equations:

$$
\begin{aligned}
P'_{n,0}(t) &= -\gamma_n P_{n,0}(t) + (1-r)\mu_1 P_{n-1,1}(t) + \gamma_{n-1} P_{n-1,0}(t) \\
P'_{n-j,j}(t) &= -(\mu_j + \gamma_{n-j}) P_{n-j,j}(t) + \gamma_{n-j+1} P_{n-j+1,j-1}(t) \\
&\quad + (1-r)\mu_{j+1} P_{n-j-1,j+1}(t) + \gamma_{n-1} P_{n-j-1,j}(t), \\
&\quad j = \overline{1, n-d} \\
P'_{d-1,n-d+1}(t) &= -\mu_{n-d+1} P_{d-1,n-d+1}(t) + \lambda_d P_{d,n-d}(t); \\
P'_{n-i,0}(t) &= -(\lambda_{n-i} + \gamma_{n-i}) P_{n-i,0}(t) + r\mu_1 P_{n-i,1}(t) \\
&\quad + \gamma_{n-i-1} P_{n-i-1,0}(t) + (1-r)\mu_1 P_{n-i-1,1}(t), \\
&\quad i = \overline{1, n-d} \\
P'_{n-i-j,j}(t) &= -(\mu_j + \lambda_{n-i-j} + \gamma_{n-i}) P_{n-i-j,j}(t) \\
&\quad + \lambda_{n-i-j+1} P_{n-i-j+1,j-1}(t) \\
&\quad + r\mu_{j+1} P_{n-i-j,j+1}(t) + \gamma_{n-i-1} P_{n-i-j-1,j}(t) \\
&\quad + (1-r)\mu_{j+1} P_{n-i-j-1,j+1}(t), \\
&\quad i = \overline{1, n-d-j}, j = \overline{1, n-i-d}; \\
P'_{d-1,n-d-i+1}(t) &= -(\mu_{n-d-i+1} + \gamma_{n-i}) P_{d-1,n-d-i+1}(t) \\
&\quad + \lambda_d P_{d,n-i-d}(t) + r\mu_{n-i-d+2} P_{d-1,n-i-d+2}(t), \\
&\quad i = \overline{1, n-d}; \\
P'_{d-1,0}(t) &= r\mu_1 P_{d-1,1}(t);
\end{aligned}
\tag{13}
$$

Based on physical considerations, we believe that:

$$
P_{n,0}(t=0) = 1; i, j \in \{\overline{0, n}\}; P_{i,j}(t=0) = 0 \tag{14}
$$

The general solution of the system can be obtained by moving on to the Laplace transformations and finding the inverse Laplace transform.

The obtained solutions of the system make it possible to determine changes in the characteristics of the TDCN over time, depending on the method of ensuring tolerance, so the average number of CM directly involved in the performance of functional tasks is equal to:

$$
n_{cp} = \sum_{i=d}^{n} i \sum_{j=0}^{n-i} j p_{i,j}(t). \tag{15}
$$

2.3 System of Ordinary Equations for Studying the Behavior of TDCN for the Stationary Case

Denote the probability that if there are $i + j$ functioning CM, i CM are in the state of performing functional tasks and j CM perform controlling functions when time t $\rightarrow \infty$ is tending.

We will determine the exact and approximate solution.

Exact Solution

To find the $P_{i,j}$ probabilities, we obtain the following system of linear equations:

Based on the state $(d-1, n-d+1)$:

$$\begin{aligned}
P_{d,n-d} &= \frac{1}{\lambda_d}\mu n - d + 1 P_{d-1,n-d+1} \\
P_{d,n-d-i} &= \frac{1}{\lambda_d}(\mu n - d + 1 + \gamma_{n-i}) P_{d-1,n-d+1-i} - \frac{r\mu_{n-d+2-i}}{\lambda_d} P_{d-1,n-d+2-i} \\
P_{d-1,0} &= \frac{r\mu_1}{\gamma_{d-1}} P_{d-1,1} \\
P_{d+j,n-d-j} &= \frac{1}{\lambda_{d+j}}(\mu_{n-d+j} + \lambda_{d+j-1}) P_{d+j-1,n-d+1-j} \\
&\quad - \frac{(1-r)\mu_{n-d+2-j}}{\lambda_{d+j}} P_{d+j-2,n-d+2-j} \\
&\quad - \frac{\gamma_{n-1}}{\lambda_{d+j}} P_{d+j-2,n-d+1-j}, j = \overline{1, n-d} \\
P_{d+j,n-i-d-j} &= \frac{(\mu_{n-i-d-j} + \lambda_{d+j-1} + \gamma_{n-i})}{\lambda_{d+j}} P_{d+j-1,n-i-d-j+1} \\
&\quad - \frac{(1-r)\mu_{n-i-d-j+2}}{\lambda_{d+j}} P_{d+j-2,n-i-d-j+2} \\
&\quad - \frac{r\mu_{n-i-d-j+2}}{\lambda_{d+j}} P_{d+j-1,n-i-d-j+2} - \frac{\gamma_{n-i-1}}{\lambda_{d+j}} P_{d+j-2,n-i-d-j+1}, \\
&\quad j = \overline{1, n-d-i}, i = \overline{1, n-d-j}
\end{aligned} \tag{16}$$

Assuming the probabilities $P_{d-1,0}, \ldots, P_{d-1,nd+1}$ as $n-d$ base probabilities, the structure of this system of equations allows us to express all other probabilities in terms of these probabilities. However, note that the equation for the state $(d-1, 1)$ follows:

$$P_{d,0} = \frac{\mu_1 + \gamma_{n-d}}{\lambda_d} P_{d-1,1} - \frac{r\mu_2}{\lambda_d} P_{d-1,2} \tag{17}$$

and the equation for the state $(d-1, 0)$ follows:

$$P_{d,0} = \frac{r\mu_1}{\lambda_d + \gamma_{n-d}} P_{d,1} + \frac{(1-r)\mu_1}{\lambda_d + \gamma_{n-d}} P_{d-1,1} + \frac{\gamma_{n-d-1}}{\lambda_d + \gamma_{n-d}} P_{d-1,0} \tag{18}$$

in this case, the probability $P_{d,1}$, included in the last equation, is determined using the base probabilities $P_{d-1,1}$ and $P_{d-1,2}$. Therefore, using the two above equations for $P_{d,0}$, one of the base probabilities can be expressed in terms of other $n-d+1$ base probabilities, which reduces the number of unknown base probabilities to $n-d+1$. Similarly, for probabilities there are also additional equations ($j = \overline{0, n-d-1}$).

$$P_{d+j,0} = \frac{1}{\lambda_{d+j} + \gamma_{d+j}}(r\mu_1)P_{d,1} + \frac{(1-r)\mu_1}{\lambda_{d+j} + \gamma_{d+j}}P_{d+j,1} + \frac{\gamma_{d+j-1}}{\lambda_{d+j} + \gamma_{d+j}}P_{d+j-1,0} \tag{19}$$

which reduces the number of unknown base probabilities to 1 base probability, i.e. all the required probabilities of the system can be expressed in terms of a single base probability, for example, in $P_{d-1,1}$.

Note that the equations obtained by for the state $(n, 0)$:

$$P_{n,0} = \frac{(1-r)\mu_1}{\lambda_n}P_{n-1,1} - \frac{\gamma_{n-1}}{\lambda_n}P_{n-1,0} \tag{20}$$

and for the state $(n-1, 1)$:

$$P_{n,0} = \frac{\lambda_{n-1} + \mu_1}{\lambda_n}P_{n-1,1} + \frac{(1-r)\mu_1}{\lambda_n}P_{n-2,2} + \frac{\gamma_{n-1}}{\lambda_n}P_{n-2,1} \tag{21}$$

it is not possible to find the remaining base probability, since they are linearly independent equations. By using this method to express all the probabilities in terms of a single base probability and substituting them into the normalization equation, we find the base probability and thus all the probabilities.

Approximate Solution

You can see from the results of this and other examples that the probabilities of being in the States of the first two upper levels of the graph are the most significant. Let's create a system of equations for a graph whose States are bounded by the first two upper levels, see Fig. 7.

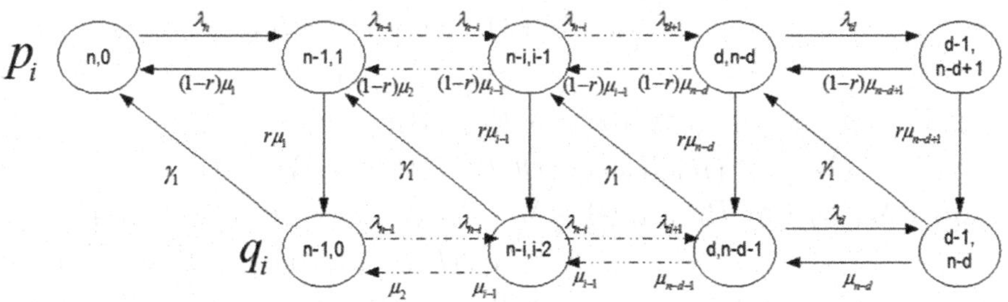

Fig. 7. .

The TDCN behavior is described by the following system of differential equations (22):

$$
\begin{aligned}
P'_{n,0}(t) &= -\lambda_n P_{n,0}(t) + (1-r)\mu_1 P_{n-1,1}(t) + \gamma_1 P_{n-1,0}(t);\\
P'_{n-j,j}(t) &= -(\mu_j + \lambda_{n-j})P_{n-j,j}(t) + \lambda_{n-j+1}P_{n-j+1,j-1}(t)\\
&\quad + (1-r)\mu_{j+1}P_{n-j-1,j+1}(t) + \gamma_1 P_{n-j-1,j}(t),\\
&\quad j = \overline{1, n-d};\\
P'_{d-1,n-d+1}(t) &= -\mu_{n-d+1}P_{d-1,n-d+1}(t) + \lambda_d P_{d,n-d}(t);\\
P'_{n-1,0}(t) &= -(\lambda_{n-1} + \gamma_1)P_{n-1,0}(t) + r\mu_1 P_{n-1,1}(t) + \mu_1 P_{n-2,1}(t)\\
P'_{n-j-1,j}(t) &= -(\mu_j + \lambda_{n-i-j} + \gamma_1)P_{n-j-1,j}(t) + \lambda_{n-j}P_{n-j,j-1}(t)\\
&\quad + r\mu_{j+1}P_{n-j-1,j+1}(t) + \mu_{j+1}P_{n-j-2,j+1}(t),\\
&\quad j = \overline{1, n-d-1};\\
P'_{d-1,n-d}(t) &= -(\mu_{n-d} + \gamma_1)P_{d-1,n-d}(t) + \lambda_d P_{d,n-d-1}(t)\\
&\quad + r\mu_{n-d+1}P_{d-1,n-d+1}(t).
\end{aligned}
\tag{22}
$$

Based on physical considerations, we believe that:

$$
P_{n,0}(t=0) = 1; i, j \in \{\overline{0, n}\}; P_{i,j}(t=0) = 0 \tag{23}
$$

The general solution of the system can be obtained by moving on to the Laplace transformations and finding the inverse Laplace transform.

To find the probabilities $P_{i,j}$ in the stationary case we obtain the following system of linear equations (24):

$$
\begin{aligned}
\lambda_n P'_{n,0}(t) &= (1-r)\mu_1 P_{n-1,1}(t) + \gamma_1 P_{n-1,0}(t);\\
(\mu_j + \lambda_{n-j})P_{n-j,j}(t) &= \lambda_{n-j+1}P_{n-j+1,j-1}(t)\\
&\quad + (1-r)\mu_{j+1}P_{n-j-1,j+1}(t)\\
&\quad + \gamma_1 P_{n-j-1,j}(t), j = \overline{1, n-d};\\
\mu_{n-d+1}P_{d-1,n-d+1}(t) &= \lambda_d P_{d,n-d}(t);\\
(\lambda_{n-1} + \gamma_1)P_{n-1,0}(t) &= r\mu_1 P_{n-1,1}(t) + \mu_1 P_{n-2,1}(t)\\
(\mu_j + \lambda_{n-i-j} + \gamma_1)P_{n-j-1,j}(t) &= \lambda_{n-j}P_{n-j,j-1}(t) + r\mu_{j+1}P_{n-j-1,j+1}(t)\\
&\quad + \mu_{j+1}P_{n-j-2,j+1}(t), j = \overline{1, n-d-1};\\
(\mu_{n-d} + \gamma_1)P_{d-1,n-d}(t) &= \lambda_d P_{d,n-d-1}(t) + r\mu_{n-d+1}P_{d-1,n-d+1}(t).
\end{aligned}
\tag{24}
$$

Assuming the probabilities $P_{d-1,n-d+1}, P_{d-1,n-d}$ as 2 basic probabilities, we will use them to express all the other probabilities. Based on the condition $(d-1, n-d)$, express the probability $P_{d,n-d-1}$, based on the state $(d-1, n-d+1)$ is the probability $P_{d,n-d}$, based on the state $(d, n-d-1)$ is the probability $P_{d+1,n-d-2}$, based on the state $(d, n-d)$ - the probability $P_{d+1,n-d-1}$, etc. based on the state $(n-2, 1)$ is the probability $P_{n-1,0}$, and based on the state $(n-2, 2)$ - probability $P_{n-1,1}$. Now, based on the state $(n-1, 0)$, we get another expression for determining the probability $P_{n-1,1}$ through the 2nd base probabilities

$P_{d-1,n-d+1}, P_{d-1,n-d}$, which reduces the number of base probabilities to one, for example, $P_{d-1,n-d+1}$. The state $(n, 0)$ gives the last equation for determining the probability $P_{n,0}$. Finally, the normalization condition provides finding the base probability $P_{d-1,n-d+1}$, and thus all probabilities. In accordance with the above algorithm for solving a system of linear equations, the paper developed its software implementation.

2.4 Evaluation of the Effectiveness of the Approximate Model

Based on the obtained solutions, the accuracy of the approximate model is estimated in relation to the exact model, in particular, for the average number of modules performing functional tasks. The results of calculations of the average number of modules that perform functional tasks for accurate and approximate models with a wide variation of parameter values (the intensities of working TSM and controlling TSM, the probability of r failure of the CM, the intensity of restoring the CM's with the number $n = 5$ TSM and $d = 3$ copies of a single task) show that the maximum inaccuracy of the approximate model when changing the parameter values in the intervals, $n = 5$, $d = 3$, $r = 0.1$ is about 4%. The speed of calculations for the approximate model increases by a factor of $O(n/2)$ in relation to the exact model.

References

1. Qu, P., Zhang, Y., Zheng, W.: High Performance simulation of spiking neural network on GPGPUs. IEEE Trans. Parallel Distrib. Syst. **31**(11), 2510–2523 (2020)
2. Pons, L., Sahuquillo, J., Selfa, V., Petit, S., Pons, J.: Phase-aware cache partitioning to target both turnaround time and system performance. IEEE Trans. Parallel Distrib. Syst. **31**(11), 2556–2568 (2020)
3. Szustak, L., Wyrzykowski, R., Olas, T., Mele, V.: Correlation of performance optimizations and energy consumption for stencil-based application on Intel Xeon scalable processors. IEEE Trans. Parallel Distrib. Syst. **31**(11), 2582–2593 (2020)
4. KhudaBukhsh, W.R., Kar, S., Alt, B., Rizk, A., Koeppl, H.: Generalized cost-based job scheduling in very large heterogeneous cluster systems. IEEE Trans. Parallel Distrib. Syst. **31**(11), 2594–2604 (2020)
5. Li, J., et al.: QWEB: high-performance event-driven web architecture with QAT acceleration. IEEE Trans. Parallel Distrib. Syst. **31**(11), 2633–2649 (2020)
6. Srinuan, P., Yuan, X., Tzeng, N.: Cooperative memory expansion via OS kernel support for networked computing systems. IEEE Trans. Parallel Distrib. Syst. **31**(11), 2650–2667 (2020)
7. Akhremtsev, Y., Sanders, P., Schulz, C.: High-quality shared-memory graph partitioning. IEEE Trans. Parallel Distrib. Syst. **31**(11), 2710–2722 (2020)
8. Losada, N., Bosilca, G., Bouteiller, A., González, P., Martín, M.J.: Local rollback for resilient MPI applications with application-level checkpointing and message logging. Future Gener. Comput. Syst. **91**, 450–464 (2019)
9. Losada, N., González, P., Martín, M.J., Bosilca, G., Bouteiller, A., Teranishi, K.: Fault tolerance of MPI applications in exascale systems: The ULFM solution. Future Gener. Comput. Syst. **106**, 467–481 (2020)

10. Tang, X., Zhai, J., Yu, B., Chen, W., Zheng, W., Li, K.: An efficient in-memory checkpoint method and its practice on fault-tolerant HPL. IEEE Trans. Parallel Distrib. Syst. **29**(4), 758–771 (2018)
11. Wang, Z., Gao, L., Gu, Y., Bao, Y., Yu, G.: A fault-tolerant framework for asynchronous iterative computations in cloud environments. IEEE Trans. Parallel Distrib. Syst. **29**(8), 1678–1692 (2018)
12. Cores, I., Rodríguez, G., Martín, M.J., González, P.: Achieving checkpointing global consistency through a hybrid compile time and runtime protocol. Procedia Comput. Sci. **18**, 169–178 (2013)
13. Luo, Y., Manivannan, D.: Hope: a hybrid optimistic checkpointing and selective pessimistic message logging protocol for large scale distributed systems. Future Gener. Comput. Syst. **28**(8), 1217–1235 (2012)
14. Castro-Le, M., Meyer, H., Rexachs, D., Luque, E.: Fault tolerance at system level based on radic architecture. J. Parallel Distrib. Comput. **86**, 98–111 (2015)
15. Panadero, J., Wong, A., Rexachs, D., Luque, E.: P3S: a methodology to analyze and predict application scalability. IEEE Trans. Parallel Distrib. Syst. **29**(3), 642–658 (2017)
16. Mohror, K., Moody, A., Bronevetsky, G., de Supinski, B.R.: Detailed modeling and evaluation of a scalable multilevel checkpointing system. IEEE Trans. Parallel Distrib. Syst. **25**(9), 2255–2263 (2014)
17. Bland, W., Bouteiller, A., Herault, T., Bosilca, G., Dongarra, J.: Post-failure recovery of MPI communication capability: design and rationale. Int. J. High Perform. Comput. Appl. **27**(3), 244–254 (2013)
18. Meyer, H., Muresano, R., Castro-León, M., Rexachs, D., Luque, E.: Hybrid message pessimistic logging. Improving current pessimistic message logging protocols. J. Parallel Distrib. Comput. **104**, 206–222 (2017)
19. Wong, A., Rexachs, D., Luque, E.: Parallel application signature for performance analysis and prediction. IEEE Trans. Parallel Distrib. Syst. **26**(7), 2009–2019 (2015)
20. Skrzypczak, J., Schintke, F., Schütt, T.: Fault-tolerant in-place consensus sequences. IEEE Trans. Parallel Distrib. Syst. **31**(10), 2392–2405 (2020)
21. Zhong, D., Bouteiller, A., Luo, X., Bosilca, G.: Runtime level failure detection and propagation in HPC systems. In: Proceedings of the 26th European MPI Users' Group Meeting, EuroMPI 2019, pp. 1–11 (2019)
22. Castro, M., Rexachs, D., Luque, E.: Radic-based message passing fault tolerance system. In: Proceedings of the the 6th International Conference on Advanced Engineering Computing and Applications in Sciences, pp. 59–64 (2012)
23. Cao, J., et al.: System-level scalable checkpoint-restart for petascale computing. In: Proceedings of the IEEE 22nd International Conference on Parallel and Distributed Systems, pp. 932–941 (2016)
24. Hassani, A., Skjellum, A., Brightwell, R.: Design and evaluation of FA-MPI, a transactional resilience scheme for non-blocking MPI. In: Proceedings of the 44th Annual IEEE/IFIP International Conference on Dependable Systems and Networks, pp. 750–755 (2014)

Asymptotic Analysis of RQ-System with Feedback and Batch Poisson Arrival Under the Condition of Increasing Average Waiting Time in Orbit

A. A. Nazarov[1], S. V. Rozhkova[1,2], and E. Yu. Titarenko[1,2(✉)]

[1] National Research Tomsk State University, Lenin Avenue 36, Tomsk 634050, Russia
nazarov.tsu@gmail.com
[2] National Research Tomsk Polytechnic University, Lenin Avenue 30, Tomsk 634050, Russia
{rozhkova,teu}@tpu.ru
https://tsu.ru/, https://tpu.ru/

Abstract. The paper studies the retrial queueing system $M^{[n]}/M/1$ with feedback and batch Poisson arrival. Customers for the system come in groups. Not more than one customer is served at once, others wait in the orbit. Having been served, the customer leaves the system or goes to re-service or into the orbit. An asymptotic analysis method is used to find the stationary distribution of the number of customers in the orbit. A long delay between customers from the orbit is proposed as an asymptotic condition. It is proved that the asymptotic probability distribution of the number of customers in the orbit is Gaussian. As a result the parameters of this distribution are obtained. The calculations to determine the range of the method applicability are carried out. The accuracy of the approximation is compared to numerical results obtained by matrix method.

Keywords: Queuing system · RQ system · Batch arrival · Feedback · Asymptotic analysis

1 Introduction

Queueing systems are widely used in various fields: call centers, intelligent transport systems, telecommunication networks, etc. [1]. Classical models of queuing theory do not appropriately describe the real technical systems, where the effect of repeated calls is observed. Therefore, to analyze these systems characteristics we use the retrial queueing system (RQ systems). Models with repeat calls are widely used for designing and optimizing information and communication systems, digital communication networks controlled by protocols of random multiple access. They are also used in many other areas. RQ systems are characterized by the fact that customers arriving in the system, go into the waiting area in

© Springer Nature Switzerland AG 2020
V. M. Vishnevskiy et al. (Eds.): DCCN 2020, CCIS 1337, pp. 327–339, 2020.
https://doi.org/10.1007/978-3-030-66242-4_26

case the server is busy, and after some random time they retries the service. Between retries customers are in the source of repeat calls (or orbit). Retrial queuing systems are considered in many works [2–4]. A review of works on this topic is given in [5,6].

However, in practice due to poor service or some external factors a customer might require re-service. Similar situations occur in multi-agent systems, where a customer having received satisfactory service requires reservice from the same agent. The behavior of such systems is accurately described by queuing systems with feedback. The application of queuing theory to optimize the performance of multi-agent systems is described in detail in [7–10].

Models of queuing systems with feedback are poorly studied. There are some works devoted to re-service and feedback [11–13]. Methods for analysis of multi-channel queueing system with instantaneous and delayed feedbacks are considered in [11]. Queueing system MMPP$|M|\infty$ with repeated service is investigated in [12].

In this paper, we study a single-channel RQ-system with exponentially distributed service time, a batch Poisson arrival, instantaneous and delayed feedbacks. We use an asymptotic analysis method under the asymptotic condition of a growing average waiting time in the orbit.

2 The Model Description and the Problem Statement

We consider the $\mathrm{M}^{[n]}/\mathrm{M}/1$ queuing system (see Fig. 1) with repeated calls and batch Poisson arrival process with parameter λ and given probabilities q_ν of occurrence of customers in the group. We assume that $\nu > 0, q_0 = 0, \sum\limits_{\nu=1}^{\infty} q_\nu = 1$. An incoming customer, which sees a server idle, occupies it, other customers from the group go to the source of the repeated calls (into orbit). Also, if the server is busy, arriving customers go into orbit. Service time is exponentially distributed with parameter μ. Having been served, the customer leaves the system with probability r_0, or goes to re-service with probability r_1 (instantaneous feedback), or into orbit with probability r_2 (delayed feedback), so $r_0 + r_1 + r_2 = 1$. In the orbit each customer independently of others waits for the time exponentially distributed with parameter σ. Then the customer occupies the device if it is idle or remains in the orbit.

We define the Markov process $\{i(t), n(t)\}$ of changing the states of the RQ-system, where $i(t)$ is the number of customers in the orbit at time t, $i(t) = 0$, 1, 2, ..., the process $n(t)$ determines the state of the server at time t, and takes one of two values: $n(t) = 0$, if the server is idle, $n(t) = 1$, if the server is busy.

Our aim is to find the stationary probability distribution of the number of customers in the orbit taking into account the state of the server $P_n(i) = P\{n(t) = n; i(t) = i\}, n = 0, 1, i = \overline{0, \infty}$.

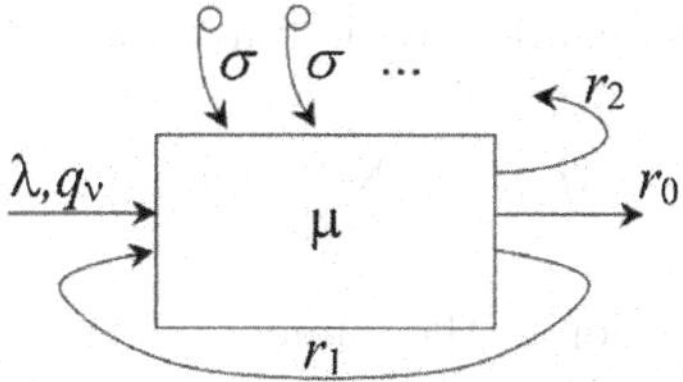

Fig. 1. System diagram.

3 Kolmogorov Equations

To obtain the probability distribution for the number of customers in the orbit, we apply the Δt-method. We fix the time t and a small time increment Δt. Then we express probability state distribution of the Markov process at time point $t + \Delta t$ by its probability state distribution at an arbitrary time point t

$$P_0(i, t+\Delta t) = P_0(i,t)(1-\lambda\Delta t)(1-i\sigma\Delta t) + \mu r_0 \Delta t P_1(i,t) + \mu r_2 \Delta t P_1(i-1,t) + o(\Delta t);$$
$$P_1(i, t+\Delta t) = P_1(i,t)(1-\lambda\Delta t)(1-\mu r_0\Delta t)(1-\mu r_2\Delta t)$$
$$+ (i+1)\sigma\Delta t P_0(i+1) + \sum_{\nu=1}^{i+1} \lambda q_\nu \Delta t P_0(i-\nu+1) + \sum_{\nu=1}^{i} \lambda q_\nu \Delta t P_1(i-\nu) + o(\Delta t).$$

The result is a system of differential equations for the probabilities $P_n(i,t)$

$$\frac{\partial P_0(i,t)}{\partial t} = -(\lambda + i\sigma)P_0(i,t) + \mu r_0 P_1(i,t) + \mu r_2 P_1(i-1,t);$$
$$\frac{\partial P_1(i,t)}{\partial t} = (i+1)\sigma P_0(i+1,t) - (\mu r_0 + \mu r_2 + \lambda)P_1(i,t)$$
$$+ \sum_{\nu=1}^{i+1} \lambda q_\nu P_0(i-\nu+1,t) + \sum_{\nu=1}^{i} \lambda q_\nu P_1(i-\nu,t).$$

From this system we derive the system of Kolmogorov equations for the stationary probabilities $P_n(i)$

$$-(\lambda + i\sigma)P_0(i) + \mu r_0 P_1(i) + \mu r_2 P_1(i-1) = 0; \tag{1}$$

$$(i+1)\sigma P_0(i+1) - (\mu r_0 + \mu r_2 + \lambda)P_1(i) + \sum_{\nu=1}^{i+1} \lambda q_\nu P_0(i-\nu+1) + \sum_{\nu=1}^{i} \lambda q_\nu P_1(i-\nu) = 0$$

and write the normalization condition

$$\sum_{i=0}^{\infty} [P_0(i) + P_1(i)] = 1. \tag{2}$$

Let us note that the stability condition has the form $\rho < 1$, where $\rho = \lambda\bar{\nu}/\mu r_0$, $\bar{\nu} = \sum_{\nu=1}^{\infty} \nu q_\nu$ is the average number of customers in the group.

We consider the partial characteristic functions of the number of customers in the orbit

$$H_n(u) = \sum_{i=0}^{\infty} e^{jui} P_n(i)$$

and the characteristic function of the number of customers in the group

$$h(u) = \sum_{\nu=1}^{\infty} e^{ju\nu} q_\nu,$$

where $j = \sqrt{-1}$. We take into account

$$\frac{\partial H_n(u)}{\partial u} = \sum_{i=0}^{\infty} ije^{jui} P_n(i),$$

$$\sum_{i=0}^{\infty}\sum_{\nu=1}^{i} q_\nu e^{jui} P_1(i-\nu) = \sum_{\nu=1}^{\infty} q_\nu e^{ju\nu} \sum_{i=0}^{\infty} e^{jui} P_1(i) = h(u) H_1(u),$$

$$\sum_{i=0}^{\infty}\sum_{\nu=1}^{i+1} q_\nu e^{jui} P_0(i-\nu+1) = e^{-ju} \sum_{\nu=1}^{\infty} q_\nu e^{ju\nu} \sum_{i=0}^{\infty} e^{jui} P_0(i) = e^{-ju} h(u) H_0(u),$$

and rewrite system (1) as

$$\sigma j \frac{\partial H_0(u)}{\partial u} - \lambda H_0(u) + \left(\mu r_0 + \mu r_2 e^{ju}\right) H_1(u) = 0; \tag{3}$$

$$-\sigma j e^{-ju} \frac{\partial H_0(u)}{\partial u} + \lambda e^{-ju} h(u) H_0(u) + \left(\lambda h(u) - \mu r_0 - \mu r_2 - \lambda\right) H_1(u) = 0.$$

The characteristic function $H(u)$ of the number of customers in the orbit for system (3) is expressed in terms of the partial characteristic functions $H_n(u)$ by the following $H(u) = H_0(u) + H_1(u)$.

Applying the inverse Fourier transform to the characteristic function, we can write the probability distribution as

$$P(i) = \frac{1}{2\pi} \int_{-\pi}^{\pi} e^{-jui} H(u) du, i = \overline{0, \infty}$$

and mathematical expectation as $E\{i(t)\} = -jH'(0)$. However, it is hardly possible to obtain an analytical expression for these integrals, therefore, it is reasonable to use numerical integration methods. But numerical calculations are computationally expensive, so we consider the asymptotic analysis method which allows us to receive an analytical approximation for the distribution $P(i)$. Using numerical experiments, we analyze the accuracy of the method.

4 Asymptotics of the First Order

We solve the equations for the characteristic functions (2) under the asymptotic condition of a growing average waiting time in the orbit, i.e. $\sigma \to 0$. We formulate the result in the following theorem.

Theorem 1. *Let $i(t)$ be the number of customers in the orbit in the $M^{[n]}/M/1$ RQ-system with a batch Poisson arrival and feedback. Then there is the following equality for a sequence of characteristic functions*

$$\lim_{\sigma\to 0} E\left\{e^{jwi(t)\sigma}\right\} = e^{jw\kappa_1}$$

where

$$\kappa_1 = \lambda\frac{\mu\bar{\nu}(r_0+r_2)}{\mu r_0 - \lambda\bar{\nu}} - \lambda, \bar{\nu} = \sum_{\nu=1}^{\infty}\nu q_\nu.$$

Proof. In the system of Eqs. (3) we use the substitutions

$$\sigma = \varepsilon, u = \varepsilon w, H_n(u) = F_n(w,\varepsilon).$$

Since

$$\frac{\partial H_0(u)}{\partial u} = \frac{1}{\varepsilon}\frac{\partial F_0(w,\varepsilon)}{\partial w},$$

the system (3) can be written as

$$j\frac{\partial F_0(w,\varepsilon)}{\partial w} - \lambda F_0(w,\varepsilon) + \left(\mu r_0 + \mu r_2 e^{jw\varepsilon}\right)F_1(w,\varepsilon) = 0; \tag{4}$$

$$-je^{-jw\varepsilon}\frac{\partial F_0(w,\varepsilon)}{\partial w} + \lambda e^{-jw\varepsilon}h(w,\varepsilon)F_0(w,\varepsilon)$$
$$+ \left(\lambda h(w,\varepsilon) - \mu r_0 - \mu r_2 - \lambda\right)F_1(w,\varepsilon) = 0.$$

Let $\varepsilon \to 0$, $F_n(w) = \lim_{\varepsilon\to 0} F_n(w,\varepsilon)$. Since $\lim_{\varepsilon\to 0} h(w,\varepsilon) = 1$, the system (3) is transformed into an equation

$$j\frac{\partial F_0(w)}{\partial w} - \lambda F_0(w) + (\mu r_0 + \mu r_2)F_1(w) = 0.$$

We find a solution to the equation of the form

$$F_n(w) = R_n e^{jw\kappa_1}, \tag{5}$$

then

$$-(\kappa_1 + \lambda)R_0 + (\mu r_0 + \mu r_2)R_1 = 0. \tag{6}$$

Then, summing the equations of system (4), we obtain

$$j\left(1 - e^{-jw\varepsilon}\right)\tfrac{\partial F_0(w,\varepsilon)}{\partial w} - \lambda\left(1 - e^{-jw\varepsilon}h(w,\varepsilon)\right)F_0(w,\varepsilon)$$
$$+ \left(\mu r_2(e^{jw\varepsilon} - 1) + \lambda(h(w,\varepsilon) - 1)\right)F_1(w,\varepsilon) = 0,$$

divide it by ε

$$j\tfrac{1-e^{-jw\varepsilon}}{\varepsilon}\cdot\tfrac{\partial F_0(w,\varepsilon)}{\partial w}-\lambda\tfrac{1-e^{-jw\varepsilon}h(w,\varepsilon)}{\varepsilon}\cdot F_0(w,\varepsilon)$$
$$+\left(\mu r_2\cdot\tfrac{e^{jw\varepsilon}-1}{\varepsilon}+\lambda\cdot\tfrac{h(w,\varepsilon)-1}{\varepsilon}\right)F_1(w,\varepsilon)=0$$

and assume $\varepsilon\to 0$

$$j\frac{\partial F_0(w)}{\partial w}+\lambda\left(\bar{\nu}-1\right)\cdot F_0(w)+\left(\mu r_2+\lambda\bar{\nu}\right)F_1(w)=0,$$

where $\bar{\nu}=\sum\limits_{\nu=1}^{\infty}\nu q_\nu$ is the average number of customers in the group. We carry out the substitution (5) and obtain the equation

$$-\left(\kappa_1+\lambda-\lambda\bar{\nu}\right)R_0+\left(\mu r_2+\lambda\bar{\nu}\right)R_1=0 \tag{7}$$

Solving the system of Eqs. (6) and (7) with the additional condition $R_0+R_1=1$, we obtain

$$R_0=\frac{\mu(r_0+r_2)}{\kappa_1+\lambda+\mu(r_0+r_2)},\quad R_1=\frac{\kappa_1+\lambda}{\kappa_1+\lambda+\mu(r_0+r_2)} \tag{8}$$

$$\kappa_1=\lambda\frac{\mu\bar{\nu}(r_0+r_2)}{\mu r_0-\lambda\bar{\nu}}-\lambda.$$

Thus, the asymptotic approximation of the characteristic function is $F(w)=exp\{jw\kappa_1\}$.

Asymptotics of the first order determines the average value of the number of customers in the orbit. For a more detailed study of the process $i(t)$, we should consider the second-order asymptotics.

5 Asymptotics of the Second Order

The main result of the analysis of the second-order asymptotics is presented in the following theorem.

Theorem 2. *Let $i(t)$ be the number of customers in the orbit in the $M^{[n]}/M/1$ RQ-system with a batch Poisson arrival and feedback. Then there is an equality as follows:*

$$\lim_{\sigma\to 0}E\left\{\exp\left\{jw\sqrt{\sigma}\left(i(t)-\tfrac{\kappa_1}{\sigma}\right)\right\}\right\}=\exp\left\{\frac{(jw)^2}{2}\kappa_2\right\}, \tag{9}$$

where

$$\kappa_2=\frac{\lambda\mu r_0}{2(\mu r_0-\lambda\bar{\nu})^2}\left(2\lambda\bar{\nu}^2+\mu\nu_2(r_0+r_2)+\mu\bar{\nu}(r_2-r_1)\right),\nu_2=\sum_{\nu=1}^{\infty}\nu^2q_\nu.$$

Proof. In the system of Eqs. (3), we use the substitutions

$$H_n(u) = H_n^{(2)}(u) \cdot e^{ju\kappa_1/\sigma}.$$

Here $H_n^{(2)}(u)$ is the partial characteristic function of the centered random variable $i(t) - \kappa_1/\sigma$. The system of equations for $H_n^{(2)}(u)$ is

$$\begin{gathered}\sigma j \frac{\partial H_0^{(2)}(u)}{\partial u} - (\kappa_1 + \lambda) H_0^{(2)}(u) + \left(\mu r_0 + \mu r_2 e^{ju}\right) H_1^{(2)}(u) = 0;\\ -\sigma j e^{-ju} \frac{\partial H_0^{(2)}(u)}{\partial u} + (\kappa_1 + \lambda h(u))\, e^{-ju} H_0^{(2)}(u)\\ + (\lambda h(u) - \mu r_0 - \mu r_2 - \lambda)\, H_1^{(2)}(u) = 0.\end{gathered}$$

Let $\sigma = \varepsilon^2$ and use the substitutions $u = \varepsilon w$, $H_n^{(2)}(u) = F_n^{(2)}(w, \varepsilon)$, then we obtain the system

$$\varepsilon j \frac{\partial F_0^{(2)}(w,\varepsilon)}{\partial w} - (\kappa_1 + \lambda) F_0^{(2)}(w,\varepsilon) + \left(\mu r_0 + \mu r_2 e^{jw\varepsilon}\right) F_1^{(2)}(w,\varepsilon) = 0; \quad (10)$$

$$\begin{gathered}-\varepsilon j e^{-jw\varepsilon} \frac{\partial F_0^{(2)}(w,\varepsilon)}{\partial w} + (\kappa_1 + \lambda h(w,\varepsilon))\, e^{-jw\varepsilon} F_0^{(2)}(w,\varepsilon)\\ + (\lambda h(w,\varepsilon) - \mu r_0 - \mu r_2 - \lambda)\, F_1^{(2)}(w,\varepsilon) = 0.\end{gathered}$$

The solution for the functions $F_n^{(2)}(w, \varepsilon)$ has the following form

$$F_n^{(2)}(w,\varepsilon) = \Phi(w) \cdot (R_n + j\varepsilon w f_n) + O(\varepsilon^2) \quad (11)$$

We substitute (11) into (10), use the approximation for $exp\{\pm jw\varepsilon\}$ and $h(\varepsilon w) = 1 + j\varepsilon w \bar{\nu} + O(\varepsilon^2)$, take into account (6) and (7), and convert the system of Eqs. (10) into an equation

$$\frac{\partial(w)}{\partial w} \frac{1}{w\Phi(w)} = \frac{1}{R_0} \left[(\kappa_1 + \lambda) f_0 - (\mu r_0 + \mu r_2) f_1 - \mu r_2 R_1\right]. \quad (12)$$

Let us denote

$$\kappa_2 = -\frac{1}{R_0} \left[(\kappa_1 + \lambda) f_0 - (\mu r_0 + \mu r_2) f_1 - \mu r_2 R_1\right] \quad (13)$$

and Eq. (12) has the form

$$\frac{1}{\Phi(w)} \frac{\partial(w)}{\partial w} = -w\kappa_2, \quad (14)$$

therefore, the function $\Phi(w)$ can be represented in the form

$$\Phi(w) = exp\left\{-\frac{1}{2} w^2 \kappa_2\right\}$$

that correlates with (9). To find unknown functions f_0, f_1 and an explicit form for κ_2, we rewrite (13) as

$$-(\kappa_1+\lambda)f_0+\mu\,(r_0+r_2)\,f_1=\kappa_2 R_0-\mu r_2 R_1.$$

Functions f_0, f_1 can be written as the sum of a general solution of the homogeneous equation and two particular solutions:

$$f_n=C\cdot R_n+g_n+\kappa_2\varphi_n,\ n=0,1. \tag{15}$$

Here $C\cdot R_n$ are the general solution of the homogeneous equation due to (6), while g_n is the solution of the equation

$$-(\kappa_1+\lambda)g_0+\mu\,(r_0+r_2)\,g_1=-\mu r_2 R_1, \tag{16}$$

and φ_n satisfies the equation

$$-(\kappa_1+\lambda)\kappa_2\varphi_0+\mu\,(r_0+r_2)\,\kappa_2\varphi_1=\kappa_2 R_0. \tag{17}$$

Differentiating (6) with respect to κ and comparing with (17), we note that

$$\varphi_0=\frac{\partial R_0}{\partial\kappa},\quad \varphi_1=\frac{\partial R_1}{\partial\kappa},\quad \varphi_0+\ \varphi_1=0.$$

Then, taking into account (8), we obtain

$$\varphi_0=-\frac{\mu(r_0+r_2)}{\left(\kappa_1+\lambda+\mu(r_0+r_2)\right)^2},\quad \varphi_1=-\varphi_0.$$

Similarly, we assume $g_0+g_1=0$ and receive from Eq. (16)

$$g_0=\frac{\mu r_2 R_1}{\kappa_1+\lambda+\mu(r_0+r_2)},\quad g_1=-g_0.$$

In order to find the explicit form κ, we sum the equations of the system (10)

$$\begin{aligned}-\varepsilon j\left(e^{-jw\varepsilon}-1\right)\tfrac{\partial F_0^{(2)}(w,\varepsilon)}{\partial w}\left[\kappa_1(e^{-jw\varepsilon}-1)+\lambda(h(w,\varepsilon)e^{-jw\varepsilon}-1)\right]F_0^{(2)}(w,\varepsilon)\\+\left[\mu r_2(e^{jw\varepsilon}-1)+\lambda(h(w,\varepsilon)-1)\right]F_1^{(2)}(w,\varepsilon)=0.\end{aligned}$$

We write the solution for $F_n^{(2)}(w,\varepsilon)$ as (11), $exp\{\pm jw\varepsilon\}=1\pm jw\varepsilon+(jw\varepsilon)^2/2+O(\varepsilon^3)$, $h(\varepsilon w)=1+j\varepsilon w\bar{\nu}+(j\varepsilon w)^2\nu_2/2+O(\varepsilon^3)$, where $\nu_2=\sum\limits_{\nu=1}^{\infty}\nu^2 q_\nu$. After some transformation and using expressions (7) and (14), we assume $\varepsilon\to 0$ and obtain

$$\begin{aligned}R_0\kappa_2 &=\frac{1}{2}(\kappa_1+\lambda+\lambda\nu_2-2\lambda\bar{\nu})R_0+\frac{1}{2}\left(\mu r_2+\lambda\nu_2\right)R_1\\ &-(\kappa_1+\lambda-\lambda\bar{\nu})f_0+(\mu r_2+\lambda\bar{\nu})\,f_1\end{aligned} \tag{18}$$

Substituting (15) into (18) and taking into account (7), (16), (17), expressions for $R_0,\ R_1,\ \varphi_1,\ g_1$, we finally obtain

$$\kappa_2=\frac{\lambda\mu r_0}{2(\mu r_0-\lambda\bar{\nu})^2}\left(2\lambda\bar{\nu}^2+\mu\nu_2(r_0+r_2)+\mu\bar{\nu}(r_2-r_1)\right).$$

Theorem 2 shows that the asymptotic probability distribution of the number of customers in the orbit in the $M^{[n]}/M/1$ RQ-system with a batch Poisson arrival and feedback is Gaussian with the parameters κ_1/σ and κ_2/σ, which allows us to make the following approximation for the distribution $P(i)$ as

$$P_{apr}(i) = \frac{G(i+0.5) - G(i-0.5)}{1 - G(-0.5)},$$

where $G(x)$ is the normal distribution function with parameters κ_1/σ and κ_2/σ.

6 Numerical Matrix Method

To evaluate the accuracy of the asymptotic analysis method we propose a numerical matrix method and compare the resulting Gaussian approximation to the distribution obtained by numerical method.

Let us consider the system (1) of infinite dimension and the Eq. (2). Then we truncate the system dimension and rewrite the system (1) and (2) for $i = 0, 1, 2, \ldots, N$ in matrix form

$$\mathbf{P}\ \mathbf{S} = \mathbf{B},$$

where the row vector $\mathbf{P}$ of dimension $2(N+1)$ is stationary probability distribution of the number of customers in the orbit

$$\mathbf{P} = (P_0(0), P_0(1), P_0(2) \ldots P_0(N), P_1(0), P_1(1), P_1(2) \ldots P_1(N)).$$

Matrix $\mathbf{S}$ of dimension $2(N+1) \times [2(N+1)+1]$ is represented in block form as

$$\mathbf{S} = \begin{pmatrix} \mathbf{S}_{11}\ \mathbf{S}_{12}\ \mathbf{S}_{13} \\ \mathbf{S}_{21}\ \mathbf{S}_{22}\ \mathbf{S}_{23} \end{pmatrix},$$

where $\mathbf{S}_{11} = \left\|s_{ij}^{11}\right\|$, $\mathbf{S}_{12} = \left\|s_{ij}^{12}\right\|$, $\mathbf{S}_{21} = \left\|s_{ij}^{21}\right\|$, $\mathbf{S}_{22} = \left\|s_{ij}^{22}\right\|$ $(i, j = 0, 1, \ldots, N)$ are matrices of dimension $(N+1) \times (N+1)$ whose nonzero elements are defined as

$$\begin{gathered} s_{ii}^{11} = -\lambda - (i-1)\sigma, \\ s_{ii}^{21} = \mu r_0,\ s_{i-1,i}^{21} = \mu r_2, \\ s_{i+1,i}^{12} = i\sigma,\ s_{ii}^{12} = \lambda q_1,\ s_{i-1,i}^{12} = \lambda q_2, \ldots,\ s_{0,i}^{12} = \lambda q_{i+1}, \\ s_{ii}^{22} = \mu r_1 - \mu - \lambda,\ s_{i-1,i}^{22} = \lambda q_1,\ s_{i-2,i}^{22} = \lambda q_2, \ldots,\ s_{0,i}^{22} = \lambda q_i. \end{gathered}$$

Blocks $\mathbf{S}_{13}, \mathbf{S}_{23}$ are unit vector columns of dimension $(N+1)$. The row vector $\mathbf{B} = \|b_n\|$ of dimension $2(N+1)+1$ is the row of free coefficients with elements $b_n = 0$ $(n = 0, 1, 2, \ldots, 2N+1), b_{2N+2} = 1$.

The vector P is easily calculated by the obvious formula

$$\mathbf{P} = \mathbf{B}\mathbf{S}^T\left(\mathbf{S}\mathbf{S}^T\right)^{-1}.$$

We use a numerical algorithm and choose N to be so large that probabilities $P_0(N)$ and $P_1(N)$ are equal to the machine zero. However, despite the high accuracy, the use of the numerical method is limited by the computer capacity.

7 Numerical Results

We consider the system with parameters $\lambda = 1, \mu = 7, r_0 = 0.5, r_1 = 0.3, r_2 = 0.2, q_1 = 0.5, q_2 = 0.3, q_3 = 0.1, q_4 = 0.1, q_5 = 0, q_6 = 0, \ldots$. Table 1 shows the results of calculating the mathematical expectation of the number of customers in the orbit for various values of σ. For each value of σ, the exact value of $E\{i(t)\}$ obtained with characteristic function, the asymptotic value of the mathematical expectation κ_1/σ, and the relative error δ are presented. The relative error is calculated by the formula

$$\delta = \frac{\left|E\{i(t)\} - \frac{\kappa}{\sigma}\right|}{E\{i(t)\}}.$$

Table 1. Approximation of the average number of customers.

σ	1	0.5	0.1	0.01	0.005
$E\{i(t)\}$	5.439	9.627	43.133	420.074	838.897
κ_1/σ	4.188	8.376	41.882	418.824	837.647
δ	0.2	0.1	0.03	0.003	0.001

The results show a sufficient accuracy of approximation of the average number of customers when $\sigma \leq 0.1$.

Figure 2 and Fig. 3 show the probability distributions obtained by the method of asymptotic analysis when $\sigma = 0.5$ and $\sigma = 0.01$, as well as a comparison with the distribution obtained by the numerical matrix method. It is noted that the approximation is quite good when parameter σ is lower.

To determine the accuracy of the approximation $P_{apr}(i)$, we use the Kolmogorov distance

$$\Delta = \max_{0 \leq k < \infty} \left| \sum_{i=0}^{k} (P_{apr}(i) - P(i)) \right|$$

that defines the difference between the asymptotic probability distribution $P_{apr}(i)$ and the probability distribution $P(i)$ obtained by the matrix method.

Table 2 shows the Kolmogorov distances for different values of the parameter σ and systems with different system load $\rho = \lambda\bar{\nu}/\mu r_0$.

Thus, if a system load is low, we can use asymptotic method to find the stationary probability distribution of customers in the orbit under asymptotic condition of a long waiting time in orbit, when $\sigma < 0.5$. If system load is high, the asymptotic method gives sufficient approximation accuracy, when $\sigma < 0.1$.

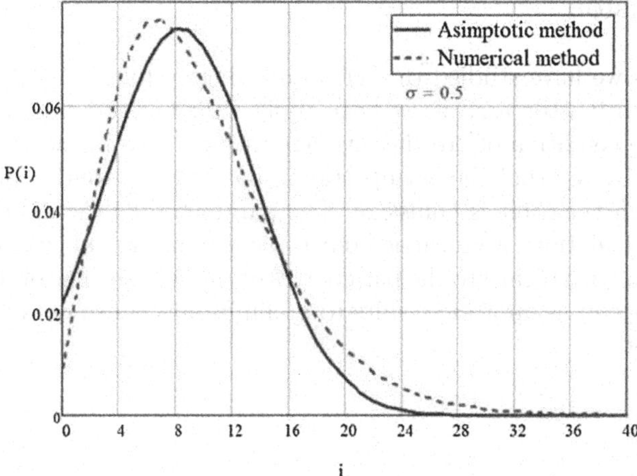

Fig. 2. Comparison of the probability distributions obtained by the asymptotic and numerical methods, when $\sigma = 0.5$.

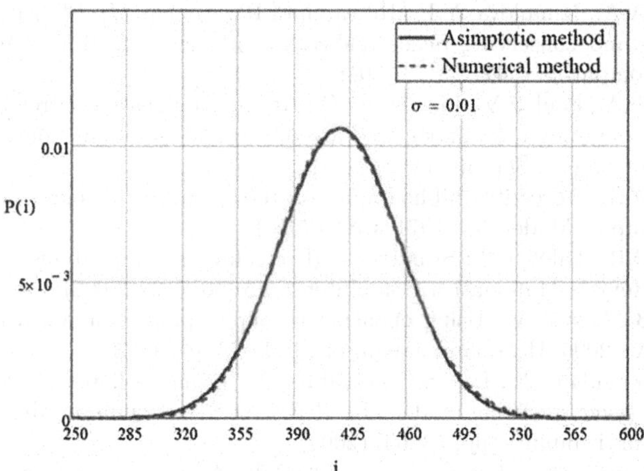

Fig. 3. Comparison of the probability distributions obtained by the asymptotic and numerical methods, when $\sigma = 0.01$.

Table 2. Kolmogorov distances.

	$\sigma = 1$	$\sigma = 0.5$	$\sigma = 0.1$	$\sigma = 0.05$	$\sigma = 0.01$
$\rho = 0.5$	0.105	0.047	0.027	0.019	0.008
$\rho = 0.6$	0.106	0.046	0.032	0.022	0.007
$\rho = 0.7$	0.107	0.045	0.039	0.027	0.008
$\rho = 0.8$	0.108	0.062	0.047	0.033	0.007
$\rho = 0.9$	0.109	0.084	0.051	0.031	0.007

8 Conclusion

In this paper, we have studied the retrial queuing system $M^{[n]}/M/1$ with feedback and batch Poisson arrival, we have applied the method of asymptotic analysis under the condition of growing average waiting time in the orbit.

It was discovered that the asymptotic probability distribution of the number of customers in the orbit is Gaussian. The parameters of the distribution were found. The calculations were carried out to determine the range of the obtained approximation with regard to the parameters of the system. The outcome showed the convergence of asymptotic results to prelimit ones which obtained using the matrix method.

References

1. Dudin, A.N., Klimenok, V.I., Vishnevsky, V.M.: The Theory of Queuing Systems with Correlated Flows. Springer, Cham (2020). https://doi.org/10.1007/978-3-030-32072-0
2. Falin, G.I., Templeton, J.G.C.: Retrial Queues. CRC Press, Boca Raton (1997)
3. Nazarov, A.A., Izmailova, Y.E.: Research of RQ-system $M|E2|1$ with request displacement and conserving phase realization of servicing. Tomsk State Univ. J. Control Comput. Sci. **42**, 72–78 (2018)
4. Nazarov, A.A., Paul, S.V., Lizyura, O.D.: Asymptotic analysis of retrial queue with n types of outgoing calls under low rate of retrials condition. Tomsk State Univ. J. Control Comput. Sci. **48**, 13–20 (2019)
5. Artalejo, J.R.: Accessible bibliography on retrial queues. Progress in 2000–2009. Math. Comput. Model. **51**, 1071–1081 (2010)
6. Artalejo, J.R., Falin, G.I.: Standard and retrial queueing systems: a comparative analysis. Revista Matematica Complutense **15**, 101–129 (2002)
7. Horling, B., Lesser, V.: Using queueing theory to predict organizational metrics. In: AAMAS 2006, Hakodate, Japan, pp. 1098–1100 (2006)
8. Gnanasambandam, N., Lee, S.C., Gautam, N., et al.: Reliable MAS performance prediction using queueing models. In: IEEE 1st Symposium on Multi-agent Security and Survivalibility, pp. 55–64. (2004)
9. Gnanasmbandam, N., Lee, S., Kumara, S.R.T.: An autonomous performance control framework for distributed multi-agent systems: a queueing theory based approach. In: AAMAS 2005, Utrecht, Netherlands, pp. 1313–1314 (2005)
10. Lee, M.H., Birukou, A., Dudin, A., Klimenok, V., Choe, C.: Quantitative analysis of single-level single-mediator multi-agent systems. In: Nguyen, N.T., Grzech, A., Howlett, R.J., Jain, L.C. (eds.) KES-AMSTA 2007. LNCS (LNAI), vol. 4496, pp. 447–455. Springer, Heidelberg (2007). https://doi.org/10.1007/978-3-540-72830-6_46
11. Koroliuk, V.S., Melikov, A.Z., Ponomarenko, L.A., Rustamov, A.M.: Methods for analysis of multi-channel queueing system with instantaneous and delayed feedbacks. Cybern. Syst. Anal. **52**(1), 58–70 (2016). https://doi.org/10.1007/s10559-016-9800-y

12. Zhidkova, L.A., Moiseeva, S.P.: Investigation of the queueing system MMPP|M|∞ with repeated service. Tomsk State Univ. J. Control Comput. Sci. **1**(26), 53–62 (2014)
13. Ayyappan, G., Gowthami, R.: Analysis of $MAP, PH_2^{OA}/PH_1^{I}, PH_2^{O}/1$ retrial queue with vacation, feedback, two-way communication and impatient customers. Soft. Comput., 1–28 (2020). https://doi.org/10.1007/s00500-020-05318-4

On Comparison of Multiserver Systems with Two-Component Mixture Distributions

Irina Peshkova[1(✉)], Evsey Morozov[1,2,3(✉)], and Maria Maltseva[1(✉)]

[1] Petrozavodsk State University, Lenin str. 33, Petrozavodsk 185910, Russia
iaminova@petrsu.ru, masha.mariam@mail.ru, maltseva@mail.ru

[2] Institute of Applied Mathematical Research of the Karelian research Centre of RAS, Pushkinskaya str. 11, Petrozavodsk 185910, Russia
emorozov@karelia.ru

[3] Moscow Center for Fundamental and Applied Mathematics, Moscow State University, Moscow 119991, Russia

Abstract. In this paper, we introduce and study the relations between parameters of the two-component Hyperexponential, Exponential-Pareto and two-component Pareto mixture distributions which admit stochastic and failure rate comparisons. Then we apply the failure rate and stochastic ordering techniques to construct the upper and lower bounds for the steady-state performance indexes of a multiserver model with Exponential-Pareto mixture service time distribution.

Keywords: Failure rate comparison · Multiserver system · Performance analysis · Finite mixture distribution

1 Introduction

Finite mixtures of distributions play an important role in various research fields, including reliability theory, actuarial science, medical and biology research, artificial neural networks and robustness studies, income analysis and queueing systems modelling [4–7], since the samples of data are often presented by quite different inputs (populations). Therefore, the finite mixture model may be an effective tool for modelling the random sample from such heterogeneous input. In the communication network modelling the application of the mixture of distributions is motivated because a randomly selected claim may be taken, for instance, from video stream, online games or from the audio stream. Moreover, a few large claims can generate a heavy-tailed distribution while some other data may be light-tailed. In general, the explicit expressions for the stationary performance measures of the mixture models are hardly available [2,3].

The main contribution of this research is as follows: using the *failure rates*, monotonicity properties (with respect to the interarrival and service times [2,18–20]) and the method of *regenerative envelops*, recently developed by the

The research is supported by Russian Foundation for Basic Research, projects No. 19-57-45022, 19-07-00303, 18-07-00156, 18-07-00147.

© Springer Nature Switzerland AG 2020
V. M. Vishnevskiy et al. (Eds.): DCCN 2020, CCIS 1337, pp. 340–352, 2020.
https://doi.org/10.1007/978-3-030-66242-4_27

authors [9–11], we compare the performance indexes in the system with the *Exponential-Pareto* mixture service time distributions with the corresponding measures in the systems having two-component service time distribution. More exactly, we use a two-component *Hyperexponential* distribution to construct a minorant system, and also use a two-component *Pareto* service time distribution to construct a majorant system. It allows us to obtain the lower and upper bounds for the mean queue size of original multiserver system with Exponential-Pareto mixture service time distribution. In contrast, in previous research, we have used comparison of the multiserver system with Exponential-Pareto mixture service time distribution with the systems having *one-component service time distributions*, either Exponential distribution (for the minorant system) or Pareto distribution (for the majorant system) [12].

The structure of the paper is as follows. In Sect. 2, we define finite mixture distributions and describe three examples. In Sect. 3, we consider conditions which allow both stochastic and failure rate comparison of the two-component mixture distribution. In Sect. 4, we describe the conditions allowing stochastic and failure rate ordering of the two-component Hyperexponential distribution, the Exponential-Pareto distribution and the two-component Pareto mixture distribution. This analysis is further applied in Sect. 5 to compare the queue sizes in the multiserver systems with infinite buffers. Finally, the obtained theoretical results are illustrated by numerical simulation presented in Sect. 6.

2 Two-Component Mixture Distributions

In this section, we discuss the two-component mixture distributions and their failure rates.

Let X_i be non-negative random variables (r.v.'s) with distribution functions (d.f.'s) $F_{X_i}(x)$, $i = 1, 2$. We say that r.v. X has *two-component (positive) mixture distribution*, if distribution F_X can be written in the form [6]

$$F_X(x) = pF_{X_1}(x) + (1-p)F_{X_2}(x), \tag{1}$$

where parameter p is called a *mixing proportion*, $p \in (0, 1)$. Suppose that the random variables X_1, X_2 are independent and let I be indicator function independent of (X_1, X_2) such that $\mathsf{P}(I = 1) = p$ and $\mathsf{P}(I = 0) = 1 - p$. Then it is said that the variable

$$X = I\,X_1 + (1 - I)\,X_2 \tag{2}$$

has *two-component mixture distribution* (1).

For each x, such that the tail distribution $\overline{F}_X(x) = \mathsf{P}(X > x) > 0$, we define the *failure rate* function as

$$r_X(x) = \frac{f_X(x)}{\overline{F}_X(x)}. \tag{3}$$

The quantity $r(x)dx$ plays an important role in the *reliability theory* and *queueing theory* representing the condition probability that a failure occurs in the interval of time $(x, x+dx)$ provided that a device is still working at instant x, see [7]. It is easy to check that the failure rate function for mixture (2) has the following form:

$$r_X(x) = \frac{pf_{X_1}(x) + (1-p)f_{X_2}(x)}{p\bar{F}_{X_1}(x) + (1-p)\bar{F}_{X_2}(x)}. \tag{4}$$

The failure rate function of a mixture of distributions tends to follow, in the long run, to the *lowest* failure rate function of the mixture components (if exists) [7], that is,

$$r_X(x) \sim \min(r_{X_1}(x), r_{X_2}(x)), \text{ as } x \to \infty, \tag{5}$$

where $a \sim b$ means $a/b \to 1$. We continue by considering three examples of mixtures.

Let X_1, X_2 have an Exponential distribution, denoted $Exp(\lambda_i)$, that is

$$F_{X_i}(x) = 1 - e^{-\lambda_i x},\ x \geq 0,\ \lambda_i > 0,$$

with constant failure rate, $r_{X_i}(x) = \lambda_i$, $i = 1, 2$. Then the mixture (2) has the Hyperexponential distribution, denoted $Hyper(\lambda_1, \lambda_2.p)$, which has the tail distribution

$$1 - F_X(x) =: \overline{F}_X(x) = pe^{-\lambda_1 x} + (1-p)e^{-\lambda_2 x},\ x > 0, \tag{6}$$

and *decreasing failure rate (DFR)* function,

$$r_X(x) = \frac{p\lambda_1 e^{-\lambda_1 x} + (1-p)\lambda_2 e^{-\lambda_2 x}}{pe^{-\lambda_1 x} + (1-p)e^{-\lambda_2 x}}. \tag{7}$$

It is easy to see that

$$r_X(x) \underset{x\to\infty}{\longrightarrow} \min(\lambda_1, \lambda_2).$$

We emphasize that the Hyperexponential distribution has DFR property, while each component in mixture (6) has constant failure rate. The explanation of this effect is given, for instance, in [14].

Let Y_1 have exponential distribution $Exp(\lambda)$ and Y_2 have Pareto type-2 distribution, denoted $Pareto(\alpha, x_0)$, with distribution

$$F_{Y_2}(x) = 1 - \left(\frac{x_0}{x_0 + x}\right)^{\alpha},\ x \geq 0,\ x_0 > 0,\ \alpha > 0, \tag{8}$$

and DFR failure rate function

$$r_{Y_2}(x) = \frac{\alpha}{x_0 + x},\ x \geq 0, \tag{9}$$

vanishing as $x \to \infty$. We say that the mixture

$$Y = IY_1 + (1-I)Y_2 \tag{10}$$

has *Exponential-Pareto mixture distribution* with tail distribution

$$\bar{F}_Y(x) = pe^{-\lambda x} + (1-p)\left(\frac{x_0}{x_0+x}\right)^{\alpha}, \quad x \geq 0. \tag{11}$$

It is easy to calculate that the failure rate of Y equals

$$r_Y(x) = \frac{p\, r_{Y_1}(x) a(x) + (1-p)\, r_{Y_2}(x)}{p\, a(x) + (1-p)}, \tag{12}$$

where

$$a(x) = e^{-\lambda x}\left(1 + \frac{x}{x_0}\right)^{\alpha}, \quad x \geq 0.$$

Note that $r_Y(x) \to 0 \quad \text{as} \quad x \to \infty$. Now we consider the mixture

$$Z = IZ_1 + (1-I)Z_2, \tag{13}$$

where Z_i have Pareto type-2 $Pareto(\alpha_i, x_0)$ distributions (8), implying

$$\bar{F}_Z(x) = p\left(\frac{x_0}{x_0+x}\right)^{\alpha_1} + (1-p)\left(\frac{x_0}{x_0+x}\right)^{\alpha_2} \tag{14}$$

It is known that the mixture of DFR distributions is a DFR distribution as well [7]. The failure rate of mixture (13) is given by

$$r_Z(x) = \frac{p r_{Z_1}(x) a(x) + (1-p) r_{Z_2}(x)}{pa(x) + (1-p)}, \tag{15}$$

where

$$a(x) = x_0^{\alpha_1 - \alpha_2}(x_0 + x)^{\alpha_2 - \alpha_1}.$$

Note that $r_Z(x) \to 0,\ x \to \infty$. Since

$$r_Z'(x) = -(1-p)\frac{pa(x)(r_{Z_1}(x) - r_{Z_2}(x))^2 + r_{Z_2}^2(x)/\alpha(pa(x) + (1-p))}{(pa(x) + (1-p))^2} < 0,$$

then $r_Z(x)$ is decreasing function and $F_Z(x)$ is DFR function.

3 Stochastic and Failure Rate Ordering of Two-Component Mixture Distributions

First we consider the connection between *stochastic ordering* and *failure rate ordering*. Let X and Y be non-negative random variables with distribution functions F_X, F_Y, and densities f_X, f_Y, respectively. We say that X *is less than* Y *stochastically (in distribution)*, and denote it as $X \underset{st}{\leq} Y$, if

$$\overline{F}_X(x) \leq \overline{F}_Y(x),\ x \geq 0; \tag{16}$$

We say that X *is less than* Y in *failure rate*, and denote it $X \underset{r}{\leq} Y$, if

$$r_X(x) \geq r_Y(x), \quad \text{for all } x \geq 0. \tag{17}$$

There are the following well-known relation between stochastic ordering and failure rate ordering [13]:

$$X \underset{r}{\leq} Y \ \Rightarrow \ X \underset{st}{\leq} Y \ . \tag{18}$$

It is evident that the latter in turns implies $\mathsf{E}X \leq \mathsf{E}Y$. The next theorem states that if the components of the mixture are ordered (stochastically or in failure rate), then the mixture has the corresponding lower and upper bounds.

Theorem 1. *If the random variable* X *has representation (2), then*

$$X_1 \underset{st}{\leq} X_2 \ \Rightarrow \ X_1 \underset{st}{\leq} X \underset{st}{\leq} X_2; \tag{19}$$

$$X_1 \underset{r}{\leq} X_2 \ \Rightarrow \ X_1 \underset{r}{\leq} X \underset{r}{\leq} X_2. \tag{20}$$

Proof. From the stochastic ordering $X_1 \underset{st}{\leq} X_2$ it follows that $\bar{F}_{X_1}(x) \leq \bar{F}_{X_2}(x)$. It is then obvious that

$$\bar{F}_{X_1}(x) \leq p\bar{F}_{X_1}(x) + (1-p)\bar{F}_{X_2}(x) \leq \bar{F}_{X_2}(x).$$

On the other hand, the failure rate ordering $X_1 \underset{r}{\leq} X_2$ means that

$$f_{X_2}(x) \leq \frac{f_{X_1}(x)\bar{F}_{X_2}(x)}{\bar{F}_{X_1}(x)},$$

and hence we obtain,

$$r_X(x) \leq \frac{pf_{X_1}(x) + (1-p)f_{X_1}(x)\bar{F}_{X_2}(x)/\bar{F}_{X_1}(x)}{p\bar{F}_{X_1}(x) + (1-p)\bar{F}_{X_2}(x)} = r_{X_1}(x),$$

that, in turn, gives the inequality $X_1 \underset{r}{\leq} X$. Further, the inequalities

$$f_{X_1}(x) \geq \frac{f_{X_2}(x)\bar{F}_{X_1}(x)}{\bar{F}_{X_2}(x)}$$

and

$$r_X(x) \geq \frac{pf_{X_2}(x)\bar{F}_{X_1}(x)/\bar{F}_{X_2}(x) + (1-p)f_{X_2}(x)}{p\bar{F}_{X_1}(x) + (1-p)\bar{F}_{X_2}(x)} = r_{X_2},$$

implies $X \underset{r}{\leq} X_2$, completing the proof. □

Now we compare two mixtures with different finite mixture distributions and with the same mixing proportion p:

$$F_X(x) = pF_{X_1}(x) + (1-p)F_{X_2}(x),$$
$$F_Y(x) = pF_{Y_1}(x) + (1-p)F_{Y_2}(x).$$

Suppose that the components are stochastically ordered in the following way

$$X_1 \underset{st}{\leq} Y_1, \quad X_2 \underset{st}{\leq} Y_2. \tag{21}$$

It is known that if the random variables X_1, Y_1, X_2, Y_2 are (mutually) independent, then the mixtures are stochastically ordered as follows [7]:

$$X \underset{st}{\leq} Y. \tag{22}$$

However, to compare the mixtures in the failure rate, some additional conditions must hold. The following statement is a modification of Theorem 2.1 from [1].

Theorem 2. *Assume the following conditions hold:*

$$X_1 \underset{r}{\leq} X_2 \text{ or } Y_1 \underset{r}{\leq} Y_2; \tag{23}$$

$$\min(X_1, Y_2) \underset{st}{\geq} \min(X_2, Y_1); \tag{24}$$

$$X_1 \underset{r}{\leq} Y_1,\ X_2 \underset{r}{\leq} Y_2. \tag{25}$$

Then

$$X \underset{r}{\leq} Y. \tag{26}$$

The results presented in (18)–(26), in particular, allow to establish the monotonicity properties of the stochastic processes in the queueing systems with *different mixture distributions* of the interarrival or/and service times. We present such results in Sect. 4 for the Hyperexponential, Exponential-Pareto mixtures and for the two-component Pareto mixture. It is important to note that the failure rate comparison of the different distributions can be often presented in an explicit form and moreover, as we mentioned above, this ordering implies the stochastic ordering. By this reason, in what follows, we use the failure rate ordering to establish, as a consequence, the stochastic ordering of the corresponding queueing systems.

4 Stochastic and Failure Rate Comparison of Some Two-Component Mixture Distributions

In this section, we demonstrate the failure rate ordering technique for mixture distributions. We consider Exponential-Pareto mixture, two-component Pareto mixture and two-component Hyperexponential mixture. Moreover we establish some auxiliary results which are used further in Sect. 5.

Suppose that X has the Hyperexponential distribution (6), Y has the Exponential - Pareto mixture distribution (11) and Z has two-component Pareto mixture distribution (14). Now we establish conditions on the basic parameters

$$\lambda,\ \lambda_1,\ \lambda_2,\ \alpha,\ \alpha_1,\ \alpha_2,$$

which imply the stochastic and failure rate ordering of the mixtures.

Theorem 3. *Assume that the distributions of $X,\ Y,\ Z$ satisfy (6), (11) and (14), respectively, and the following relations hold:*

$$\lambda_1 \geq \lambda \geq \frac{\alpha_1}{x_0}; \tag{27}$$

$$\lambda_2 \geq \frac{\alpha}{x_0} \geq \frac{\alpha_2}{x_0}. \tag{28}$$

Then

$$X \underset{st}{\leq} Y \underset{st}{\leq} Z. \tag{29}$$

Proof. It is easy to check that if the following relations between parameters hold,

$$\lambda_1 \geq \lambda \text{ and } \lambda_2 \geq \frac{\alpha}{x_0}, \tag{30}$$

then the components of mixtures are ordered in the failure rate as follows:

$$X_1 \underset{r}{\leq} Y_1,\quad X_2 \underset{r}{\leq} Y_2. \tag{31}$$

But as we know, (31) implies

$$X_1 \underset{st}{\leq} Y_1,\quad X_2 \underset{st}{\leq} Y_2, \tag{32}$$

and thus it follows from (21) and (22) that $X \underset{st}{\leq} Y$.

If the parameters of the Exponential-Pareto mixture and two-component Pareto mixture are connected by the following relations,

$$\lambda \geq \frac{\alpha_1}{x_0},\quad \alpha \geq \alpha_2, \tag{33}$$

then the following implications hold:

$$r_{Y_i}(x) \geq r_{Z_i}(x) \Rightarrow Y_i \underset{r}{\leq} Z_i \Rightarrow Y_i \underset{st}{\leq} Z_i,\ i = 1, 2 \Rightarrow\ Y \underset{st}{\leq} Z.$$

Thus, under conditions (27) and (28) the mixtures X, Y, Z are stochastically ordered as in (29). □

Theorem 4. *If the random variables $X,\ Y,\ Z$ have the (tail) distributions (6), (11), and (14), respectively, and the following relations hold*

$$\frac{\alpha_1}{x_0} \leq \frac{\alpha_2}{x_0} \leq \frac{\alpha}{x_0} \leq \lambda \leq \lambda_1 \leq \lambda_2, \tag{34}$$

then the following failure rate ordering holds:

$$X \underset{r}{\leq} Y \underset{r}{\leq} Z. \tag{35}$$

Proof. First we mention that relation (34) imply conditions (27) and (28). Hence, it follows from the proof of Theorem 3, that

$$X_i \underset{r}{\leq} Y_i,\ Y_i \underset{r}{\leq} Z_i,\ i = 1, 2,$$

and assumption (24) is fulfilled.

In order to check whether the mixtures are ordered by failure rate, it remains to check the expressions (23) and (24). Let $\lambda_1 \leq \lambda_2$, then $X_1 \underset{r}{\geq} X_2$. As it is easy to see, condition (24) for mixtures X and Y can be written as

$$\left(\frac{x_0}{x_0 + x}\right)^{\alpha} \geq e^{-(\lambda + \lambda_2 - \lambda_1)x},\ x \geq 0. \tag{36}$$

To verify this inequality, we note that if $\lambda \geq \alpha/x_0$ then $Y_1 \underset{r}{\leq} Y_2$ and condition (23) is fulfilled, which in turn implies

$$Y_1 \underset{st}{\leq} Y_2.$$

But the latter inequality can be written as

$$\left(\frac{x_0}{x_0 + x}\right)^{\alpha} \geq e^{-\lambda x},\ x \geq 0,$$

and then (36) follows because $e^{-\lambda x} \geq e^{-(\lambda + \lambda_2 - \lambda_1)x}$. Thus, the following implication holds:

$$\lambda_2 \geq \lambda_1 \geq \lambda \geq \frac{\alpha}{x_0} \ \Rightarrow\ X \underset{r}{\leq} Y. \tag{37}$$

It remains to check inequality (24), which is equivalent to the inequality

$$e^{-\lambda x} \geq \left(\frac{x_0}{x_0 + x}\right)^{\alpha_1 + \alpha - \alpha_2},\ x \geq 0. \tag{38}$$

Let $\alpha_1 \leq \alpha_2$, then we have $Z_1 \underset{r}{\geq} Z_2$. Because $Y_1 \underset{st}{\leq} Z_1$, then it follows that

$$e^{-\lambda x} \geq \left(\frac{x_0}{x_0 + x}\right)^{\alpha_1},\ x \geq 0$$

and we can conclude that (24) holds since

$$\left(\frac{x_0}{x_0 + x}\right)^{\alpha_1} \geq \left(\frac{x_0}{x_0 + x}\right)^{\alpha_1 + \alpha - \alpha_2},\ x \geq 0.$$

Thus the following implication takes place:

$$\alpha_1 \leq \alpha_2 \leq \alpha \leq \lambda x_0 \ \Rightarrow\ Y \underset{r}{\leq} Z. \tag{39}$$

Now we conclude that conditions (34) for the parameters of mixtures guarantee that the mixtures are ordered in the failure rate as in (35). □

5 Comparison of Multiserver Systems with Different Service Distributions

In this section, we demonstrate how the failure rate comparison allows to stochastically compare the steady-state performance of the multiserver queueing systems with the infinite buffers and a renewal input flow. We consider two queueing systems, denoted by $\Sigma^{(1)}$ and $\Sigma^{(2)}$, with the same number N servers working in parallel. (In what follows, the superscript (i) denotes the number of the system.) The service discipline is assumed to be First-Come-First-Served. We denote by $S_n^{(i)}$ the service time of customer n, and by $t_n^{(i)}$ his arrival instant. The sequence of the independent identically distributed (iid) interarrival times $\tau_n^{(i)} = t_{n+1}^{(i)} - t_n^{(i)}$, $n \geq 1$, and the sequence of the iid service times $\{S_n^{(i)}, n \geq 1\}$ are assumed to be independent, $i = 1, 2$. Denote $S^{(i)}$ the generic service time, and $\tau^{(i)}$ the generic interarrival time, $i = 1, 2$. Now we compare the steady-state queue-size processes in the systems $\Sigma^{(1)}$ and $\Sigma^{(2)}$. At the arrival instant of customer n in in the system $\Sigma^{(i)}$, denote by $\nu_n^{(i)}$ the number of customers, by $Q_n^{(i)}$ the *queue size* and by $W_n^{(i)}$ the *waiting time* of this customer. Denote, when exists, the limits (in distribution)

$$Q_n^{(i)} \Rightarrow Q^{(i)},\ \nu_n^{(i)} \Rightarrow \nu^{(i)},\ W_n^{(i)} \Rightarrow W^{(i)},\ n \to \infty,\quad i = 1, 2.$$

These limits exists, in particular, when the interarrival times $\tau^{(i)}$, $i = 1, 2$ are *non-lattice* and the following negative drift assumption holds [2]:

$$\mathsf{E}S^{(i)} < N\mathsf{E}\tau^{(i)}. \tag{40}$$

The following statement, which is a modification of Theorem 5 from [20], contains conditions implying ordering of the queue sizes in the infinite buffer systems.

Theorem 5. *Assume the following conditions hold in the queueing systems:*

$$\nu_1^{(1)} \underset{st}{=} \nu_1^{(2)} = 0,\ \tau^{(1)} \underset{st}{=} \tau^{(2)},\ S^{(1)} \underset{r}{\leq} S^{(2)}. \tag{41}$$

Then, with probability 1 (w.p.1),

$$\nu_n^{(1)} \leq \nu_n^{(2)},\ Q_n^{(1)} \leq Q_n^{(2)},\ W_n^{(1)} \leq W_n^{(2)},\ n \geq 1. \tag{42}$$

It is worth mentioning that, indeed the stochastic ordering of service times in (41) is sufficient for inequalities (42) . (Indeed we use a coupling to obtain ordering (42) w.p.1.) From this point of view, the failure rate ordering seems to be rather restrictive. However, for the most of distributions the failure rate comparison can be expressed as an explicit relation between given parameters, which allows to compare the performance indexes in the systems with *different (mixture) distributions* of the interarrival and/or service times.

To simulate and compare the described above multiserver systems, we apply the well-developed *regenerative simulation*, which is a powerful and efficient method of the output analysis of complex queueing systems. We note that our systems are regenerative. A detailed analysis of the regenerative method can be found in the following basic works [2,3,8,15–17].

6 Numerical Examples

In this section we illustrate, by the numerical simulation, the comparison results obtained for the multiserver systems in Sects. 4 and 5.

Denote by Σ the original system with Exponential-Pareto (mixture) service times with parameters λ, α, x_0, p. Then we construct two new systems $\Sigma^{(1)}$ and $\Sigma^{(2)}$. The system $\Sigma^{(1)}$ has two-component Hyperexponential service times $S^{(1)}$, with parameters λ_1, λ_2, p, and the system $\Sigma^{(2)}$ has two-component Pareto service times $S^{(2)}$, with parameters $\alpha_1, \alpha_2, x_0, p$. The Poisson input flow with parameter λ_τ is the same in all three systems.

It follows from results, obtained in Sects. 4 and 5, that, under condition (34), the following ordering between the basic performance indexes must take place:

$$Q_n^{(1)} \le Q_n \le Q_n^{(2)}, \;\; W_n^{(1)} \le W_n \le W_n^{(2)}, \; n \ge 1. \tag{43}$$

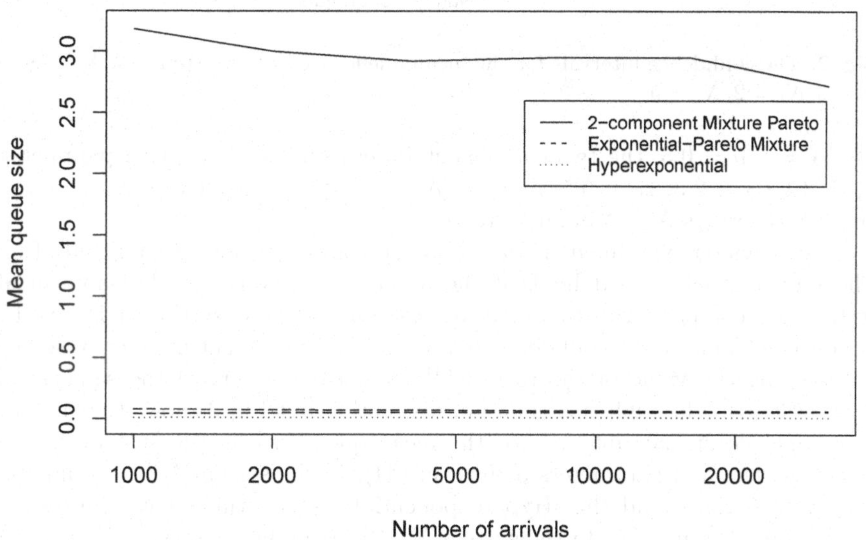

Fig. 1. The confidence intervals for the mean queue size in the systems Σ, $\Sigma^{(1)}$, $\Sigma^{(2)}$, $N = N_1 = N_2 = 4$

Figure 1 demonstrates the confidence interval bounds for the mean queue size in the basic system Σ with Exponential-Pareto service time with parameters:

$$p = 0.5, \; x_0 = 0.5, \; \lambda = 7, \; \alpha = 3, \; x_0 = 0.5. \tag{44}$$

It follows from (43) that $\Sigma^{(1)}$ is *minorant* system for Σ and $\Sigma^{(2)}$ is *majorant* system for Σ. We compare regenerative simulation results for all three systems. We take the Hyperexponential service time with parameters $\lambda_1 =$

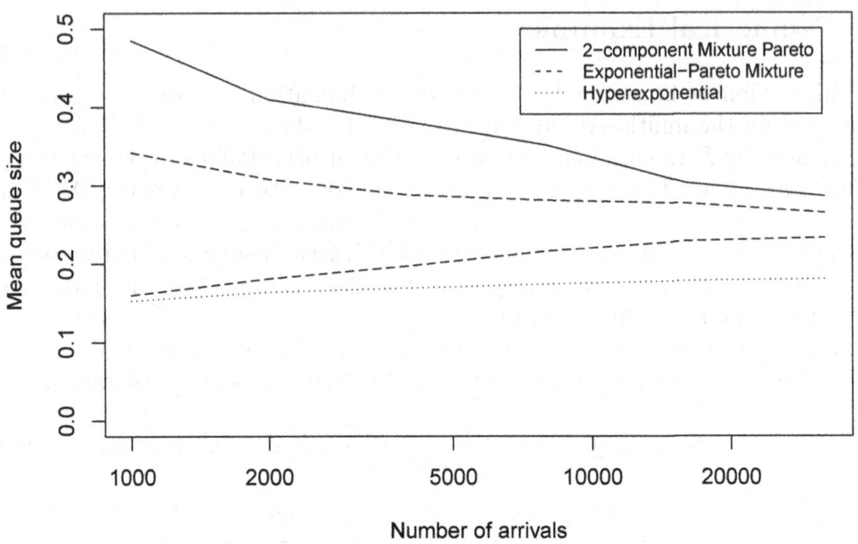

Fig. 2. The confidence intervals for the mean queue size in the systems Σ, $\Sigma^{(1)}$, $\Sigma^{(2)}$, $N = 3, N_1 = 2, N_2 = 6$.

$7.5, \lambda_2 = 8, p = 0.5$, the two-component Pareto service times with parameters $\alpha_1 = 2.1, \alpha_2 = 2.3$, $x_0 = 0.5$, $p = 0.5$. Also we use the input rate $\lambda_\tau = 10$ and number of servers $N = 4$ in all systems.

The previous experiment is based on the correct monotonicity results from Theorem 5 which indeed holds if the number of servers are the same in all systems we compare [20]. In two next experiments we show that it is possible to construct more tight confidence bounds allowing different number of servers but keeping the value of the ratio $\mathsf{E}S^{(i)}/N$ (load per server) the same in all three systems. Although this is an approximation but Fig. 2 shows that it gives more tight confidence interval for the mean queue size in the system Σ with $N = 3$ servers and parameters defined in (44). In the system $\Sigma^{(1)}$, the number of servers $N_1 = 2$ and the (Hyperexponential) service times have parameters $\lambda_1 = 8.5, \lambda_2 = 9, p = 0.5$. In the system $\Sigma^{(2)}$, the number of servers $N_2 = 6$ and (two-component Pareto) service times have parameters $\alpha_1 = 2.1, \alpha_2 = 2.5$, $x_0 = 0.5$, $p = 0.5$. The input rate is $\lambda_\tau = 10$ in all systems. We stress that in both experiments the selected parameters of the service time distributions satisfy the failure rate ordering.

Figure 3 demonstrates the confidence intervals (also obtained by regenerative simulation) for the mean queue size in the system Σ with parameters $p = 0.5$, $x_0 = 0.5$ and $\lambda = 5.1$, $\alpha = 2, x_0 = 0.5$, cf. (44), for the system $\Sigma^{(1)}$ with parameters $\lambda_1 = 6, \lambda_2 = 5.7, p = 0.5$, and for the system $\Sigma^{(2)}$ with parameters $\alpha_1 = 2.5, \alpha_2 = 1.9, x_0 = 0.5, p = 0.5$. We use the input rate $\lambda_\tau = 10$ in all systems, and the number of servers (in an evident notation) $N_1 = 2, N = 4, N_2 = 5$. We again note that the selected parameters satisfy conditions (27) and (28) which imply stochastic ordering of the service times.

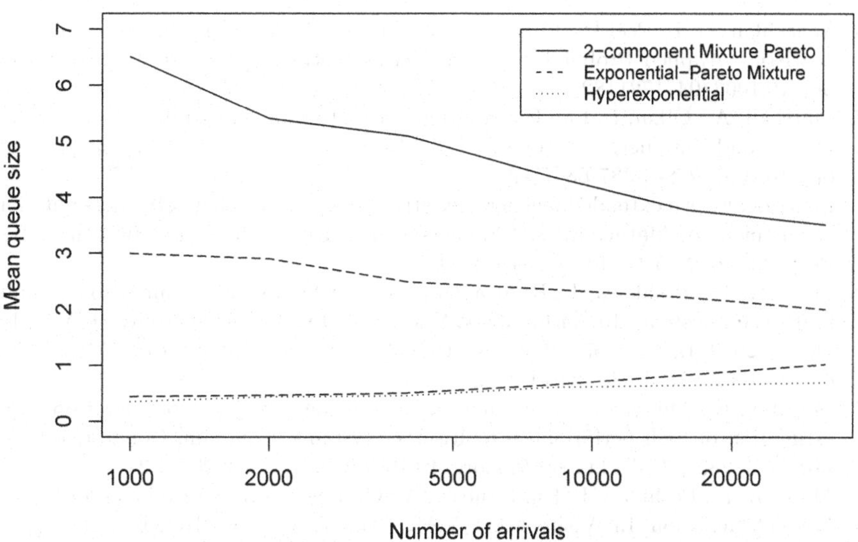

Fig. 3. The confidence intervals for the mean queue size in the systems $\Sigma, \Sigma^{(1)}$, $\Sigma^{(2)}$, $N = 4, N_1 = 2, N_2 = 5$.

7 Conclusion

In this paper, we study the applicability the *failure rate comparison* of the steady-state performance measures in the multiserver system with two-component mixture distributions. We demonstrate a few refined monotonicity results for the systems in which the properties of service time distributions imply specific monotonicity properties of the failure rate functions. Some numerical results based on regenerative simulation are presented. The obtained results may be useful for the estimation, by the regenerative simulation, the performance indexes for a wide class of the queueing systems with *two-component mixture distributions* of service time.

References

1. Amini-Seresht, E., Zhang, Y.: Stochastic comparisons on two finite mixture models. Oper. Res. Lett. **45**, 475–480 (2017). https://doi.org/10.1016/j.orl.2017.07.009
2. Asmussen, S.: Applied Probability and Queues, vol. 2, p. 439. Springer, New York (2003). https://doi.org/10.1007/b97236
3. Asmussen, S., Glynn, P.: Stochastic Simulation: Algorithms and Analysis, p. 476. Springer, New York (2007). https://doi.org/10.1007/978-0-387-69033-9
4. Aven, T., Jensen, U.: Stochastic Models in Reliability, p. 297. Springer, New York (2013). https://doi.org/10.1007/978-1-4614-7894-2
5. Al-Hussaini, E.K., Sultan, K.S.: Reliability and hazard based on finite mixture models. In: Handbook of Statistics. Advances in Reliability, vol. 20, pp. 139–183 (2001). https://doi.org/10.1016/S0169-7161(01)20007-8

6. Mclachlan, G. I., Peel D.: Finite Mixture Models. Wiley Series in Probability and Statistics. Applied Probability and Statistics Section, p. 439 (2001). https://doi.org/10.1002/0471721182
7. Marshall, A., Olkin, I.: Life Distributions: Structure of Nonparametric, Semiparametric, and Parametric Families, p. 783. Springer, New York (2007). https://doi.org/10.1007/978-0-387-68477-2
8. Morozov, E.: An extended regenerative structure and queueing network simulation. Department of Mathematics, Chalmers University, Gothenburg, Sweden (1995). Preprint No 1995–08/ISSN 0347–2809
9. Morozov, E., Peshkova, I., Rumyantsev, A.: On failure rate comparison of finite multiserver systems. In: Vishnevskiy, V.M., Samouylov, K.E., Kozyrev, D.V. (eds.) DCCN 2019. LNCS, vol. 11965, pp. 419–431. Springer, Cham (2019). https://doi.org/10.1007/978-3-030-36614-8_32
10. Morozov, E., Nekrasova, R., Peshkova, I., Rumyantsev, A.: A regeneration-based estimation of high performance multiserver systems. Commun. Comput. Inf. Sci. **608**, 271–282 (2016). https://doi.org/10.1007/978-3-319-39207-3_24
11. Morozov, E., Peshkova, I., Rumyantsev, A.: On Regenerative Envelopes for Cluster Model Simulation. In: Vishnevskiy, V.M., Samouylov, K.E., Kozyrev, D.V. (eds.) DCCN 2016. CCIS, vol. 678, pp. 222–230. Springer, Cham (2016). https://doi.org/10.1007/978-3-319-51917-3_20
12. Peshkova, I., Morozov, E., Maltseva, M.: On comparison of multiserver systems with Exponential-Pareto mixture distribution. Comput. Netw. **1231**, 141–152 (2020). https://doi.org/10.1007/978-3-030-50719-0_11
13. Ross, S., Shanthikumar, J., Zhu, Z.: On increasing-failure-rate random variables. J. Appl. Probab. **42**, 797–809 (2005). https://doi.org/10.1239/jap/1127322028
14. Shaked, M., Spizzichino, F.: Chap. 6. Mixtures and monotonicity of failure rate functions. In: Handbook of Statistics, vol. 2, pp. 185–198 (2001). https://doi.org/10.1016/S0169-7161(01)20008-X
15. Shedler, G.S.: Regeneration and Networks of Queues. Springer, New York (1987). https://doi.org/10.1007/978-1-4612-1050-4
16. Shedler, G.S.: Regenerative Stochastic Simulation, p. 400. Academic Press Inc. (1992). https://www.elsevier.com/books/regenerative-stochastic-simulation/shedler/978-0-08-092572-1
17. Sigman, K., Wolff, R.W.: A review of regenerative processes. SIAM Rev. **35**, 269–288 (1993). https://doi.org/10.1137/1035046
18. Sonderman, D.: Comparing multi-server queues with finite waiting rooms, I: same number of servers. Adv. Appl. Probab. **11**, 439–447 (1979). https://doi.org/10.2307/1426848
19. Sonderman, D.: Comparing multi-server queues with finite waiting rooms, II:different number of servers. Adv. Appl. Probab. **11**(2), 448–455 (1979). https://doi.org/10.2307/1426849
20. Whitt, W.: Comparing counting processes and queues. Adv. Appl. Probab. **13**(1), 207–220 (1981). https://doi.org/10.2307/1426475

On Splitting Scenarios for Effective Simulation of Reliability Models

Alexandra Borodina[1,2(✉)] and Vitaliy Tishenko[2]

[1] Institute of Applied Mathematical Research of the Karelian Research Centre of RAS, Petrozavodsk, Russia
borodina@krc.karelia.ru
[2] Petrozavodsk State University, Petrozavodsk, Russia
musen@cs.pertsu.ru, vitalik1tishenko@gmail.com

Abstract. This article discusses split-based simulation algorithms for the reliability analysis of systems with rare failures. We consider a regenerative degradation process with independent increments, where a two-level management policy is introduced. The evaluation of the failure probability and other characteristics of the regeneration cycle is critical for optimal control of such systems. However, real degradation processes are usually described by rather complicated models, for which it is difficult to obtain any analytics or heuristic solutions. Accelerated simulation techniques provide a more efficient way to investigate models associated with rare events than the naive Monte Carlo method. The paper considers two scenarios for constructing regeneration cycles using the splitting procedure. One algorithm uses splits at degradation stages of the process and is a nested procedure that runs after the stage when failure becomes possible. Another algorithm implements splitting at levels that are calculated adaptively from the value of the total time spent in the degradation stages. For both scenarios, a series of experiments were performed and both algorithms were compared with the naive Monte Carlo method. In the case of exponential degradation stages, the estimates were compared with the analytical solution. The key criteria for comparing the effectiveness of the methods were the simulation time, the relative error, and the relative experimental error. Moreover, the questions of the efficiency of the optimal choice of levels and the number of splits were investigated.

Keywords: Failure probability · Reliability analysis · Degradation process · Regenerative splitting · Speed-up simulation · Relative error

1 Introduction

Analysis of degradation and shock multi-stage processes is a key step in the development and implementation of modern highly reliable technologies. It is

The study was carried out under state order to the Karelian Research Centre of the Russian Academy of Sciences (Institute of Applied Mathematical Research KarRC RAS) and supported by the Russian Foundation for Basic Research, project 18-07-00187.

© Springer Nature Switzerland AG 2020
V. M. Vishnevskiy et al. (Eds.): DCCN 2020, CCIS 1337, pp. 353–366, 2020.
https://doi.org/10.1007/978-3-030-66242-4_28

quite expected that the emergence of new models forces to modernize research methods for shock models with changing degradation rate, with soft and hard failures with a natural or predetermined threshold level, which is generally a random variable [9,14]. In particular, effective methods for calculating the realistic model parameters and system reliability are in demand.

Nevertheless, the *naive Monte Carlo method* is still popular in a large number of modern works, despite its well-known inefficiency, for example, see [6,7,11,17]. For instance, as it proposed in [15], a standard solution is based on the stationary distribution of the built-in Markov chain for the simplest cases. Then, for the generalized model, Monte Carlo simulation is used to approximate the cost of maintenance.

Nevertheless, the models become more complex, and analytical methods are less available. In particular, the Wiener process with independent and normally distributed increments is widely used for non-monotonic degradation. The Gamma process is useful in the stochastic modeling of monotonic and gradual degradation, characterized by the sequence of tiny increments, such as wear, fatigue, corrosion, crack growth, erosion, consumption, degrading health index.

In addition, more complex models of two-stage or multi-stage degradation include Gamma-Gamma, Wiener-Wiener, Gamma-Wiener degradation models are also considered (for example, see [10,16]). In this regard, we suggest looking for more effective alternatives for the naive simulation method that can be used to analyze the reliability of modern systems.

For a homogeneous case of exponential degradation stages of the system with gradual and instantaneous failures, analytical formulas were obtained and an advanced simulation technique was proposed in [3].

In [2] a heterogeneous degradation process was investigated analytically for exponential stages. Numerical experiments confirmed that using the standard Monte Carlo method impairs the accuracy of the probability estimates and other characteristics of the degradation process.

In [1] the variance reduction technique has been extended to estimate the failure probability that a random sum exceeds a random variable V. In [4] a variance reduction technique using a special variant of *conditional Monte Carlo* approach proposed for heterogeneous degradation process. It was shown by numerical examples that relative error is bounded and even is vanishing when the degradation stage has heavy-tailed distribution. On the other hand, our experiments show that this method has not an advantage for the light-tailed stages.

A variance reduction technique based on the *Importance sampling* with an exponential change of measure for light-tailed degradation stages was introduced in [5].

All these techniques were tested on a model of the degradation process, which describes the thickness of the anti-corrosion coating and is described in [3]. Since it is possible to obtain analytical results in the simplest cases, this model is convenient for analyzing the effectiveness of accelerated simulation and variance reducing methods. In addition, the process has a regenerative structure, which is typical for degradation models. The proposed accelerated algorithm can also be extended for more complex models by replacing the procedure of simulation the time spent at the degradation stages.

2 Crude Simulation of the Degradation Process

Following [3] consider the degradation process $X := \{X(t),\, t \geq 0\}$ with a finite state space $E = \{0, 1, \ldots, L, \ldots, M, \ldots, K; F\}$ describing the degradation stages of the system. Two-threshold policy (K, L) is considered, which means that the system is restored in the stage K and then proceeds to stage L (see Fig. 1a).

Let T_i be the transition time from i to $i+1$ stage. Random variables (r.v.) T_i are independent but not necessarily identically distributed. Note that

$$S_{i,j} := \sum_{k=i}^{j-1} T_k, \quad 0 \leq i \leq K-1,\, j > i,$$

is the transition time from state j to state i. We define the following distribution functions (d.f.):

$$F_i(t) = \mathbb{P}(T_i \leq t); \;\; F_V(t) = \mathbb{P}(V \leq t); \tag{1}$$

$$F_{U_F}(t) = \mathbb{P}(U_F \leq t); \;\; F_{U_{K,L}}(t) = \mathbb{P}(U_{K,L} \leq t); \tag{2}$$

$$F_{i,j}(t) = \mathbb{P}(S_{i\,j} \leq t) = F_{i,\,j-1} * F_j(t) = \int_{0-}^{t} F_{i,\,j-1}(t-v)dF_j(v); \tag{3}$$

$$F_{i,i+1}(t) = F_i(t), \quad F_{i,i}(t) \equiv 0, \tag{4}$$

where $*$ means convolution. Starting in state $X(0) = 0$ the process successively passes $K-1$ intermediate degradation stages and reaches the state M. We denote by V a random time after which a failure can occur. Thus, after the stage M either event $\{S_{M,K} \geq V\}$ (*instantaneous failure*) may happen during a random period V, or event $\{S_{M,K} < V\}$ (*starting the preventive repair stage*) occurs during the time

$$S_{M,K} = \sum_{i=M}^{K-1} T_i.$$

The process X is strongly regenerative with moments

$$\tau_{n+1} = \inf\{Z_i > \tau_n :\, X(Z_i^+) = M\},\; n \geq 0,\; \tau_0 := 0,$$

where Z_k is the the hitting time of the stage $k \geq 1$, and cycle lengths $Y_k = \tau_{k+1} - \tau_k$, $k \geq 1$ are i.i.d.

There are two types of regeneration cycles for degradation process: with and without failure (see the illustration Fig. 1a)

$$Y = \begin{cases} Y_F = V + U_F + S_{0,M}, & \text{if } S_{M,K} \geq V \\ Y_{NF} = S_{M,K} + U_{K,L} + S_{L,M}, & \text{if } S_{M,K} < V, \end{cases} \tag{5}$$

where r.v. V, U_F, $S_{0,M} = \sum_{i=0}^{M-1} T_i$ with known distributions are independent as well as r.v. $S_{M,K}$, $U_{K,L}$, $S_{L,M} = \sum_{i=L}^{M-1} T_i$ After failure and repair, the process returns to the initial state 0. Thus (unconditional) regeneration cycle length Y can be written as

$$Y = Y_F \cdot I_{\{V \leq S_{M,K}\}} + Y_{NF} \cdot I_{\{S_{M,K} < V\}}. \tag{6}$$

where I_A denotes indicator function. The variable Y plays an important role in the analysis of the degradation process [3].

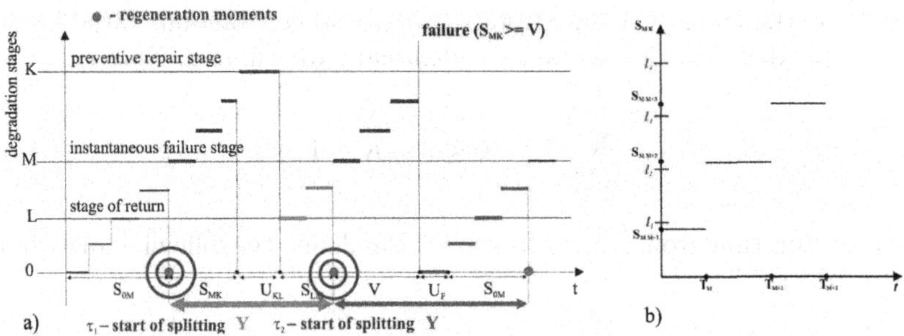

Fig. 1. Two splitting schemes: a) by stages b) by the value of $S_{M,K}$

The main target is to find the probability of instantaneous failure within the regeneration cycle, that is

$$p_F = \mathbb{P}(S_{M,K} \geq V) = \mathsf{E}[F_V(S_{M,K})], \tag{7}$$

where F_V is the distribution function of the random variable V. But the simulation also necessary for other characteristics like the mean lifetime T

$$\mathbb{E}[T] = \mathbb{E}[Y_{NF}](\mathbb{E}[N] - 1) + \mathbb{E}[V | V \leq S_{M,K}],$$

where $\mathbb{E}[N] = 1/p_F$ is the mean number of cycles until complete failure; mean cycle length $\mathbb{E}[Y_F]$ with failure or $\mathbb{E}[Y_{NF}]$ without failure; reliability function

$$R(t) = \mathbb{P}[T > t | X(0) = 0], \ t \geq 0,$$

where T stands for the lifetime of the system.

Note that, for more complex shock models simulation allows us to estimate parameters of the model like the damage threshold, the intensity of random shock, critical shock inter-arrival time, scale, and shape parameters in Gamma models, etc.

If the failure is not a rare event and it is easy to simulate the r.v. V, U_F, $S_{0,M}$, $S_{M,K}$, $U_{K,L}$, $S_{L,M}$ on a computer and the performance function is computationally inexpensive to construct the regeneration cycles, then the p_F and other characteristics of degradation process can be approximated by *crude Monte Carlo* unbiased estimators. However, for highly reliable systems, such an assumption seems to be naive.

3 Splitting Scenarios for the Degradation Process

The splitting procedure for a homogeneous degradation process was firstly proposed in [3], where the method showed the effectiveness of the estimation for exponential degradation stages. Denote $\{l_i\}, \{R_i\}, i \geq 0$ the sequence of levels and the corresponding factors. After crossing the level l_i, the trajectory is split (produces independent copies) into R_i trajectories, which further develop independently. Taking into account the generalization of the method for other models, let's now compare *two possible splitting scenarios:*

a) the process splits at each degradation stage $i \in [M, K-1]$ (Fig. 1a);
b) the levels of splitting l_i depend on the value of the accumulated amount of time $S_{M,K}$ (Fig. 1b), their total number is determined by the pilot run.

In the first case a), it is impossible to go through several levels of splitting, while in the second b), the trajectory of the process can cross several trajectories at once in one step during event simulation.

In both cases, the splitting of the trajectories occurs only in the area after stage M, when instantaneous failure becomes possible.

Unlike processes with negative drift (which is typical for problems of evaluating rare events), it is impossible to perform optimal leveling due to the randomness of the threshold time to failure V for the degradation process. However, all methods for rare events probabilities simulation are designed to solve problems with a **constant value** of failure threshold [12,13].

For the standard splitting procedure, an optimal distance between thresholds $\{l_i\}$ and splitting factors $\{R_i\}$ at each threshold are defined by the pilot run [12]. The pilot run defines threshold partition in accordance with the requirement that conditional probabilities of transition between thresholds p_i **is not so rare** but gives the biased estimator itself.

From the point of view of the regeneration theory, the problem of rare event probability (and other characteristics of the process) estimating can be reduced to the problem of constructing regeneration cycles using the accelerated simulation technique. In comparison with the naive Monte Carlo method, the splitting method will allow building regeneration cycles much faster. Yet, it should be noted that the cycles constructed in this way cannot be considered independent.

Recall that, as it was proposed in [3], to speed-up the simulation procedure of waiting for a rare system failure, we introduce the embedded splitting procedure. The splitting started strictly in regeneration moments τ_i after which an instantaneous failure becomes possible, see Fig. 1. Further, we will use splitting only in the region between stages M and K and construct a mathematical model of cycles based on the branching process.

3.1 Fully Branching Regeneration Splitting - Scenario a)

The first scheme a) employs fully branching regeneration, where the splitting levels l_i are fixed and strictly correspond to the degradation stages. For this reason, initially, we cannot change the partition into levels. Thus, it is impossible to influence the estimation variance by optimizing the levels, unlike the standard splitting method. However, this algorithm is the easiest to implement regardless of the choice of programming language.

Besides, given the artificial branching of the process, it is necessary to take into account the dependency between cycles. At each level, we generate R_i copies of r.v. T_i, $M \le i \le K-1$. So, each original path generates

$$D = R_M \cdots R_{K-1}$$

subpaths called *group of cycles.* Each process trajectory started from the initial threshold l_M gives the group of D dependable regeneration cycles. The cycles from different groups are independent by construction. The total number of groups is R_{M-1}. The dependence is generated by the same pre-history of the random sum S_{MK} realizations before the splitting point at each stage.

For convenience, we will assume that after the moment of splitting at the level $l_i, i \in [M+1, K-1]$, all R_i trajectories of brunching degradation process X^* develop **simultaneously**. (This can be realistic if the algorithms are implemented using parallel programming tools.)

Let us consider in detail the construction of traditional regenerative cycles from the first group X_1^* generated by the first trajectory starting from the stage $M-1$. For the remaining groups $X_i^*, i \in [2, R_{M-1}]$, the reasoning is similar. Let's introduce the following notation:

g_i - generation of $l_i, i \in [M, K-1]$ threshold;
n_i - the number of trajectories starting from the level l_{i-1} and reaching the level l_i, $i \in [M+1, K-1]$, $n_M = R_{M-1}$;
N^* - the maximum stage number reached by the trajectories before the failure (if it occurs) $N^* = \max\{i : n_i > 0, i \in [M, K-1]\}$;
$t_{k_i}^{(i)}$ - the splitting moment of the bunch number $k_i \in [1, n_i]$ at the level $l_i, i \in [M+1, K-1]$;
$H_{k_i,j}^{(i)}(t)$ - accumulated time spent in degradation stages at time t by jth process copy that started from l_i threshold at the moment $t_{k_i}^{(i)}$;
$\eta_{k_i,j}^{(i)}$ - the instantaneous failure moment for jth copy of the process after the hitting l_i threshold;
$Z_{k_i,j}^{(i+1)}$ - the moment of the next event, namely, either a transition to the next stage or a failure occur at the level l_i.

We can conclude that

$$H_{k_i,j}^{(i)}(t_{k_i}^{(i)}) = \sum_{p=M}^{i-1} T_p, \quad i \ge M+1; \tag{8}$$

$$\eta_{k_i,j}^{(i)} = \min\{t > t_{k_i}^{(i)} : H_{k_i,j}^{(i)}(t) \ge V\}; \tag{9}$$

$$t_{k_i,j}^{(i+1)} = t_{k_i}^{(i)} + T_{i,j}, \;\; T_{i,j} = {}_{st}T_i \sim F_i; \tag{10}$$

$$Z_{k_i,j}^{(i+1)} = \min\{\eta_{k_i,j}^{(i)}, t_{k_i,j}^{(i+1)}\}. \tag{11}$$

We'll say that the trajectory *belongs to the generation* g_i if it started from a level l_i and either reached the level l_{i+1} or the failure occurred, so denote:

$$g_i = \{H^{(i)}_{k_i,j},\ j \geq M\}, \tag{12}$$

$$H^{(i)}_{k_i,j} = \{H^{(i)}_{k_i,j}(t),\ t^{(i)}_{k_i} \leq t \leq Z^{(i+1)}_{k_i,j}\}. \tag{13}$$

Recall that D cycles in each group $X^*_i, i \in [1, R_{M-1}]$ will be formed from trajectories that met failure at one of the levels (cycles with the failure Y_F), or reached stage K (cycles without failure Y_{NF}).

Below we give an algorithm for sequentially constructing cycles in the group X^*_1 assuming that $n_{M+1} > 0$. The output of the Algorithm 1 is the set of cycles in first group

$$X^*_1 = \{G^{(M)}, \ldots, G^{(N^*)}\}.$$

Remark 1. As a rule, for degradation models with rare failures, it is most often performed that $N^* = K - 1$ and thus splitting occurs at each stage of degradation, therefore this algorithm becomes **fully branching.**

Further, each group must be supplemented to a completed regeneration cycle according to the type of cycle (6) by simulation. Thus, each set of cycles $G^{(i)}$ has its own tail of the regeneration cycle. The proposed Algorithm 1 is started R_{M-1} times at regeneration moments τ_i. The cycles from different groups X^*_i are independent, but the trajectories in one group cycles have a construction dependency. Then, the total number of the failures in the ith group is

$$A_i = \sum_{j=(i-1)\cdot D+1}^{i\cdot D} I^{(j)}, i = 1, \ldots, R_{M-1},$$

where $I^{(j)} = 1$ for the cycle with failure ($I^{(j)} = 0$, otherwise). Sequence $\{I^{(j)}, j \geq 1\}$ is discrete D-dependent regenerative with constant cycle length D and regeneration instants are $\{i \cdot D\}$, $i \in [1, R_{M-1}]$.

Remark 2. Note that constructing trajectories according to the algorithm is necessary for estimating performance measures such as the mean life-time $\mathsf{E}T$, the average length of the cycles $\mathsf{E}Y$, $\mathsf{E}Y_F$, the reliability function $R(t)$, etc. If it is required to estimate only the failure probability, then the regeneration cycle can be represented by an indicator, and the Algorithm 1 can be simplified.

Moreover, the regenerative interpretation is successful in terms of interval estimation. The regenerative structure gives the following unbiased and strongly consistent estimator $\widehat{p}_F$:

$$\widehat{p}_F = \frac{\sum_{j=1}^{R_{M-1}} A_j}{R_{M-1} \cdot D} \to \frac{\mathbb{E}\sum_{j=1}^{D} I^{(j)}}{D} = p_F \tag{15}$$

Algorithm 1. Algorithm for sequential design of degradation trajectories

Draw an i.i.d. R_M-sample $T_{M,j}, j \in [1, R_M]$ from df F_M and $V \sim F_V$. Denote the next-stage time $t_j^{(M+1)} = T_{M,j}$, the failure time $\eta_j^{(M)} = V$. Let $H_j^{(M)} = t_j^{(M+1)}$ until time $Z_j^{(M)} = \min\{\eta_j^{(M)}, t_j^{(M+1)}\}$, where index j means the ordinal number of the trajectory in the generation g_M;

Compute $n_{M+1} = \#\{j : Z_j^{(M)} = t_j^{(M+1)}, j \in [1, R_M]\}$;

$N^* \leftarrow M + 1$;

$i \leftarrow M + 1$;

$n_M \leftarrow 1$;

while $i \leq N^*$ and $i < K$ **do**

Draw an i.i.d. R_i-sample $T_{i,j}, j \in [1, n_i R_i]$ from df F_M and $V \sim F_V$. Draw the trajectories $H_{k_i,j}^{(i)}, k_i \in [1, n_i], j \in [1, n_i R_i]$ after time $t_{k_i}^{(i)}$. **Add** common pre-history to the next generation

$$H_{k_i,j}^{(i)}(t) \equiv H_{k_i}^{(i-1)}(t), \tau_1 \leq t \leq t_{k_i}^{(i)}, k_i \in [1, n_i]$$

Denote the failure time, next-stage time and choose the next event-time:

$$\eta_{k_i,j}^{(i)} = \min\{t > t_{k_i}^{(i)} : H_{k_i,j}^{(i)}(t) \geq V\}, \; k_i \in [1, n_i], \; j \in [1, n_i R_i];$$
$$t_{k_i,j}^{(i+1)} = t_{k_i}^{(i)} + T_{i,j}, \; T_{i,j} =_{st} T_i \sim F_i;$$
$$Z_{k_i,j}^{(i+1)} = \min\{\eta_{k_i,j}^{(i)}, t_{k_i,j}^{(i+1)}\},$$

Define the trajectories of generation g_i (their number is $n_i R_i$)

$$H_{k_i,j}^{(i)} = \{H_{k_i,j}^{(i)}(t), \tau_1 \leq t \leq Z_{k_i,j}^{(i+1)}, \; j \in [1, n_i R_i]\}$$

Omit the index k_i for g_i-trajectories and use notation $H_j^{(i)}$, $\eta_j^{(i)}$, $t_j^{(i+1)}$, $Z_j^{(i+1)}$ for all $j \in [1, n_i R_i]$.

Compute $n_{i+1} = \#\{j : Z_j^{(i+1)} = t_j^{(i+1)}, j \in [1, n_i R_i]\}$;

Reorder the trajectories and assign numbers $[1, n_i R_i - n_{i+1}]$ to the paths with the failure at the stage l_i, and numbers $(n_i R_i - n_{i+1}, n_i R_i]$ to the paths without failure.

For every $j : Z^{(i+1)} = \eta^{(i)}$, **generate** $D_i = R_{i+1} \cdots R_{K-1}$ stochastically identical paths

$$G_{j,d}^{(i)} =_{st} H_j^{(i)}, d \in [1, D_i].$$

Define the paths from generation g_i

$$G^{(i)} = \{G_{j,d}^{(i)}, j \in [1, n_i R_i - n_{i+1}], d \in [1, D_i]\} \tag{14}$$

if $n_{i+1} > 0$ and $i < K - 1$ **then**

$N^* \leftarrow i + 1$

end if

$i \leftarrow i + 1$

end while

if $N^* = K - 1$ **then**

there are n_K cycles without failure. Every path $H_{k_{K-1},j}^{(K-1)}$ generates the only trajectory in stage N^*, $k_{K-1} \in [1, n_{K-1}]$, $j \in [1, n_{K-1} R_{K-1}]$. So, the paths from g_{K-1} are:

$$G^{(K-1)} = \{G_{j,d}^{(K-1)} \equiv H_{k_{K-1},j}^{(K-1)}, j \in [1, n_{K-1} R_{K-1} - n_K], d \in [1, D_{K-1}]\}$$

else

define $G^{(N^*)}$ as in formula (14)

end if

as $R_{M-1} \to \infty$ w.p.1. The following $100(1-\delta)\%$ confidence interval for p_F based on the *regenerative variant of Central Limit Theorem*, that is well known from [8]

$$\left[\widehat{p}_F \pm \frac{z(\delta)\sqrt{v_n}}{\sqrt{n}}\right] \tag{16}$$

where quantile $z(\delta)$ satisfies $\mathbb{P}[N(0,1) \leq z(\delta)] = 1 - \delta/2$, and

$$v_n = \frac{n^{-1}\sum_{i=1}^{n}[A_i - \widehat{p_F}D]^2}{D^2} \tag{17}$$

is a weakly consistent estimator of $\sigma^2 = \mathbb{E}[A_1 - p_F D]^2/D^2$ if $\mathbb{E}(A_1 - \gamma\alpha_1)^2 < \infty$. Under moment assumptions, $\mathsf{E}A_1^2 < \infty$, the estimate (17) is strongly consistent.

3.2 Standard Splitting Procedure - Scenario b)

Another splitting scenario can be proposed based on the standard splitting method [12,13]. Recall that in standard spitting procedure the state space E of process is divided into $M+1$ nested subsets C_i, which defined by *importance function* f:

$$C_i = \{x \in E \mid f(x) \geq l_i\},\ i \in [1, M+1].$$

For the degradation process the splitting levels l_i selected according to the value of $S_{M,K}$, see the illustration in Fig. 1b), so the cumulative residence time in the degradation stages is chosen as importance function f. Since the standard algorithm arranges the levels $\{l_i\}$ based on the fixed value (level $l_{M+1} = l$), it is proposed to use the expectation value $\mathsf{E}V$ as a threshold value instead of r.v. V. Denote the following notation for conditional probabilities of reaching the next level $i+1$ while being at the current level i:

$$p_1 = \mathsf{P}(C_1), \ldots, p_{i+1} = \mathsf{P}(C_{i+1}|C_i), i \in [1, M].$$

Thus, the standard point estimator $\widehat{p}_F$ given by splitting scenario b) is

$$\hat{p}_F = \prod_{i=1}^{M+1} \hat{p}_i = \frac{n_{M+1}}{R_0} \prod_{i=1}^{M} \frac{1}{R_i} = \frac{n_{M+1}}{R_0 D}, \tag{18}$$

where $\widehat{p}_i$ is the estimate of transition probability p_i, R_i is the splitting factor, n_{M+1} is the number of failures, R_0 is the number of simulation starts.

It is known that the splitting **simulation effort** is defined as follows:

$$\sum_{i=1}^{M+1} R_i \mathsf{E}[n_{i-1}] = R_0 \sum_{i=1}^{M+1} \frac{1}{p_i} \prod_{j=1}^{i} p_j R_j, \tag{19}$$

where n_i is the number of reaching the level l_i. It is optimal if the number of branching trajectories does not grow exponentially on the one hand and the process is not damped on the other. It is obvious that the case $p_j R_j > 1$ implies the explosion of the algorithm effort, the case $p_j R_j < 1$ gives $n_{M+1} = 0$ with high probability, and the case $R_j = 1/p_j$ **is optimal**. Thus, the optimal values for simulation are related by the ratio $R_i = 1/\widehat{p}_i$, but for the degradation process, it should be expected that $\widehat{p}_i \approx 1$. Thus, the optimal option is not suitable for accumulation processes and does not give an effect.

Generally an optimal levels $\{l_i\}$ and optimal splitting factors $\{R_i\}$ at each threshold determined from the so-called **adaptive pilot run** of splitting procedure (for instance, see [12]). Pilot run gives biased estimator but defines threshold partition according to requirement that p_i **is not rare event probabilities**. Thus, the input parameters to the adaptive algorithm include sample size N, $p_i = p \in (0,1)$, fixed level $l_{M+1} = l$ and importance function $f(x)$. The key points are:

1. to get the optimal levels $l_i = \min(l_{M+1}, \hat{l}_i)$, where

$$\hat{l}_i = \underset{l \in \{f(X_1),\ldots,f(X_N)\}}{arg\,min} \left\{ \frac{1}{N} \sum_{i=1}^{N} I[f(X_i) \geq l] \leq p \right\}$$

2. to get the splitting factors

$$R_i = \mathsf{Ber}(p) + \left\lfloor \frac{N}{n_i} \right\rfloor$$

.

Due to the randomness of the parameter V the adaptive pilot run for average level system, where $l = E[V]$ was applied to calculate thresholds l_i and factors R_i.

Remark 3. In addition, the degradation process due to arbitrary increments can cross several levels at once, so the following **splitting condition** was introduced: if any trajectory starting from level l_i crossed some level $l_{i+k}, i + k \leq M + 1$, then it splits into $\prod_{j=1}^{k} R_{i+j}$ copies.

4 Simulation Results

An experimental comparison of two splitting scenarios was made with respect to the following evaluation quality criteria: relative error RE and relative experimental error RER (if p_F is analytically available)

$$RE[\widehat{p}_F] = \frac{\sqrt{Var[\widehat{p}_F]}}{\mathbb{E}[\hat{p}_F]}, \quad RER[\widehat{p_F}] = |\hat{p_F} - p_F| \cdot 100/p_F.$$

To estimate the variance, samples of 100 values were constructed. The tables show the average computation time for one sample element. In some cases, Monte Carlo estimation for probabilities less than 10^{-7} was too long to be complete. An invariant for comparing the three algorithms MC, $RS_{a)}$ and $RS_{b)}$ is to construct the same number n of cycles.

All numerical tests were executed on ultrabook HP ENVY Intel(R) Core(TM) i3 7100U 2.4 GHz processor with 4 GB of RAM, running Windows 10. Tables 1, 2, and 3 show the results of running a programs in Python3 and C++ for the Monte Carlo method (MC) and the regenerative splitting method (RS) for $a)$ and $b)$ scenarios.

Table 1. Time (s) estimator: MC vs. $RS_{a)}$ and $RS_{b)}$: $T_j \sim Exp(\lambda_j)$, $V \sim Exp(\nu)$

λ_{K-1}, s	n	p_F	t_{MC}	$t_{RS_{a)}}$	$t_{RS_{b)}}$
10^3, 50	10^4	$8.75 \cdot 10^{-3}$	0.318	0.019	0.228
10^4, $5 \cdot 10^2$	10^4	$8.79 \cdot 10^{-4}$	0.312	0.018	0.222
10^5, $5 \cdot 10^3$	10^5	$8.79 \cdot 10^{-5}$	3.06	0.180	2.16
10^6, $5 \cdot 10^4$	10^7	$8.79 \cdot 10^{-6}$	253	17.4	218

Table 2. RER estimator: MC vs. $RS_{a)}$ and $RS_{b)}$: $T_j \sim Exp(\lambda_j)$, $V \sim Exp(\nu)$

λ_{K-1}, s	n	p_F	RER_{MC}	$RER_{RS_{a)}}$	$RER_{RS_{b)}}$
10^3, 50	10^4	$8.75 \cdot 10^{-3}$	3.02	2.06	3.91
10^4, $5 \cdot 10^2$	10^4	$8.79 \cdot 10^{-4}$	4.59	3.46	4.17
10^5, $5 \cdot 10^3$	10^5	$8.79 \cdot 10^{-5}$	2.85	4.43	1.46
10^6, $5 \cdot 10^4$	10^7	$8.79 \cdot 10^{-6}$	0.79	1.23	4.79

Table 3. RE estimator: MC vs. $RS_{a)}$ and $RS_{b)}$: $T_j \sim Exp(\lambda_j)$, $V \sim Exp(\nu)$

λ_{K-1}, s	n	p_F	RE_{MC}	$RE_{RS_{a)}}$	$RE_{RS_{b)}}$
10^3, 50	10^4	$8.75 \cdot 10^{-3}$	0.18	3.02	1.5
10^4, $5 \cdot 10^2$	10^4	$8.79 \cdot 10^{-4}$	0.36	2.16	4.60
10^5, $5 \cdot 10^3$	10^5	$8.79 \cdot 10^{-5}$	0.31	3.23	4.65
10^6, $5 \cdot 10^4$	10^7	$8.79 \cdot 10^{-6}$	0.11	4.75	1.53

Let's fix the parameters of the model $\nu = 0.5$, $\mu_F = 1.5$, $\mu = 2$, $L = 1$, $M = 5$, $K = 17$. To observe both RE and RER values we give an example of exponential degradation periods $T_j \sim Exp(\lambda_j)$ for the heterogeneous case where the analytical formulas are known from [2]. So, we will vary number of regeneration cycles n and sequence of values

$$\lambda_j = \lambda_{K-1} - (K - j - 1)s, \; j \in [0, K-2],$$

where λ_{K-1} will be initialized before starting the splitting procedure, and the other values will be shifted by step s, thus the condition

$$\lambda_0 < \cdots < \lambda_{K-1}, \ \nu < \lambda_j, \ j = [0, K-1],$$

for increasing the degradation rate is guaranteed.

We compare all methods with the same number n of regeneration cycles and track the corresponding time t_{MC}, $t_{RS_{a)}}$, $t_{RS_{b)}}$ of each method in seconds. The number of cycles n is chosen for practical reasons so that the RER does not exceed 5% for all methods. For variance estimation, a sample of 100 values was constructed.

The Table 1 shows that the $RS_{a)}$ algorithm is significantly superior in time to the others MC and $RS_{b)}$. Despite the higher RE values the $RS_{a)}$ splitting scenario provides a lower RER and is, therefore, closer to the analytical value than $RS_{b)}$.

It should be noted that for the scenario $RS_{b)}$, cases with a fixed optimal number of splits R_i and a random factors with exponential $R_i \sim Exp(0.33)$ (RS_{exp}) and uniform $R_i \sim Uni[1, 5]$ (RS_{uni}) were also studied.

Table 4. RER and time (s) results for various R_i types: $T_j \sim Exp(\lambda_j)$, $V \sim Exp(\nu)$

p_F	RER_{MC}	RER_{RS}	$RER_{RS_{exp}}$	$RER_{RS_{uni}}$	t_{MC}	t_{RS}	$t_{RS_{exp}}$	$t_{RS_{uni}}$
$8,79 \cdot 10^{-3}$	2.29	3.07	0.7	3.23	0.39	0.11	0.19	0.18
$8,79 \cdot 10^{-4}$	0.57	2.82	0.31	0.4	3.48	1.02	1.68	1.68
$8,79 \cdot 10^{-5}$	0.34	4.35	2.46	4.08	36.6	10.1	16.2	17.4
$8,79 \cdot 10^{-6}$	–	0.73	–	1.33	–	108	–	168

The Table 4 demonstrates the values for comparison RER and time for $RS_{b)}$ algorithm in these cases with MC method. Nevertheless, no significant differences between the types of splitting and MC estimates are observed. That fact shows again that the standard optimal conditions have no effect in the case of the degradation process. Empty spaces in the table mean that the computation time for one element of the sample was too long to build a full sample of size 100.

Table 5. RE and time (s) estimator in C++: MC vs. $RS_{a)}$: $T_i \sim exp(\lambda)$, $V \sim Exp(\nu)$

n	p_F	t_{MC}	$t_{RS_{a)}}$	RE_{MC}	$RE_{RS_{a)}}$
10^4	$8,79 \cdot 10^{-4}$	0.037	0.010	0.318	2.167
10^5	$8,79 \cdot 10^{-5}$	0.389	0.095	0.366	3.254
10^7	$8,79 \cdot 10^{-6}$	**37.83**	**6.56**	0.125	1.967
10^8	$8,79 \cdot 10^{-7}$	378.0	25.06	0.166	3.188

The Table 5 shows the results of the algorithm in the C++ language. Expectedly, the running time for the same number of cycles is more comfortable in comparison with the Python language, so that we can calculate the estimate by MC for a lower probability. Note that the MC method gives a smaller RE in some cases. That fact can be explained due to the direct dependence of variance $\widehat{p}_F$ on ν and it is rather big here.

Table 6. Var and time (s) estimator in C++: MC vs. $RS_{a)}$ and $RS_{b)}$: $T_i \sim$ Weibull

$\hat{p}_{F_{MC}}$	$\hat{p}_{F_{RS_{a)}}}$	$\hat{p}_{F_{RS_{b)}}}$	Var_{MC}	$Var_{RS_{a)}}$	$Var_{RS_{b)}}$	t_{MC}	$t_{RS_{a)}}$	$t_{RS_{b)}}$
0.156	0.153	0.155	4.24e−09	8.28e−05	3.56e−06	137.7	13.67	89.3
5.43e−03	5.29e−03	5.33e−03	2.38e−10	1.33e−05	7.31e−07	69.64	6.53	25.3
6.43e−05	6.02e−05	6.12e−05	1.29e−12	1.58e−10	4.59e−10	234.9	17.33	210.1
–	2.07e−07	–	–	3.99e−15	–	>2 h	628.33	>2 h

In the next example, we try to avoid the influence of the variance r.v. V and assume that it is fixed $V = 1/\nu$. We use ν as a rarity parameter to change the probability value in the experiment. Consider i.i.d $T_i = T$ with light tail Weibull d.f. $F_T(x) = 1 - e^{-3x^4}, x \geq 0$. It is difficult to get the analytics for the Weibull degradation stages, so we compare simulation methods with each other. This is clear from Table 6 that it is possible to repeat the observation that splitting at the degradation stages is more beneficial than scenario b).

5 Conclusion

We performed a series of experiments that demonstrated the effectiveness of the two splitting schemes for the accelerated simulation of regeneration cycles and estimation of the degradation process characteristics. Both variants of the splitting procedure were compared with the cycle constructing procedure via naive Monte Carlo method in terms of simulation time, relative error, and relative experimental error estimates. Based on numerical experiments, it can be summarized that the first scenario a) is the most efficient for simulation of the degradation process with rare failures. At the same time, the organization of branching at the moments of transition between the stages of degradation is simpler and more intuitively clear. In addition, it turns out that the choice of the optimal splitting parameters is difficult due to the fact that the time to failure is a random variable. In particular, the use of the optimal splitting rates does not give an effect for processes with independent increments, even compared to the Monte Carlo method.

References

1. Borodina, A., Lukashenko, O., Morozov, E.: On conditional Monte Carlo for the failure probability estimation. In: Proceedings of 2018 10th International Congress on Ultra Modern Telecommunications and Control Systems, ICUMT 2018, pp. 202–207. IEEE (2018)
2. Borodina, A., Tishenko, V.: Simulation of a heterogeneous degradation process in a system with gradual and sudden failures. Proc. Karelian Res. Cent. Russ. Acad. Sci. **7**, 3–13 (2018). https://doi.org/10.17076/mat836. (in Russian)
3. Borodina, A., Efrosinin, D., Morozov, E.: Application of splitting to failure estimation in controllable degradation system. In: Vishnevskiy, V.M., Samouylov, K.E., Kozyrev, D.V. (eds.) DCCN 2017. CCIS, vol. 700, pp. 217–230. Springer, Cham (2017). https://doi.org/10.1007/978-3-319-66836-9_18
4. Borodina, A., Lukashenko, O., Morozov, E.: A rare-event estimation of heterogeneous degradation process. In: 2019 11th International Congress on Ultra Modern Telecommunications and Control Systems and Workshops (ICUMT), pp. 1–6. IEEE (2019)
5. Borodina, A., Lukashenko, O., Morozov, E.: On estimation of rare event probability in degradation process with light-tailed stages. In: Proceedings the 22nd International Conference on Distributed Computer and Communication Networks: Control, Computation, Communications, DCCN 2019, pp. 477–483 (2019)
6. Dong, Q., Cui, L.: A study on stochastic degradation process models under different types of failure thresholds. Reliab. Eng. Syst. Saf. **181**, 202–212 (2019)
7. Gao, H., Cui, L., Kong, D.: Reliability analysis for a wiener degradation process model under changing failure thresholds. Reliab. Eng. Syst. Saf. **171**, 1–8 (2018)
8. Glynn, P.W., Iglehart, D.L.: Conditions for the applicability of the regenerative method. Manage. Sci. **39**(9), 1108–1111 (1993)
9. Jensen, H., Papadimitriou, C.: Reliability analysis of dynamical systems. Substructure Coupling for Dynamic Analysis. LNACM, vol. 89, pp. 69–111. Springer, Cham (2019). https://doi.org/10.1007/978-3-030-12819-7_4
10. Kong, D., Balakrishnan, N., Cui, L.: Two-phase degradation process model with abrupt jump at change point governed by Wiener process. IEEE Trans. Reliab. **66**(4), 1345–1360 (2017)
11. Lin, Y.H., Li, Y.F., Zio, E.: A comparison between Monte Carlo simulation and finite-volume scheme for reliability assessment of multi-state physics systems. Reliab. Eng. Syst. Saf. **174**, 1–11 (2018)
12. Rubinstein, R.Y., Kroese, D.P.: Simulation and the Monte Carlo Method. Wiley, New Jersey (2017)
13. Rubinstein, R.Y., Ridder, A., Vaisman, R.: Fast Sequential Monte Carlo Methods for Counting and Optimization. Wiley, New Jersey (2014)
14. Shahraki, A., Yadav, O.P., Liao, H.: A review on degradation modelling and its engineering applications. Int. J. Perform. Eng. **13**(3), 299–314 (2017)
15. Sun, Q., Ye, Z.S., Zhu, X.: Managing component degradation in series systems for balancing degradation through reallocation and maintenance. IISE Trans. **52**(7), 797–810 (2020)
16. Wang, X., Jiang, P., Guo, B., Cheng, Z.: Real-time reliability evaluation for an individual product based on change-point gamma and Wiener process. Qual. Reliab. Eng. Int. **30**(4), 513–525 (2014)
17. Yousefi, N., Coit, D.W., Song, S., Feng, Q.: Optimization of on-condition thresholds for a system of degrading components with competing dependent failure processes. Reliab. Eng. Syst. Saf **192**, 106547 (2019)

Ergodicity and Polynomial Convergence Rate of Generalized Markov Modulated Poisson Processes

Galina Zverkina(✉)

V. A. Trapeznikov Institute of Control Sciences of Russian Academy of Sciences, 65 Profsoyuznaya street, 117997 Moscow, Russia
ldan@ipu.ru
https://www.ipu.ru/

Abstract. Generalization of the Lorden's inequality is an excellent tool for obtaining strong upper bounds for the convergence rate for various complicated stochastic models. This paper demonstrates a method for obtaining such bounds for some generalization of the Markov modulated Poisson process (MMPP). The proposed method can be applied in the reliability and queuing theory.

Keywords: Markov modulated Poisson process · Lorden's inequality · Convergence rate · Strong upper bounds · Coupling method · Successful coupling.

1 Introduction

As well-known, a Markov modulated Poisson Process (MMPP) is a Poisson process whose rate varies according to some Markov process. There is some Markov process X_t, and the behaviour of the other process Y_t at the time t depends on the state of X_t at this time. Because Poisson process are considered, the processes X_t and Y_t are Markov chains in continuous time, obviously.

Thus, the "rate" or *intensity* and transition function of the process Y_t depends on the state of X_t. The paired process (X_t, Y_t) is Markov, and it has the constant intensities between the changes of the state of (X_t, Y_t). This is a multi-dimensional Markov chain in continuous time.

Obviously, the processes X_t and Y_t are one-dimensional, but they may be multi-dimensional.

The behavior of a large class of queueing systems, reliability systems, networks may be described in terms of MMPP under the assumption about exponential distribution of all random variables (r.v.'s).

Here, the generalization of MMPP is given; we suppose that random variables in the model can have arbitrary distributions. Moreover, we suppose, that the processes X_t and Y_t can be multi-dimensional, and the intensities of everyone components of the paired process (X_t, Y_t) can be depended on everyone. Such a

The work is supported by RFBR, project No 20-01-00575A.

© Springer Nature Switzerland AG 2020
V. M. Vishnevskiy et al. (Eds.): DCCN 2020, CCIS 1337, pp. 367–381, 2020.
https://doi.org/10.1007/978-3-030-66242-4_29

process describes a wide class of models of queueing systems, reliability systems, and networks.

So, we consider the process $X_t = (X_t^{(1)}, X_t^{(2)}, X_t^{(3)}, \dots X_t^{(n)})$, where the processes $X_t^{(i)}$ are one-dimensional Markov chains in continuous time. For simplicity, we suppose that all these Markov chains in continuous time have only one state of embedded Markov chain (i.e. some renewal processes). In other words, the processes are flows of events (not Poisson!).

In general, this multidimensional process is non-regenerative.

It is well known that in a queuing theory and related disciplines it is very important to establish the ergodicity of the systems, as well as their stationary distribution. Also it is important to know what is the rate of convergence of the distribution of the system to the limit distribution. This is important because when we use some systems in practice, we need to know when the stationary distribution can be used in calculations and for estimations instead of the time-varying distribution of the system.

Our aim is to prove the ergodicity of the process X_t under some assumptions about the distributions of random times between the events of all processes $X_t^{(i)}$.

2 Some Preliminary Considerations

Definition 1 (Regenerative Process). *A random process is called regenerative if there exists an increasing sequence* $\{t_i\}_{i=0,1,2,\dots}$*, such, that the random elements* $\Theta_i \stackrel{\text{def}}{=} \{X_t,\, t \in [t_{i-1}; t_i]\}$ *are i.i.d.* $\forall\, i = 1, 2, \dots$.

Times t_i *are named regeneration times.*

Denote $\tau_i \stackrel{\text{def}}{=} t_{i+1} - t_i$*, and let* $\mathcal{P}_t$ *be a distribution of regenerative process at the time* t. ▷

It is well-known that:

1. If $\mathbf{E}\,\tau_i < C < \infty$, then the regenerative process is ergodic, i.e. there exists the probability distribution $\mathcal{P}$, such that $\mathcal{P}_t \Longrightarrow \mathcal{P}$.
2. If $\mathbf{E}\,(\tau_i)^k < \infty$, then $\exists\, K(\mathcal{P}_0)$:

$$\|\mathcal{P}_t - \mathcal{P}\|_{TV} \le \frac{K((k-1), \mathcal{P}_0)}{t^{k-1}}. \tag{1}$$

3. If $\mathbf{E}\,\exp(\alpha\tau_i) < \infty$, then $\forall\, \beta < \alpha\ \exists\, K(\mathcal{P}_0, \beta)$:

$$\|\mathcal{P}_t - \mathcal{P}\|_{TV} \le K(\mathcal{P}_0, \beta) \exp(-\beta t).$$

These results are the classic results, but they make it impossible to estimate the value K – see, e.g., [1,2,7,19].

The convergence rate can be estimated explicitly in a fairly limited number of cases.

These are situations when, for example, in the QS the incoming flow is Poisson, and the service times are distributed exponentially. The same approach is applied to the study of reliability for systems consisting of restorable elements: the rate of convergence of the distribution of such a reliability system can be estimated when the work and repair times are exponential (see, e.g., [6]). It is always assumed that all switching (transitions between operating and repair modes) occur instantly.

The goal of the present paper is to estimate the convergence rate of the QS distribution, consisting of *dependent* servers, and the connection of which may be delayed.

Previously, such systems were studied using the special Lyapunov function – see, e.g., [16,17].

Note, that the regenerative process have an embedded renewal process, and the bounds of the convergence rate of this renewal process in total variation metrics are also the bounds of the convergence rate of studied regenerative process. This fact is the basis for obtaining strong upper bounds for the convergence rate of regenerative processes. Also, this general method for obtaining an upper bounds for the constant K was invented in [14]. However, special methods are desirable for specific QS.

All known upper bounds for the constants for the convergence rate are based on the coupling method, invented in [4].

2.1 Some Information About the Coupling Method

Consider two *independent Markov processes* with different initial states X_0 and $\widehat{X}_0$ and with the same transition function. Denote these processes by X_t and $\widehat{X}_t$ correspondingly. Suppose we can find the time τ where they are coincided. The time τ is called *coupling epoch* and it depends on X_0 and $\widehat{X}_0$. After the time $\tau(X_0,\widehat{X}_0)$, the distributions of the processes X_0 and $\widehat{X}_0$ are coincided by Markov property. Thus, for all $t \geq \tau(X_0,\widehat{X}_0)$, and for all set $\mathcal{A} \in \sigma(\mathcal{X})$, $\mathbf{P}\{X_t \in \mathcal{A}\} = \mathbf{P}\{\widehat{X}_t \in \mathcal{A}\}$. It implies the basic coupling inequality:

$$|\mathbf{P}\{X_t \in \mathcal{A}\} - \mathbf{P}\{\widehat{X}_t \in \mathcal{A}\}| = |\mathbf{P}\{X_t \in \mathcal{A} \,\&\, \tau > t\} - \mathbf{P}\{\widehat{X}_t \in \mathcal{A} \,\&\, \tau > t\}$$

$$+ |\mathbf{P}\{X_t \in \mathcal{A} \,\&\, \tau \leq t\} - \mathbf{P}\{\widehat{X}_t \in \mathcal{A} \,\&\, \tau \leq t\}|$$

$$= |\mathbf{P}\{X_t \in \mathcal{A} \,\&\, \tau > t\} - \mathbf{P}\{\widehat{X}_t \in \mathcal{A} \,\&\, \tau > t\}| \leq \mathbf{P}\{\tau > t\}.$$

Then, if it is possible to find the increasing positive function $\varphi(\tau)$ such that $\mathbf{E}\,\varphi(\tau(X_0,\widehat{X}_0)) < \infty$, then by Markov inequality,

$$\mathbf{P}\{\tau(X_0,\widehat{X}_0) \geq t\} = \mathbf{P}\{\varphi(\tau(X_0,\widehat{X}_0)) \geq \varphi(t)\} \leq \frac{\mathbf{E}\,\varphi(\tau(X_0,\widehat{X}_0))}{\varphi(t)}.$$

From the last inequality the bounds for convergence of the distribution $\mathcal{P}_t$ can be obtained. Let's explain it.

If the process $\widehat{X}$ starts from the stationary distribution $\mathcal{P}$ of X_t, i.e. at any time, the distribution of $\widehat{X}_t$ is the same as the one of $\widehat{X}_0$, then

$$|\mathbf{P}\{X_t \in \mathcal{A}\} - \mathcal{P}(\mathcal{A})\}| \leq \frac{\int\limits_{\mathcal{X}} \varphi(\tau(X_0, \widehat{X}_0))\mathcal{P}(\mathrm{d}\widehat{X}_0)}{\varphi(t)},$$

where $\mathcal{X}$ is the state space of X_t

This schema can be used for discrete Markov chain and for Markov chain in continuous time. But for the process X_t in continuous time, the "direct" coupling method is impossible, because for different values $X_0 \neq \widehat{X}_0$, $\mathbf{P}\{\tau(X_0, \widehat{X}_0) < \infty\} = 0$. Thus, the modification of coupling method, or *successful coupling* will be used.

2.2 Successful Coupling (See [8]).

Let X_t and $\widehat{X}_t$ be two independent Markov processes with the same transition function, but with different initial states at time $t = 0$.

Suppose that (*dependent*) processes Y_t and $\widehat{Y}_t$ are constructed on some probability space, in such a way that:

1. $Y_t \stackrel{\mathcal{D}}{=} X_t$ and $\widehat{Y}_t \stackrel{\mathcal{D}}{=} \widehat{X}_t$ for all *non-random* t;

2. $\mathbf{P}\{\tau(X_0, \widehat{X}_0) < \infty\} = 1$, where $\tau(X_0, \widehat{X}_0) = \tau(Y_0, \widehat{Y}_0) = \inf\{t > 0 : \ Y_t = \widehat{Y}_t\}$.

This pair of processes Y_t and $\widehat{Y}_t$ is called *successful coupling* for the processes X_t and $\widehat{Y}_t$, and $\tau(X_0, \widehat{X}_0)$ is called *coupling epoch.*

For successful coupling, the basic coupling inequality can be applied as:

$$\begin{aligned}
|\mathbf{P}\{X_t \in \mathcal{A}\} - \mathbf{P}\{\widehat{X}_t \in \mathcal{A}\}| &= |\mathbf{P}\{Y_t \in \mathcal{A}\} - \mathbf{P}\{\widehat{Y}_t \in \mathcal{A}\}| \\
&= |\mathbf{P}\{Y_t \in \mathcal{A} \,\&\, \tau > t\} - \mathbf{P}\{\widehat{Y}_t \in \mathcal{A} \,\&\, \tau > t\} \\
&\quad + |\mathbf{P}\{Y_t \in \mathcal{A} \,\&\, \tau \leq t\} - \mathbf{P}\{\widehat{Y}_t \in \mathcal{A} \,\&\, \tau \leq t\}| \qquad (2) \\
&= |\mathbf{P}\{Y_t \in \mathcal{A} \,\&\, \tau > t\} - \mathbf{P}\{\widehat{Y}_t \in \mathcal{A} \,\&\, \tau > t\}| \leq \mathbf{P}\{\tau > t\}
\end{aligned}$$

for any set $\mathcal{A} \in \sigma(\mathcal{X})$. Here, identical distribution of pairs $Y_t \stackrel{\mathcal{D}}{=} X_t$ and $\widehat{Y}_t \stackrel{\mathcal{D}}{=} \widehat{X}_t$ means that only marginal distributions coincide in any time, but not finite-dimensional distributions of these processes.

Now, our goal is a construction of the successful coupling and an estimation of polynomial moments of a random variable $\tau(X_0, \widehat{X}_0)$. For this construction the Basic Coupling Lemma is needed.

2.3 Basic Coupling Lemma (see, e.g., [13,18,20])

Here the simplest formulation of the Basic Coupling Lemma is given.

Lemma 1. *If the random variable* ϑ_1 *and* ϑ_2 *have c.d.f.* $\Phi_1(s)$ *and* $\Phi_2(s)$ *correspondingly, and their common part* $\varkappa \stackrel{\text{def}}{=} \int\limits_{\mathbf{R}} \min\{\Phi_1'(s), \Phi_2'(s)\}\,\mathrm{d}\,s > 0$, *then the random variables* $\widehat{\vartheta}_1$ *and* $\widehat{\vartheta}_2$ *can be constructed (on some probability space) such, that*

1. $\widehat{\vartheta}_1 \stackrel{\mathcal{D}}{=} \vartheta_1$, $\widehat{\vartheta}_2 \stackrel{\mathcal{D}}{=} \vartheta_2$; *2.* $\mathbf{P}\{\widehat{\vartheta}_1 = \widehat{\vartheta}_2\} = \varkappa$. ▷

The statement of Lemma 1 has been extended to any finite number of random variables.

Lemma 2. *Let* ϑ_1, ϑ_2, …, ϑ_n *be the random variable with probability densities* $\varphi_1(s)$, $\varphi_2(s)$, …, $\varphi_n(s)$ *correspondingly, and* $\varkappa \stackrel{\text{def}}{=} \int\limits_{\mathbf{R}} \min\limits_{i=1,\dots,n} \{\varphi_i(s)\}\,\mathrm{d}\,s > 0$. *Then on some probabilistic space it is possible to construct the random variables* $\widehat{\vartheta}_1(s)$, $\widehat{\vartheta}_2(s)$, …, $\widehat{\vartheta}_n(s)$ *such that*

1. $\widehat{\vartheta}_i \stackrel{\mathcal{D}}{=} \vartheta_i$, $i = 1, 2, \dots n$; 2. $\mathbf{P}\{\widehat{\vartheta}_1 = \widehat{\vartheta}_2 = \ldots = \widehat{\vartheta}_n\} = \varkappa$. ▷

Proof. Let $\varkappa < 1$ (the proof for the case $\varkappa = 1$ is very simple). Consider a probability space $(\Omega, \sigma(\Omega), \mathbf{P})$, where $\Omega = [0;1)^{n+1} = [0;1)_1 \times [0;1)_2 \times \cdots \times [0;1)_{n+1}$, $\sigma(\Omega)$ is its Borel σ-algebra, and $\mathbf{P}$ is Lebesgue measure on Ω.

Let $\mathcal{U}_i$ be the random variable with continuous uniform distribution on $[0;1)_i$, $i = 1, \dots, (n+1)$.

Let $\varphi(s) \stackrel{\text{def}}{=} \min\limits_{i=1,\dots,n} \varphi_i(s)$, and

$$\Psi(s) \stackrel{\text{def}}{=} \frac{1}{\varkappa} \int\limits_{-\infty}^{s} \varphi(u)\,\mathrm{d}\,u, \qquad \Psi_i(s) \stackrel{\text{def}}{=} \frac{1}{1-\varkappa} \int\limits_{-\infty}^{s} (\varphi_i(u) - \varphi(u))\,\mathrm{d}\,u.$$

Ψ and Ψ_i are the distribution functions.

Put $\widehat{\vartheta}_i \stackrel{\text{def}}{=} \Psi_i^{-1}(\mathcal{U}_i) \times \mathbf{1}(\mathcal{U}_{n+1} > \varkappa) + \Psi^{-1}(\mathcal{U}_i) \times \mathbf{1}(\mathcal{U}_{n+1} \le \varkappa)$, $i = 1, \dots, n$. It is easy to see that the random variables $\widehat{\vartheta}_i$ satisfy the conditions 1 and 2 of Lemma 2. ▷

The Lorden's inequality has been used in [13–15] for coupling method application to find upper bounds of convergence rate for regenerative processes.

Now, we need to use some *generalization* of the Lorden's inequality.

3 Lorden's Inequality

Consider the renewal process $N_t \stackrel{\text{def}}{=} \sum\limits_{i=1}^{\infty} \mathbf{1}\left\{\sum\limits_{k=1}^{i} \xi_k \le t\right\}$, where $\{\xi_1, \xi_2, \dots\}$ is the of i.i.d. positive random variables. N_t is a counting process that changes its value at the times $t_k = S_k \stackrel{\text{def}}{=} \sum\limits_{j=1}^{k} \xi_j$. The times t_k are the renewal times.

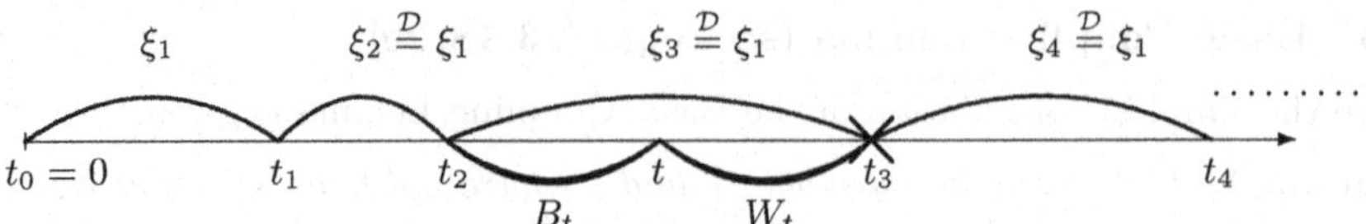

Fig. 1. B_t is a backward renewal time, and W_t is a forward renewal time at the fixed time t; $B_t = t - \sum\limits_{1}^{N_t} \xi_i$.

In Fig. 1, we see the *backward renewal time (or overshoot)* B_t, and the *forward renewal time (or undershot)* W_t at the **fixed** time t (See Fig. 1):

$$B_t \overset{\text{def}}{=} t - S_{N_t}; \qquad W_t \overset{\text{def}}{=} S_{N_t+1} - t.$$

Theorem 1 (Lorden, G. (1970) *[10]*; **(see, e.g.** *[3]*). *Lorden's inequality states that the expectation of this overshoot is bounded as*

$$\mathbf{E}\, B_t \leq \frac{\mathbf{E}\, \xi_1^2}{\mathbf{E}\, \xi_1}.$$ ▷

This inequality is the tool for upper bounds for convergence rate in total variation metrics.

However, we need to use some generalization of the Lorden's inequality.

3.1 Generalized Lorden's Inequality

Definition 2 (Quasi-renewal process). *Consider the counting process* $N_t \overset{\text{def}}{=} \sum\limits_{i=1}^{\infty} \mathbf{1}\left\{ \sum\limits_{k=1}^{i} \xi_k \leq t \right\}$, *where* $\{\xi_1, \xi_2, \ldots\}$ *are positive random variables is named quasi regenerative if the r.v.'s* ξ_i *are not i.i.d.*

As for a classic renewal process, N_t *is a counting process that changes its value at the times* $t_k = S_k \overset{\text{def}}{=} \sum\limits_{j=1}^{k} \xi_j$. *The times* t_k *are the renewal times.* ▷

The generalized Lorden's inequality estimates the expectation of the backward renewal time of quasi-renewal process – in some condition, naturally.

3.2 Some Preliminary Considerations About Generalized Intensities

The random variables can be defined by distribution function, by its density, and by its *intensity*. Obviously, the intensity is a means for study the absolutely continuous distributions, and for d.f. $F(x)$ the intensity is equal $\lambda(x) = \dfrac{F'(x)}{1 - F(x)}$.

This formula is correct:

$$F(x) = 1 - \exp\left(\int_0^x -\lambda(s)\,\mathrm{d}\,s\right). \tag{3}$$

But we are interested in mixed random variables, i.e. their distribution functions can have jumps.

If $F(a-0) \neq F(a+0)$, the **we put** $\lambda(a) \stackrel{\text{def}}{=} -\ln\big(F(a+0) - F(a-0)\big)\delta(0)$, where $\delta(\cdot)$ is a "classic" δ-function. So, **we put**

$$f(s) = \begin{cases} F'(s), \text{ if } \exists\, F'(s); \\ 0, \qquad \text{otherwise.} \end{cases}$$

Finally, (generalized) intensity is

$$\lambda(s) \stackrel{\text{def}}{=} \frac{f(s)}{1-F(s)} - \sum_i \delta(s-a_i)\ln\big(F(a_i+0) - F(a_i-0)\big),$$

where $\{a_i\}$ is a set of the discontinuous points of d.f. $F(s)$.

It is easy to see, that the formula (3) remains true for generalized intensity.

Let $Int_\xi(s)$ be an intensity of r.v. ξ.

We skip very easy proof of the next Lemma 3.

Lemma 3. $Int_{\min\{\xi;\eta\}} = Int_\xi + Int_\eta$. ▷

3.3 Notations and Assumptions

Consider the sequence $\{\xi_1, \xi_2, \ldots\}$ of random variables.

Assumptions

1. $\xi_j = \min\{\zeta_j; \eta_j\}$, *were* $\{\zeta_j\}$ *are i.i.d. r.v.'s defined by (generalized) intensities* $\varphi_i(s) \equiv \varphi(s)$, *and* $\zeta_i \perp\!\!\!\perp \eta_j$ *for all* i, j; η_j *are defined by (generalized) intensities* $\mu_j(s)$;
2. *There exists some (generalized) measurable function* $Q(s)$ *such that for all* $s \geq 0$, $\varphi(s) + \mu_j(s) = \lambda_i(s) \leq Q(s)$;
3. $\int_0^\infty \varphi(s)\,\mathrm{d}\,s = \infty$, *and* $\int_0^\infty \left(x^{k-1}\exp\left(-\int_0^x \varphi(s)\,\mathrm{d}\,s\right)\right)\mathrm{d}\,x < \infty$ *for some* $k \geq 2$;
4. $Q(s)$ *is bounded in some neighborhood of zero;*
5. $\varphi(s) > 0$ *a.s. for* $s > T \geq 0$. ▷

Definition 3. *If conditions 1–4 are satisfied, then the counting process*

$$N_t \stackrel{\text{def}}{=} \sum_{i=1}^\infty \mathbf{1}\left\{\sum_{k=1}^i \xi_k \leq t\right\} \tag{4}$$

is named generalized renewal process. ▷

For this quasi-renewal process the r.v.'s $B_t \stackrel{\text{def}}{=} t - S_{N_t}$ and $W_t \stackrel{\text{def}}{=} S_{N_t+1} - t$ are backward renewal time and forward renewal time accordingly.

Remark 1. The r.v. W_t also can be represented as $W_t = \min\{\zeta(B_t), \eta(B_t)\}$, where $\zeta(B_t)$ has the distribution function $\mathbf{P}\{\zeta(B_t) \le s\} = \dfrac{\mathbf{P}\{\zeta_{N_t+1} - B_t \le s\}}{1 - \mathbf{P}\{\zeta_{N_t+1} > B_t\}}$, and $\eta(B_t)$ has the distribution function $\mathbf{P}\{\eta(B_t) \le s\} = \dfrac{\mathbf{P}\{\eta_{N_t+1} - B_t \le s\}}{1 - \mathbf{P}\{\eta_{N_t+1} > B_t\}}$. ▷

Remark 2. The Assumption 1 holds: the r.v.'s ξ_i and ξ_j are dependent, and this dependency is "weak" dependency in some sens.

Also the intensity of renewal time of considered quasi-renewal process is a sum of two independent processes: one of them is a classic renewal process (with renewal times ζ_i) and quasi-renewal process with (dependent) renewal times η_i.

Such a process describes some process that has some minimal intensity and additional increase in intensity due to the influence of some external factors. ▷

Remark 3. The Assumptions 3 and 4 hold: $\mathbf{E}\,\xi_i > 0, \qquad \operatorname{Var}\xi_i^2 > 0.$ ▷

Remark 4. The Assumptions 1 and 2 hold:

$$F_i(t) = \mathbf{P}\{\xi_i \le t\} = 1 - \exp\left(\int_0^t -\lambda_i(s)\,\mathrm{d}\,s\right) \ge 1 - \exp\left(\int_0^t -Q(s)\,\mathrm{d}\,s\right)$$

$\Rightarrow \ \exists\, \mathbf{E}\,\xi_i^2 < \infty.$ ▷

Remark 5. The Assumption 5 reports that the renewal process under study is a delay renewal process, and a delay time does not exceed T. ▷

Theorem 2 (Generalized Lorden's inequality – see [9]). *If the conditions 1–4 are satisfied, then for the process (4) the inequality*

$$\mathbf{E}\,B_t \le \mathbf{E}\,\zeta + \frac{\mathbf{E}\,\zeta^2}{2\mathbf{E}\,\xi} = \Xi(= \Xi(1)), \tag{5}$$

is true, where $\mathbf{E}\,\zeta = \int\limits_0^\infty x\,\mathrm{d}\,\Phi(x)$; $\mathbf{E}\,\xi = \int\limits_0^\infty x\,\mathrm{d}\,G(x)$, *and*

$G(x) = 1 - \exp\left(-\int\limits_0^x Q(t)\,\mathrm{d}\,t\right)$, *and* $\Phi(x) = 1 - \exp\left(-\int\limits_0^x \varphi(t)\,\mathrm{d}\,t\right)$. ▷

Remark 6. In the proof of this Theorem, there is an intermediate result:

If $\mathbf{E}\,\zeta^k < \infty$, then for $N \in (0; k-1]$,

$$\mathbf{E}\,(B_t)^N \le = \mathbf{E}\,\zeta^N + \frac{\mathbf{E}\,\zeta^{N+1}}{(N+1)\mathbf{E}\,\xi} \stackrel{\text{def}}{=} \Xi(N).$$ ▷

4 Considered Generalized Markov-Modulated Poisson Process

Consider the process $X_t \stackrel{\text{def}}{=} (X_t^{(1)}, X_t^{(2)}, X_t^{(3)}, \dots X_t^{(m)})$, where the processes $X_t^{(i)}$, $i = 1, 2, \dots m$ are generalized renewal processes $X_t^{(i)}$, $i = 1, 2, \dots m$.

For these processes, the backward renewal times $B_t^{(i)}$ and forward renewal times $W_t^{(i)}$ are for processes $X_t^{(i)}$ accordingly defined in the Remark 1.

At the time t, the state of the process $X_t^{(i)}$ is $x_t^{(i)} = t - \sum_{j=1}^{N_t^{(i)}} \xi_i^{(j)} (= B_t^{(i)}$ - elapsed renewal time or backward renewal time of process $X_t^{(i)}$) – see Fig. 1.

The intensities of the processes $X_t^{(i)}$, $i = 1, 2, \dots m$ are $\varphi(B_t^{(i)}) + \mu_i(B_t^{(i)})$.

Renewal times of quasi-renewal process $X_t^{(i)}$ are $\xi_i^{(j)} = \min\{\zeta_i^{(j)}, \eta_i^{(j)}\}$, where $\zeta_i^{(j)}$ has intensity $\varphi(x_t^{(i)})$, and $\eta_i^{(j)}$ has intensity $\mu_i(X_t)$.

So, the "full" intensity of $X_t^{(i)}$ (which corresponds to $\zeta_i^{(j)}$ in Assumption 1) is $\lambda_i(X_t) = \varphi(x_t^{(i)}) + \mu_i(X_t)$. "Additional" intensity (which corresponds to $\eta_i^{(j)}$ in Assumption 1) is $\mu_i(t) = \mu(x_t^{(1)}, x_t^{(2)}, x_t^{(3)}, \dots x_t^{(m)})$.

For simplicity, here we replace inequalities $\lambda_i(x) \geq \varphi_i(x)$ and $\lambda_i(x) + \mu_i(x) \leq Q_i(x)$ by homogeneous inequalities $\lambda_i(x) \geq \varphi(x)$ and $\lambda_i(x) + \mu_i(x) \leq Q(x)$.

The processes $X_t^{(i)}$ are dependent, and the intensity of $X_t^{(i)}$ depends on the state of the processes $X_t^{(j)}$, $j \neq i$.

The process $X_t \stackrel{\text{def}}{=} (X_t^{(1)}, X_t^{(2)}, X_t^{(3)}, \dots X_t^{(m)})$ is not regenerative in general.

It can be considered as a kind of variation on the topic of the Markov modulated Poisson process (see, e.g., [5,11])

Emphasize, that for simplicity, here we suppose, that the functions $\varphi(s)$ and $Q(s)$ are the same for all processes.

Additional Assumption

6. For all $t > 0$, $\int_0^s Q(s)\,\mathrm{d}\,s < \infty$, and there exists non-negative function $q(s)$ such that $\mu_i(s) = \lambda_i(s) - \varphi(s) \geq q(s)$.

Remark 7. From Assumptions 1 and 6 we have the formulae for distributions of ζ and η.

$$\mathbf{P}\{\zeta_i \leq s\} = \Phi(s) > 0 \text{ for } s > T,$$

and

$$\mathbf{P}\{\eta_i \leq s\} = \widetilde{G}(s) \stackrel{\text{def}}{=} 1 - \exp\left(-\int_0^s (Q(t) - \varphi(t))\,\mathrm{d}\,t\right);$$

$$\mathbf{P}\{\eta_i > s\} = 1 - \widetilde{G}(s) \geq \exp\left(-\int_0^s Q(t)\,\mathrm{d}\,t\right) = 1 - G(s).$$

Note, that $\mathbf{P}\{\zeta < \infty\} = 1$, and $\mathbf{P}\{\eta < \infty\} \leq 1$.

Assumptions 1–6 hold: the probability densities of the r.v.'s $\xi_j^{(i)}$ can be bounded as:

$$F_j(x) = \mathbf{P}\{\xi_j^{(i)} \leq x\} = 1 - \exp\left(-\int\limits_0^x (\lambda_j(s) + \mu_j(s))\,\mathrm{d}\,s\right) \leq \\ 1 - \exp\left(-\int\limits_0^x \varphi(s)\,\mathrm{d}\,s\right); \qquad (6)$$

$$f_j(x) = \frac{\mathrm{d}\,F_j(x)}{\mathrm{d}\,x} \geq \exp\left(-\int\limits_0^x Q(s)\,\mathrm{d}\,s\right)\varphi(s) > 0 \text{ a.s. for } t > T.$$

For forward renewal times $W_t^{(i)}(a)$ of the processes $X_t^{(i)}$ given $B_t^{(i)} > a$ have the probability densities:

$$f_t^{(i)}(x,a) \geq \varphi(a+x)\exp\left(-\int\limits_a^{x+a} Q(s)\,\mathrm{d}\,s\right) \stackrel{\text{def}}{=\!=} f(a,x) > 0 \text{ a.s. for } t > T, \quad (7)$$

see the Assumption 5. ▷

$$\mathbf{P}\{\zeta_t^{(i)} > \eta_t^{(i)} | B_t^{(i)} < \Theta\} \geq A(\Theta).$$

4.1 Ergodicity of the Multi-dimensional Process X_t

Here we give *only* the schema for the proof of the ergodicity of the multi-dimensional process X_t.

Suppose, $X_0 = (0, 0, 0, \ldots, 0)$.

Put $\theta_0 = 0$. This is markov moment of X_t.

At these times, the backward renewal time $B_{\theta_1}^{(i)}$ satisfies the inequality (5). Put constant $\Theta > \Xi$. By Markov inequality,

$$\mathbf{P}\{B_{\theta_1}^{(i)} \geq \Theta\} \leq \frac{\Xi}{\Theta}, \text{ and } \mathbf{P}\{B_{\theta_1}^{(i)} < \Theta\} \geq 1 - \frac{\Xi}{\Theta} = \pi_0(\Theta).$$

These bounds are uniform for all the processes $X_t^{(i)}$, they don't depend on the number i of the process $X_t^{(i)}$.

So, at the time θ_ℓ for all the processes $X_t^{(i)}$, $B_{\theta_1}^{(i)} < \Theta$, with probability greater than $p_0 \stackrel{\text{def}}{=\!=} (\pi_0(\Theta))^m$.

At this time, the forward renewal times $W_{\theta_1}^{(i)}$ of the processes $X_t^{(i)}$ are the residual times of $\zeta^{(i)}$ given the elapsed time is less then Θ for all $\zeta^{(i)}$.

Now, we can use the generalization of the Basic Coupling Lemma, and we can create (in some probability space) the prolongation of the processes $X_t^{(i)}$ by such a way, that the next renewal times of both these processes coincide with probability greater then

$$\pi_1 \stackrel{\text{def}}{=} \inf_{a_j \in [0;\Theta], j=1,2,\dots,m} \int_0^\infty \min\{f(x,a_j)\}\,\mathrm{d}\,x > 0 \quad \text{– see (7).}$$

If at the time $\theta_1 = \theta_0 + \max\{\xi_1^{(j)}\} = \theta_0 + \max\{W_{\theta_0}^{(j)}\}$ all the processes $X_{\theta_0}^{(i)} = 0$, we say that θ_1 is the first regeneration point of the constructed version for X_t.

In this case $X_{\theta_1} = X_{\theta_0}$, and we repeat the procedure descripted above. Otherwise consider $W_{\theta_1}^{(i)}$. Again with probability greater then π_1 at the time $\theta_2 \stackrel{\text{def}}{=} \theta_1 + \max_i\{W_{\theta_1}^{(i)}\}$ constructed versions of the processes $X_t^{(i)}$ coincide.

Note that $\mathbf{E}\,(\theta_2 - \theta_1) \le \Xi$ (Theorem 1).

Anew we use the generalization of the Basic Coupling Lemma, and we can create (in some probability space) the prolongation of the processes $X_t^{(i)}$ by such a way, that the next renewal times of both these processes coincide with probability greater then π_1, and so on.

This operation is repeated at every times $\theta_\ell \stackrel{\text{def}}{=} \theta_{\ell-1} + \max_i\{W_{\theta_{\ell-1}}^{(i)}\}$.

By this way, on some probability space we construct the new process $\widetilde{X}_t \stackrel{\text{def}}{=} (\widetilde{X}_t^{(1)}, \widetilde{X}_t^{(2)}, \widetilde{X}_t^{(3)}, \dots, \widetilde{X}_t^{(m)})$, such that the marginal distributions of $\widetilde{X}_t$ and X_t are equal, and after any time θ_ℓ the process $\widetilde{X}_t$ hits to the state $(0,0,0,\dots,0)$ with probability greater then $p_0\pi_1$.

The time between the hits to this zero-state are distributed as geometrical sum of constants M and the finishing residual time of the last period. So, it has the finite expectation, and the process $\widetilde{X}_t$ is regenerative. Therefore, it has a limit probability distribution.

In this situation, it is natural to call the process X_t as "quasi-regenerative process".

In fact, here we used the scheme of successful coupling, but in a slightly different way.

Remark 8. The limit distribution of the process $\widetilde{X}_t$ can be estimated by the same schema as for a "classic" renewal process (see [12]):

$$\mathbf{P}\{\widetilde{X}_t^{(i)} > s\} \le \Psi(s) \stackrel{\text{def}}{=} \frac{\int_0^s (1-\Phi(u))\,\mathrm{d}\,u}{\int_0^\infty (1-G(u))\,\mathrm{d}\,u}. \qquad \triangleright$$

4.2 About Polynomial Convergence of the Process X_t

Now, consider two independent version of multi-dimensional process X_t: X_t and $\widehat{X}_t = (\widehat{X}_t^{(1)}, \widehat{X}_t^{(2)}, \widehat{X}_t^{(3)}, \dots, \widehat{X}_t^{(m)})$.

For simplicity, $x_0 = 0$, $y_0 = 0$. The second process has an arbitrary initial state: $\widehat{x}_0^{(i)} = a_i$.

So, the residual times $\widehat{\xi}_1^{(i)}(a_i)$ of the processes $\widehat{X}_t^{(i)}$ have the densities greater then $\varphi(a_i + x)\exp\left(-\int\limits_{a_i}^{x+a_i} Q(s)\,\mathrm{d}\,s\right)$ (see (7)).

So, if there exists finite $\int\limits_0^\infty x^k \exp\left(-\int\limits_0^x Q(s)\,\mathrm{d}\,s\right)\mathrm{d}\,x < \infty$, then there exist finite expectations of $\mathbf{E}\,(\widehat{\xi}_1^{(i)}(a_i))^{k-1}$.

After the time $\theta_0 = S_0(a_1, a_2, \ldots, a_m) \stackrel{\text{def}}{=} \max\limits_{i=1,2,3,\ldots,m}\{\widehat{\xi}_1^{(i)}(a_i)\} \leq \sum\limits_{i=1}^m \widehat{\xi}_{N_t+1}^{(i)}(a_i)$, we can use the generalization of Lorden's inequality.

Denote $\theta_\ell \stackrel{\text{def}}{=} \theta_{\ell-1} + \max\limits_{i=1,2,3,\ldots,m}\{W_{\theta_{\ell-1}}^{(i)}, \widehat{W}_{\theta_{\ell-1}}^{(i)}\}$.

At the time θ_0, consider probability

$$\Pi \stackrel{\text{def}}{=} \mathbf{P}\{B_{\theta_\ell}^{(i)} < \Theta,\ \widehat{B}_{\theta_\ell}^{(i)} < \Theta,\ i = 1, 2, 3, \ldots, m\}\ -$$

here the letters with caps refer to the process $\widehat{X}_t$.

Similarly to the calculations in the Subsect. 4.1 we have $\Pi \geq (p_0)^2$.

Now the generalized Basic Coupling Lemma gives the probability of coincidence of the prolonged versions of both processes after any time θ_ℓ – this probability $\geq \pi_1$.

Thus, after any time θ_0 the construction of successful coupling leads to the moment of coincidence of both processes in zero-state $(0, 0, 0, \ldots, 0)$ with probability greater then $(p_0)^2\pi_1$ – this is a *coupling epoch* for the processes X_t and $\widehat{X}_t$. Otherwise, we use the generalized Basic Coupling Lemma at the time θ_1, and so on.

So, after any time θ_ℓ with probability $\pi \geq (p_0)^2\pi_1$ (see Lemma 2) we can continue the versions of all processes X_t and $\widehat{X}_t$ hit to the zero-states $(0, 0, 0, \ldots 0)$ contemporaneously is a coupling epoch of created on some probability space versions of X_t and $\widehat{X}_t$ – see Subsect. 2.2.

Let $\mathcal{E}_\ell$ be the event {*coupling epoch happened right* after *time* θ_ℓ}.

So, the coupling epoch

$$\tau(a_1, a_2, \ldots, a_m) \leq S_0 + \sum_{j=1}^{\nu+1}(\max_i\{W_{\theta_j}^{(i)}\}) + \{\xi_{N_{\theta_\ell}}^{(1)}|\mathcal{E}_\ell\},$$

with probability $\mathbf{P}\,(\mathcal{E}_\ell)$, and $\mathbf{P}\{\nu - 1 > n\} \geq (1-\pi)^n$; denote $\widetilde{W}_j \stackrel{\text{def}}{=} \max\limits_i\{W_{\theta_j}^{(i)}\}$. $\mathbf{E}\,(\widetilde{W}_j)^N \leq \Xi(N)$ for $N \leq k-1$ – see Remark 6.

(Note that $\mathbf{E}\{\varsigma|B\} \times \mathbf{P}\{B\} \leq \mathbf{E}\,\varsigma$ for all r.v. ς and event B.)

Thus, using the Jensen's inequality in the form

$$\left(\sum_{i=1}^{j} r_i\right)^N = j^N\left(\sum_{i=1}^{j} r_i/j\right)^N \le m^{N-1}\sum_{i=1}^{j}(r_i)^N, \quad j>1,\ j\in\mathbf{N},$$

so, for coupling epoch we have for $N \le k-1$:

$$(\tau(a_1,a_2,\dots,a_m))^N \le (m+2)^{N-1} \times \left(\sum_{r=1}^{m}(\widehat{\xi}_i^{(r)}(a_i))^N + \left(\sum_{r=1}^{\nu+1}\widetilde{W}_r\right)^N + (\{\xi^{(1)}_{N_{\Theta_\ell}}|\mathcal{E}_\ell\})^N\right)$$

with probability $\mathbf{P}\{\mathcal{E}_\ell\}$, and the last term is a conditional r.v., provided that the coupling epoch is on the end of this r.v.

So,

$$\mathbf{E}\,(\tau(a_1,a_2,\dots,a_m))^N \le (m+2)^{N-1} \times \left(\sum_{r=1}^{m}\mathbf{E}\,(\widehat{\xi}_i^{(r)}(a_i))^N + \mathbf{E}\,(\nu+1)^N \times \varXi(N)^N + \mathbf{E}\,(\xi^{(1)}_{N_{\Theta_\ell}})^N\right) \stackrel{\text{def}}{=} \mathrm{T}(a_1,a_2,\dots,a_m)_N,$$

where ν is geometric r.v. with parameter π; here $\mathbf{E}\,(\xi^{(1)}_{N_{\Theta_\ell}})^N \le \mathbf{E}\,(\zeta_1)^N$.

Now, integrating $\mathrm{T}(a_1,a_2,\dots,a_m)_N$ over a stationary measure bounded in (8) and considering Remark 8, we can find an upper bound $K(N)$ in formula (1) for initial zero-state of the process X_t.

Remark 9. For arbitrary initial state of the process X_t, the same schema can be used. ▷

Remark 10. The value of $K(N)$ depends on the value Θ. It can be optimized. ▷

5 Application for One Reliability System

Consider two elements reliability system. Let the intensities of the failure and of the repair of the main element depend only on the state of this element. And let the intensities of the failure and of the repair of the reserve element depend on the full state of this reliability system.

The state of the main element is (i,x), where i is an indicator of work/failure state, and x is the elapsed time in this state. So, the intensities of failure and of repair are $\lambda(i,x)$.

The state of the reserve element also is is (j,y), where j is an indicator of work/failure state, and y is the elapsed time in this state. The intensities of failure and of repair of this reserve element are $\widehat{\lambda}(j,y)$.

We suppose that these intensities satisfy the Assumptions 1–7.

In this case, the full periods (work+repaire) of the main element are i.i.d. random variables. The full period (work+repaire) of the reserve element are the periods of embedded generalized renewal process.

Thus, we can prove the ergodicity of this reliability system, and calculate the upper bounds for the convergence rate of its distribution.

If we take into account the peculiarities of the distributions of work and repair, and consider not only the times of work beginnings of the main element, these estimates can be improved.

6 Conclusion

Here we give only the schema of use of the generalized Lorden's inequality for multi-dimensional process and one application to the reliability theory. This is only beginning of development of this approach to the finding of strong bounds for various stochastic models.

Acknowledgments. The author is grateful to E. Yu. Kalimulina for the great help in preparing this paper. The work is supported by RFBR, project No. 20-01-00575A.

References

1. Afanasyeva, L.G., Tkachenko, A.V.: On the convergence rate for queueing and reliability models described by regenerative processes. J. Math. Sci. **218**(2), 119–36 (2016)
2. Asmussen, S.: Applied Probability and Queues. SMAP, vol. 51, 2nd edn. Springer, New York (2003). https://doi.org/10.1007/b97236
3. Chang, J.T.: Inequalities for the overshoot. Ann. Appl. Probab. **4**(4), 1223 (1994). https://doi.org/10.1214/aoap/1177004913
4. Doeblin, W.: Exposé de la théorie des chaînes simples constantes de Markov á un nombre fini d'états. Rev. Math. de l'Union Interbalkanique **2**, 77–105 (1938)
5. Fischer, W., Meier-Hellstern, K.: The Markov-modulated Poisson process (MMPP) cookbook. Perform. Eval. **18**(2), 149–171 (1993)
6. Gnedenko, B.V., Belyayev, Y., Solovyev, A.D.: Mathematical Methods of Reliability Theory. Academic Press, Cambridge (2014)
7. Gnedenko, B.V., Kovalenko, I.N.: Introduction to Queuing Theory. Mathematical Modeling. Birkhäeuser Boston, Boston (1989). https://doi.org/10.1007/978-1-4615-9826-8
8. Griffeath, D.: A maximal coupling for Markov chains. Zeitschrift für Wahrscheinlichkeitstheorie und Verwandte Gebiete **31**(2), 95–106 (1975). https://doi.org/10.1007/BF00539434
9. Kalimulina E., Zverkina G.: On some generalization of Lorden's inequality for renewal processes. Cornell university library, Cornell, pp. 1–5 (2019). arXiv:1910.03381v1
10. Lorden, G.: On excess over the boundary. Ann. Math. Stat. **41**(2), 520 (1970). https://doi.org/10.1214/aoms/1177697092. JSTOR 2239350
11. Rydén, T.: Parameter estimation for Markov modulated Poisson processes Communications in Statistics. Stoch. Models **10**(4), 795–829 (1994)

12. Smith, W.L.: Renewal theory and its ramifications. J. Roy. Statist. Soc. Ser. B **20**(2), 243–302 (1958)
13. Zverkina, G.: On Strong Bounds of Rate of Convergence for Regenerative Processes. In: Vishnevskiy, V.M., Samouylov, K.E., Kozyrev, D.V. (eds.) DCCN 2016. CCIS, vol. 678, pp. 381–393. Springer, Cham (2016). https://doi.org/10.1007/978-3-319-51917-3_34
14. Zverkina, G.: Lorden's inequality and coupling method for backward renewal process. In: Proceedings of XX International Conference on Distributed Computer and Communication Networks: Control, Computation, Communications, DCCN-2017, Moscow, pp. 484–491 (2017)
15. Zverkina, G.: A system with warm standby. In: Gaj, P., Sawicki, M., Kwiecień, A. (eds.) CN 2019. CCIS, vol. 1039, pp. 387–399. Springer, Cham (2019). https://doi.org/10.1007/978-3-030-21952-9_28
16. Veretennikov, A.Y., Zverkina, G.A.: Simple proof of Dynkin's formula for single-server systems and polynomial convergence rates. Markov Process. Relat. Fields **20**(3), 479–504 (2014)
17. Veretennikov, A.Y.: On polynomial recurrence for reliability system with a warm reserve. Markov Process. Relat. Fields **25**(4), 745–761 (2019)
18. Kato, K.: Coupling lemma and its application to the security analysis of quantum key distribution. Tamagawa Univ. Quant. ICT Res. Inst. Bull. **4**(1), 23–30 (2014)
19. Thorisson, H.: Coupling, Stationarity, and Regeneration. Springer, New York (2000)
20. Veretennikov, A., Butkovsky, O.A.: On asymptotics for Vaserstein coupling of Markov chains. Stoch. Process. Appl. **123**(9), 3518–3541 (2013)
21. Zverkina, G. About some extended Erlang-Sevast'yanov queueing system and its convergence rate (English and Russian versions) (2018). https://arxiv.org/abs/1805.04915. Fundamentalnaya i Prikladnaya Matematika **22**(3), 57–82

The Mathematical Model for Calculating Physical Entity of DPI Analyser

Boris Goldstein(✉) and Vadim Fitsov

The Bonch-Bruevich Saint-Petersburg State University of Telecommunications, St. Petersburg, Russia
bgold@niits.ru, noldi@iks.sut.ru

Abstract. This article describes the specialized servers that build up the DPI system architecture. Some initial data for calculating DPI system based on traffic statistics have been formalized. The develop formalization of the ratio of flows and packets. This paper describes a mathematical model for calculating the analysis time for a given number of processors in a specialized server of the DPI system. A mathematical model for calculating physical entity of DPI analyser in the DPI system, based on the model by Ventcel-Ovcharov, is provided. The concept of the Ventcel-Ovcharov model with equal mutual assistance is to combine channels into groups for the joint service of requests. The formula of the final processing time of requests of the physical entity of DPI analyser is presented. The DPI simulation model in GPSS World is briefly described. The general architecture of the DPI system is taken into account when building a simulation model. The results of the mathematical and simulation modeling are compared.

Keywords: DPI · QoS · Queuing system (QS) · Queuing network (QN) · Mathematical model

1 Introduction

Many telecom operators use deep packet inspection (DPI) systems to manage and offload their networks, analyze user interests, behavioral targeting, implement personal tariffs, protect copyrighted content, provide lawful interception (LI) according to the laws of their country, additional network protection against hacker attacks.

However, DPI requires a significant investment in hardware resources. Meanwhile the efficiency of using hardware resources in DPI systems remains understudied due to the complexity and novelty of the issue. There is a lack of necessary mathematical and simulation models for determining the parameters of a DPI architecture.

This paper describes a mathematical model for calculating the analysis time for a given number of processors in a specialized server of the DPI system. DPI mathematical and simulation models would reduce the purchase cost of DPI systems and to avoid overloads.

© Springer Nature Switzerland AG 2020
V. M. Vishnevskiy et al. (Eds.): DCCN 2020, CCIS 1337, pp. 382–393, 2020.
https://doi.org/10.1007/978-3-030-66242-4_30

2 Related Work

There is a fairly large number of works dedicated to the questions of mathematical description, analysis and classification of network traffic. In this article, when describing and formalizing the process of transferring flows for analysis, the studies [3,14,17,30] were used. The closest is the work [14], in which a mathematical model of DPI interaction with additional servers is presented and the traffic flow is taken as an example (which will be discussed below). In Russia, the mathematical description of packet traffic is developed by Stepanov S.N. [21], Samuilov K.E., Gaidamaka Y.V. [2], Levakov A.K., Sokolov N.A., Zaitsev V.S. [18], and others.

There are various mathematical models, but when applying, one should take into account the peculiarities of a packet traffic. Modern western research suggests that network traffic is similar to itself or fractal in structure (pulsating on a wide time scale). This kind of traffic is most successfully described by the Pareto and Weibull distributions or as fractal Brownian motion (FBM) [5,13]. Also, traffic often has the feature of packets arriving in batches with service requests, which is described in [21]. There are promising models for representing packet traffic as fractal Brownian motion (FBM), which are detailed in [5].

When there are several interacting queuing systems (QS), they make up a queuing network (QN). In a QN, the interest is the parameters of the output after processing in QS1, which determine the models that can be used to describe the subsequent QS (QS2). For the mathematical description of the QS as a part of the QN which receives packet traffic, the models G/M/1, G/G/1 described in [1,5,19] and others can be used. However, they restrict the model to one device. The G/G/V model should be used to overcome this limitation and to describe modern systems, but it cannot be calculated [5]. Most of the known models suggest an arriving Poisson flow of requests. For example, M/G/1, which is suitable for calculating the QS with one device. M/M/V and M/G/V with an infinite queue, which do not take into account the possibility of simultaneous processing of a request by several devices. As well as processor sharing models [12] and so-called Ventcel-Ovcharov model with an equal mutual assistance (where several devices work to serve one request) [15,16,23,24].

The processor sharing model, described by Kleinrock in 1967 [12] is widely known. PS or EPS (egalitarian processor sharing) is a service policy in which all requests are served simultaneously. Each newly arrived request receives an equal share of the bandwidth. This does not imply the presence of a queue. Significant results in the systems with fair processor sharing research were introduced in theory by Yashkov S.F. [25–28]. ESP is very close to the so-called Ventcel-Ovcharov model with full mutual assistance.

A queuing system in which several devices jointly serve one request is called QS with mutual assistance. In the works of Ventcel and Ovcharov [16,23,24], a description of a model with several service devices, a finite queue and equal mutual assistance is given.

According to Burke's theorem for QS1 (M/M/V and M/M/1), the distribution of time intervals between outgoing requests, as well as the time intervals between incoming requests, are distributed exponentially with the same parameter. Which was proved mathematically by Burke, based on the following: when QS1 becomes empty after it's done with the query, the time interval when the next request leaves the QS1 will be the sum of the time until the next request arrives at the QS1 and the service time of the next request; when there is a next request in the QS1 queue, after the previous one is finished, then the time interval when it leaves QS1 is distributed in the same way as the service time.

The mathematical description of the output flow of requests after processing in QS1 (of M/G/1 type, with Poisson input flow of requests) is described in [19]. At the same time, note that the value of the coefficient of the interval duration of requests moving between systems cannot be considered as a sufficient condition for the correspondence of the distribution function of the output flow to the exponential distribution law. But it is a necessary condition for the exponential approximation. The coefficient of variation of the output flow is close to 1 in two cases: when the primary server load is very low or when the variation coefficient of the request service time is very close to 1. It was shown for the M/M/1 model in [19] that the output flow is also a Poisson flow. In [1], it is said that for a primary server G/G/1 with an unlimited capacity storage unit operating without overloads, the intensity of the outgoing flow of requests is equal to the intensity of the incoming flow (since the mathematical expectations of the intervals between successive requests at the exit and the entrance coincide). In addition, it is said in [1] that for the M/M/1 model, the variation coefficient of the outgoing flow is equal to one.

According to the study [29], the value of the variation coefficient of the resulting flow of requests (packet traffic) on the input of QS1 of a large number of sources with similar distributions for certain cases approaches one. The subsequent verification carried out in [29] using the Kolmogorov-Smirnov criterion showed that the addition of a large number (more than 100) of flows with the Weibull distribution gives the resulting flow a manner similar to Poisson flow. In this case, one should take into account the magnitude of errors in such a simplification.

Let us consider the cases in which a forced assumption will be made about the exponential distribution of the flow exiting QS1 and entering the QS2, in order to compare the results of the QS2 calculations, which can be given by the so-called Ventcel-Ovcharov model with the results obtained in the DPI simulation model described in [4]. Proceeding from the requirement to avoid packet loss, the QN (consisting of QS1 and QS2) is designed in such a way that it can be represented as a system with an infinite queue, which is important for the cases considered below.

In the first case, QS1 receives aggregated packet traffic received from more than 100 sources, where each traffic from each source can be approximately described by the Weibull distribution. Then, according to [29], we assume that the distribution of the flow of requests entering the QS1 may be close to the Poisson distribution, but it should be borne in mind that there are some errors associated with this assumption. Then, according to [1], let us assume for one

QS1 service device that the output flow entering QS2 is also close to the Poisson distribution.

In the second case, if we assume that QS1 processes requests according to an exponential distribution, and is in a mode when the average intensity of requests is equal to the average intensity of request processing, but the system remains in a stable state and the waiting time in the queue does not become infinite. Then, according to Burke's theorem, the intensity of the output flow of requests from QS1 to QS2 will be distributed in the same way as the service time in QS1. This means that if we assume that requests in QS1 are treated exponentially, then the output flow will have an exponential distribution.

The third case is less interesting, since imposes even greater restrictions on the QS1. First, according to the conditions and assumptions of the first case, it is assumed that a flow close to Poisson arrives at the QS1. And secondly, QS1 must process requests according to an exponential distribution. Then, according to Burke's theorem, the distribution of time intervals between outgoing requests, as well as the time intervals between incoming requests, are distributed exponentially. In this case, there is no limitation per 1 device as in the first case, and there is no limitation of the mode necessarily operating in a limited but stable state, as in the second case.

Further mathematical calculation of the QS2 (DPI analyser) is carried out based on these three cases.

3 DPI

In 2012, the International Telecommunication Union (ITU) officially approved the Y.2770 [6] document. This recommendation defines the requirements for DPI objects and describes aspects such as application identification, flow identification, signature management, interaction with the decision functional entity according to network policy, and more. In subsequent years, a series of recommendations for DPI came out [7–11].

A flow (associated with a specific user application or service protocol) is usually identified using address information from layers 2–4 of the OSI model. So DPI analyzes the first packets of a traffic flow or all packets passing through it [3,17,22].

The basis of the DPI system is the Bypass server, the hardware filter (HWF: DPI scanner and DPI action execution function), DPI analyser (DPI-An or Front-End), PCRF (Policy and Charging Rules Function) and Back-End. Each of the DPI servers performs its own tasks and actively interacts with the rest.

DPI analyser - is the main element of the system, as it analyzes data from the traffic flow previously identified by the hardware filter. It uses signature analysis, statistical methods, behavioral analysis, and other approaches [20]. Having recognized the application that generated the traffic flow, DPI analyser asks the PCRF server for a decision on what to do with this traffic. Further, based on this decision, it receives more detailed instructions on filtering from the Back-End server. Then it gives the flow and instructions for execution on the hardware filter.

Back-End - is a repository of information about policy rules, collected statistics, signatures (analysis criteria), mirroring and redirection rules, AAA (Authentication, Authorization, Accounting). PCRF - is a server for real-time policy enforcement and policy management. Contains policy ID (Identifier) for each case. DPI analyser sends the ID of the subscriber or application, and PCRF returns the required policy number.

Thus, the DPI system can be represented as a QN, consisting of several QS. Where the hardware filter acts as QS1, which receives the packet data, and the output flow from it goes to QS2 (DPI analyser). Subsequent DPI servers are not taken into account in this work, due to the relatively smaller number of request sentering them.

4 The Mathematical Model

4.1 Formalization DPI Input Parameters

For the practical use of mathematical models, it is necessary to determine methods for obtaining quantitative characteristics of the operating conditions of the DPI system and formalize some features of the operation of the system itself. This article describes the formalization of some common initial parameters and the number of requests processed on the physical entity of DPI analyser (3).

To calculate, you need to know or set the intensity of incoming requests for the DPI system. Therefore, it is rational to use the statistics of the transmitted traffic on the network where you plan to install the DPI system. A peculiarity of DPI is that QS1 (hardware filter) processes all incoming packets and identifies traffic flows from them, and QS2 (DPI analyser) receives a request to analyze a specific flow. For analysis, a certain number of packets of the flow (n_f) is transmitted, about which studies have been carried out [3,22]. Only new unknown flows are analyzed, already known flows are processed according to the previously set rules and are not sent to the DPI analyser.

From the traffic statistics collected by wireshark or cisco NetFlow, you can get the number of packets and the number of flows during statistical analysis, and then the average number of packets in the flow (n_{af}) and the rate of arrival of the packets (λ_0). In this case, you can get the number and frequency of occurrence of new unknown flows, the probability that the flow was previously known (P_{kn}), the time of occurrence of the flow, the time of the end of the flow and the average duration of the flow.

The rate of arrival of unknown packets will be determined by the product of the probability of an unknown flow ($1 - P_{kn}$) and λ_0. Based on the above description of the interaction of DPI servers, only a part of the packets of the flow as a request is sent for analysis, which is set by the ratio of the number of analyzed packets (n_f) to the n_{af} (coefficient of the number of requests to the Front-End K_n formula (1)).

$$K_n = \frac{n_f}{n_{af}} \tag{1}$$

At the same time, for each analyzed flow, DPI analyser sends a response to the hardware filter, which is set by the ratio of 1 to the n_{af} (the coefficient for recalculating flows and packets (K_{fe}) formula (2)).

$$K_{fe} = (1 - P_{kn}) \times (K_n + \frac{1}{n_{af}}) \tag{2}$$

From these components, the number of requests processed at the DPI analyser is obtained, determined by the formula (3).

$$\lambda_{fe} = (1 - P_{kn}) \times (\frac{n_f + 1}{n_{af}}) \times \lambda_0 \tag{3}$$

If necessary, formula (3) can be supplemented to take into account the interaction with the PCRF and Back-End servers. In formula (4), these interactions are partially given.

$$\lambda_{fefull} = (1 - P_{kn}) \times (\frac{n_f + 1 + 2 \times P_p \times (1 + P_b)}{n_{af}}) \times \lambda_0 \tag{4}$$

The parameters defined here, as well as other parameters of traffic characteristics and DPI system features, are presented in Table 1.

4.2 Ventcel-Ovcharov Model and DPI Analyser

Previously, a forced assumption was made in the cases considered in which the exponential distribution of the flow leaving the hardware filter and entering the DPI analyser, in order to compare the results of DPI analyser calculations, which can provide the so-called Ventcel-Ovcharov model with the results obtained in the DPI simulation model.

Ventcel and Ovcharov distinguish two types of mutual assistance: full and partial (equal). This article discusses equal mutual assistance. Since the DPI system must process all requests, to simplify the calculations of the DPI analyser mathematical model, it is advisable to take a model with an infinite queue and with equal mutual assistance [15] mentioned, but not completely described, in the works of Ventcel.

The concept of the model with equal mutual assistance is to combine channels into groups for the joint service of requests. In this case, a system with equal mutual assistance will have 3 modes of operation, shown in Fig. 1 below: I - the number of requests is less than the maximum number of groups (like a classical QS), II - the number of requests is greater than the maximum number of groups, but less than the number of channels (transient mode), III - the number of requests is greater than the number of channels (like a classical QS). One of the advantages of the considered mathematical model is the use of all possible resources of the system before the number of requests equals the number of channels.

Mode I of operation implies the formation of channel groups. In this mode, the system operates as a classical QS, in which a group of channels is taken as a

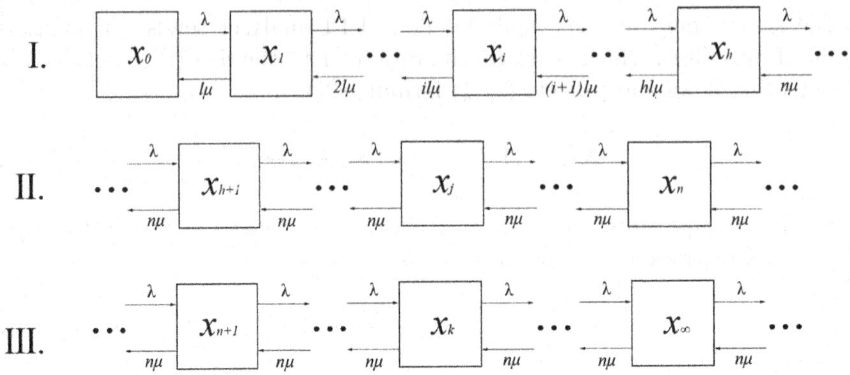

Fig. 1. The state graph of the service model with equal mutual assistance

service device. Mode II of operation, when all possible groups of channels have already been formed, and the system begins to gradually disband them as new requests arrive. Mode III implies the placing of newly received requests in the queue. The system switches to the classic QS operation mode. Accordingly, the operating mode of the system is determined by the number of requests in it.

The Ventcel-Ovcharov model with equal mutual assistance describes the operation of servers (in our case, we are talking about physical entity of DPI analyser), when many service processors can distribute computing power to simultaneously work on a single request in the system, or evenly to work on several requests received in the system. However, in the works of Ventcel and Ovcharov [16,23,24] there is no description of a model with an infinite queue and equal mutual assistance. Therefore, the Ventcel-Ovcharov mathematical model had to be transformed for the case with an infinite queue.

Let us assume that the intensity of servicing one request by a group of channels will be directly proportional to the number of involved channels. Taking into account the above information about the system, you can get the relevant formulas. Let us denote V - the number of devices in the system, l - the number of devices in one group, h - the maximum possible number of groups.

For this model, the probability of system downtime (P_{0fe}) (5):

$$P_{0fe} = \left[\sum_{i=0}^{h} \frac{\alpha^i}{i!} + \frac{\alpha^h}{h!} \times \frac{\beta^{h+1}}{1-\beta} \right]^{-1}, \tag{5}$$

where the ratio of the intensity of incoming requests to the intensity of processing by one group is defined in (6) and the ratio of the intensity of incoming requests to the intensity of processing by all devices of the DPI analyser is defined in (7).

$$\alpha = \left(\frac{\lambda}{l \times \mu}\right) \tag{6}$$

$$\beta = (\frac{\lambda}{V \times \mu}) \tag{7}$$

Using the definitions of the average waiting time and service, as well as P_{0fe} (5) [15], we obtain the formula for the average time spent in the queue (8) and the formula for the average service time (9):

$$\bar{t_w} = \frac{\beta}{V \times \mu} \times \frac{\alpha^h}{h!} \times \beta \times P_{0fe} \times \frac{1}{(1-\beta)^2} \tag{8}$$

$$\bar{t_s} = \frac{1}{\sum_{i=1}^{h} i \times l \times \mu + \sum_{j=h+1}^{V} j \times \mu} \tag{9}$$

The main indicator of the performance of the DPI system is the average time spent by a request in the system (10). We get it by adding the average service time (9) and the average waiting time (8) (with substituting (5)):

For this model, the probability of system downtime (P_{0fe}), the ratio of the intensity of incoming requests to the intensity of processing by one group (α), the ratio of the intensity of incoming requests to the intensity of processing by all devices of the DPI analyser (β), the formula for the average time spent in the queue ($\bar{t_q}$), the formula for the average service time ($\bar{t_w}$) is defined in [15]. The main indicator of the performance of the DPI system is the average time spent by a request in the system (10). We get it by adding the average service time ($\bar{t_q}$) and the average waiting time ($\bar{t_w}$) (with substituting P_{0fe}):

$$\bar{T_{fe}} = \frac{(\frac{1}{\sum_{i=1}^{h} i \times l \times \mu + \sum_{j=h+1}^{V} j \times \mu} + \frac{\beta}{V \times \mu} \times \frac{\alpha^h}{h!} \times \beta \times \frac{1}{(1-\beta)^2})}{(\sum_{i=0}^{h} \frac{\alpha^i}{i!} + \frac{\alpha^h}{h!} \times \frac{\beta^{h+1}}{1-\beta})} \tag{10}$$

The mathematical model for the physical entity of DPI analyser was shown. Part of the initial data for calculating the DPI system has been determined. The formula of the final processing time of requests (10) of the physical entity of DPI analyser is presented.

4.3 Data Sets and Calculation

In this section, we will briefly present the calculated average time spent by a request on the DPI analyser obtained for a given set of initial data. All values are summarized in Table 1.

To obtain the initial data for the calculations, it was necessary to study the statistics of network traffic that is supposed to be passed through the DPI system. One must find the total number of flows, n_{af} and $1 - P_{kn}$. To determine $1 - P_{kn}$, it is necessary to divide the number of new flows by the total number of flows received during a given period of time. For the calculations presented, we used the traffic collected in the dormitories of SPbSUT using Cisco NetFlow equipment. DPI analyser performance was taken to comply with system stability factors. For simplicity, the calculations were performed for one DPI analyser

Table 1. Initial and calculated data.

Actions	Name	Value
Determining the average duration of one flow, s	T_f	300
Determining the number of packets for the period of traffic collection, packets and their rate of arrival, packets per second	N_p	245925000
Determining the intensity of arrival packets, packets per second	λ_0	409875
Determining the probability that a flow is known	P_{kn}	0.78
Determination of the average number of packets in a flow, packets	n_{af}	1093
Specifying the number of packets required for analysis on QS2	n_f	10
Specifying the preliminary number of service devices on QS2	V_{fe}	1
Specifying the intensity of processing requests (QS2)	μ_{fe}	917
Calculation of the value of the stability condition	ρ_{fe}	0.99
Calculation of the intensity of incoming requests for QS2 from QS1 (3)	λ_{fe}	908
Calculation of the average processing time of a request on QS2 (10), s	$\bar{T_{fe}}$	0.10847

server (which reduces the visibility of the Ventcel-Ovcharov model). Average time spent by a request on the DPI analyser server ($\bar{T_{fe}}$) is the sum of the waiting time in the queue and the processing time on the DPI analyser. The result of the calculation showed that equipment with a given performance successfully copes with processing the load with a stability coefficient of 0.99. However, a temporary increase in the number of requests to the DPI analyser can lead to a significant increase in the time spent by a request on the DPI analyser server. With the obtained and given in Table 1 value of the average time of finding a request on the DPI analyser server, there is no need to change the server performance.

5 Simulation

For the DPI system, a simulation model (SM) was created in GPSS [4]. GPSS use a discrete-event approach and a set of distribution laws to describe incoming traffic and how it is processed. In the SM, the initial parameters are set, presented in Table 1. However, to describe the traffic arrival, the Weibull distribution (for the HWF) is used, and for the processing law, the exponential distribution (for the HWF and DPI-An). It is possible to apply a SM in GPSS to obtain the

Table 2. Time characteristics of hardware filter and DPI analyser.

Model type	Total time ms	HWF req., items	HWF av. time req. in system, ms	DPI-An req., items	DPI-An av. time req. in system, ms
SM	108.9	884128	24.278	908	84.593
MM	142.8	847983	33.306	908	108.47

probabilistic-temporal characteristics of the DPI system and compare with the results of the calculation using the mathematical model.

The simulation model of the hardware filter describes the arrival of a request, its spot in a queue, the marking of new requests in the system, the distribution with a given probability for already known flows and for flows requiring analysis. Receiving instructions from DPI analyser, processing new requests and sending requests to the network, how the end of processing is indicated with different requirements for processing capabilities by a hardware filter and are set when describing the filter. Likewise, on DPI analyser - different times are indicated for the process of analyzing a new flow and for processing other requests.

As a result of the simulation model, you can get the number of requests and data on queues. The number of requests received and processed by the system, requests to the hardware filter and to the DPI analyser, and responses. The total time spent by all requests in the system (waiting time and processing time), the total number of requests in the queue (without waiting, with waiting). The simulation results are shown in Table 2.

The simulation modeling results showed that DPI system hardware can handle the load with a stability coefficient of 0.99 in this case. Changes in the characteristics of the distribution of incoming traffic to the DPI system, which was described in the simulation model by the Weibull distribution, significantly affects the size of the queue, and through it, the processing time of requests in the hardware filter. Comparison of the calculation results based on the mathematical and simulation models given in Table 2 indicates the possibility of their use. To obtain more accurate results, it is necessary to clarify the parameters of the distribution of the arrival processes and processing of requests in the simulation model.

6 Conclusion

The aforementioned work formalized the initial data for calculating the DPI system based on traffic statistical data. The mathematical model of Ventcel-Ovcharov is presented. The possibility of practical application of this model for calculating specialized DPI servers is shown.

This paper describes the need for a mathematical model to determine the parameters of the DPI analyser in the DPI architecture. The mathematical model is used to calculate the average analysis time for a given number of processors in a dedicated DPI server. The use of the DPI mathematical model will help reduce the purchase cost of DPI equipment and avoid overloads on communication networks early. The developed formalization of the ratio of flows and packets can be useful in calculating DPI systems and other similar systems.

References

1. Aliev, T.: Basics of Modeling Discrete Systems. SPbGU ITMO, Saint Petersburg (2009)
2. Basharin, G., Gaidamaka, Y., Samuilov, K., Yarkina, N.: Models for Analyzing the Quality of Service in Next Generation Communication Networks. RUDN, Moscow (2008)
3. Dainotti, A., Pescape, A., Sansone, C.: Early classification of network traffic through multi-classification. In: Traffic Monitoring and Analysis III, pp. 122–135 (2011)
4. Fitsov, V.: A simulation model of the DPI system based on the GPSS world software. In: Actual Problems of Information and Telecommunications in Science and Education V Conference, pp. 539–545 (2016)
5. Grimm, C., Schluchtermann, G.: IP Traffic Theory and Performance. Springer, Heidelberg (2008). https://doi.org/10.1007/978-3-540-70605-2
6. ITU-T: Recommendation ITU-T Y.2770 requirements for deep packet inspection in next generation networks (2012)
7. ITU-T: Recommendation ITU-T Y.2771 framework for deep packet inspection (2014)
8. ITU-T: Recommendation ITU-T Y.2772 mechanisms for the network elements with support of deep packet inspection (2016)
9. ITU-T: Recommendation ITU-T Y.2773 performance models and metrics for deep packet inspection (2017)
10. ITU-T: Recommendation ITU-T Y.2774 functional requirements of deep packet inspection for future networks (2019)
11. ITU-T: Recommendation ITU-T Y.2775 functional architecture of deep packet inspection for future networks (2019)
12. Kleinrock, L.: Time-shared system: a theoretical treatment. J. Assoc. Comput. **14**(2), 242–251 (1967)
13. Lozhkovsky, A., Kaptur, V., Verbanov, O.: Mathematical model of packet traffic. Bull. Natl. Polytech. Univ. KhPI **9**, 113–119 (2011)
14. Niang, B.: Bandwidth management - a deep packet inspection mathematical model. In: ICUMT 2014, pp. 169–175 (2014)
15. Novikov, A., Fitsov, V.: Application of the Ventcel-Ovcharov mathematical model with uniform mutual assistance for modern NFV systems. In: Actual Problems of Information and Telecommunications in Science and Education VIII Conference, vol. 1, pp. 705–709 (2019)
16. Ovcharov, L.: Applied Problems of the Theory of Queuing. Mashinostroenie, Moscow (1969)
17. Park, J., Yoon, S., Kim, M.: Software architecture for a lightweight payload signature-based traffic classification system. In: Traffic Monitoring and Analysis III, pp. 136–149 (2011)
18. Samuilov, K., Levakov, A., Sokolov, N., Zaitcev, V.: Method of arrival process description for packet switch. In: Proceedings of the Workshop on ITTMM, pp. 47–56 (2020)

19. Sokolov, N.: Telecommunication Networks Planning Tasks. BHV, Saint Petersburg (2012)
20. Song, W., et al.: A software deep packet inspection system for network traffic analysis and anomaly detection. Sensors **20**(6), 1637 (2020)
21. Stepanov, S., Thu, D.: The system of equilibrium equations for the model of channel resource allocation on traffic concentration lines. In: Telecommunication and Computing Systems, pp. 24–25 (2010)
22. Wang, Z., et al.: SymTCP: eluding stateful deep packet inspection with automated discrepancy discovery. In: NDSS-2020 Symposium (2020)
23. Ventcel, E.: Probability Theory. Nauka, Moscow (1969)
24. Ventcel, E.: Operations Research. Soviet Radio, Moscow (1972)
25. Yashkov, S.: A derivation of response time distribution for an M/G/1 processor-sharing queue. Problem. Control Inf. Theory. **12**(2), 133–148 (1983)
26. Yashkov, S.: Analysis of Queues in a Computer. Radio i Svyaz, Moscow (1989)
27. Yashkov, S., Yashkova, A.: Egalitarian separation of the processor. Inform. Process. **6**(4), 396–444 (2006)
28. Yashkov, S., Yashkova, A.: Processor sharing: a survey of the mathematical theory. Automat. Remote Control **68**(9), 1662–1731 (2007)
29. Zaitcev, V.: Characteristics of the total flow of IP packets at the input of the switching node of a multiservice network. In: Proceedings of the XII Conference Technologies of the Information Society, pp. 178–182 (2018)
30. Zeng, Y., Guo, S.: Deep packet inspection with delayed signature matching in network auditing. In: Naccache, D., et al. (eds.) ICICS 2018. LNCS, vol. 11149, pp. 75–91. Springer, Cham (2018). https://doi.org/10.1007/978-3-030-01950-1_5

The Overflow Probability Asymptotics in a Multiclass Single-Server Retrial System

Evsey Morozov[1,2,3] and Ksenia Zhukova[1,2(✉)]

[1] Institute of Applied Mathematical Research, Petrozavodsk, Russia
emorozov@karelia.ru, kalininaksenia90@gmail.com
[2] Petrozavodsk State University, Petrozavodsk, Russia
[3] Moscow Center for Fundamental and Applied Mathematics, Moscow State University, 119991 Moscow, Russia

Abstract. We discuss the asymptotic of the large deviation probability in a single-server retrial queue with K classes of customers. The constant retrial rate policy is assumed. The input is assumed to be a general renewal process and the retrial attempts follow an exponential distribution. The system is described with a regenerative process. We are interested in the stationary probability that the orbit size reaches a level N within regeneration cycle, as $N \to \infty$, called large deviation probability. We interpret the original multiclass retrial system as a buffered system with the service time of special type to construct the upper and lower bounds of the large deviation probability. Some numerical examples are also included.

Keywords: Retrial queue · Large deviation · Overflow probability · Virtual orbits · Stochastic ordering

1 Introduction

The large deviation analysis is closely connected with the evaluation of the quality of service indexes of the telecommunication and computer systems. One of such an important parameter is the stationary overflow probability that the buffer content (the number of customers in the system or the orbit size) exceeds a given threshold N, as N infinitely grows. The problem of calculating and estimating the overflow probability in the multiserver classic buffered system has been well-studied previously, see for instance, [2,12]. For our research, the approach developed in the paper [10], where this problem for the tandem system has been considered, plays an important role. The retrial models are highly motivated by practical applications in the modern wireless communications systems.

The research was carried out under state order to the Karelian Research Centre of the Russian Academy of Sciences (Institute of Applied Mathematical Research KarRC RAS) and supported by the Russian Foundation for Basic Research, projects 18-07-00147 and 19-57-45022.

© Springer Nature Switzerland AG 2020
V. M. Vishnevskiy et al. (Eds.): DCCN 2020, CCIS 1337, pp. 394–406, 2020.
https://doi.org/10.1007/978-3-030-66242-4_31

In this paper, we consider a single-server retrial system with two orbits and constant retrial rate. The large deviation analysis of the overflow probability in the multi-server retrial queue has been studied in particular, in [3,5,6]. In the paper [3], using the approach developed in [10], the lower and upper bounds for the decay rate of the overflow probability during a regeneration cycle in retrial system with constant and classical retrial rates have been constructed. These results, based on the analysis developed in the fundamental monograph [9], have been extended in [7] to the probability of the overflow during *full busy cycle* of the *multiserver* retrial system.

To analyze a multiclass retrial queue with several orbits, we will apply the approach suggested in [3]. The key idea is to represent the original retrial system as an equivalent *buffered single-server system* with dependent service times. This interpretation allows further to compare the original queueing system with the well-studied buffered classic systems. More exactly, this approach allows to construct the upper and lower bounds for the large deviation (overflow) probability that the number of customers in the system reaches level N within regeneration cycle, as $N \to \infty$. The main contribution of this research is an extension of the approach from [3] to the multiclass retrial queueing system with stochastically ordered service times.

The paper is organized as follows. In Sect. 2 description of the model is given and interpretation of the original multiclass retrial queue is discussed. Moreover this section contains the analysis of the overflow probability, which is the main contribution of this research. Numerical examples are further presented in Sect. 3, which demonstrate the tightness of the constructed bounds for the retrial system with exponential and Weibull service times.

2 An Associated Buffered Multiclass System

In this section we discuss an interpretation of a retrial queue and apply it to construct the lower and upper bounds of asymptotic of the overflow probability.

2.1 Description of a Model

We consider a single-server K-class retrial system with a renewal input of customers arriving at the instants $\{t_n\}$, with independent identically distributed (iid) interarrival times $\tau_n := t_{n+1} - t_n$, $t_1 = 0$, and K classes of the customers. It is assumed that with probability p_i the n-th arrival is class-i customer with the service time $S_n^{(i)}$, $n \geq 1$. The service times $\{S_n^{(i)}, n \geq 1\}$, being class-dependent, are assumed to be iid, with generic element $S^{(i)}$, $i = 1, \dots, K$. (In what follows we omit the serial index to denote a generic element of an iid sequence.) We introduce the service time S_n of customer n as

$$S_n = \sum_{i=1}^{K} S_n^{(i)} \, 1_n^{(i)}, \tag{1}$$

where $1_n^{(i)}$ is the indicator function defined by

$$1_n^{(i)} = \begin{cases} 1, & \text{with probability } p_i \\ 0, & \text{with probability } 1-p_i, \end{cases} \tag{2}$$

We note that $\sum_{i=1}^K p_i = 1$. Also we note that, for each fixed i, the sequence $\{1_n^{(i)}, n \geq 1\}$ is iid by construction, with generic element $1^{(i)}$. It is easy to see that the service times $\{S_n, n \geq 1\}$ defined in (1) are iid as well. Also we introduce the input rate and the service rates, respectively,

$$\lambda = \frac{1}{\mathsf{E}\tau}, \quad \mu_i = \frac{1}{\mathsf{E}S^{(i)}}, \quad i = 1, \ldots, K.$$

It is assumed that there are K infinite-capacity virtual orbits. It is assumed that, if a new customer finds the server busy, he joins a class-dependent orbit and attempts to occupy the server after an exponential time with class-dependent rate. We consider a *constant retrial rate policy*, when the retrial rate does not depend on the orbit size. Denote by $\{\xi_n^{(i)}\}$ the iid retrial times of class-i customers with generic time $\xi^{(i)}$ and retrial rate $\gamma_i = 1/\mathsf{E}\xi^{(i)}$, $i = 1, \ldots, K$.

We will study the *stationary overflow probability* P_N, that the number of customers in the retrial system reaches a level N during a regeneration cycle. (Below we give more precise definition of P_N.) The regenerations of the system occur when an arrival meets completely empty system. (A detailed discussion the regenerative structure of this system can be found in [3].) Our purpose is to construct the upper and lower bounds of the logarithmic asymptotics of the *large deviation probability* P_N as $N \to \infty$. To construct these bounds, we compare the probability P_N in the original system with the similar probabilities in the classic buffered systems with the same input flow, in which service times satisfy some stochastic ordering. This idea has been applied previously in the large deviation analysis for the tandem networks in the paper [10], and for the retrial multi-server queue in the paper [3].

In the retrial queue, after each departure, the server stays idle until the next customer, either a new arrival or a retrial customer, occupies it. In the further analysis we interpret this idle time as a part of the service time of the *next customer*. This allows us to define a new sequence of the service times, denoted by $\{\hat{S}_n, n \geq 1\}$, written as

$$\hat{S}_n = \sum_{i=1}^{K} \tilde{S}_n^{(i)}\, \tilde{1}_n^{(i)} + \zeta_n =: \tilde{S}_n + \zeta_n, \; n \geq 1, \tag{3}$$

where the indicators function $\tilde{1}_n^{(i)}$ takes the values

$$\tilde{1}_n^{(i)} = \begin{cases} 1, & \text{if the } n-\text{th customer in the server is class} - i \\ 0, & \text{otherwise,} \end{cases} \tag{4}$$

and ζ_n is the mentioned above idle time of the server between the departure of customer $n-1$ and the next service initiation. Moreover, the random variables

$\tilde{S}_n^{(i)}$ are selected from the same basic sequence $\{S_n^{(i)}, n \geq 1\}$, that is $\tilde{S}^{(i)}$ is distributed as $S^{(i)}$, however $\{\tilde{S}_n^{(i)}, n \geq 1\}$ are not iid in general. It is because we assign the service times in the order in which service initiations occur, and it destroys the order of arrivals. As a result, the actual fractions of the class-i customers in the representation (1) may differ with the given proportions p_i at least on a finite interval of the observation. The interpretation of the idle time of the server as a part of service time does not change distributional properties of the the number of customers in the retrial queue. Moreover, it allows us to compare the original system to classical systems with the same input and service times which are (stochastically) less or bigger than the service times $\hat{S}_n$.

2.2 Analysis of a Large Deviation Probability

The interpretation described above allows to consider the original retrial system as a buffered system with the same input flow and the service time of type (3) which in general are not iid because the summands $\{\tilde{S}_n\}$ are not iid. Moreover the random variable ζ_n depends on which event occurs first: a retrial customer or a new arrival captures the server. Thus the exact distribution of the variable ζ_n is unknown, but we know that by construction,

$$0 < \zeta_n \leq_{st} \max(\xi^{(1)}, \ldots, \xi^{(K)}),\ n \geq 1, \tag{5}$$

where $\leq_{st}$ means the *stochastic inequality.* We exclude the value $\zeta_n = 0$ in (5) because consider the dynamics of the system within a regeneration cycle. For the further analysis we accept the following *ordering* additional assumptions:

$$S^{(1)} \leq_{st} S^{(2)} \leq_{st} \cdots \leq_{st} S^{(K)}, \tag{6}$$

$$\xi^{(1)} \leq_{st} \xi^{(2)} \leq_{st} \cdots \leq_{st} \xi^{(K)}. \tag{7}$$

By assumption (7), class-1 orbital customers are "the most insist" to occupy server. Note that the order (7) means the ordering

$$\gamma_1 \geq \cdots \geq \gamma_K,$$

among the parameters of the exponential retrial times. Also note that condition (6) holds, for instance, for exponential and Weibull service times. Now we introduce the iid sequences of the random variables, $\{S_n^{(l)}\}$ and $\{S_n^{(u)}\}$, as the lower and upper bounds for the service times $\{\hat{S}_n\}$:

$$S^{(l)} =_{st} S^{(1)}, \tag{8}$$

$$S^{(u)} =_{st} S^{(K)} + \xi^{(K)}. \tag{9}$$

Note that indeed the lower bound is distributed as the *minimal* service time. By assumptions (6), (7) and by (5), it follows that, for $n \geq 1$,

$$S_n^{(l)} \leq_{st} \hat{S}_n \leq_{st} S_n^{(u)},\ n \geq 1.$$

By the stochastic monotonicity property [14], the buffered system with the interarrival time τ and service time $S^{(l)}$ is less loaded than the buffered system with service time $\{\hat{S}_n\}$ and the same input process. (For this monotonicity property the independence of the service times is not required. Some additional details can be found in [3].) By this reason, the buffered system is a *minorant* for the original retrial system. In particular the overflow probability in the minorant system is dominated by the same probability in the original retrial queue. Analogously, the buffered system with the same interarrival time τ and the iid service times $\{S_n^{(u)}\}$ dominates the number of customers in the original retrial system. This observation allows us to construct the lower and upper bounds for the asymptotics of the overflow probability P_N as $N \to \infty$.

In the subsequent analysis we apply the large deviation asymptotics in the classic single-server system adapted from the paper [2]. This result assumes that the basic processes describing the dynamics of the original system are *stationary.* To be more precise, we consider the basic process $\mathcal{Q} = \{Q(t),\, t \geq 0\}$, where $Q(t)$ is the number of customers in the system, in orbits and server, at instant t. We remind that the regenerations of the process $\mathcal{Q}$, occur when an arrival finds the system completely empty. The regenerative process $\mathcal{Q}$ (and the system) is called *positive recurrent* if the mean regeneration cycle length is finite. Since the constructed buffered system with service time $S^{(u)}$ dominates the original retrial queue, the finitness of the mean regeneration cycle length in this dominating system implies that the mean cycle length in the original queue is finite as well. In the classic system the positive recurrence implies the existence of the stationary distribution of the corresponding regenerative process (if, additionally, τ is non-lattice) [1]. Recall that the stability criterion of the *dominating* single-server classic buffered system introduced above is [1,15]:

$$\lambda \mathsf{E} S^{(u)} < 1. \tag{10}$$

Since

$$\mathsf{E} S^{(u)} = \mathsf{E}(S^{(K)} + \xi^{(K)}) = \frac{1}{\mu_K} + \frac{1}{\gamma_K},$$

then the stability criterion (10) can be rewritten as

$$\frac{1}{\mu_K} + \frac{1}{\gamma_K} < \frac{1}{\lambda}. \tag{11}$$

Moreover, condition (11) guarantees the positive recurrence of the minorant system with the service time $S^{(l)}$. Noe that the analysis of a single-server system with classic and constant retrials can be found in particular in the works [4,13,17].

Now we define the main object of our analysis. First of all, for an arbitrary random variable X, we define the *logarithmic (log) moment generating function:*

$$\Lambda_X(\theta) = \log \mathsf{E} e^{\theta X}, \;\; \theta > 0.$$

Now we return to the service times in our system and define the value

$$\hat{\theta} = \sup(\theta > 0 : \mathsf{E} e^{\theta S} < \infty), \tag{12}$$

which delimits the zone of the existence of the exponential moment generation function of the variable S. Taking into account (1) and (2), we obtain

$$\mathsf{E}e^{\theta S} = \mathsf{E}e^{\theta \sum_{i=1}^{K} S^{(i)} 1^{(i)}} = \sum_{i=1}^{K} p_i \mathsf{E}e^{\theta S^{(i)}}. \tag{13}$$

Then the parameter $\hat{\theta}$ is defined as

$$\hat{\theta} = \sup\Big(\theta > 0 : \sum_{i=1}^{K} p_i \mathsf{E}e^{\theta S^{(i)}} < \infty\Big). \tag{14}$$

It is then evident that the component service times $S^{(i)}$, $i = 1, \ldots, K$, must have finite exponential moments to satisfy assumption $\hat{\theta} < \infty$.

We will study the stationary overflow probability P_N that the number of customers in the stationary retrial system exceeds a fixed level N during regenerative cycle. More exactly we consider the logarithmic asymptotics of this probability as $N \to \infty$. It means that, for each value of N, we consider the stationary probability P_N, then increase the value of N and again consider stationary probability, etc. The logarithmic asymptotic lower and upper bounds for the probability P_N are given in the following statement.

Theorem 1. *Assume that conditions (6), (7) and (11) hold true. Then the overflow probability* P_N *satisfies*

$$\Lambda_\tau(-\theta_*) \le \limsup_{N\to\infty} \frac{1}{N} \log \mathsf{P}_N \le \Lambda_\tau(-\theta^*), \tag{15}$$

where θ_* *and* θ^* *are defined, respectively, as*

$$\theta_* = \sup\Big(\theta \in (0,\, \hat{\theta}) : \Lambda_\tau(-\theta) + \Lambda_{S^{(l)}}(\theta) \le 0\Big) \tag{16}$$

and

$$\theta^* = \sup\Big(\theta \in (0,\, \min(\hat{\theta}, \gamma_K)) : \Lambda_\tau(-\theta) + \Lambda_{S^{(u)}}(\theta) \le 0\Big). \tag{17}$$

Proof. The proof of the theorem repeats the main steps of the proof of Lemmas 1 and 3 from [3]. In particular, under condition (11) which provides the positive recurrence of the classic systems with service times $S^{(l)}$ and $S^{(u)}$ respectively, we can apply the basic result from the paper [2] concerning the stationary overflow probabilities, denoted by P^l_N and P^u_N, in the corresponding buffered systems, and then show that

$$\lim_{N\to\infty} \frac{1}{N} \log \mathsf{P}^l_N = \Lambda_\tau(-\theta_*), \tag{18}$$

$$\lim_{N\to\infty}\frac{1}{N}\log \mathsf{P}_N^u = \Lambda_\tau(-\theta^*), \tag{19}$$

where parameters θ_* and θ^* are defined in (16) and (17), respectively. It follows from discussion following formulas (8) and (9) that

$$\mathsf{P}_N^l \le \mathsf{P}_N \le \mathsf{P}_N^u,$$

and then we obtain the statement (15).

Remark. Note that lim sup in (18) can be replaced by lim inf. Also we note that, by construction, the obtained bounds do not depend on the probabilities p_i, $i = 1, \ldots, K$.

3 Numerical Examples

In this section we present some numerical examples which show the estimated overflow probability obtained by means of the regenerative simulation in comparison with the theoretical bounds (15).

Example 1. The exponential service times and three orbits. We consider $M/GI/1$ retrial system with the input rate $\lambda = 1$. There are three classes of customers with exponential service times $S^{(1)}$, $S^{(2)}$, $S^{(3)}$ with parameters $\mu_1 = 1.5$, $\mu_2 = 1.2$, $\mu_3 = 1.25$, respectively. This implies the required condition

$$S^{(1)} \le_{st} S^{(2)} \le_{st} S^{(3)}.$$

Also we take $p_i = 1/3$, $i = 1, 2, 3$; the retrial rates are $\gamma_1 = 20$, $\gamma_2 = 15$, $\gamma_3 = 10$, implying

$$\xi^{(1)} \le_{st} \xi^{(2)} \le_{st} \xi^{(3)}.$$

The chosen parameters satisfy condition (10), and thus the statements of Theorem 1 hold true. To estimate the overflow probability P_N we sample 10000 regeneration cycles. Then we calculate the upper and lower bounds (15) for various values of N. The simulation results together with the bounds are presented in Fig. 1. To calculate the bounds (15), we need to find functions $\Lambda_{S^{(l)}}(\theta) = \Lambda_{S^{(1)}}(\theta)$ and

$$\Lambda_{S^{(u)}}(\theta) = \log \mathsf{E}e^{\theta(S^{(3)}+\xi^{(3)})} = \Lambda_{S^{(3)}}(\theta) + \Lambda_{\xi^{(3)}}(\theta).$$

For the exponential τ and $S^{(i)}$ we easily find that:

$$\Lambda_\tau(\theta) = \log \mathsf{E}e^{\theta\tau} = \log\frac{\lambda}{\lambda-\theta},\ \ \theta \in (0,\, \lambda), \tag{20}$$

and

$$\Lambda_{S^{(i)}}(\theta) = \log\frac{\mu_i}{\mu_i-\theta},\ \ i = 1, 2, 3;\ \ \theta \in (0,\, \min \mu_i). \tag{21}$$

Using (14) we find that $\hat{\theta} = 1.25$, and then using (20) and (21), we obtain the equation

$$\begin{aligned}\theta_* &= \sup\Big(\theta \in (0,\,\hat{\theta}) : \Lambda_\tau(-\theta) + \Lambda_{S^{(1)}}(\theta) \le 0\Big) \\ &= \sup\Big(\theta \in (0,\,\hat{\theta}) : \frac{\lambda}{\lambda+\theta}\cdot\frac{\mu_1}{\mu_1-\theta} \le 1\Big),\end{aligned} \tag{22}$$

which has solution $\theta_* = 0.5$. This gives $\Lambda_\tau(-\theta_*) = -0.41$.

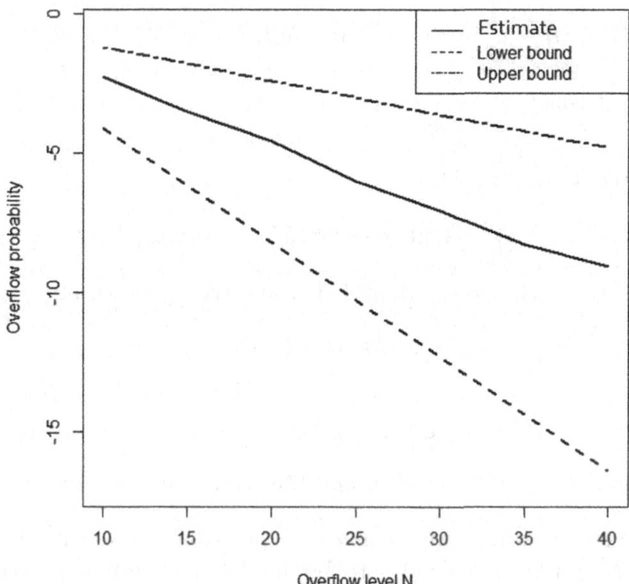

Fig. 1. The estimated overflow probability in M/GI/1 retrial system and theoretical asymptotics vs. overflow level N; exponential service times, logarithmic scale.

Then we calculate parameter θ^*:

$$\begin{aligned}\theta^* &= \sup\left(\theta \in (0,\,\min(\hat{\theta},\gamma_3)) : \Lambda_\tau(-\theta) + \Lambda_{S^{(3)}}(\theta) + \log\frac{\gamma_3}{\gamma_3-\theta} \le 0\right) \\ &= \sup\left(\theta \in (0,\,\hat{\theta}) : \frac{\gamma_3}{\gamma_3-\theta}\cdot\frac{\lambda}{\lambda+\theta}\cdot\frac{\mu_3}{\mu_3-\theta} \le 1\right).\end{aligned} \tag{23}$$

and obtain $\theta^* = 0.13$. This implies $\Lambda_\tau(-\theta^*) = -0.126$. Figure 1 shows the estimated overflow probability and theoretical bounds (15).

Example 2. Weibull service times and three orbits. Again we consider an $M/GI/1$ retrial system with the input rate $\lambda = 1/3$ and three classes of customers with Weibull service times, denoted by *Weibull(a, b)*, with the density,

$$f(x) = \frac{a}{b}\left(\frac{x}{b}\right)^{a-1} e^{-(x/b)^a},\ x \geq 0,\, b > 0,\, a \geq 1. \tag{24}$$

Note that the rate of *Weibull(a, b)* service time is

$$\mu = \frac{1}{\mathsf{E}S} = \frac{1}{b\,\Gamma(1+\frac{1}{a})}, \tag{25}$$

where Γ is the *Gamma function.* We take

$$S^{(1)} =_{st} \mathit{Weibull}\big(2, \frac{1}{3}\big),\ S^{(2)} =_{st} \mathit{Weibull}\big(2, \frac{1}{2}\big),\ S^{(3)} =_{st} \mathit{Weibull}(2, 1),$$

so that the ordering

$$S^{(1)} \leq_{st} S^{(2)} \leq_{st} S^{(3)},$$

holds true. According to (25),

$$\mu_1 = 3.39,\ \mu_2 = 2.258,\ \mu_3 = 1.129.$$

Also we take $p_i = 1/3$, $i = 1, 2, 3$, and choose retrial rate parameters as

$$\gamma_1 = 10,\ \gamma_2 = 5,\ \gamma_3 = 2,$$

implying

$$\xi^{(1)} \leq_{st} \xi^{(2)} \leq_{st} \xi^{(3)}.$$

Note that condition (10) is satisfied and the statements of Theorem 1 hold.

As in previous example, we estimate the probability P_N using 100000 regeneration cycles, and then calculate the bounds of the overflow probability using (15). In general, it is a hard problem to derive function $\Lambda_{S^{(i)}}(\theta)$ in an explicit form for the non-exponential $S^{(i)}$. However for Weibull distribution this function is available in some cases. In our setting, $\Lambda_{S^{(l)}} = \Lambda_{S^{(1)}}$, so we obtain by [11],

$$\Lambda_{S^{(1)}}(\theta) = \log \int_0^\infty 18x e^{-9x^2+\theta x} dx = \log\left[1 + \frac{\theta}{6}\sqrt{\pi} e^{\frac{\theta^2}{36}}(1 - \Phi(-\theta/6))\right], \tag{26}$$

where

$$\Phi(x) = \frac{2}{\sqrt{\pi}}\int_0^x e^{-t^2} dt.$$

Using (20) and (26) we must calculate

$$\begin{aligned}\theta_* &= \sup\Big(\theta \in (0, \hat{\theta}) : \Lambda_\tau(-\theta) + \Lambda_{S^{(1)}}(\theta) \leq 0\Big)\\ &= \sup\Big(\theta > 0 : \frac{\lambda}{\lambda+\theta}\cdot\left[1 + \frac{\theta}{6}\sqrt{\pi} e^{\frac{\theta^2}{36}}(1 - \Phi(-\theta/6))\right] \leq 1\Big).\end{aligned} \tag{27}$$

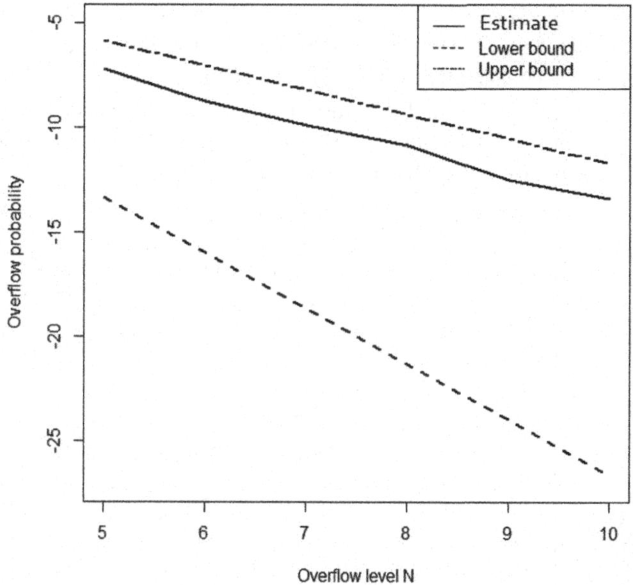

Fig. 2. The estimated overflow probability in $M/GI/1$ retrial system and theoretical asymptotics vs. overflow level N; Weibull service times; logarithmic scale.

Note, that in this example $\hat{\theta} = +\infty$. After some algebra, we find that $\theta_* = 4.23$, implying $\Lambda_\tau(-\theta_*) = -2.66$. Next we derive the upper bound $\Lambda_{S^{(u)}} = \Lambda_{S^{(3)}}$ and obtain

$$\Lambda_{S^{(3)}}(\theta) = \log \int_0^\infty 2xe^{-x^2+\theta x} dx = \log \left[1 + \frac{\theta}{2}\sqrt{\pi}e^{\frac{\theta^2}{4}}(1 - \Phi(-\theta/2))\right]. \quad (28)$$

It remains to calculate parameter θ^* satisfying the equation

$$\begin{aligned}\theta^* &= \sup\left(\theta \in (0, \min(\hat{\theta}, \gamma_3)) : \Lambda_\tau(-\theta) + \Lambda_{S^{(3)}}(\theta) + \log\frac{\gamma_3}{\gamma_3 - \theta} \le 0\right) \\ &= \sup\left(\theta \in (0, \gamma_3) : \frac{\gamma_3}{\gamma_3 - \theta} \cdot \frac{\lambda}{\lambda + \theta} \cdot \left[1 + \frac{\theta}{2}\sqrt{\pi}e^{\frac{\theta^2}{4}}(1 - \Phi(-\theta/2))\right] \le 1\right).\end{aligned}$$

After some algebra we find that $\theta^* = 0.73$, and it gives $\Lambda_\tau(-\theta^*) = -1.17$. As in the previous experiment, Fig. 2 shows that the estimated probability is located between the theoretical bounds being closer to the upper bound.

Example 3. The exponential service times and two orbits. Now we consider an $M/GI/1$ retrial system with *two classes* and the input rate $\lambda = 1/2$. Service times are exponential and ordered as $S^{(1)} \le_{st} S^{(2)}$ with the rates $\mu_1 = 10$, $\mu_2 = 1.25$. We also take the retrial rates $\gamma_1 = 20$, $\gamma_2 = 2$, implying the ordering $\xi^{(1)} \le_{st} \xi^{(2)}$. As above, the chosen parameters satisfy stability condition (10).

First, we estimate the overflow probability P_N for two sets of the probabilities p_i, $i = 1, 2$. The set $p_1 = 0.2$, $p_2 = 0.8$ represents the scenario when class-2 customers have (stochastically) bigger service times, that is $S^{(2)} \geq_{st} S^{(1)}$, and rarer retrial attempts, $\xi^{(2)} \geq_{st} \xi^{(1)}$, and thus attempt to enter the server more frequently than class-1 customers. In this case the system is more loaded than that under the opposite case with $p_1 = 0.8$, $p_2 = 0.2$.

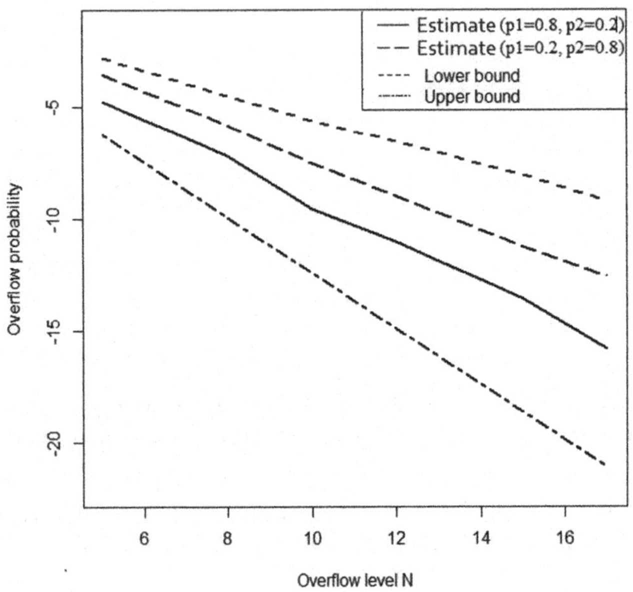

Fig. 3. The estimated overflow probabilities and the theoretical asymptotics vs. the overflow level N; exponential service times; two orbits; logarithmic scale.

Then we calculate the theoretical result (15). We recall that, by construction, the theoretical bounds do not depend on p_1, p_2. As in example 1, we obtain $\hat{\theta} = 1.25$. Using (20) and (21) we obtain

$$\theta_* = \sup\Big(\theta \in (0,\, \hat{\theta}) : \frac{\lambda}{\lambda+\theta}\, \frac{\mu_1}{\mu_1 - \theta} \leq 1\Big), \tag{29}$$

and solution $\theta_* = 0.37$. This gives $\Lambda_\tau(-\theta_*) = -0.56$. Finally, we calculate θ^* using the following equation:

$$\theta^* = \sup\left(\theta \in (0,\, \hat{\theta}) : \frac{\gamma_2}{\gamma_2 - \theta}\, \frac{\lambda}{\lambda+\theta}\, \frac{\mu_2}{\mu_2 - \theta} \leq 1\right). \tag{30}$$

Its solution is $\theta^* = 1.25$, implying $\Lambda_\tau(-\theta^*) = -1.23$. The results based on 10000 regeneration cycles are given in Fig. 3. The simulation results indicate an expected shift of the estimate to the upper bound for the more loaded system.

4 Conclusion

A single-server multiclass retrial system with constant class-dependent retrial rates is considered in the stationary regime which is defined in the terms of the regeneration theory. The stationary overflow probability that the number of customers in the system reaches a high level N during a regeneration cycle is studied. It is shown that the original retrial system can be treated as buffered system with the dependent service times of a special type. This interpretation allows further to compare the original retrial system with the classic buffered systems, which in turn can be analysed using the large deviation result from [2]. Based on this approach, the lower and upper bounds for the asymptotic of the large deviation probability are constructed as $N \to \infty$. The tightness of the bounds is illustrated by a few numerical examples.

References

1. Asmussen, S.: Applied Probability and Queues, 2nd edn. Springer, New York (2003). https://doi.org/10.1007/b97236
2. Sadowsky, J.S.: Large deviations theory and efficient simulation of excessive backlogs in a GI/GI/m queue. IEEE Trans. Autom. Control **36**(12), 1383–1394 (1991)
3. Morozov, E., Zhukova, K.: A large deviation analysis of retrial models with constant and classic retrial rates. Perform. Eval. **135**, 102021 (2019)
4. Morozov, E., Rumyantsev, A., Dey, S., Deepak, T.G.: Performance analysis and stability of multiclass orbit queue with constant retrial rates and balking. Perform. Eval. (2019). https://doi.org/10.1016/j.peva.2019.102005
5. Kim, J., Kim, B.: Tail asymptotics for the queue size distribution in the MAP/G/1 retrial queue. Queueing System. **66**, 79–94 (2010)
6. Kim, J., Kim, B., Ko, S.-S.: Tail asymptotics for the queue size distribution in an M/G/1 retrial queue. J. Appl. Probab. **44**, 1111–1118 (2007)
7. Morozov, E., Zhukova, K.: An upper bound of the large deviation probability in multi-server constant retrial rate system. In: Vishnevskiy, V.M., Samouylov, K.E., Kozyrev, D.V. (eds.) DCCN 2019. CCIS, vol. 1141, pp. 325–337. Springer, Cham (2019). https://doi.org/10.1007/978-3-030-36625-4_26
8. Artalejo, J.R., Resing, J.A.C.: Mean value analysis of single server retrial queues. Asia-Pac. J. Oper. Res. **27**(3), 335–345 (2010)
9. Thorisson, H.: Coupling, Stationarity, and Regeneration. Springer, New York (2000)
10. Buijsrogge, A., de Boer, P.T., Rosen, K., Scheinhardt, W.: Large deviations for the total queue size in non-Markovian tandem queues. Queueing Syst. **85**(3–4), 305–312 (2017). https://doi.org/10.1007/s11134-016-9512-z
11. Gradshteyn, I.S., Ryzhik, I.M.: Table of Integrals, Series and Products. Elsevier, Amsterdam (2007)
12. Chang, C.-S.: Performance Guarantees in Communication Networks. Springer, London (2000). https://doi.org/10.1007/978-1-4471-0459-9
13. Morozov, E.: A multiserver retrial queue: regenerative stability analysis. Queueing Syst. **56**, 157–168 (2007)
14. Whitt, W.: Comparing counting processes and queues. Adv. Appl. Probab. **13**, 207–220 (1981)

15. Altman, E., Borovkov, A.A.: On the stability of retrial queues. Queueing Syst. **26**, 343–363 (1997)
16. Nekrasova, R.: On verification of stability of multi-orbit system with general retrials: simulation approach. In: Vishnevskiy, V.M., Samouylov, K.E., Kozyrev, D.V. (eds.) DCCN 2019. CCIS, vol. 1141, pp. 401–412. Springer, Cham (2019). https://doi.org/10.1007/978-3-030-36625-4_32
17. Avrachenkov, K., Morozov, E., Nekrasova, R., Steyaert, B.: Stability analysis and simulation of N-class retrial system with constant retrial rates and Poisson inputs. Asia-Pacific J. Oper. Res. **31**(2) (2014) https://doi.org/10.1142/S0217595914400028

Wireless Channel Modeling and Simulation with K Distribution

S. G. Shorokhov(✉)

Peoples' Friendship University of Russia (RUDN University),
6 Miklukho-Maklaya St, Moscow 117198, Russian Federation
shorokhov-sg@rudn.ru

Abstract. We study modeling of wireless channel with fading and shadowing effects using K distribution with modified Bessel function of the second kind with half integer order. This allows us to obtain probability density function and cumulative distribution function in closed form in terms of elementary functions and simplifies the calculation of exact average bit error rate. The problem of Monte Carlo simulation using random variables with K distribution is also addressed.

Keywords: Wireless channel · K distribution · Bessel function · Bit error rate · Monte Carlo method

1 Introduction

K distribution was introduced in [10] for describing the statistics of radiation scattered by media with a wide range of length scales. K distribution can be derived from the product of two random variables, where one variable has a chi distribution and another variable has a complex Gaussian distribution [24]. K distribution is widely used for modeling diverse scattering phenomena such as tropospheric propagation of radio waves, various types of radar clutter, optical scintillation from the atmosphere [3], in synthetic-aperture radar (SAR) imagery [25], radiative heat transfer [15] and also in wireless communication to model composite fading and shadowing effects [18,20].

One of the most important approaches to stochastic simulation is Monte Carlo method. Problems in communication theory form one of the most important domains of application for Monte-Carlo method [19].

We study the problem of wireless channel modeling with K distribution and Monte Carlo method, when modified Bessel function in probability density of K distribution is of half integer order.

2 Model of Shadowed Fading Channel

Basically, the model for shadowed fading channel can be described by the following equation [18]

$$r = A\,B\,s + n, \tag{1}$$

The publication has been prepared with the support of the "RUDN University Program 5–100". The research was funded by RFBR, grant No. 19-08-00261.

© Springer Nature Switzerland AG 2020
V. M. Vishnevskiy et al. (Eds.): DCCN 2020, CCIS 1337, pp. 407–421, 2020.
https://doi.org/10.1007/978-3-030-66242-4_32

where r is the received signal, s is the transmitted signal, n is the Gaussian distributed noise with zero mean, A and B represent the fluctuations in the channel due to fading and shadowing.

Fading in wireless channel (1) can be described using various stochastic models, such as Rayleigh fading, Rician fading, Nakagami fading, etc. [18].

When the envelope of the signal is Rayleigh distributed, its power has an exponential probability density function (PDF), given by

$$f_F(p) = \frac{1}{p_0} \exp\left(-\frac{p}{p_0}\right) U(p), \tag{2}$$

where p_0 is the average power of the received signal, $U(p) = \mathbb{1}_{\{p>0\}} = \begin{cases} 1, p > 0, \\ 0, p \leq 0. \end{cases}$

In wireless communications average power often varies randomly due to the existence of shadowing by surrounding terrain, mountains, buildings, etc. The density function of the average power can be modeled in terms of lognormal or gamma distribution [1,2,17]. In model with gamma distribution, the PDF of the shadowing power is equal to

$$f_S(z) = \frac{z^{c-1}}{y_0^c \Gamma(c)} \exp\left(-\frac{z}{y_0}\right) U(z), \; c > 0, \tag{3}$$

where parameters c and y_0 are related to the average power and standard deviation of shadowing.

Taking into account the simultaneous effect of fading and shadowing on the received signal, the PDF of the received signal power in (1) can be expressed as

$$f(p) = \int_0^\infty f_F(p \mid z) f_S(z) \, dz, \tag{4}$$

where f_F is the PDF of channel power, f_S is the PDF of the mean power.

Rayleigh-lognormal distribution [8], which is a mixture of Rayleigh and lognormal distributions, is probably the most appropriate description of signal envelope in fading-shadowing wireless channels [22]. But the complicated form of its PDF motivates to approximate lognormal distribution by gamma distribution and proceed with K distribution, which is a mixture of Rayleigh (2) and gamma (3) distributions.

3 K Distribution and Modified Bessel Functions

Random variable has K distribution with shape $\alpha > 0$ and scale $\lambda > 0$, if its PDF and cumulative distribution function (CDF) are equal to [3]

$$f_K(x) = \frac{2}{\lambda \Gamma(\alpha)} \left(\frac{x}{\lambda}\right)^{(\alpha-1)/2} K_{\alpha-1}\left(2\sqrt{\frac{x}{\lambda}}\right) U(x), \tag{5}$$

$$F_K(x) = 1 - \frac{2}{\Gamma(\alpha)} \left(\frac{x}{\lambda}\right)^{\alpha/2} K_\alpha\left(2\sqrt{\frac{x}{\lambda}}\right) U(x), \tag{6}$$

respectively, where Γ is the gamma function, $K_\nu(x)$ is the modified Bessel function of the second kind of order ν. Mean and variance of K distributed random variable X are equal to

$$\mathbb{E}[X] = \alpha\lambda, \; \mathbb{V}[X] = \lambda^2\alpha(\alpha+2).$$

The modified Bessel function of the second kind $K_\nu(x)$ in (5) and (6) is a solution of the modified Bessel's ordinary differential equation

$$x^2 \frac{d^2w}{dx^2} + x\frac{dw}{dx} - \left(x^2 + \nu^2\right) w = 0 \tag{7}$$

and, basically, can't be expressed in terms of elementary functions [12].

But in the case of half integer order ν $\left(\nu = \pm\frac{1}{2}, \pm\frac{3}{2}, ...\right)$ functions $K_\nu(x)$ can be expressed through elementary functions, for example:

$$K_{-\frac{1}{2}}(x) = K_{\frac{1}{2}}(x) = \sqrt{\frac{\pi}{2x}} e^{-x}, \; K_{\frac{3}{2}}(x) = \sqrt{\frac{\pi}{2x}} \left(\frac{1}{x} + 1\right) e^{-x}, ... \tag{8}$$

Modified Bessel functions of the second kind of higher half integer orders keep the same structure and are represented in the following form [7]

$$K_{n+\frac{1}{2}}(x) = \sqrt{\frac{\pi}{2x}} e^{-x} \sum_{k=0}^{n} \frac{(n+k)!}{k!\,(n-k)!\,(2x)^k}, \; n \geq 0. \tag{9}$$

For $x > 0$ and $\nu > 0$ function $K_\nu(x)$ is positive and monotonically decreasing (see plots of modified Bessel functions of the second kind of half integer orders in Fig. 1).

4 K-Fading Channel in Half Integer Case

If we set $b = \frac{2}{\sqrt{\lambda}}$ and $c = \alpha$ in PDF (5), we receive the PDF of power in shadowed fading channel (1) in the following form [18]

$$f_K(p) = \frac{2}{\Gamma(c)} \left(\frac{b}{2}\right)^{c+1} p^{\frac{c+1}{2}-1} K_{c-1}\left(b\sqrt{p}\right) U(p), \tag{10}$$

where b is a parameter related to the average power, c is a positive parameter related to the effective number of scatterers.

Rayleigh-lognormal and K distributions are similar, but K distribution has a simpler form, which makes it possible to obtain closed-form solutions in the calculation of bit error rates, diversity effects, etc. The calculations can be further simplified in half integer case $c = n + \frac{1}{2}$, where $n \in \mathbb{N}$ or $n = 0$.

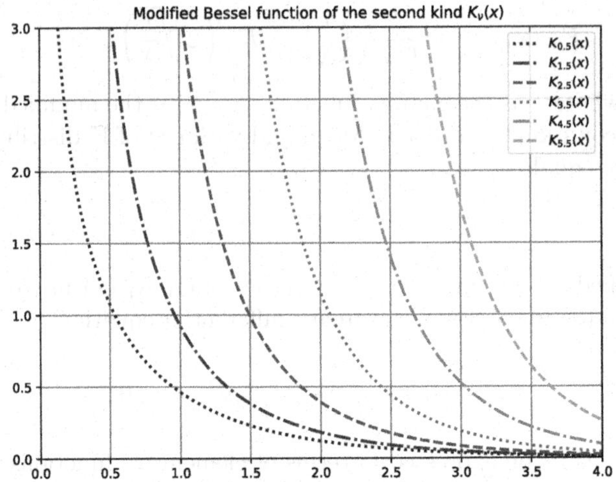

Fig. 1. Modified Bessel functions of the second kind $K_\nu(x)$.

In half integer case PDF (10) takes the form

$$f_K^{(n+\frac{1}{2})}(p) = \frac{2}{\Gamma\left(n+\frac{1}{2}\right)}\left(\frac{b}{2}\right)^{n+\frac{3}{2}} p^{\frac{n}{2}-\frac{1}{4}} K_{n-\frac{1}{2}}\left(b\sqrt{p}\right) U(p). \tag{11}$$

When $n = 0$, we take into account equalities (8) and value $\Gamma\left(\frac{1}{2}\right) = \sqrt{\pi}$ to receive expression for the first half-integer PDF

$$f_K^{(\frac{1}{2})}(p) = \frac{b}{2\sqrt{p}} e^{-b\sqrt{p}} U(p). \tag{12}$$

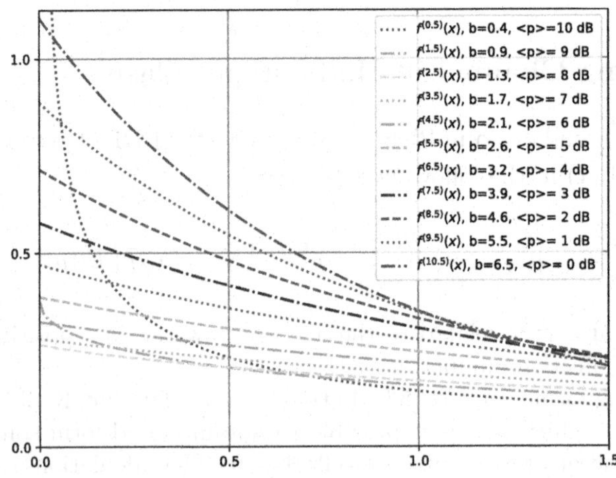

Fig. 2. Probability density functions $f_K^{(n+\frac{1}{2})}(p)$ in half integer case.

When $n \in \mathbb{N}$, gamma function for half integer argument is equal to

$$\Gamma\left(n+\frac{1}{2}\right)=\frac{(2n-1)!!}{2^n}\sqrt{\pi},$$

so from (9) we obtain that

$$K_{n-\frac{1}{2}}\left(b\sqrt{p}\right)=\sqrt{\frac{\pi}{2b\sqrt{p}}}e^{-b\sqrt{p}}\sum_{k=0}^{n-1}\frac{(n+k-1)!}{k!\,(n-k-1)!\left(2b\sqrt{p}\right)^k},$$

and PDF (11) can be expressed in terms of elementary functions as follows

$$f_K^{\left(n+\frac{1}{2}\right)}(p)=\frac{1}{(2n-1)!!}e^{-b\sqrt{p}}\sum_{k=0}^{n-1}\frac{(2n-k-2)!\,b^{k+2}}{(n-k-1)!\,k!\,2^{n-k}}p^{\frac{k}{2}}\,U(p)\,. \tag{13}$$

In particular, from (13) we receive the following representations for half integer PDF:

$$f_K^{\left(\frac{3}{2}\right)}(p)=\frac{b^2}{2}e^{-b\sqrt{p}}\,U(p)\,, \tag{14}$$

$$f_K^{\left(\frac{5}{2}\right)}(p)=\frac{1}{3}\left[\frac{b^2}{2}+\frac{b^3}{2}\sqrt{p}\right]e^{-b\sqrt{p}}\,U(p)\,. \tag{15}$$

Plots of functions $f_K^{\left(n+\frac{1}{2}\right)}(p)$ for various values of n and b are shown in Fig. 2. The value of b is determined by selected value $\langle p\rangle$ of average power in the channel.

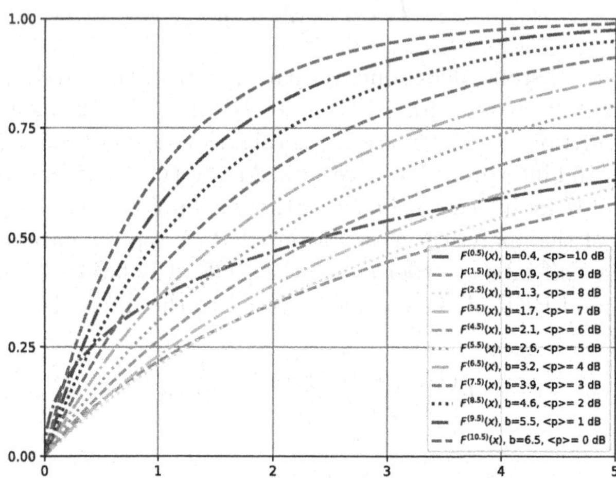

Fig. 3. Cumulative distribution functions $F_K^{\left(n+\frac{1}{2}\right)}(p)$ in half integer case.

In half integer case CDF of K distribution can also be essentially simplified. Using representation (9) we receive CDF of power in shadowed K-faded channel for half integer case ($n \geq 0$):

$$F_K^{\left(n+\frac{1}{2}\right)}(p) = 1 - \frac{1}{(2n-1)!!} e^{-b\sqrt{p}} \sum_{k=0}^{n} \frac{(2n-k)!b^k}{k!\,(n-k)!2^{n-k}} p^{\frac{k}{2}}\, U(p)\,. \tag{16}$$

Plots of distribution functions $F_K^{\left(n+\frac{1}{2}\right)}(p)$ for various values of n and b are demonstrated in Fig. 3.

5 Bit Error Rate for K-Fading Channel in Half Integer Case

Closed form expressions for power PDF and CDF in half integer case simplify calculation of miscellaneous performance criteria, for instance, the capacity of wireless channel [13].

Average bit error rate (BER) is an important measure for performance analysis of a digital wireless channel. Evaluation of BER for K-fading channel under DPSK (differential phase shift keying) scheme requires calculation of the integral

$$P_e = \int_0^{\infty} P_e(x)\, f_K(x)\, dx$$

with signal dependent BER $P_e(x)$ given by expression

$$P_e(x) = \frac{1}{2} e^{-s\,x},$$

where $s > 0$ is a constant, depending on parameters of the wireless channel.

Explicit exact expression for BER under DPSK [2,11,16] includes sophisticated special functions and do not allow straightforward determination of channel parameters and interpretation of results. An alternative is the approximation of power PDF of K-fading channel using orthogonal polynomials [21] with further determination of the analytical closed form BER of approximate model.

For the first half integer order $c = \frac{1}{2}$ the PDF of K-fading channel is determined by (12) and BER of the channel is equal to the integral

$$P_e^{\left(\frac{1}{2}\right)} = \frac{b}{4} \int_0^{\infty} \frac{1}{\sqrt{p}} e^{-s\,p - b\sqrt{p}} dp$$

Using the substitution

$$p = y^2,\ dp = 2ydy \tag{17}$$

we obtain

$$P_e^{\left(\frac{1}{2}\right)} = \frac{b}{2} \int_0^{\infty} e^{-s\,y^2 - by} dy = \frac{b}{2} e^{\frac{b^2}{4s}} \int_0^{\infty} e^{-s\left(y+\frac{b}{2s}\right)^2} dy$$

The next substitution

$$u = \sqrt{2s}\left(y + \frac{b}{2s}\right),\ du = \sqrt{2s}dy \tag{18}$$

results in equality

$$P_e^{\left(\frac{1}{2}\right)} = \frac{1}{\sqrt{2s}} e^{\frac{b^2}{4s}} \int\limits_{\frac{b}{\sqrt{2s}}}^{\infty} e^{-\frac{u^2}{2}} du.$$

Let us denote by J_0 the integral

$$J_0 = \int\limits_{\frac{b}{\sqrt{2s}}}^{\infty} e^{-\frac{u^2}{2}} du = \sqrt{2\pi}\Phi\left(u\right)\Big|_{\frac{b}{\sqrt{2s}}}^{\infty} = \sqrt{2\pi}\,\Phi\left(-\frac{b}{\sqrt{2s}}\right), \tag{19}$$

where $\Phi\left(x\right) = \frac{1}{\sqrt{2\pi}} \int_{-\infty}^{x} e^{-\frac{u^2}{2}} du$ is standard normal CDF.

Thus BER of wireless channel for the first half integer order is represented by the following simple formula

$$P_e^{\left(\frac{1}{2}\right)} = \frac{b}{2}\sqrt{\frac{\pi}{s}} e^{\frac{b^2}{4s}} \Phi\left(-\frac{b}{\sqrt{2s}}\right). \tag{20}$$

Dependence of average BER on parameters b and s for the first half integer order is characterized by the surface in Fig. 4.

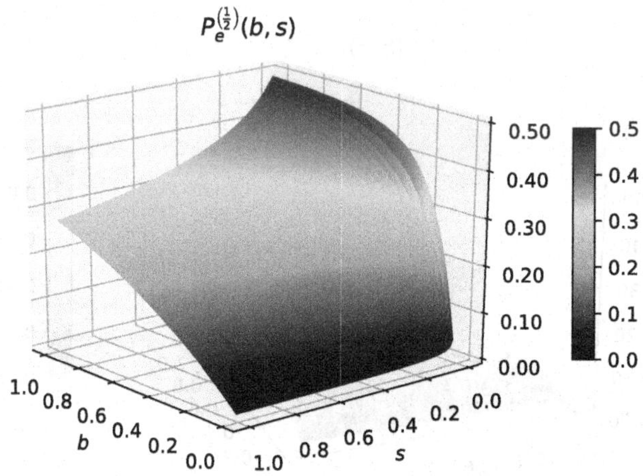

Fig. 4. Average BER surface for K-fading channel when $c = \frac{1}{2}$.

For the next half integer order $c = \frac{3}{2}$ the PDF of K-fading channel is determined by (14) and BER is equal to

$$P_e^{\left(\frac{3}{2}\right)} = \frac{b^2}{4} \int\limits_0^{\infty} e^{-s\,p - b\sqrt{p}} dp.$$

Using substitutions (17) and (18) we receive that

$$P_e^{\left(\frac{3}{2}\right)} = \frac{b^2}{4s} \int_{\frac{b}{\sqrt{2s}}}^{\infty} e^{-\frac{u^2}{2}} \left(u - \frac{b}{\sqrt{2s}}\right) du$$

Using notation J_1 for the integral

$$J_1 = \int_{\frac{b}{\sqrt{2s}}}^{\infty} e^{-\frac{u^2}{2}} u du = \int_{\frac{b}{\sqrt{2s}}}^{\infty} d\left(-e^{-\frac{u^2}{2}}\right) = -e^{-\frac{u^2}{2}}\Big|_{\frac{b}{\sqrt{2s}}}^{\infty} = e^{-\frac{b^2}{4s}}, \qquad (21)$$

we obtain that

$$P_e^{\left(\frac{3}{2}\right)} = \frac{b^2}{4s} \int_{\frac{b}{\sqrt{2s}}}^{\infty} d\left(-e^{-\frac{u^2}{2}}\right) - \frac{1}{2}\left(\frac{b}{\sqrt{2s}}\right)^3 J_0$$

$$= \frac{b^2}{4s} e^{-\frac{b^2}{4s}} - \sqrt{\frac{\pi}{2}} \left(\frac{b}{\sqrt{2s}}\right)^3 \Phi\left(-\frac{b}{\sqrt{2s}}\right)$$

Average BER surface for half integer order $c = \frac{3}{2}$ in Fig. 5 differs from average BER surface for previous half integer order $c = \frac{1}{2}$.

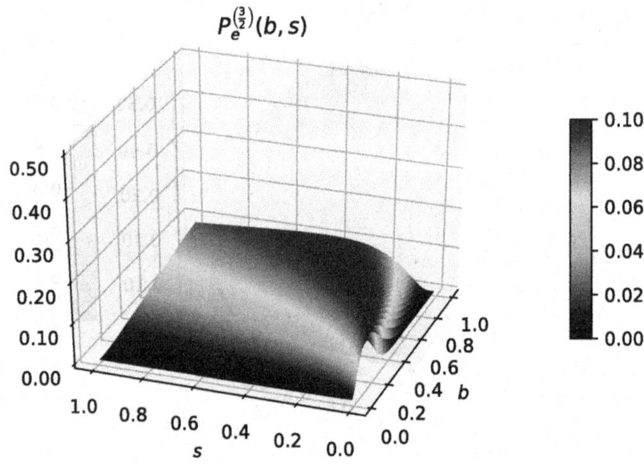

Fig. 5. Average BER surface for K-fading channel when $c = \frac{3}{2}$.

For the general half integer case $c = n + \frac{1}{2}$, $n \in \mathbb{N}$ the PDF of K-fading channel is determined by (13) and BER is equal to

$$P_e^{\left(n+\frac{1}{2}\right)} = \int_0^{\infty} \frac{1}{2} e^{-sp} f_K^{\left(n+\frac{1}{2}\right)}(p)\, dp$$

$$= \frac{1}{2}\frac{1}{(2n-1)!!}\sum_{k=0}^{n-1}\frac{(2n-k-2)!\,b^{k+2}}{(n-k-1)!\,k!\,2^{n-k}}\int_0^\infty e^{-s\,p-b\sqrt{p}}p^{\frac{k}{2}}\,dp \tag{22}$$

Application of substitutions (17) and (18) to the integrals in right hand side of (22) results in equality

$$\int_0^\infty e^{-s\,p-b\sqrt{p}}p^{\frac{k}{2}}\,dp = \frac{e^{\frac{b^2}{4s}}}{2^{\frac{k}{2}}s^{\frac{k}{2}+1}}\int_{\frac{b}{\sqrt{2s}}}^\infty e^{-\frac{u^2}{2}}\left(u-\frac{b}{\sqrt{2s}}\right)^{k+1}du$$

The integrand in the right hand side of equality is the product of an exponent and a polynomial, so the calculation of BER for general half integer order is reduced to the calculation of integrals

$$J_m = \int_{\frac{b}{\sqrt{2s}}}^\infty e^{-\frac{u^2}{2}}u^m du,\ m\geq 0.$$

For $m=0$ and $m=1$ the integrals J_m are already calculated in (20) and (21). For $m>1$ the integral J_m can be calculated using the following recurrent formula:

$$J_m = \int_{\frac{b}{\sqrt{2s}}}^\infty d\left(-e^{-\frac{u^2}{2}}u^{m-1}\right) + (m-1)\int_{\frac{b}{\sqrt{2s}}}^\infty e^{-\frac{u^2}{2}}u^{m-2}du = -e^{-\frac{u^2}{2}}u^{m-1}\Big|_{\frac{b}{\sqrt{2s}}}^\infty$$

$$+(m-1)\int_{\frac{b}{\sqrt{2s}}}^\infty e^{-\frac{u^2}{2}}u^{m-2}du = e^{-\frac{b^2}{4s}}\left(\frac{b}{\sqrt{2s}}\right)^{m-1} + (m-1)\,J_{m-2} \tag{23}$$

For instance, for half integer order $c=\frac{5}{2}$ the PDF is determined by (15), so BER is equal to

$$P_e^{(\frac{5}{2})} = \frac{b^2}{12}\int_0^\infty e^{-s\,p-b\sqrt{p}}\,(1+b\sqrt{p})\,dp.$$

After substitutions (17) and (18) the integral is reduced to

$$P_e^{(\frac{5}{2})} = \frac{b^2}{12s}\int_{\frac{b}{\sqrt{2s}}}^\infty e^{-\frac{u^2}{2}}\left(\frac{bu^2}{\sqrt{2s}}+\left(1-\frac{b^2}{s}\right)u-\left(1-\frac{b^2}{2s}\right)\frac{b}{\sqrt{2s}}\right)du.$$

According to (23)

$$J_2 = \int_{\frac{b}{\sqrt{2s}}}^\infty e^{-\frac{u^2}{2}}u^2du = e^{-\frac{b^2}{4s}}\frac{b}{\sqrt{2s}} + J_0 = e^{-\frac{b^2}{4s}}\frac{b}{\sqrt{2s}} + \sqrt{2\pi}\,\Phi\left(-\frac{b}{\sqrt{2s}}\right),$$

so finally average BER of K-fading channel for half integer order $c = \frac{5}{2}$ is represented by the formula

$$P_e^{\left(\frac{5}{2}\right)} = \frac{b^2}{12s}\left(e^{-\frac{b^2}{4s}} + \sqrt{2\pi}\,\Phi\left(-\frac{b}{\sqrt{2s}}\right)\left(\frac{b}{\sqrt{2s}}\right)^3\right)$$

with average BER surface in Fig. 6.

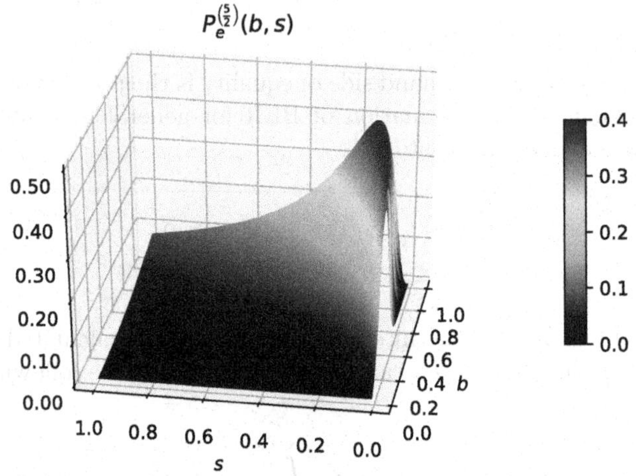

Fig. 6. Average BER surface for K-fading channel when $c = \frac{5}{2}$.

6 Monte Carlo Simulation with K Distribution

Monte Carlo method implies generation of random numbers with given probability distribution [14]. The most straightforward way to generate a non-uniform random variable is by inversion of its CDF: if given distribution is characterised by CDF F with known closed form inverse function (quantile function) F^{-1} and X is a random variable with continuous uniform distribution in the interval $(0, 1)$, then $Y = F^{-1}(X)$ is a random variable with given probability distribution.

For distributions with no closed-form inverse CDF one has to solve equation $F(x) = y$ for $y \in (0, 1)$ using approximations or numerical methods [6].

To find the quantile function of K distribution in general case, one has to solve transcendental equation

$$F_K(x) = 1 - \frac{2}{\Gamma(c)}\left(\frac{b}{2}\right)^c x^{\frac{c}{2}} K_c\left(b\sqrt{x}\right) = y,\ y \in (0, 1)\,. \tag{24}$$

Equation (24) can be solved using various approaches [9].

The basic version of Newton's method for solving (24) gives the following iterative process:

$$x_{l+1} = x_l - \frac{F_K(x_l) - y}{\frac{d}{dx}F_K(x_l)} = x_l - \frac{F_K(x_l) - y}{f_K(x_l)}$$

$$= x_l - \frac{(1-y)\,\Gamma(c) - 2\left(\frac{b}{2}\right)^c x_l^{\frac{c}{2}} K_c\left(b\sqrt{x_l}\right)}{2\left(\frac{b}{2}\right)^{c+1} x_l^{\frac{c-1}{2}} K_{c-1}\left(b\sqrt{x_l}\right)},\ l \geq 0. \quad (25)$$

The iterative process (25) stops when the required accuracy $\varepsilon > 0$ is achieved, i.e. $|F_K(x_l) - y| < \varepsilon$, or when absolute or relative error approximation is below the given tolerance $\epsilon > 0$, i.e. $|x_{l+1} - x_l| < \epsilon$ or $\left|\frac{x_{l+1}}{x_l} - 1\right| < \epsilon$.

The Newton's method of quantile function construction requires an initial starting point $x_0 = x_0(y)$ with good accuracy and a sufficiently efficient algorithm for modified Bessel function K_c and K_{c-1} calculation in (25).

Existence of approximations to the quantile function of K distribution has not been investigated, so the determination of an initial guess requires either application of general considerations [26], or approximation of K distribution by similar distribution with known quantile function. For instance, general K distribution can be approximated by Rayleigh or Weibull distribution, or K distribution with parameter $c = \frac{1}{2}$. In the latter case CDF is equal to $F_K(x) = 1 - be^{-b\sqrt{x}}$ and the quantile function is equal to $x = F_K^{-1}(y) = \left(\frac{1}{b}\ln\left(\frac{1}{b}(1-y)\right)\right)^2$.

Nowadays digital signal processing is often performed in Python [5], so the simulation with K distribution is carried out in Python.

Efficient calculation of modified Bessel functions of the second kind $K_\nu(x)$ in half integer case $\nu = n + \frac{1}{2}$ can be implemented in Python using formula (9). As an alternative, modified Bessel functions of the second kind are implemented in SciPy library [23] as wrappers for AMOS package routines in Fortran [4].

Performance of modified Bessel function calculation for half integer case is tested by execution of 1 million modified Bessel function calls with results in Table 1 (execution time in milli-seconds).

Table 1. Execution time of modified Bessel function calls for half integer case.

ν	1.5	2.5	3.5	4.5	5.5	6.5	7.5	8.5	9.5	10.5
$kv_\nu(x)$	107.7	195.5	181.7	194.5	210.4	197.2	187.6	197.1	198.3	205.7
$K_\nu(x)$	26.4	29.7	49.3	72.4	98.9	116.9	138.0	153.9	175.1	209.2

In Table 1 ν is the order of the modified Bessel function of the second kind, $kv_\nu(x)$ corresponds to modified Bessel function implementation in SciPy and $K_\nu(x)$ corresponds to modified Bessel function implementation in Python using formula (9).

Table 1 demonstrates that for small half integer orders ν implementation, based on formula (9), is essentially more efficient than implementation in SciPy, but for larger half integer orders execution time is almost the same.

If the order of modified Bessel function of the second kind is not half integer, execution time of calls is essentially larger as it follows from Table 2.

Table 2. Execution time of modified Bessel function calls for non half integer case.

ν	1.51	2.51	3.51	4.51	5.51	6.51	7.51	8.51	9.51	10.51
$kv_\nu(x)$	373.0	433.7	445.7	446.7	438.3	441.6	441.3	448.6	456.7	459.4

Thus, use of formula (9) for modified Bessel function implementation can bring essential gain in efficiency of computing.

In half integer case the problem of Monte Carlo simulation with K distribution can be easily implemented using CDF inversion approach and Newton's iterative method. Algorithm 1) describes Monte Carlo simulation with K distribution and subsequent construction of empirical density function from the simulated values.

Algorithm 1 for Monte Carlo simulation with K distribution and construction of empirical density function is implemented in Python and produces empirical density function in Fig. 7 similar to theoretical density function of K distribution.

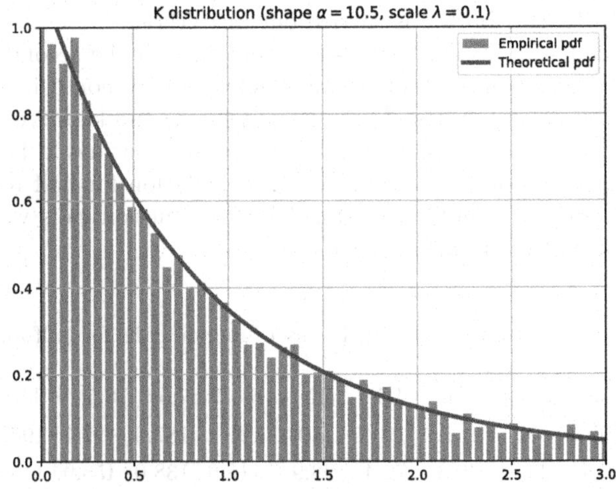

Fig. 7. Empirical and theoretical densities for K distribution

Since samples in Monte Carlo method are generated independently of each other, Monte Carlo simulation is naturally suited to parallel computing. So Monte Carlo simulation with K distribution can be implemented using modern multi-core processors or graphics processing units (GPU).

Algorithm 1: Construction of empirical density with K distribution.

Data: number of trials N, number of bins M, order $c = n + \frac{1}{2}$, $n \in \mathbb{N}$, accuracy $\varepsilon > 0$ (or error tolerance $\epsilon > 0$);

Result: empirical PDF $\rho_K^{(n+\frac{1}{2})}(x)$ of random variable with K distribution

for $i \leftarrow 1$ **to** N **do**

  create a pseudo-random number y_i, uniformely distributed on $(0, 1)$;

  `/* solve equation` $F_K^{(n+\frac{1}{2})}(x_i) = y_i$ `for` $x_i \in (0, +\infty)$ `*/` $\tilde{x}_0 \leftarrow x_0(y_i)$, $l \leftarrow 0$; `/* an initial approximation */`

  repeat `/* Newton's method */`

    $l \leftarrow l + 1$;

    $$\tilde{x}_l \leftarrow \tilde{x}_{l-1} - \frac{(2n-1)!!(1-y)e^{b\sqrt{\tilde{x}_{l-1}}} - \sum_{k=0}^{n} \frac{(2n-k)!b^k}{k!(n-k)!2^{n-k}} \tilde{x}_{l-1}^{\frac{k}{2}}}{\sum_{k=0}^{n-1} \frac{(2n-k-2)!\, b^{k+2}}{(n-k-1)!\, k!\, 2^{n-k}} \tilde{x}_{l-1}^{\frac{k}{2}}};$$

  until $|F_K(\tilde{x}_l) - y| < \varepsilon$ $(|\tilde{x}_l - \tilde{x}_{l-1}| < \epsilon)$; `/* accuracy/error tolerance */`

  $x_i \leftarrow \tilde{x}_l$;

end

$m \leftarrow \max \{x_i\}_{i=1}^{N}$;

divide the half-interval $(0, m]$ into M equal parts of length $h = \frac{m}{M}$ and calculate ρ_j as the number of values from the set $\{x_i\}_{i=1}^{N}$, that belong to half-intervals of the form $((j-1)\,h,\, j\,h]$ for $j = \overline{1, M}$;

$$\rho_K^{(n+\frac{1}{2})}(x) \leftarrow \begin{cases} 0, & x \le 0, \\ \frac{1}{N}\rho_j, & (j-1)\,h < x \le j\,h,\ j = \overline{1, M}, \\ 0, & x > m \end{cases}$$

7 Conclusion

Investigation of wireless channel, modeled by K distribution with modified Bessel function of half integer order, gives an opportunity to receive the framework with closed form PDF and CDF expressions and simplifies calculation of wireless channel performance measures and simulation of wireless channel performance. Consideration of Bessel functions of half integer orders may be useful in other models of wireless channel with fading and shadowing effects.

References

1. Abdi, A., Kaveh, M.: K distribution: an appropriate substitute for rayleigh-lognormal distribution in fading-shadowing wireless channels. Electron. Lett. **34**(9), 851–852 (1998). https://doi.org/10.1049/el:19980625
2. Abdi, A., Kaveh, M.: Comparison of DPSK and MSK bit error rates for k and rayleigh-lognormal fading distributions. IEEE Commun. Lett. **4**(4), 122–124 (2000). https://doi.org/10.1109/4234.841317
3. Abraham, D.A.: Underwater Acoustic Signal Processing. MASP. Springer, Cham (2019). https://doi.org/10.1007/978-3-319-92983-5

4. Amos, D.E.: Algorithm 644: a portable package for Bessel functions of a complex argument and nonnegative order. ACM Trans. Math. Softw. **12**(3), 265–273 (1986). https://doi.org/10.1145/7921.214331
5. Charbit, M.: Digital Signal Processing with Python Programming. Wiley, Hoboken, December 2016. https://doi.org/10.1002/9781119373063
6. Dagpunar, J.S.: Simulation and Monte Carlo. Wiley, Hoboken, January 2007. https://doi.org/10.1002/9780470061336
7. Gradshteyn, I., Ryzhik, I.: Table of Integrals, Series, and Products, 7th edn. Elsevier (2007). https://doi.org/10.1016/c2010-0-64839-5
8. Hansen, F., Meno, F.: Mobile fading–rayleigh and lognormal superimposed. IEEE Trans. Veh. Technol. **26**(4), 332–335 (1977). https://doi.org/10.1109/t-vt.1977.23703
9. Hauser, J.R.: Numerical Methods for Nonlinear Engineering Models. Springer, Netherlands (2009). https://doi.org/10.1007/978-1-4020-9920-5
10. Jakeman, E., Pusey, P.N.: Significance of K distributions in scattering experiments. Phys. Rev. Lett. **40**(9), 546–550 (1978). https://doi.org/10.1103/physrevlett.40.546
11. Kiasaleh, K.: Performance of coherent DPSK free-space optical communication systems in k-distributed turbulence. IEEE Trans. Commun. **54**(4), 604–607 (2006). https://doi.org/10.1109/tcomm.2006.873067
12. Korenev, B.G.: Bessel Functions and Their Applications. CRC Press, Boca Raton, July 2002. https://doi.org/10.1201/b12551
13. Laourine, A., Slim Alouini, M., Affes, S., Stephenne, A.: On the capacity of generalized-k fading channels. IEEE Trans. Wireless Commun. **7**(7), 2441–2445 (2008). https://doi.org/10.1109/twc.2008.070103
14. Metropolis, N., Ulam, S.: The Monte Carlo method. J. Am. Stat. Assoc. **44**(247), 335–341 (1949). https://doi.org/10.1080/01621459.1949.10483310
15. Modest, M.: Radiative Heat Transfer, 3rd edn. Elsevier (2013). https://doi.org/10.1016/c2010-0-65874-3
16. Samimi, H.: Performance analysis of free-space optical links with transmit laser selection diversity over strong turbulence channels. IET Commun. **5**(8), 1039–1043 (2011). https://doi.org/10.1049/iet-com.2010.0075
17. Shankar, P.M.: Error rates in generalized shadowed fading channels. Wireless Pers. Commun. **28**(3), 233–238 (2004). https://doi.org/10.1023/B:wire.0000032253.68423.86
18. Shankar, P.M.: Fading and Shadowing in Wireless Systems. Springer, Cham (2017). https://doi.org/10.1007/978-3-319-53198-4
19. Shreider, Y. (ed.): The Monte Carlo Method: The Method of Statistical Trials. Elsevier (1966). https://doi.org/10.1016/c2013-0-01870-1
20. Simon, M.K., Alouini, M.S.: Digital Communication over Fading Channels, 2nd edn. Wiley, Hoboken, November 2004. https://doi.org/10.1002/0471715220
21. Singh, R.K., Kumar, S.K.: A novel approximation for k distribution: closed-form BER using DPSK modulation in free-space optical communication. IEEE Photonics J. **9**(5), 1–14 (2017). https://doi.org/10.1109/jphot.2017.2746763
22. Stüber, G.L.: Principles of Mobile Communication. Springer, Cham (2017). https://doi.org/10.1007/978-3-319-55615-4
23. Virtanen, P., et al.: SciPy 1.0: fundamental algorithms for scientific computing in python. Nature Methods **17**(3), 261–272 (2020). https://doi.org/10.1038/s41592-019-0686-2
24. Ward, K.: Compound representation of high resolution sea clutter. Electron. Lett. **17**(16), 561 (1981). https://doi.org/10.1049/el:19810394

25. Ward, K., Tough, R., Watts, S.: Sea Clutter: Scattering, the K Distribution and Radar Performance. Institution of Engineering and Technology, 2nd edition edn. (2013). https://doi.org/10.1049/pbra025e
26. Yu, C., Zelterman, D.: A general approximation to quantiles. Commun. Stat. Theory Methods **46**(19), 9834–9841 (2016). https://doi.org/10.1080/03610926.2016.1222433

Melanoma Detection Computer System Development with Deep Neural Networks

Eugene Yu. Shchetinin[1], Leonid A. Sevastianov[2,3], Dmitry S. Kulyabov[2,3](✉), Edik A. Ayryan[3], and Anastasia V. Demidova[2]

[1] Financial University, Government of the Russian Federation, Moscow, Russian Federation
riviera-molto@mail.ru

[2] Peoples' Friendship University of Russia (RUDN University), Moscow, Russian Federation
{sevastianov-la,kulyabov-ds,demidova-av}@rudn.ru

[3] Joint Institute for Nuclear Research, Dubna, Russian Federation
ayrjan@jinr.ru

Abstract. This paper examines the problem of detecting skin malignancies, in particular, melanoma, from the analysis of dermoscopic images using deep learning methods. For this purpose, a deep convolutional neural network architecture was developed, which was used to process dermoscopic images of various skin lesions contained in the HAM10000 data set. The studied images were previously cleared of noise and other contaminants for processing by neural networks. In addition, since the disease classes are unbalanced, a number of transformations have been made to balance them. At the first stage, the images were divided into two classes: melanoma and benign tumor. At the second stage, all images of skin injuries were grouped into seven classes. Computer experiments on the use of the constructed deep neural network on the data obtained in this way have shown that the proposed approach provides 91%.

Keywords: Melanoma · Classification · Deep learning · Convolutional neural networks

1 Introduction

Melanoma is a deadly form of skin cancer that is often undiagnosed or misdiagnosed as a benign skin lesion. Its early and accurate detection is extremely important, because patients lives depend on it. In their practice, doctors are accustomed to rely on their professional experience and evaluate the injuries of each patient on the basis of a personal examination. However, such a system for detecting skin lesions is time-consuming and requires time and resources [7,18,24]. In addition, the process of manual dermatoscopy is more prone to diagnosis errors, requires many years of experience in complex situations and a huge number of visual studies of similarities and differences between various skin lesionss [13,20].

© Springer Nature Switzerland AG 2020
V. M. Vishnevskiy et al. (Eds.): DCCN 2020, CCIS 1337, pp. 422–434, 2020.
https://doi.org/10.1007/978-3-030-66242-4_33

Clinical studies allow us to obtain an accuracy of the diagnosis of melanoma from 65 to 80%, which was a good result for some time [9,16]. However, modern research claims that the use of dermoscopic images in diagnosis significantly increases the accuracy of diagnosis of skin lesions. However, the visual differences between melanoma and benign skin lesions can be very small, making diagnosis difficult even for an expert doctor. Recent advances in the use of artificial intelligence methods in the analysis of medical images have allowed us to consider the development of intelligent medical diagnostics systems based on visualization as a very promising direction that will help the doctor in making more effective decisions about the health of patients and making a diagnosis at an early stage and in adverse conditions [5].

In this paper, we investigate an approach to solving the problem of classification of skin diseases, namely, detection of melanoma, based on deep learning methods. For this purpose, the architecture of a deep convolutional neural network was developed, which was applied to the processing of dermoscopic images of various skin lesions contained in the set of dermoscopic images HAM10000 [6]. The studied data was previously processed to eliminate noise, contamination, and change the size and format of images. In addition, since the disease classes are unbalanced, a number of transformations were performed to balance them. The data obtained in this way were divided into two classes: Melanoma and Benign. Computer experiments on the use of a built deep neural network on the data obtained in this way have shown that the proposed approach provides an accuracy of 91% on the test sample, which exceeds similar results obtained by other algorithms.

2 Modern Results in Computer Recognition of Dermatoscopic Images

Most classical methods in the field of melanoma classification rely on manual selection of features such as the type of lesion (primary morphology), lesion configuration (secondary morphology), color, distribution, shape, texture, and uneven borders of the pigment spot [21] then, after extracting the main characteristics of images, machine learning methods such as the K-nearest neighbor (k-NN) algorithm, logistic regression, decision trees, and others are used to solve the classification problem [11].

Modern computer research on the diagnosis of skin diseases in order to detect melanoma actively implements deep learning methods and is aimed at improving existing and developing new models of deep neural networks, primarily convolutional neural networks (CNN) [14,27]. Esfahani et al. [19] proposed a CNN architecture for the diagnosis of melanoma, where clinical images were pre-processed in such a way as to reduce the illumination of the image. Research results have shown that the proposed method is able to diagnose cases of melanoma in 70% of cases.

Mahbod et al. showed that convolutional neural networks are superior to traditional machine learning methods [17]. The authors proposed a hybrid automated computerized method for classifying skin diseases using three pre-trained deep networks (AlexNet, VGG16, ResNet-18) to extract the features. The features extracted in this way are then used to train the support vector machine on 150 images from the ISIC 2017 dataset. Chelebi et al. [10] proposed an ensemble of threshold methods for determining the boundaries of skin lesions.

Heckler et al. [15] applied deep learning methods to classify histopathological diagnosis of melanoma and compared the result with qualified histopathologists. Esteva et al. [25] implemented a pre-trained INCEPTIONV3 network to classify nine classes, where they used a labeled set of dermatological data that has 3374 dermoscopic images, 129,450 clinical images, and reaches an accuracy of 72%. Harangi and co-authors [8] used the ensemble method DCNN (deep convolutional neural network), where they combined the result of four different architectures, improving the accuracy of melanoma classification on the 2017 ISBI dataset.

XI and Bovik [26] proposed a method for segmentation of skin lesions in which the CNN model is combined with a genetic algorithm. In [2], a method for segmentation of skin lesions on dermoscopic images from the ISIC 2017 data set and their classification of various types of skin cancer using deep neural network models Mask R-CNN and U-net is proposed. The proposed method consists of preprocessing and segmentation using a hybrid learning algorithm. The goal of the first stage is to remove noise using the filtering method. In the second stage, images are segmented based on the clustering method. In [1], it was proposed to use the deep convolutional network ResNet50 for recognition of melanoma.

3 Data Description and Their Pre-processing

In clinical dermatology, there are relatively few data sets with digital images of skin lesions. Most of these sets are too small and/or not publicly available, which creates an additional barrier to research in this area. Examples of such dermatology-related image datasets are: the Dermofit image Library [23] – a dataset containing 10 different classes, including 1,300 high-quality images of skin lesions collected worldwide. Dermnet [22] – the website-enabled Atlas of skin diseases contains more than 23,000 skin images divided into 23 classes. In early 2016, the international biomedical imaging Symposium (ISBI) published a set of data for analyzing skin lesions for early detection of melanoma [4].

In order to support the training of clinical dermatologists and the development of new information technologies, the International society for skin imaging (ISIC) has developed an international repository of dermoscopic images, known as the ISIC archive. This data set contains pigmented skin lesions obtained using standard Dermoscopy. Every year, ISIC adds new images to its archive and promotes the task of implementing computer methods for detecting melanoma and other skin diseases. For example, the HAM10000 data set created within this organization served as data for the ISIC-2018 Challenge [3]. In 2019, the number of samples already numbered more than 25,000 images for Dermoscopy, available for training in 7 different categories. In Table 1 examples of images from the studied data are given. The lesion classes in the HAM10000 dataset are listed below.

1. nv: Melanocytic nevi-benign neoplasms of melanocytes [6705 images];
2. Mel: Melanoma-malignancy [1113 images];
3. bkl: Benign keratosis - a common class that includes seborrheic keratosis, solar lentigo, and lichen-squamous as keratosis [1099 images];
4. bcc: Basal cell carcinoma is a common variant of epithelial skin cancer that rarely metastasizes, but grows if untreated [514 images];
5. akiec: Actinic keratosis and intraepithelial carcinoma are common non-invasive variants of squamous cell carcinoma [327 images];
6. Vasc: Vascular lesions of the skin of cherry angiomas to angiokeratoma and pyogenic granulomas [142 images];
7. Df: Dermatofibroma-a benign skin lesion [115 images].

Table 1. Image examples from ISIC archive

Melanoma	Benign

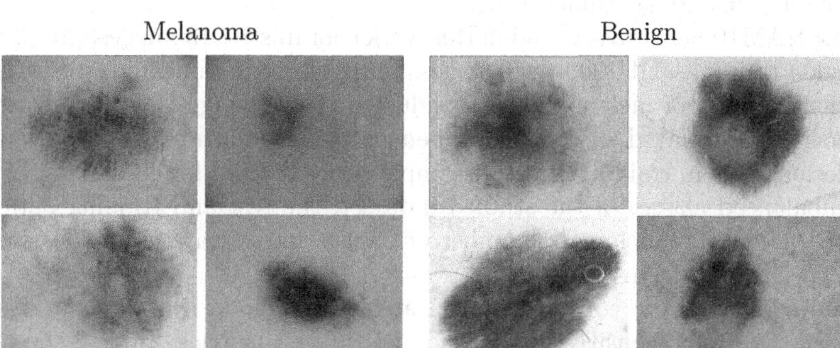

4 Building a Deep Neural Network Model and Computer Experiments

In this paper, a model of the CNN deep convolutional neural network was constructed to analyze dermoscopic images and detect melanoma in them [23]. The CNN model is initialized as a sequence of layers using the Sequential class. Next, a convolutional Conv2D layer was added, with the input parameters of the feature map `input_shape` = (32, 32, 3), where 32 is the size of the spatial features of the input map, and 3 is the number of color channels (in this case, the color of the image in RGB format). Convolution is defined by the following parameters: the size of templates extracted from input data is (3×3); the depth of the output feature map is the number of filters calculated by the convolution. In this model, the first convolutional layer outputs a feature map of size (30,30,32) and calculates 32 filters based on input data. Each of these 32 output channels contains a 30×30 value grid-a map of filter responses to input data that defines the response of this filter template for different sections of input data. The last parameter of the convolutional layer is the activation function that we use to activate neurons in the neural network, specifically `'relu'`. In the third step, the pooling layer (MaxPooling2D) with the map (3 × 3) was used. The main purpose of using this layer is to reduce the number of coefficients in the feature map for processing, as well as to implement spatial filter hierarchies by creating successive convolution layers for viewing larger and larger windows. Then create a vector for the fully connected `flatten()_layer`. The flatten layer serves as a link between the data received by the network and the output vector, converting the multidimensional output of the preceding layer into a one-dimensional (vector). In the last step, a fully connected layer is built – the density layer. The density function has 2 parameters – the number of nodes for the output layer(128) and the '*relu*' activation function. The output layer has 1 node where the `'sigmoid'` activation function is used. Next, you need to compile the model and optimize the weight coefficients and loss function for evaluating the model. The Adam algorithm is selected as the optimizer for our model. The loss function is `binary_crossentropy`, since we have two classes (1 or 0: Benign (Benign) or Malignant (Melanoma) tumors).

The HAM10000 data set used in this work contains 10013 images, pre-divided into 8000 training and 2000 test images. Computer experiments in research on the identification of melanoma was conducted in two stages. In the first stage, all images were placed as containing benign Benign and malignant Melanoma skin lesions. Then, training and test samples were created for each class. Due to the unbalanced classes of the studied data set, the test and training data sets were reduced by reducing the number of images in separate classes, as well as using special algorithms [22].

The quality of the constructed neural network model was evaluated using the basic Accuracy, Sensitivity, Precision and Recall metrics. Table 2 shows the confusion matrix obtained as a result of the neural network operation on the test sample. From the Table 2 we get Accuracy=91.4%, $Precision = TP/(TP+FP) = 92\%$, $Recall = TP/(TP+FN) = 0.91\%$. Similar computer experiments were performed with the architecture of convolutional networks U-net [8] and LeNet [19]. Their results are shown in Table 3.

All experiments were performed using a computer equipped with a core i5 processor, 8 GB SDRAM, and an NVIDIA GeForce 920 M graphics card. Keras and TensorFlow were used to develop a neural network model and train it [4] (Fig. 1).

Table 2. Confusion Matrix for proposed DCNN

Matrix		Actual class	
		Melanoma	Non-Melanoma
Prediction	Melanoma	69	6
Class	Non-Melanoma	7	68

Table 3. Accuracy metric values obtained by various neural network models in the test sample

Neural network model	Accuracy, %
DCNN	91
U-net [30]	74
LeNet [14]	78.2

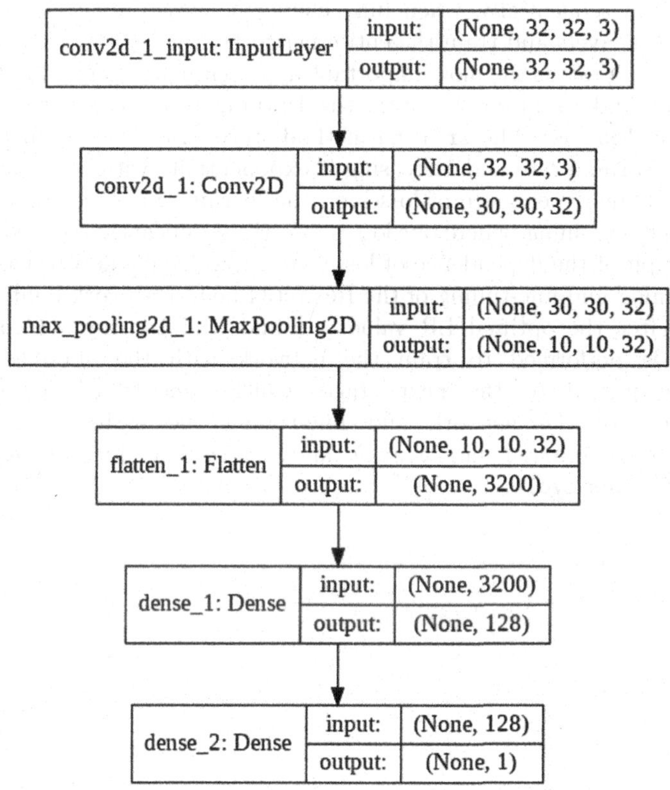

Fig. 1. The architecture of the deep convolutional neural network (DCNN)

At the second stage of research, computer experiments were conducted on the multiclass classification of the HAM10000 dataset described in this paper above. To solve this problem, we used models of deep neural networks, such as ResNet50, MobileNet, InceptionV3, and VGG16. The choice of these models is due to their popularity among researchers and their effectiveness in applying to the processing of complex images. In this paper, we also propose a strategy for training and fine-tuning the parameters of the ResNet34 deep neural network model, previously trained on ImageNet data [4]. For its implementation, we used the FastAI deep learning library [3], one of the most effective neural networks for training. The FastAI library provides strategies for improving the learning efficiency of deep neural networks based on varying the learning rate by setting different values for different parts of the network. Its feature and advantage is the use of the method of freezing network layers, when by default only the last fully connected layers are trained first. To do this, use the `fit` and `fit_one_cycle` algorithms. The `fit` method is a way to train a neural network at a constant LR learning rate, while the `fit_one_cycle` method uses the `one_cycle` principle, where the learning rate changes over time to achieve better accuracy. After applying them sequentially, when fully connected layers are trained, you can unfreeze other layers and train the entire network.

The learning rate is the most important hyperparameter for training a neural network. To find its optimal values, the learning process uses the strategies `lr_find_and_learner`. The `lr find` method starts trial training first, starting with a low learning rate and increasing it exponentially with each batch. From the graph of the dependence of losses on the learning rate, we can choose the highest level of training, where our losses are still significantly reduced. Figure 2 shows a graph of the dependence of losses on the value of the LR learning rate during training and fine-tuning of the ResNet34 neural network built by us.

Then, after the optimal LR value is found, the `learner.fit_one_cycle` procedure is performed to train the network with the obtained LR. Its results are passed to the next stage, which consists of defrosting the upper layers of the network and fine-tuning the entire network using the procedures `learner.unfreeze()` and `learner.lr_find()`, as well as `learner.fit_one_cycle`.

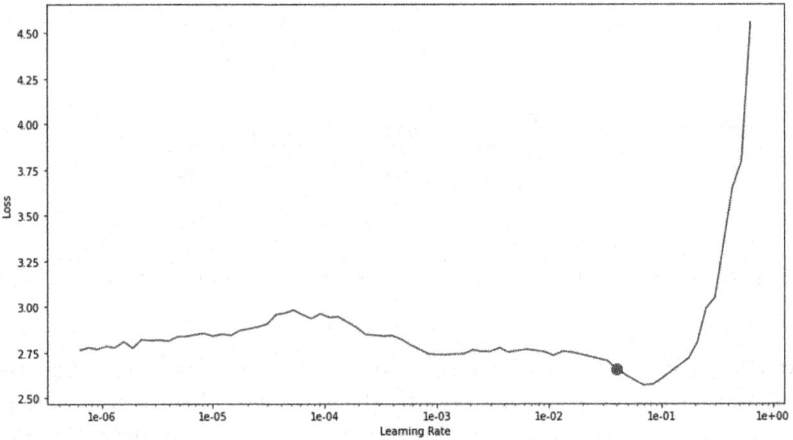

Fig. 2. Graph of the loss dependence on the learning rate using the `lr_find()` procedure

Recall that the graph of the ROC curve represents the dependence of the TPR indicator on the FPR. The true positive rate of the classification (TPR) shows which percentage of objects of the positive class the algorithm predicted correctly. The false positive rate of the classifications (FPR) shows which percentage of objects of the negative class the algorithm predicted incorrectly. The area under the ROC curve AUC (Area Under Curve) is an effective characteristic of classification quality. This indicator is often used for comparative analysis of several classification models [12]. For the ResNet34 model, it was 98.4%. The AUC-ROC criterion is more resistant to unbalanced classes and can be interpreted as the probability that a randomly selected positive object will be ranked higher by the classifier (it will have a higher probability of being positive) than a randomly selected object from the negative class.

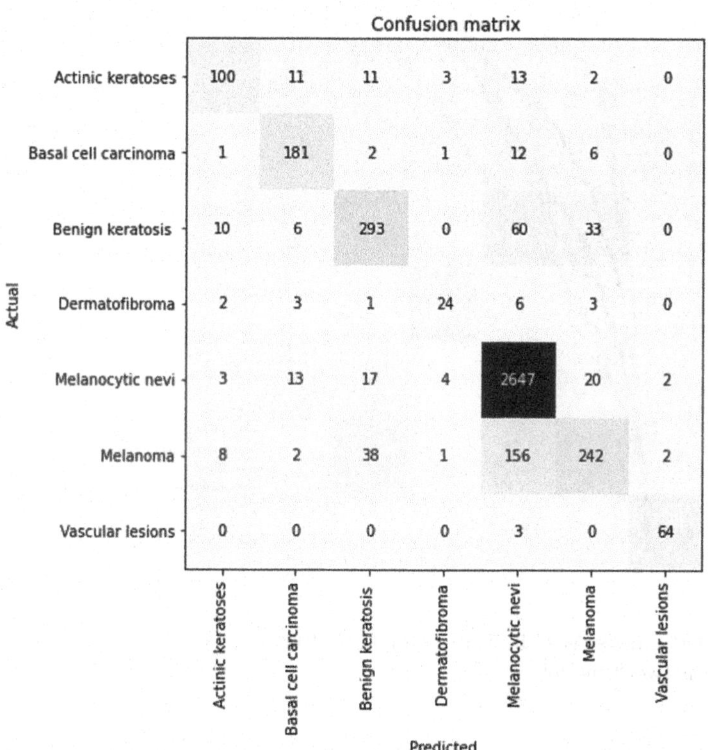

Fig. 3. Confusion matrix of the ResNet34 neural network on HAM10000

Another indicator of the performance of the neural network model is the confusion matrix. Figure 3 shows the confusion matrix for the ResNet34 model, from which the accuracy value was obtained for the test sample equal to 88.75%. Figures 4 and 5 show graphs of ROC curves, as well as AUC indicators for all classes of skin damage from HAM1000, obtained using the ResNet34 and

Table 4. Accuracy metric values obtained by various neural network models in the test sample.

Neural network model	Accuracy, %
ResNet-34_FastAI	88.75
InceptionV3 [21]	72.2
MobileNet	78.2
VGG16 [18]	84.3

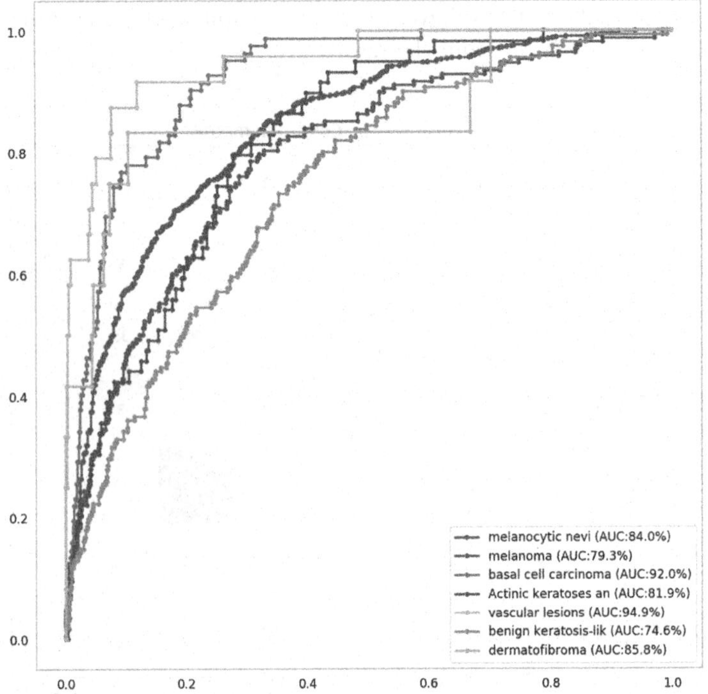

Fig. 4. ROC curves-AUC indicators for HAM1000 dataset classes obtained using the ResNet34 network model

MobileNet models, respectively. Table 3 shows the results of applying various neural network models to classify images of skin lesions from HAM10000. It follows that the ResNet34 model is significantly more accurate than the other neural network models we have studied (Table 4).

All experiments were performed using a computer equipped with an Intel Core i5 processor, 8 GB SDRAM, and an NVIDIA GeForce 920M graphics card. The estimated time spent on getting the achieved accuracy in the first case was from a few minutes to half an hour, depending on the number of epochs (from 25 to 100). In the second case, the specified accuracy required the use of the

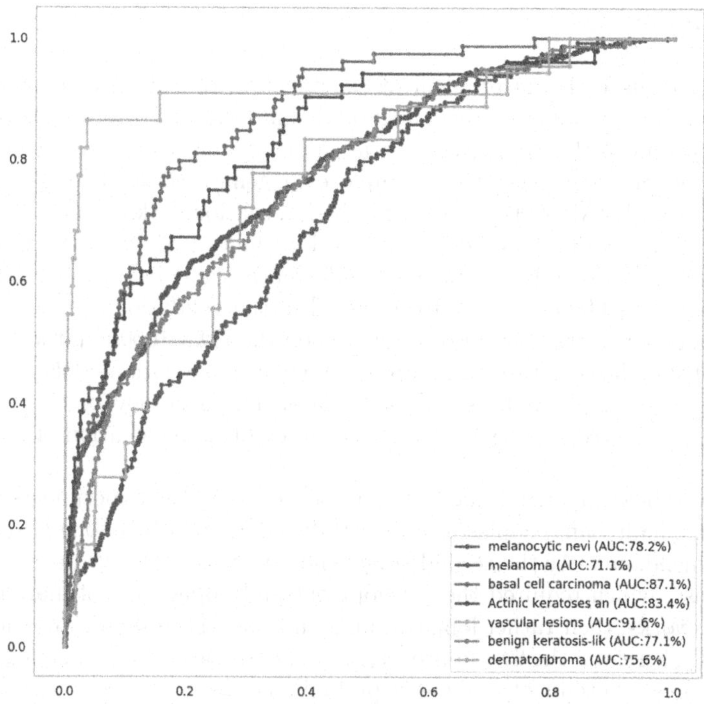

Fig. 5. ROC curves-AUC indicators for HAM1000 dataset classes obtained using MobileNet

Colab.Research.Google.com environment and 11 GB k80s or 14 GB Tesla T4S GPUs. The calculation time was about 11 h.

Dermoscopic images alone are not sufficient for a dermatological diagnosis or for reliable remote triage of a patient in a teledermatology setting. Dermoscopic images have no context. To include it in research, you need to create an image acquisition Protocol that includes panoramic images of the entire patient's body, as well as approximative images of each lesion, which are taken using a ruler or other reference frame visible in the image to provide contextual information about the size of the lesion. The images obtained with the ruler are also important for the patient who is already being treated to allow the accompanying doctor to monitor the development of the lesion. Both panoramic and approximation images must be performed in accordance with a protocol that ensures that the images are in focus, taken from the correct distance, and with the correct illumination. There are also details that cannot be reliably detected using the standard Dermoscopy technique currently used, and in some cases a biopsy will be required to confirm the diagnosis.

5 Conclusion

The paper deals with the problem of classification of skin diseases and recognition of melanoma among other diagnosed skin lesions. To solve this problem, a deep convolutional neural network was implemented, which was then applied to solve the problem using the example of a generated set of digital images extracted from HAM10000 provided by the international Skin Imaging Collaboration (ISIC). The developed neural network includes various methods of image preprocessing in order to obtain more informative and balanced training and control samples. The accuracy of the model in the validation sample was 91%. Computer experiments were also conducted on the HAM10000 multiclass classification of skin diseases using a group of deep neural network models pre-trained on ImageNet and tuned to HAM10000 data. The accuracy of the ResNet34 model we configured was 88%, and the accuracy of its recognition of melanoma was 79%.

It should be emphasized that the first stage of computer experiments is aimed at detecting malignant neoplasms in the studied data, which, among other things, includes melanoma. At the second stage, the task was to detect melanoma as a skin disease, which required the development and application of more complex neural networks, their rather long training, and the cost of significant computing resources. Therefore, the first approach can be used, for example, in telemedicine, as an Express method of making a preliminary diagnosis, as a signal to the patient to contact specialists for further clarification of the diagnosis. The second approach has several advantages and disadvantages. First of all, despite the focus on detecting melanoma, it is limited by computational costs. On the other hand, in the absence of the possibility of personal consultation with a professional dermatologist, obtaining a diagnosis with high accuracy based on the proposed approach remains almost the only way to solve the problem.

We have shown that the use of deep neural networks is a promising approach to solving the problem of automated recognition of various dermatological diseases, including such dangerous ones as melanoma. Our further research will focus on improving the architecture of deep neural networks to improve accuracy, and exploring the possibility of using other models of deep neural networks. In addition, efforts will be made to make this model available and usable as a mobile device application in telemedicine systems.

Acknowledgments. The publication has been prepared with the support of the "RUDN University Program 5–100".

References

1. Dermnet – skin disease atlas. http://www.dermnet.com/
2. Dermofit image library. https://licensing.eri.ed.ac.uk/i/software/dermofit-image-library.html
3. Github.com. https://github.com/fastai/fastai/
4. Imagenet database. http://www.image-net.org/

5. Isbi'16. https://biomedicalimaging.org/2016/
6. Isic archive. https://www.isic-archive.com/
7. American cancer society, cancer facts and figures 2019 (2020). https://www.cancer.org
8. Codella, N.C.F., et al.: Deep learning ensembles for melanoma recognition in dermoscopy images. IBM J. Res. Dev. **61**(4/5), 51–515 (2017). https://doi.org/10.1147/jrd.2017.2708299
9. Dreiseitl, S., Ohno-Machado, L., Kittler, H., Vinterbo, S., Billhardt, H.,Binder, M.: A comparison of machine learning methods for the diagnosis of pigmented skin lesions. J. Biomed. Informat. **34**(1), 28–36 (2001). https://doi.org/10.1006/jbin.2001.1004,http://www.sciencedirect.com/science/article/pii/S1532046401910044
10. Emre Celebi, M., Wen, Q., Hwang, S., Iyatomi, H., Schaefer, G.: Lesion border detection in dermoscopy images using ensembles of thresholding methods. Skin Res. Technol. **19**(1), 252–258 (2013). https://doi.org/10.1111/j.1600-0846.2012.00636.x
11. Esteva, A., Kuprel, B., Thrun, S.: Deep networks for early stage skin disease and skin cancer classification. Project Report, Stanford University (2015)
12. Fawcett, T.: An introduction to ROC analysis. Pattern Recogn. Lett. **27**(8), 861–874 (2006). https://doi.org/10.1016/j.patrec.2005.10.010
13. Gandhi, S.A., Kampp, J.: Skin cancer epidemiology, detection, and management. Med. Clin. North Am. **99**(6), 1323–1335 (2015). https://doi.org/10.1016/j.mcna.2015.06.002
14. Harangi, B.: Skin lesion classification with ensembles of deep convolutional neural networks. J. Biomed. Inform. **86**, 25–32 (2018). https://doi.org/10.1016/j.jbi.2018.08.006
15. Hekler, A., Utikal, J.S., Enk, A.H., Berking, C., Klode, J., Schadendorf, D., Jansen, P., Franklin, C., Holland-Letz, T., Krahl, D., von Kalle, C., Fröhling, S., Brinker, T.J.: Pathologist-level classification of histopathological melanoma images with deep neural networks. Eur. J. Cancer **115**, 79–83 (2019). https://doi.org/10.1016/j.ejca.2019.04.021
16. Kittler, H., Pehamberger, H., Wolff, K., Binder, M.: Diagnostic accuracy of dermoscopy. Lancet Oncol. **3**(3), 159–165 (2015). https://doi.org/10.1016/S1470-2045(02)00679-4
17. Mahbod, A., Schaefer, G., Wang, C., Ecker, R., Ellinger, I.: Skin lesion classification using hybrid deep neural networks. In: ICASSP 2019–2019 IEEE International Conference on Acoustics, Speech and Signal Processing (ICASSP), pp. 1229–1233 (2019). https://doi.org/10.1109/ICASSP.2019.8683352
18. Marks, R.: An overview of skin cancers. Cancer **75**(S2), 607–612 (1995). https://doi.org/10.1002/1097-0142(19950115)75:2+⟨607::AID-CNCR2820751402⟩3.0.CO;2-8
19. Nasr-Esfahani, E., et al.: Melanoma detection by analysis of clinical images using convolutional neural network. In: 2016 38th Annual International Conference of the IEEE Engineering in Medicine and Biology Society (EMBC), pp. 1373–1376 (2016). https://doi.org/10.1109/EMBC.2016.7590963
20. Rogers, H.W., Weinstock, M.A., Feldman, S.R., Coldiron, B.M.: Incidence estimate of nonmelanoma skin cancer (keratinocyte carcinomas) in the us population, 2012. JAMA Dermatol. **151**(10), 1081–1086 (2015). https://doi.org/10.1001/jamadermatol.2015.1187
21. Salerni, G., et al.: Meta-analysis of digital dermoscopy follow-up of melanocytic skin lesions: a study on behalf of the international dermoscopy society. J. Eur. Acad. Dermatol. Venereol. **27**(7), 805–814 (2013). https://doi.org/10.1111/jdv.12032

22. Sevastyanov, L.A., Shchetinin, E.Y.: On methods of increasing the accuracy of multiclass classification based on unbalanced data. Informat. Appl. **14**(1), 67–74 (2020). https://doi.org/10.14357/19922264200109
23. Shollet, F.: Deep Learning with Python. Manning Publications Co. (2017)
24. Siegel, R.L., Miller, K.D., Jemal, A.: Cancer statistics, 2019. CA: A Cancer J. Clin. **69**(1), 7–34 (2019). https://doi.org/10.3322/caac.21551, https://acsjournals.onlinelibrary.wiley.com/doi/abs/10.3322/caac.21551
25. Xie, F., Bovik, A.C.: Automatic segmentation of dermoscopy images usingself-generating neural networks seeded by genetic algorithm. Pattern Recogn. **46**(3), 1012–1019 (2013).https://doi.org/10.1016/j.patcog.2012.08.012,http://www.sciencedirect.com/science/article/pii/S0031320312003585
26. Yu, L., Chen, H., Dou, Q., Qin, J., Heng, P.A.: Automated melanoma recognition in dermoscopy images via very deep residual networks. IEEE Trans. Med. Imaging **36**(4), 994–1004 (2017). https://doi.org/10.1109/TMI.2016.2642839
27. Yuan, Y., Chao, M., Lo, Y.C.: Automatic skin lesion segmentation using deep fully convolutional networks with jaccard distance. IEEE Trans. Med. Imaging **36**(9), 1876–1886 (2017). https://doi.org/10.1109/tmi.2017.2695227

Distributed Systems Applications

Comparison of Different Smoothing Methods for Initial Data of the DSN-PC Sensor Network

Ádám Vas[1](✉) and László Tóth[2]

[1] University of Debrecen, Debrecen, Hungary
vas.adam@inf.unideb.hu
[2] SciTech Műszer Kft, Debrecen, Hungary
laszlo.toth@scitechmuszer.com

Abstract. Our Distributed Sensor Network for Prediction Calculations (DSN-PC) is a surface-based observational and computational network which is currently capable of calculating the change of an upper-air atmospheric parameter. The number of actually installed stations is limited thence we include data from the NOAA GFS database to create the 2D field of initial values for the prediction calculations. This hybrid application leads to numerical instability because some grid points get values from DSN-PC stations while others from NOAA GFS data. Previously we applied a smoothing algorithm based on moving average calculations. As presented in this paper, we applied two other methods. Each algorithm improved the numerical stability and prediction reliability. The type of the algorithm and the adjacency distance had a significant impact on the goodness of the forecasts. Below we compare these new results with the moving average method and the raw, unsmoothed case.

Keywords: Sensor network · Distributed computing · Weather prediction · Data assimilation · Data smoothing · DSN-PC

1 Introduction

Since the 1990s, distributed sensor networks have been an actively researched and developed field. There have been a trend of moving from centralized to distributed systems consisting of cheap nodes which together are often capable of more complex tasks compared to the centralized ones. Such distributed systems are displacing the traditional systems at a significant rate [10].

A distributed sensor network is a collection of nodes that are distributed by a logical, spatial or geographical aspect and are connected to each other through wired or wireless networks. These nodes can be equipped with many types of

This work was supported by the construction EFOP-3.6.3-VEKOP-16-2017-00002. The project was co-financed by the Hungarian Government and the European Social Fund.

© Springer Nature Switzerland AG 2020
V. M. Vishnevskiy et al. (Eds.): DCCN 2020, CCIS 1337, pp. 437–449, 2020.
https://doi.org/10.1007/978-3-030-66242-4_34

sensors (temperature, wind, air pressure, sound, light, magnetic field, acceleration etc.) and always contain a central unit used for signal processing, communication control and computational tasks. With the emergence and wild availability of high-speed networks, distributed sensor networks are used extensively in aerospace industry, automation, medical imaging, geology, weather prediction etc. The purpose of interconnecting these nodes can be the improvement of the data collection speed and reliability or the coverage of wider areas where central network availability is limited.

Recent technological advances have enabled the development of low-cost, low-power and multi-functional sensor devices. These are autonomous nodes with sensing, processing, and communication capabilities. Looking at a weather sensor network, the processing capabilities usually do not include mathematical computations, only signal processing and communication tasks. However, they can be used for distributed calculations of differential equation systems which generally are executed on central supercomputers of meteorological agencies.

Our goal with the DSN-PC system is to build a large distributed sensor network not only for the measurement of certain atmospheric parameters but also for weather prediction calculations [20]. This way the central supercomputer can be omitted from the whole system. Previously we followed a mixed approach by integrating our own DSN-PC nodes and NOAA GFS data into a hybrid sensor- and computational network [23]. After that we increased the involvement of our own measurements by directly inserting DSN-PC node measurements instead of the simultaneous interpolation [22]. This lead to problems due to significant differences (spikes) between adjacent grid points which caused numerical instability and incorrect results. This is a known issue in meteorology and several data assimilation [1,13,14] and data smoothing [11,16,25] techniques have been developed to address it. The spline methods have been the most widely used [2,5–7,15,19] and distributed algorithms have been developed based on them [17]. Their applications in atmospheric and geosciences have shown their viability [8,9,12]. Before moving to these advanced methods we tried one simpler approach based on 2-dimensional moving average calculations [22] to see whether smoothing is enough to maintain numerical stability.

In this paper we compare the previous smoothing method with two others: a distributed moving median algorithm [11] and a Savitzky-Golay filter [16] - the latter not yet implemented in a distributed way thence executed in MATLAB. Below the results of the numerical weather prediction calculations are shown.

2 System and Model Description

2.1 Geographical Properties

Currently our system implements a 20×20 size hybrid distributed sensor network which consist of 5 pieces of DSN-PC weather stations with the rest being simulated as Java threads on a server computer. This network forms a regular grid over Europe using polar stereographic projection [3]. Figure 1 shows the locations of the grid points. The detailed properties of the grid are:

- lower-left grid point coordinates: 39 °N, 2.6 °W
- upper-right grid point coordinates: 54.1371 °N, 38.6715 °E
- grid step at North Pole: 150 km
- central angle of the map: 0°

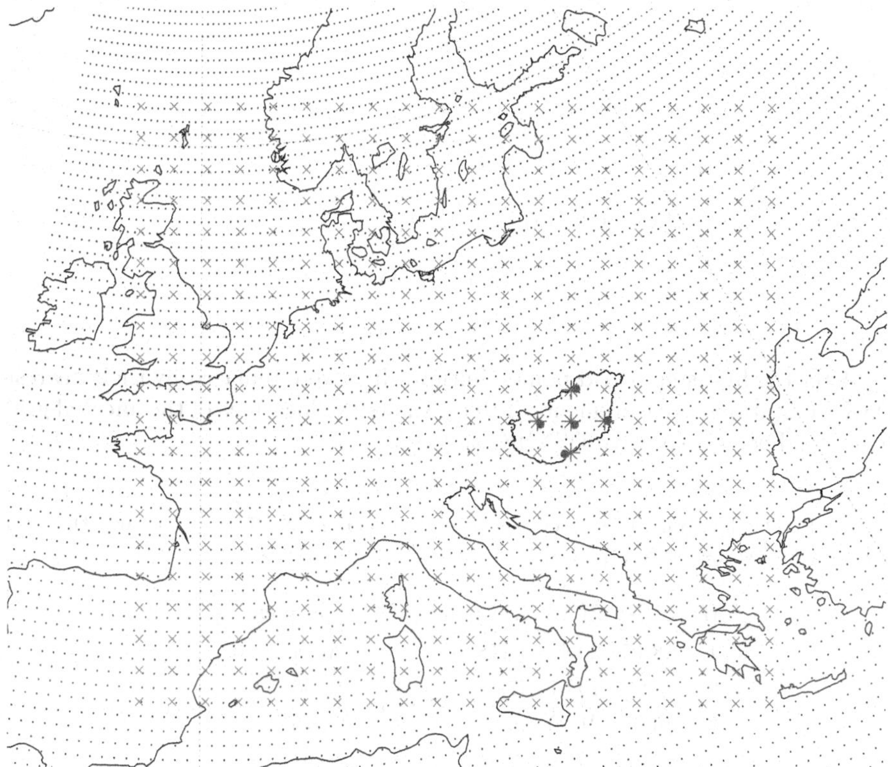

Fig. 1. The regular grid of the 20 × 20 computational network (x), the locations of the NOAA GFS dataset points (·), the locations of our DSN-PC weather stations (o) and the 5 grid points whose data were replaced with the nearest DSN-PC stations' measurements (*).

2.2 Input Data Sources and Their Assimilation

The initial values for the forecast calculations are taken from 2 sources: our 5 pieces of existing DSN-PC weather stations and publicly available data from the NOAA GFS-ANL database [4]. The DSN-PC weather stations in their present state can measure temperature, pressure and relative humidity [21]. The currently used forecast variable (500 hPa geopotential height) can be approximated based on these parameters using the hypsometric equation [23, 24]. From the NOAA GFS-ANL database we downloaded and applied the 0.5° resolution dataset which already contains the 500 hPa geopotential height values.

To calculate the initial data of the 20×20 grid, as a first step, natural neighbor interpolation [18,23] was performed considering only the GFS-ANL grid points. Before the interpolation the latitude(φ[°]) and longitude(λ[°]) coordinates of the grid points and the GFS-ANL points were converted to (x,y) coordinates based on polar stereographic map projection [3]:

$$r = \frac{\cos(\varphi)}{1 + \sin(\varphi)} \cdot 2a, \tag{1}$$

where

$$a = \frac{4 \cdot 10^7}{2\pi} \tag{2}$$

is the radius of the Earth (m). Then

$$x = r \cdot \sin(\lambda) \tag{3}$$

$$y = -r \cdot \cos(\lambda) \tag{4}$$

After the first step 5 grid points' initial values were replaced by their nearest DSN-PC stations' measurements. Table 1 shows the locations of the affected grid points and their respective DSN-PC stations.

Table 1. The locations of our currently operational DSN-PC stations and the grid points whose values were replaced by DSN-PC measurements.

ID	φ[°N]	λ[°E]	grid point φ[°N]	grid point λ[°E]
1	48.17	20.42	48.18	20.14
2	46.92	19.67	47.08	19.55
3	46.65	21.29	46.67	21.15
4	47.31	18.01	47.46	17.91
5	46	18.68	45.98	18.99

2.3 The Applied Numerical Weather Prediction Algorithm

In its present state the DSN-PC runs a relatively simple weather forecast model that is based on the barotropic vorticity equation originally developed by Charney, Fjørtoft, and von Neumann (CFvN) [3]. The original algorithm was refactored to a distributed form so that the nodes of the DSN-PC network are able to solve the equations in a fully distributed way [20]. In this article we cover the period between 21 March 2019 and 27 March 2019. Each day the 00:00 UTC measurements were chosen as initial values for the forecasts. We investigated the goodness of the forecasts by calculating the Mean Absolute Error (MAE):

$$MAE = \frac{1}{18 \cdot 18} \sum_{i=1}^{18} \sum_{j=1}^{18} |z_{500,i,j} - z'_{500,i,j}|, \quad (5)$$

where $z'_{500,i,j}$ is the predicted and $z_{500,i,j}$ is the measured 500 hPa geopotential height (m) 24 h later.

2.4 Smoothing the Initial Data

During our forecast calculations numerical instability can occur. This happens because we use GFS-ANL and DSN-PC data simultaneously and GFS-ANL contains data that are already initialized, smoothed and interpolated onto a grid. However, DSN-PC measurements contain raw data. This hybrid approach leads to large differences between adjacent grid points which our simple CFvN model is not able to handle. Another reason is that it was originally designed for a larger geographical area, larger-scale atmospheric movements and larger distance between grid points. Also, DSN-PC measurements may contain measurement errors and may be affected by local weather phenomena.

Trying to address this problem we previously applied a 2-dimensional moving average algorithm which could smooth those large differences between adjacent points [22]. That approach lead to satisfactory results in terms of numerical stability and the goodness of the predictions. Presented in this paper, we applied two other methods: moving median and Savitzky-Golay filter.

We implemented the moving median algorithm in a distributed form. First, each node queries the initial values from its adjacent nodes over TCP/IP. Then based on the queried values they calculate the median of the whole dataset that also includes their own measurements. The adjacency distance (Np) varies between 1–3 hops. On Fig. 2 these adjacency distances are visualized on a 20×20 grid. On Fig. 3 the flowchart diagram of the moving median algorithm is shown.

The Savitzky-Golay filter [16] is yet to be refactored to a distributed form which will be applicable to DSN-PC nodes. In the meantime we used MATLAB's sgolayfilt function to calculate the filtered matrix from the raw grid data. As the function is not capable of 2-dimensional filtering, we applied two consequent steps by first going row-by-row then column-by-column:

```
matrixOut = sgolayfilt(matrixIn', degree, Np*2+1);
matrixOut = sgolayfilt(matrixOut', degree, Np*2+1);
```

Np=2 and Np=3 cases were applied in the case of the Savitzky-Golay filter. The degree of the fitted polynomial varied between 2 and (2*Np)-1.

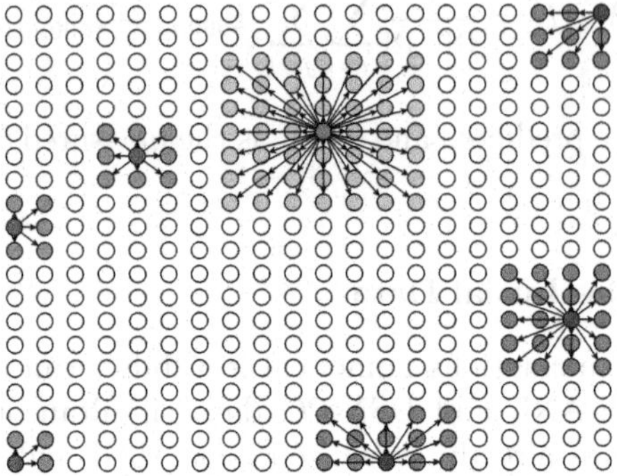

Fig. 2. Examples of moving median smoothing areas for different nodes with Np=1 (red), Np=2 (green) and Np=3 (blue) (Color figure online)

area_values = new List<double>()
area_values[0] = initial_value
count = 0
x = max(0, i-Np)
x ≤ min(i+Np, num_rows-1)
true
y = max(0, j-Np)
false
y ≤ min(j+Np, num_columns-1)
false
x++
area_values = arraySort(area_values)
true
x==i && y==j
true
false
count%2 == 1
false
true
y++
area_values[++count] = tcpGetNeighborValue(x,y)
smoothed_value = area_values[floor(count/2)]
smoothed_value = (area_values[count/2] + area_values[count/2-1]) / 2

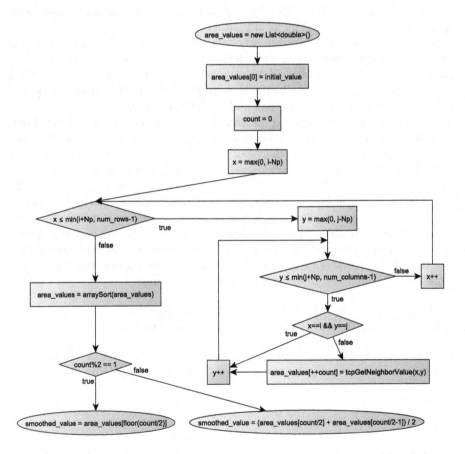

Fig. 3. The moving median algorithm as executed on a node at position (i,j)

3 Results

The MAE values of the forecast calculations are summarized in Table 2 and Table 3. Smoothing seems to be necessary at our current state of development to maintain numerical stability. The adjacency distance highly affects the goodness of the prediction calculations. A minimum of Np = 2 seems to be necessary in the case of moving median. The Savitzky-Golay method produced satisfactory results in both cases (Np = 2 and Np = 3). Increasing the degree of the fitted polynomial doesn't result in significantly better forecasts in the investigated time period. An example is shown on Fig. 4 containing initial fields based on unsmoothed and smoothed values.

Table 2. Mean Absolute Error (m) values of the forecast calculations performed by the CFvN algorithm on initial data smoothed using moving median method.

Date	No smoothing	Np = 1	Np = 2	Np = 3
21 March	NaN	NaN	25.55	54.63
22 March	NaN	191.65	43.44	65.23
23 March	81.19	60.83	95.01	NaN
24 March	NaN	64.47	111.53	71.29
25 March	89.95	210.41	145.99	58.42
26 March	NaN	194.47	76.89	100.65
27 March	NaN	199.53	78.78	50.23

Table 3. Mean Absolute Error (m) values of the forecast calculations performed by the CFvN algorithm on initial data smoothed using Savitzky-Golay filter with different polynomial degrees values.

Date	No smoothing	Np = 2		Np = 3			
		2nd	3rd	2nd	3rd	4th	5th
21 March	NaN	22.65	48.71	20.36	34.21	48.94	56.13
22 March	NaN	45.66	44.84	46.07	44.75	44.02	44.58
23 March	81.19	64.95	70.72	65.22	61.12	72.77	73.17
24 March	NaN	59.96	31.38	59.24	42.00	37.38	51.07
25 March	89.95	158.96	81.24	158.37	113.86	79.86	98.40
26 March	NaN	38.85	47.39	37.05	42.39	48.53	50.03
27 March	NaN	56.72	44.26	50.39	47.03	35.40	44.18

Comparing the moving average, moving median and Savitzky-Golay methods show the advantage of the Savitzky-Golay filter as it produces the smallest MAE values generally. However, this needs more review in the future by involving much more datasets as the current results containing 7 cases are not definite.

The initial, the analysis and the forecast height fields of the calculations using initial data from 24 March 2019 are shown based on smoothed initial data using moving average (see Fig. 5), moving median (see Fig. 7) and Savitzky-Golay (see Fig. 9) methods. The error maps of the forecasts are shown on Fig. 6, 8, 10. In these examples, for the moving average and the moving median methods adjacency distance of Np=1 was set. For the Savitzky-Golay method Np=2 was set with 3rd degree fitted polynomials.

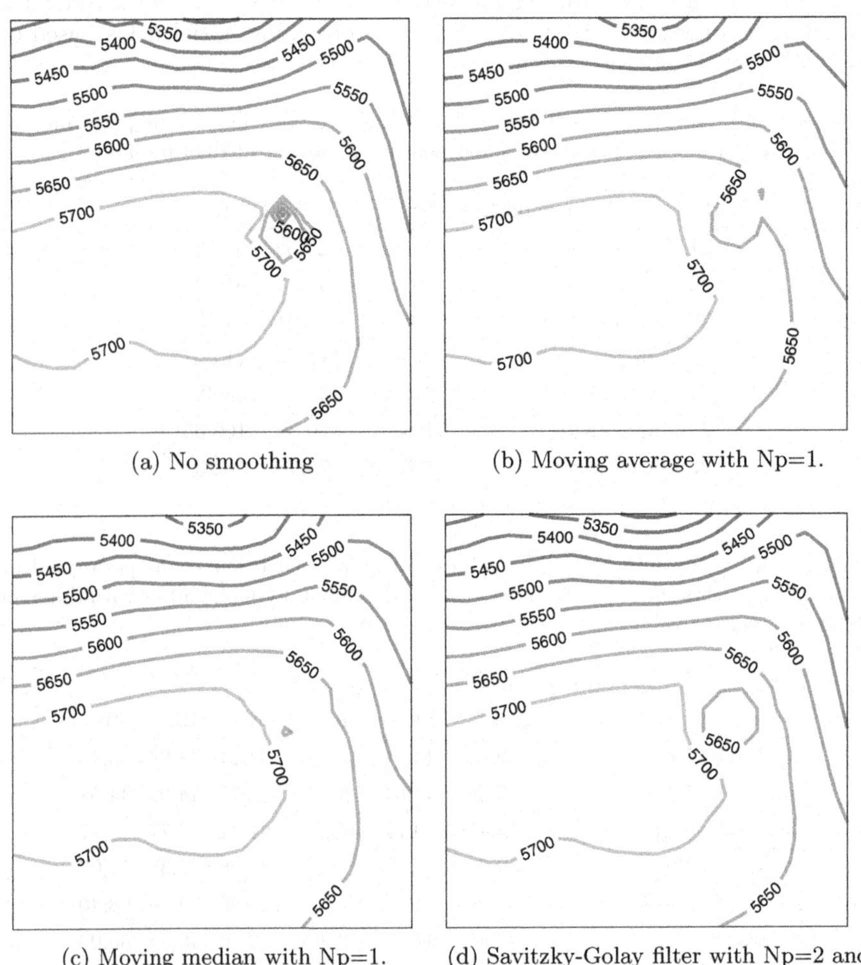

(a) No smoothing

(b) Moving average with Np=1.

(c) Moving median with Np=1.

(d) Savitzky-Golay filter with Np=2 and 3rd degree polynomials.

Fig. 4. Initial height fields using different smoothing methods performed on 24 March 2019 00:00 UTC data.

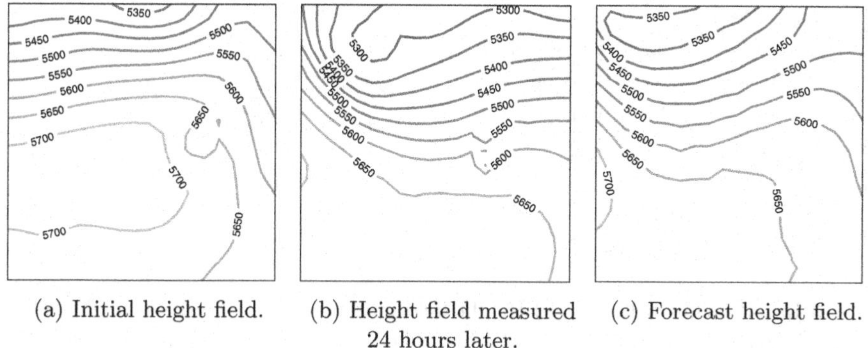

(a) Initial height field. (b) Height field measured 24 hours later. (c) Forecast height field.

Fig. 5. The result of the forecast based on initial data from 24 March 2019 00:00 UTC smoothed using moving average method with Np = 1.

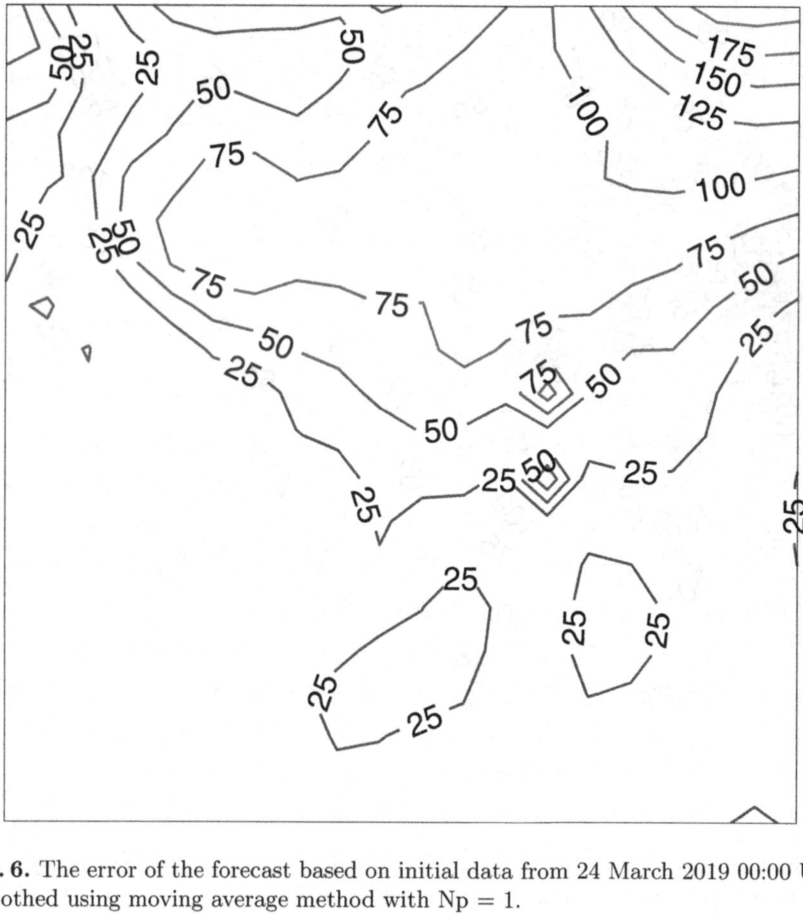

Fig. 6. The error of the forecast based on initial data from 24 March 2019 00:00 UTC smoothed using moving average method with Np = 1.

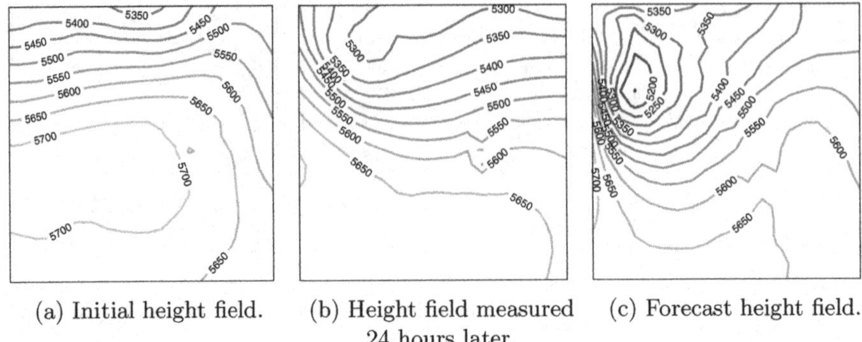

(a) Initial height field. (b) Height field measured 24 hours later. (c) Forecast height field.

Fig. 7. The result of the forecast based on initial data from 24 March 2019 00:00 UTC smoothed using moving median method with Np = 1.

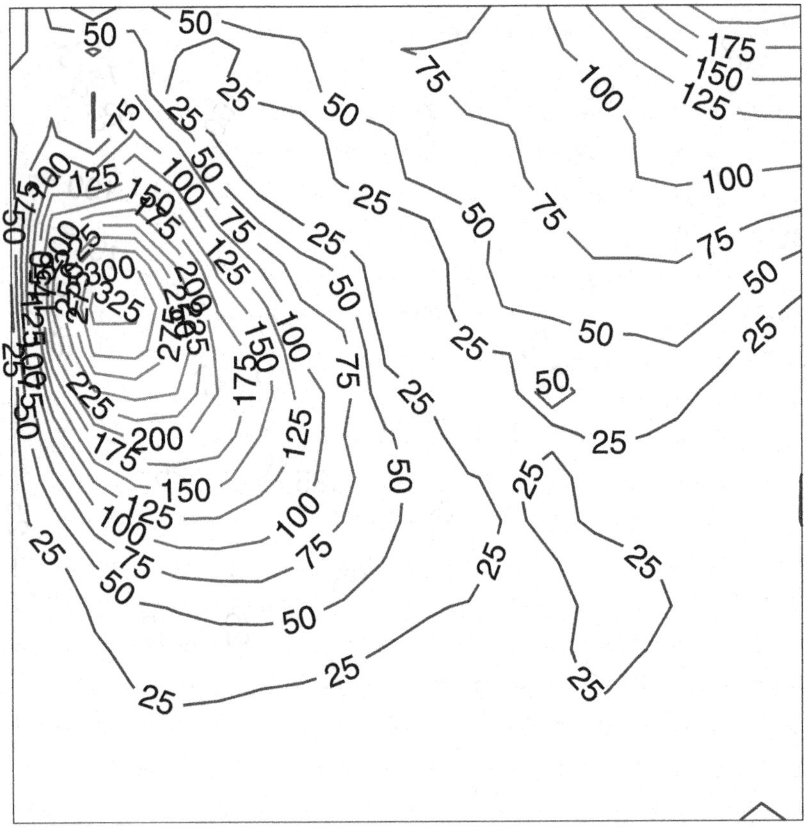

Fig. 8. The error of the forecast based on initial data from 24 March 2019 00:00 UTC smoothed using moving median method with Np = 1.

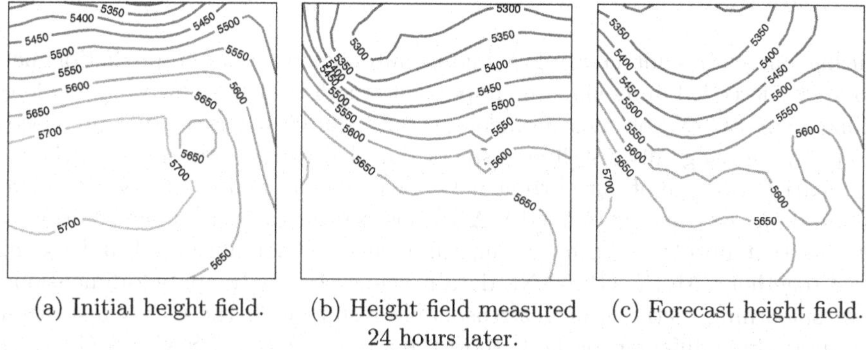

(a) Initial height field. (b) Height field measured 24 hours later. (c) Forecast height field.

Fig. 9. The result of the forecast based on initial data from 24 March 2019 00:00 UTC smoothed using Savitzky-Golay filter with Np = 2 and 3rd degree polynomials.

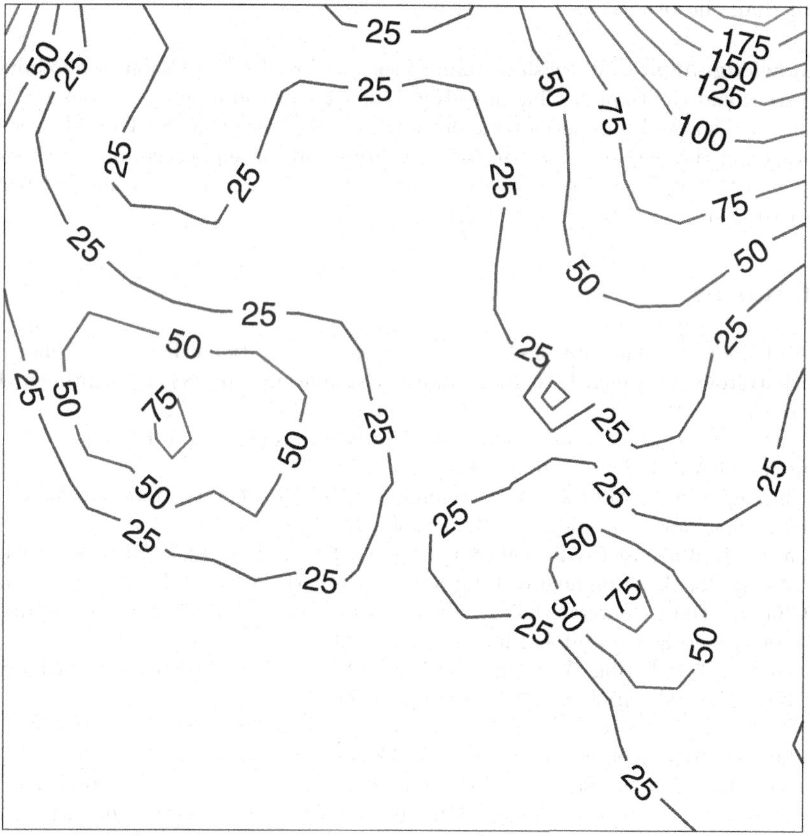

Fig. 10. The error of the forecast based on initial data from 24 March 2019 00:00 UTC smoothed using moving median method with Np = 2 and 3rd degree polynomials.

4 Conclusion

During our efforts for creating a highly autonomous weather prediction network we succeeded in moving one important step forward as now we are able to utilize DSN-PC station measurements as data sources for a weather prediction model by applying fully distributed smoothing algorithms to the nodes. Although the spatial coverage of our current network is not optimal as of now, involving public databases like NOAA GFS-ANL has proven to be a viable solution to address that problem. Since our measurements are not initialized and assimilated together with the GFS-ANL data it was necessary to apply 2-dimensional data smoothing methods to get reasonable results with the CFvN algorithm. Applying three different methods show the advantage of the Savitzky-Golay filter in terms of prediction reliability but its computational needs are the highest. Further investigation and comparison of these methods seems to be necessary to draw a general conclusion. Also, it's worth investigating whether there are some particularities in the certain initial fields that make one smoothing method more useful than the others.

Acknowledgements. We wish to thank Ficsor Endre, Perlaki Csaba, Szabó Sándor, Vas Ferenc and the Baptist Church of Kecskemét for providing place for our DSN-PC weather stations and thus supporting our research. We also thank SciTech Műszer Kft. for providing the possibility to use their facilities and for supporting our work financially; and Gábor Nagy for his contribution in designing, manufacturing and testing our weather stations' electronic circuits.

References

1. Andersson, E., Thépaut, J.-N.: Assimilation of Operational Data. In: Lahoz, W., Khattatov, B., Menard, R. (eds.) Data Assimilation, pp. 283–299. Springer, Heidelberg (2010). https://doi.org/10.1007/978-3-540-74703-1_11
2. Baramidze, V., Lai, M.J., Shum, C.K.: Spherical splines for data interpolation and fitting. SIAM J. Sci. Comput. **28**(1), 241–259 (2006)
3. Charney, J.G., Fjørtoft, R., Neumann, J.V.: Numerical integration of the barotropic vorticity equation. Tellus **2**(4), 237–254 (1950)
4. for Environmental Prediction, N.C.: Global Forecast System (GFS) — National Centers for Environmental Information (NCEI) formerly known as National Climatic Data Center (NCDC). https://www.ncdc.noaa.gov/data-access/model-data/model-datasets/global-forcast-system-gfs
5. Freeden, W., Törnig, W.: On spherical spline interpolation and approximation. Math. Methods Appl. Sci. **3**(1), 551–575 (1981)
6. Freeden, W., Nashed, M.Z., Schreiner, M.: Spherical Harmonics Interpolatory Sampling. In: Spherical Sampling, pp. 267–300. Birkhäuser Basel (2018)
7. Freeden, W., Schreiner, M.: Special functions in mathematical geosciences: an attempt at a categorization. In: Handbook of Geomathematics, pp. 2455–2482. Springer, Heidelberg (2015). https://doi.org/10.1007/978-3-642-01546-5_31
8. Hutchinson, M.F.: Interpolation of rainfall data with thin plate smoothing splines -part II: analysis of topographic dependence. J. Geographic Inf. Decis. Anal. **2**(2), 152–167 (1998)

9. Hutchinson, M.F.: Interpolation of rainfall data with thin plate smoothing splines. Part I: Two dimensional smoothing of data with short range correlation. J. Geographic Inf. Decis. Anal. **2**(2), 139–151 (1998)
10. Iyengar, S.S., Brooks, R.R.: Distributed Sensor Networks : Sensor Networkingand Applications (Volume 2). Chapman and Hall/CRC, Boca Raton (2016)
11. Justusson, B.I.: Median Filtering: Statistical Properties. In: Two-Dimensional Digital Signal Prcessing II, pp. 161–196. Springer, Heidelberg (1981). https://doi.org/10.1007/BFb0057597
12. Kiani, M.: Template-based smoothing functions for data smoothing in Geodesy. Geodesy Geodyn. **11**(4), 300–306 (2020)
13. Lahoz, W., Khattatov, B., Ménard, R.: Data assimilation and information. In: Lahoz, W., Khattatov, B., Menard, R. (eds.) Data Assimilation, pp. 3–12. Springer, Heidelberg (2010). https://doi.org/10.1007/978-3-540-74703-1_1
14. Lynch, P., Huang, X.-Y.: Initialization. In: Lahoz, W., Khattatov, B., Menard, R. (eds.) Data Assimilation, pp. 241–260. Springer, Heidelberg (2010). https://doi.org/10.1007/978-3-540-74703-1_9
15. S., L.L., Wahba, G.: Spline Models for Observational Data. Mathematics of Computation 57(195), 444 (jul 1991)
16. Savitzky, A., Golay, M.J.E.: Smoothing and differentiation of data by simplified least squares procedures. Anal. Chem. **36**(8), 1627–1639 (1964)
17. Shang, Z., Cheng, G.: Computational limits of a distributed algorithm for smoothing spline. J. Mach. Learn. Res. **18**(108), 1–37 (2017)
18. Sibson, R.: A Brief Description of Natural Neighbour Interpolation. In: Interpreting multivariate data, pp. 21–36. Wiley, Hoboken (1981)
19. Van Assche, W., Freeden, W., Gervens, T., Schreiner, M.: Constructive approximation on the sphere, with application to geomathematics. J. Approximation Theory **112**(2), 324–325 (2001)
20. Vas, Á., Fazekas, Á., Nagy, G., Tóth, L.: Distributed sensor network for meteorological observations and numerical weather prediction calculations. Carpathian J. Electr. Comput. Eng. **6**(1), 56–63 (2013)
21. Vas, Á., Nagy, G., Tóth, L.: Networkable sensor station for DSN-PC system. Carpathian J. Electr. Comput. Eng. **8**(2), 37–40 (2015)
22. Vas, Á., Owino, O.J., Tóth, L.: Improving the simultaneous application of the DSN-PC and noaa GFS datasets. Annales Mathematicae et Informaticae **51**, 77–87 (2020)
23. Vas, Á., Tóth, L.: Investigation of a hybrid sensor- and computational network for numerical weather prediction calculations. In: Vishnevskiy, V.M., Samouylov, K.E., Kozyrev, D.V. (eds.) DCCN 2019. CCIS, vol. 1141, pp. 510–523. Springer, Cham (2019). https://doi.org/10.1007/978-3-030-36625-4_41
24. Wallace, J.M., Hobbs, P.V.: Atmospheric Thermodynamics. In: Atmospheric Science, pp. 63–111. Elsevier (2006)
25. Woodford, C.H.: An algorithm for data smoothing using spline functions. BIT **10**(4), 501–510 (1970)

NFC Vulnerabilities Investigation

S. Japertas[1](✉) and R. Jankūnienė[2](✉)

[1] Kaunas University of Technology, K. Donelaičio St. 73, 44249 Kaunas, Lithuania
saulius.japertas@ktu.lt

[2] University of Applied Engineering Sciences, Tvirtovės av. 35, 50155 Kaunas, Lithuania
ruta.jankuniene@edu.ktk.lt

https://www.ktu.edu/, http://www.ktk.lt/

Abstract. The evolving Internet of Things is based on such short-range technologies as Bluetooth, RFID, etc. However, one of the biggest drawbacks of these technologies is their vulnerabilities. This paper addresses the vulnerabilities of NFC technology which belong to the RFID class, namely, signal interception and jamming. Near Field Communication (NFC) defines a short-range wireless standard allowing data to be transmitted between two devices at a maximum distance of 20 cm (usually, 4 cm). NFC technology is widely used including banking and access control. Therefore, loss of data or jammed data on an NFC chip may result in a variety of extremely negative consequences for the users of this technology. The signal produced by the NFC scanner and tag is analyzed in this research. The maximum distance is investigated at which it is possible to record a signal of 13.56 MHz. Experiments show that data can be intercepted at a distance of up to 10 m by using an improvised passive antenna without any amplification or signal filtering circuits and by employing publicly available spectrum analyzers. Jamming of the signal was performed with a specially designed generator and another NFC reader. NFC signal attenuation results indicate that an attenuator with a very precise frequency is required. When another NFC scanner is used as the damper, the maximum distance at which the transmission between the scanner and the tag is successfully jammed is about 9 cm.

Keywords: Near field communications · Jamming · Signal interception

1 Introduction

The use of Near Field Communication (NFC) technology is expanding. It is used in bank payment cards, identity documents, as well as in access control and mobile phones. A major challenge, however, is the security gaps in the employed technology. These security disadvantages are multi-layered and multifunctional.

© Springer Nature Switzerland AG 2020
V. M. Vishnevskiy et al. (Eds.): DCCN 2020, CCIS 1337, pp. 450–463, 2020.
https://doi.org/10.1007/978-3-030-66242-4_35

1.1 RFID Technologies

NFC belongs to the group of RFID technologies, but uses different frequencies (operates at 13.56 MHz) and is significantly more popular than other RFID technologies [1]. Radio Frequency Identification (RFID) is a system which transmits object identification by using radio waves. One of the simplest RFID systems consists of the following components: an RFID tag that stores information, a scanner (reader) receives information from an RFID tag and transmits it for further processing, a transmitter generates radio signals, and an antenna transmits a radio signal between the scanner and the RFID tag.

NFC technology is similar to other RFID technologies, because it consists of two devices featuring antennas, which are used to exchange information due to electromagnetic induction (not by radio waves). Information is usually transmitted over short distances, stored on chips and transmitted to the end-user. In general, three main RFID technologies are distinguished: low-frequency, high-frequency, and ultra-high-frequency technologies.

Low frequency RFID technology operates at 125 kHz and 134 kHz. The specificity of such technologies is that they are typically used with no additional cryptographic protection. Such technologies are needed when rapid access to the information stored on microchips is required, but strong protection of such information is not necessary. For example, such a RFID technology is used to tag animals.

High frequency technology is NFC. It is the most popular RFID technology and it operates at 13.56 MHz. As was mentioned above, with the help of this technology, various operations are performed that require a sufficiently high level of security. Therefore, such technologies feature additional cryptographic protection. NFC works when NFC devices touch each other so that to establish connection between the devices. The NFC system consists of two modes: active and passive devices. The difference between NFC and other RFID technologies is that other RFID technologies can only work when one device is active and the other is passive.

Most UHF RFID systems operate between 860–960 MHz bands. However, there are UHF RFID systems operating at 433 MHz and 2.45 GHz. A distinctive feature of the UHF RFID technology is that it does not communicate through magnetic coupling but instead uses passive backscatter modulation according to ISO-18000-6C Standard [2]. The information contained in these tags can be scanned over relatively long distances. Therefore, such passive UHF RFID systems are used in hundreds of various applications, such as tool tracking, IT asset tracking, race timing, and so on. However, such systems have the advantage that UHF RFID information, as mentioned above, can be scanned over long distances. At the same time, this advantage becomes the biggest disadvantage of UHF RFID. That is, since a UHF RFID reader can emit a signal of up to 2 W, it is not a complex problem to receive such a signal [3]. Therefore, this variation of RFID technology is also used relatively less frequently than NFC.

NFC Operation Modes. According to NFC Forum [1], NFC can communicate either in two-way communication, or in peer-to-peer communication. NFC devices are unique in that they can change their operating mode depending on the situation. A total of three NFC operating modes are distinguished: *Peer-to-Peer* (P2P), *Reade rWriter*, and *Card Emulation.*

Peer-to-Peer Mode. This mode allows two NFC devices to communicate with each other, exchange information and share files, i.e., both devices work as active devices.

Reader Writer Mode: A device that supports NFC (the active device) can read the information stored in the NFC tag (the passive device). Such tags can be placed in various locations, such as smart posters, advertisements, and so on. The user can read the tag information stored on the passive device chip for further use.

Card Emulation Mode. This mode allows the NFC device to act as a smart card: to perform transactions such as shopping, ticketing and controlling access to public transportation.

1.2 NFC Technology Vulnerabilities

Although NFC is a fairly mature technology, there is a significant likelihood of encountering various attacks which could affect NFC because of the free space data channel. Interception of the communication channel by a third party which uses or modifies the captured data for its own purposes is possible without knowing that this third party is intervening (man-in-the-middle attack) Fig. 1. Especially difficult to detect such an attack on an NFC channel when communicating in the peer-to-peer mode and when the two mobile devices are communicating actively.

Fig. 1. Man-in-the-middle pattern. *Left:* Attacker captures data sended by transponder to reader. *Right:* Attacker is participating in both sides of the communication.

Table 1 reveals information about some other threats, such as Eavesdropping, Ticket Cloning, Use Only Single ID, Spoofing, Phishing, Tag Tampering, Relay Attack, Data Corruption, Data Modification, Data Insertion, and Denial of Service (DOS) [4]. Such threats occur in touch-and-go NFC applications, particularly when the user needs to touch the NFC device to the NFC reader [4].

Table 1. Threats in NFC

Threats	Description
Eavesdropping	Signal scanning is possible up to 1 meter for passive signals and up to 10 m for active signals. Active signals are much stronger. It is hard to prevent eavesdropping because signals have to be received reliably by the receiver and to be strong enough. There is no need for the attacker to receive all messages; it is commonly sufficient to get some of them. Attacker can use big and complex antennas, and logistics in the receiver terminal is commonly used; thus, the size of antennas and the level of performance can be restricted. This is relevant for the card emulation and peer-to-peer modes
Ticket Cloning	The purpose is to share data with the ticket until its expiry (in the e-ticket system). This is manifested if the tickets were copied and transferred to others before checking. Everybody can use such a ticket as a new one. If the cloned ticket has already been checked, it can be used until its expiry date
Use Only Single ID	Every contactless smart card chip has a unique ID for non-collision wireless reading, and this ID is used for identification. There is neither encryption nor authentication to read an ID with an NFC reader. An attacker can hijack or copy a unique ID
Spoofing	Tag content is counterfeited by providing false information, such as a fake domain name, a false URL, or an email. Smart poster URI spoofing allows attacks on the web browser, URLs and mobile services by using SMS URI, telephony URI, etc.
Phishing	Phishing in NFC is possible with social engineering. Disclosing personal information is directed to malware
Tag Tampering	Attackers use passive tags to create an attack vector. The most common way is to simply destroy the original passive tag or to block the connection to it, and then replace it with a malicious tag
Relay Attack	These attacks are impossible to recognize both with a card and with a reader. This attack is performed by using Application Protocol Data Unit (APDU) commands. Malicious apps can get APDU commands from the network socket. The main relay attack system uses two devices: ghost or leech. A ghost is a device which fakes a card for the reader; meanwhile, a leech is a device which fakes the reader's card. Bidirectional connection is created between the ghost and the leech by using a transparent communication channel through the reader and the tag
Data Corruption	If the attacker has modified the NFC interface, the data may be corrupted. This leads to denial of service because the attacker has changed data to an unrecognized format, and this causes disruption of communication between the user and the receiver. If the data in the NFC tag is corrupted, the tag will be useless, and the device must recover the data again. Data corruption can be done by malicious software running on a smartphone
Data modification	Data may be changed to incorrect values when the data is being transmitted between NFC devices. This type of vulnerability requires the attacker's experience in the wireless and radio technologies because transmission amplitude modulation needs to be processed
Data Insertionk	An attacker in the form of messages can insert any unwanted data, especially during the exchange of data between NFC devices. The attacker must respond to the device before the legitimate device makes a connection. The obtained data would be corrupted if both legitimate devices and impersonators transmit data at the same time
Denial of Service (DOS)	A DOS attack occurs when the request for access to a NFC security chip or malicious mobile phone application continues. The access to the security chip will also be locked until the attack stops sending mass requests and therefore may lose its function. The entire application process will be terminated. Additionally, DOS can occur when the NFC scanner is touched with an empty tag. Error messages that may affect NFC devices or services may then be stopped

The vulnerability described in Table 1 is only possible if malware acts remotely. However, there are only a few works dealing with the investigation of the distance of NFC signal propagation. According to the NFC Forum, such a small amount of researches in this area stems from the fact that the NFC signal usually travels only a short distance (< 20 cm) which is physically controlled [1]. Therefore, it is argued that real signal interception is practically impossible and that NFC technology is safe [4].

However, NFC can have EM fields above this range that would definitely be vulnerable for a potential attack, especially when devices are performing in the active mode. Thus, the NFC signal can be intercepted at distances greater than 20 cm, i.e., there are no major problems in capturing the signal at these distances with a scanner [5–13].

Paper [5] observes that physical attacks on NFC technology such as remove, shield, clone, and high distance read are possible, but the details were not revealed.

Some works described NFC security systems regarding prevention of man-in-the-middle and relay attacks [6,7]. The authors [6] provided an example how criminals can intercept important data (in this particular case, information stored in a citizen's passport) using this technology.

Two different signals are necessary to achieve successful data transfer: the first when the scanner sends a signal to the NFC tag, and the second when the NFC tag sends a response to the scanner [6]. The signals were captured at a distance of one meter by using the following equipment: an H-field antenna, a radio frequency receiver, an oscilloscope, and a computer (for storing the intercepted data).

The security of NFC payments is analyzed in [7]. The authors propose a method to attack mobile-based NFC payment methods by making a wormhole through which a payment could be made with the card physically being in a different location.

As it was shown in [8], it is possible to receive the signal at a distance of up to 30 cm and demodulate the command by using a tag passive antenna.

Paper [9] states that the NFC passive element signal can be intercepted at short distances, while the active element signal can be intercepted at distances of up to 10 m. However, this work does not provide any evidence or references to other papers. Therefore, it is not clear under what conditions such signal interception is actually possible.

It is possible to receive the signal at a distance of up to 100 cm when making near-field contactless payments [10]. However, such experiments were performed with HF RFID technology only.

Another research states that the NFC signal can be transmitted at a distance of several meters from the reader [11]. Hence, this signal can be picked up at the same distance. However, again, there are only statements without clear experimental results or references to other works.

All the weaknesses inherent in radio communication are essentially inherent in NFC technology as well, including signal jamming. Paper [12] experimentally showed that the presence of NFC transponders can be determined at a distance of up to the 2.5 m. Paper [12] also states that it would be useful to investigate NFC technology jamming separately for the reader and the tag. Reflective and pulse jamming can be used for attacks on NFC technologies. In order to avoid such physical effects, the authors suggest using the EnGarde solution. However, this paper does not reveal how NFC technology is actually jammable.

Paper [13] strives to analyze jamming during a cyber-attack session.

Signals at 13.56 MHz, correctly chosen for the NFC transmitter, can jam an NFC signal very easily [14,15]. However, such attacks are very rare due to their low effectiveness.

Keeping in mind the information listed above, it is evident that studies on NFC signal interception are incomplete. Therefore, the question of the real possibilities of intercepting the NFC signal still remains open.

Despite all the strengths of NFC technology regarding such security threats as data pre-encoding, synchronization between devices before communication and short distance for data transmission (from 4 to 20 cm), there are some NHC threats that need to be kept in mind. The most common vulnerabilities are data modification, eavesdropping (or interception), man-in-the-middle attack, and data corruption. The latter one is more vulnerable for an attack at a data rate of 106 Kbps. In order to overcome this issue, it makes sense to have more power at the NFC receiver. This is because the attacker's transmitters always use more power compared to NFC transmission (equalling to about 10 to 20 dBm). NFC is particularly vulnerable for an attack when the devices are performing in the active mode because RF fields are stronger in the active mode. The active-passive mode for communication is harder to overcome by eavesdropping. The use of a secure channel for communication is another alternative. This provides protection against eavesdropping and data modification attacks. For example, Diffie-Hellman as the key agreement protocol because of its inherent protection is the proper decision against the man-in-the-middle attack [16]. The shared key can be used to derive a symmetric key for the NFC secure channel, which assures confidentiality, integrity and authenticity of the data transferred between the involved devices.

2 Methodology of Experiments

Two vulnerabilities of the NFC technology are of interest: NFC signal interception distance and scanner suppression.

Reader. ACR122U NFC Reader. This device is compliant with ISO/ IEC18092 Standard. It supports not only MIFARE ®and ISO 14443 A and B cards but also all the four types of NFC tags. The operating distance of ACR122U is up to 5 cm depending on the type of the tag in use. Readers ACR1222L, ACR128U and NFC integrated to Nexus 7 Tablet were investigated in the jamming investigations.

Tag. A plastic card with an integrated NFC tag based on a *Mifare* 4k chip.

Equipment Used for Measurement. This device is capable of measuring the characteristics of signals from 9 kHz to 8 GHz. It also allows performing many other functions; the one that is relevant in this study is the averaging function.

Jammer. A special attenuator operating at 13.56 MHz was designed and manufactured to investigate the possibilities of NFC signal attenuation. For the most accurate frequency of 13.56 MHz, the quartz generator scheme was chosen. 13.56 MHz signal was amplified with an amplifier based on IRFP450 transistors and 1.7 nH, 570 nH, and 88 nH inductors (the latter were specially designed for this research). The power at the output of such an amplifier is 4 W.

Measurement Errors. Three measurements were made at each point in order to determine possible measurement errors. The obtained results were averaged, and these means were used in graphs and mathematical models. As a result, the following relative error δ (1), variance D (2), and standard deviation RMS (3) values were determined:

$$\delta = \left| \frac{x_i - \mu}{x_i} \right| \cdot 100\%; \tag{1}$$

$$D = \frac{\sum_1^n (x_i - \mu)}{n - 1}; \tag{2}$$

$$RMS = \sqrt{\frac{1}{n} \sum_1^n x_1^2}; \tag{3}$$

where, x_i is the measured value, μ is the average, and n is the number of measurements. The error estimate showed that the maximum relative error was $\delta < 5\%$; maximum variance was D < 7 dBm, and the maximum measurement deviation from the RMS was < 5%.

Figure 2 reveals the signal variation of the produced 13.56 MHz jammer depending on the polarization of the antenna in the closed space (corridor) and the calculated signal variation in the free space. The study was performed by fixing the specially manufactured 13.56 MHz frequency generator, and signal measurements were performed at every 1 m further away from the generator. The results were approximated with the logarithmic formula (4):

$$P_Rx = -14.6d + 14.6lg(f) - 85.25; \tag{4}$$

where d is the distance between the transmitter and the receiver, m; f is the frequency, MHz. These results are plotted in Fig. 2 as Model.

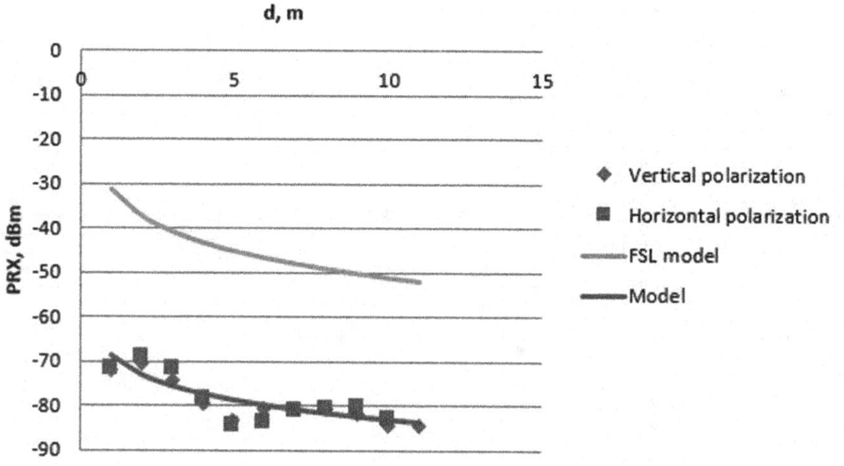

Fig. 2. Variation of jammer signal with distance.

One can see that the signal strength slightly increases between about 6 and 9 m from the location of the transmitter. This could be explained by the waveguide effect in the case of higher frequencies. However, in this case, if it is the waveguide effect, it occurs significantly earlier than in the case of higher frequencies, although the measurements were made under identical conditions of construction and geometry of the premises [17].

Antenna. The first experiments were performed with a 20 cm long omnidirectional antenna. However, with the help of such an antenna it was possible to receive the signal at a distance of 1.5 m only from the reader. Therefore, a 1.7 m long omnidirectional antenna was used for further experiments.

Measurements. It was assumed that equipment selection and/or its manufacturing could be sufficiently freely available to the attacker. In this case, the NFC signal strength was measured only to the level of −85 dBm. Therefore, the experimental distance between the NFC reader and the spectrum analyzer was increased until the signal level of −85 dBm was reached.

Two scenarios were used to study the propagation distance of the NFC signal: when the reader was operating in the passive mode, i.e., no tag was being used, and when the tag was covering the reader. The latter one was placed 1 m above the floor. A 1.7 m long antenna was connected to a spectrum analyzer. Measurements were made by varying the distance between the instrument and the spectrum analyzer at every one meter. The radiation of the reader signal was measured at various positions of the reader antenna relative to the receiver antenna Fig. 3.

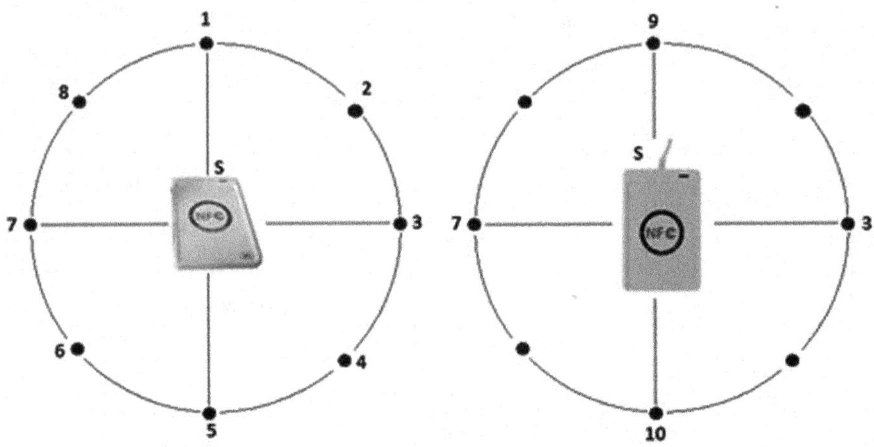

Fig. 3. ACR122U NFC Reader's positional orientation to the antenna of the spectrum analyzer. The positions of the spectrum analyzer antenna are numbered.

3 Results

3.1 NFC Signal Interception Investigation

Scenarios 1 (Reader's passive mode). During the experiments, the signal strength of the reader was measured in the standby (passive) mode of the reader. The signal strength was recorded from all the sides of the reader. The results are shown in Fig. 4.

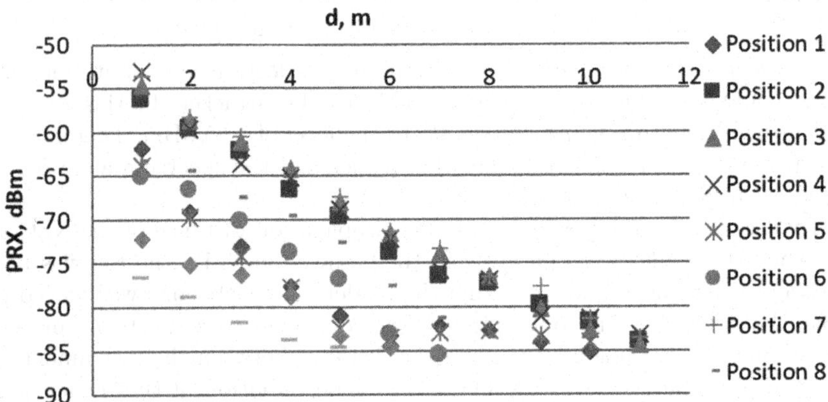

Fig. 4. Changes of the reader signal level vs distance when the reader is in the passive mode.

The results show that the maximum distance radiated by the signal is observed when the scanner is positioned and measured from the left, right, or by turning to the right at an angle of 45°. It is obvious that there are some orientations of such an antenna for the optimal position of signal reception. In all other cases, with the increasing distance, the signal strength decreases sharply, and it is difficult to detect it at 5 m (when the scanner is on the plane from the back). The maximum radiating distance of the scanner was recorded at up to 11 m away.

It can be observed that the variation of these signals follows the linear law $y = ax + b$, and slope coefficient a of these lines (which describes the physics of wave propagation) is approximately the same. Coefficient b depends on the orientation angle of the antennas. Although, as shown above, the variation of 13.56 MHz signal is close to the logarithmic scale, it is still convenient to summarize the obtained results by linear relationship Eq. (5):

$$P_Rx = -2.715 \cdot d + b \tag{5}$$

where PRx is the signal level at the receiver, dBm; d is the distance between the transmitter and the receiver, m; b is a free member which depends on the orientation angle of the transmitter and the receiver antennas.

Scenarios 2 (with the Reader Being in the Active Mode). This scenario investigated the variation of the NFC signal level by increasing the distance between the reader and the receiver (spectrum analyzer) during tag scanning, i.e., when the tag covers the scanner. The tag ID number was scanned. The tag ID was provided by the manufacturers provided the tag ID and it was fixed.

The signal strength was recorded from all the sides of the reader. The results are shown in Fig. 5.

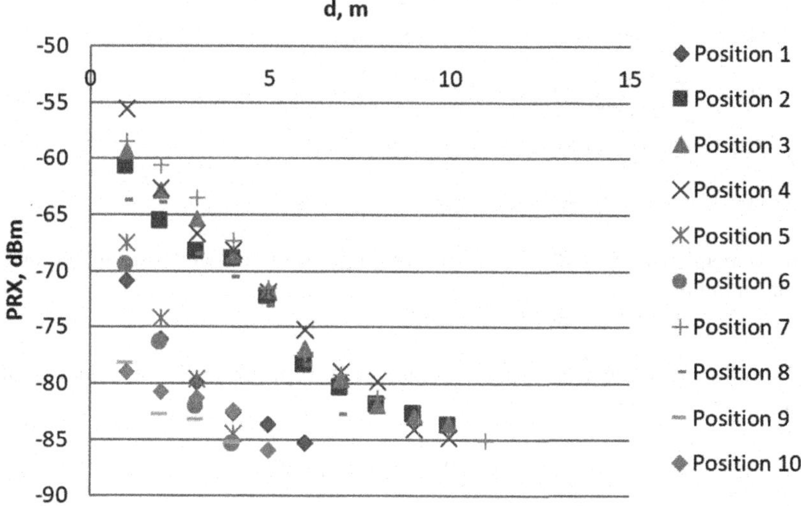

Fig. 5. Changes of the reader signal level vs distance (the reader is in the active mode).

One can observe that the variation of these signals, as in the previous case Fig. 4, follows the linear law $y = ax + b$, and slope coefficient a of these lines (which describes the physics of wave propagation) is approximately the same. Therefore, these results can be reflected with the same relation Eq. (5).

By summarizing Figs. 4 and 5, the following regularities of NFC signal changes can be observed:

- By looking at the signal strength of the reader in different directions of orientation of the transmitter antenna towards the receiver antenna, we see that the signal strength decreases linearly rather than logarithmically with varying distances (as in the case of UHF frequency ranges). The shortest distance between the transmitter and the receiver was fixed when the reader was placed on a plane from the back. However, in this case, the signal strength was recorded even at a distance of up to 4 m from the transmitter.
- Coefficient b depends on the angle between the antennas of the reader and the receiver, i.e., from the angle between axis 2–4 Fig. 3 and the location of the receiver Fig. 6.

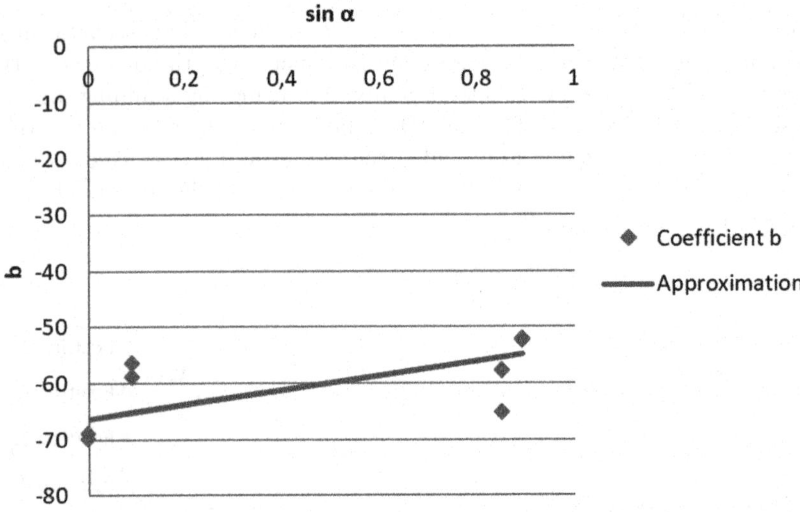

Fig. 6. Dependence of coefficient b on the angle between the transmitter and the receiver antennas.

It should be noted Fig. 6 that coefficient b approximates formula (6) values well with some error:

$$b = 12.90|sin(\alpha)| - 66.33; \tag{6}$$

where α is the angle between the transmitter and the receiver antennas. When inserting this expression into formula (5), it yields:

$$P_R x = -2.715d + 12.90|sin(\alpha)| - 66.33; \quad (7)$$

However, this is not a very accurate approximation as the purpose of this work was not to determine the regularities of signal changes, but rather to evaluate certain safety considerations.

3.2 NFC Signal Jamming Investigation

Firstly, the signal strengths of the readers and the jammer were determined at a distance of about 4 cm between them. The results (presented in Table 2) reveal the attenuation of the signals of all the three readers' dependency on the attenuator's signal strength.

Table 2. Signal strengths of the readers and the jammer at a distance of about 4 cm from them.

Passive mode	
Reader	Signal strength, dBm
ACR1222L	−77.72
Nexus 7 tablet	−81.55
ACR128U	−68.98
Jammer	−35.1

The signal strength of the NFC reader increases during tag scanning (the tag is placed on the reader). Therefore, this needs to be evaluated (Table 3).

Table 3. Signal strengths of the readers in the active mode at a distance of about 4 cm from them.

Active mode	
Reader	Signal strength, dBm
ACR1222L	−46.03
Nexus 7 tablet	−48.56
ACR128U	−45.84

Although it would seem that the attenuator must suppress the NFC signal, the effect of the attenuator was minimal in the real experiment, and the signal emitted by the scanner was not actually affected. Accurate measurements of the attenuator frequency revealed that the attenuator frequency differed by

6 kHz from the NFC scanner frequency. When the attenuator frequency was tuned exactly to the NFC scanner frequency, the NFC readers' signals were successfully suppressed, i.e., the data was not read from the card when the card was being laid on the reader. An attempt was made to read the NFC tag ID number during the investigation. However, the distances at which the silencer operated effectively were short (Table 4). For the jamming of the NFC tag in this case, the jammer had to be close enough to the reader: from 5.5 to 9 cm away.

Table 4. Maximum distance jammer from the reader for successful tag jamming.

Jamming of NFC tag	
Reader	Distance, cm
ACR1222L	5.5
Nexus 7 tablet	6.5
ACR128U	9.5

By using various types of tags, the reader was also jammed at the same distances. It was obvious that the jammer acted directly on the reader, but not on the tag.

4 Conclusions

Analysis of literature shows that there has been little work done for the investigation of real jamming of NFC technology, i.e., insufficient attention has previously been paid to this issue although the consequences would be particularly grave if successful jamming were achieved.

Our investigation has shown that the tag signal of the NFC reader can be intercepted at a significantly longer distance than officially announced – it was actually greater than 10 m. It should be noted that the experiments were performed until the signal level dropped below −85 dBm. The outcomes of such experiments revealed the possibility to jam the signal with a jammer. The jammer used in this investigation can jam the NFC reader's signal at a distance of 9.5 cm between the scanner and the attenuator, depending on the type of the scanner.

For the jamming of the signal, it is necessary to generate an electromagnetic wave of exactly the same frequency as that of the NFC reader. A small deviation (6 kHz in our case) from this frequency does not allow jamming the NFC signal anymore.

The proposed models evaluate the regularity of 13.56 MHz signal strength variation in small indoor distances as well as the dependence of 13.56 MHz signal variation on the angle between the transmitter and the receiver. The accuracy of the latter model is not high. However, this research did not set the objective of establishing strict regularities.

The research of this problem will likely take place in two directions in the future: investigation of the regularities of 13.56 MHz signal propagation over short distances under various environmental conditions as well as the structure of the intercepted NFC signal regarding the feasibility of detecting such a signal.

References

1. NFC forum. https://nfc-forum.org/. Accessed 20 Sep 2020
2. Farsens. http://www.farsens.com/en/2016/03/31/how-backscattering-affects-read-range-in-uhf-rfid-systems/. Accessed 20 Sep 2020
3. Garcia-Alfaro, J., Herrera-Joancomartí, J., Melià-Seguí, J.: Security and privacy concerns about the RFID layer of EPC gen2 networks. In: Navarro-Arribas, G., Torra, V. (eds.) Advanced Research in Data Privacy. SCI, vol. 567, pp. 303–324. Springer, Cham (2015). https://doi.org/10.1007/978-3-319-09885-2_17
4. Singh, M.M., Adzman, K.A.A.K., Hassan, R.: Near field communication (NFC) technology security vulnerabilities and countermeasures. Int. J. Eng. Technol. **7**, 298–305 (2018)
5. Masyuk, M.A.: Information security of RFID and NFC technologies. In: Journal of Physics: Conference Series, vol. 1399, pp. 1–6 (2019)
6. Kavya, S., Pavithra, K., Rajaram, S., Vahini, M., Harini, N.: Vulnerability analysis and security system for NFC-enabled mobile phones. Int. J. Sci. Technol. Res. **3**(6), 207–210 (2014)
7. Giese, D., Liu, K., Sun, M., Syed, T., Zhang, L.: Security analysis of Near Field Communication (NFC) payments, arXiv preprint (2019)
8. Kortvedt, H. S., Mjølsnes, S. F.: Eavesdropping near field communication. In: The Norwegian Information Security Conference (NISK), pp. 57–68 (2019)
9. Jin, R., Zeng, K.: Secure inductive-coupled near field communication at physical layer. IEEE Trans. Inf. Forensics Secur. **13**(12), 3078–3093 (2018)
10. Diakos, T.P., Briffa, J.A., Brown, T.W.C., Wesemeyer, S.: Eavesdropping near-field contactless payments: a quantitative analysis. J. Eng. **7**, 1–7 (2013)
11. Near Field Communications and mobile threats (2015). https://fedvte.usalearning.gov/courses/MSDS2015/course/videos/pdf/ MS_2015_D01_S07_T01_STEP.pdf
12. Gummeson J., Priyantha B., Ganesan D., Thrasher D., Zhang P. EnGarde: protecting the mobile phone from malicious NFC interactions.In: proceeding of the 11th Annual International Conference on Mobile Systems, Applications and Services, pp. 445–458. Association for Computing Machinery (2013). https://doi.org/10.1145/2462456.2464455
13. Oh S., Doo T., Ko T., Kwak J., Hong M.: Countermeasure of NFC relay attack with jamming. In: 12TH International Conference & Expo on Emerging Technologies for a Smarter World (CEWIT), pp. 1–4. IEEE (2015). https://doi.org/10.1109/CEWIT.2015.7338165
14. Ramos, A., Scott, W., Scott, W., Lloyd, D., O'Leary, K.: A threat analysis of RFID passports. Queue **7**(9), 1–5 (2009)
15. Nelson, D., Qiao, M., Carpenter, A.: Security of the near field communication protocol: an overview. J. Comput. Sci. Coll. **29**(2), 94–104 (2013)
16. Biswas, B., Basuli, K., Sen, S.S.: On a key exchange technique, avoiding man-in-the-middle attack. J. Glob. Res. Comput. Sci. **3**(9), 28–30 (2012)
17. Japertas, S., Orzekauskas, E., Slanys, R.: Investigation of Wi-Fi indoor signals under LOS and NLOS conditions. Int. J. Digit. Inf. Wire. Commun. (IJDIWC) **2**(1), 26–32 (2012)

Data Migration Rate of CRUSH-Based Distributed Object Storage with Dynamic Topology

Alexey Vanin[1,2(✉)], Vladimir Bogatyrev[3], and Stanislav Bogatyrev[2]

[1] ITMO University, Saint Petersburg, Russia
[2] NEO Saint Petersburg Competence Center, Saint Petersburg, Russia
{alexey,stanislav}@nspcc.ru
[3] Saint Petersburg State University of Aerospace Instrumentation, Saint Petersburg, Russia

Abstract. Distributed systems are widely used to solve problems that require large computational or storage resources. The scalability of such systems makes them cheaper. However, the overhead for maintaining the system's performance or operation ability may be significant. This paper considers a distributed P2P storage system in uncontrolled dynamic environment. The change of the structure or the topology in such system can lead to data migration that may consequently cause system overload. This paper examines the intensity and the amount of these migrations for different CRUSH-based data placement approaches.

Keywords: Distributed system · Storage system · Simulation modeling · Data migration · CRUSH · DHT · P2P

1 Introduction

The amount of generated and stored data is increasing every year [9]. This allows to develop new big data processing algorithms and train accurate neural networks. These research areas continue to improve because there is a place where all this big data can be stored. Due to technical limitations and economic reasons, storage systems virtualize their resources in hardware level: hard disks or magnetic tapes combined in clusters; as well as at operation level when data is spread across several storage nodes. These storage nodes may be geographically distributed [8,11] to increase reliability [4].

Data should be securely stored and accessible to a variety of agents that interact with it. Such agents are databases, web applications, raw data parsers, neural networks, etc. In order to unify interaction, modern storage systems store data in the form of objects. These objects contain data as a payload and provide additional meta-information, some context to it. Amazon S3 object storage interface has become a de facto standard for clients working with such data. Other cloud storage providers [2] implement this interface to be competitive on the market.

© Springer Nature Switzerland AG 2020
V. M. Vishnevskiy et al. (Eds.): DCCN 2020, CCIS 1337, pp. 464–471, 2020.
https://doi.org/10.1007/978-3-030-66242-4_36

This paper considers a distributed object storage system as a set of storage peers. This approach allows to develop cheap, horizontally scaling P2P systems [7,10]. However, this imposes additional costs for maintaining data consistency in the storage. It must remain available, replicated, while workload on storage nodes should be uniformly distributed.

Data load is controlled by a distribution algorithm (or distribution function) that selects nodes on which data must be stored. If network topology changes, data should migrate in order to remain available for clients. DHT-based algorithms are widely used in P2P network [5,12]. CEPH [14] object storage uses Controlled, Scalable, Decentralized Placement algorithm of Replicated Data (CRUSH) [13]. This paper evaluates how effective these approaches can be in a distributed decentralized object storage. Efficiency criterion is the rate of data migration. Migration is a necessary overhead to maintain the consistency of the storage system. A large number of migrations can significantly lower the efficiency of the entire system by reducing data availability or increasing the probability of storage node overload. The more migrations are in the system, the less likely the client is to meet the declared quality of service.

2 Program Simulation of Data Migration

The intensity and amount of migrated data are examined in a simulation model. The model defines a set of nodes that have different attributes, or buckets in CRUSH terminology. Nodes are connected with a logical topology that determines how the system places objects in the storage nodes. Hash-ring [6] topology was used to examine DHT approach. In CRUSH, nodes are connected in a graph, where the vertices correspond to the attributes of the nodes, and the leaves are to the nodes themselves, see Fig. 1. Implementation of the CRUSH algorithm called network map [3].

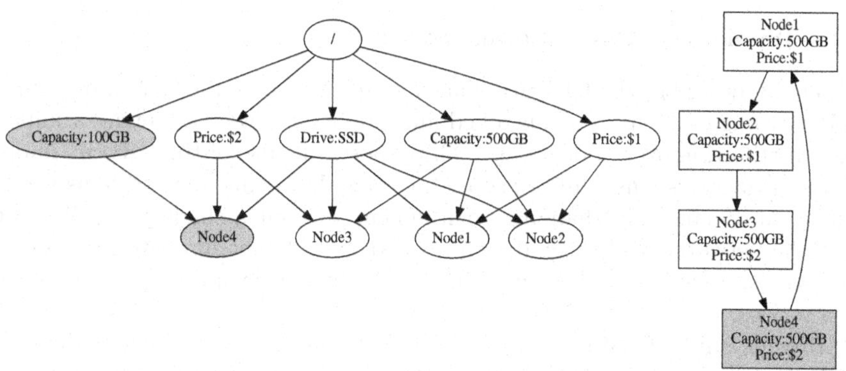

Fig. 1. Topology representation for CRUSH and hash ring

Simulation is performing in two stages. The whole procedure is briefly described in Algorithm 1. At the first stage, the model generates a stream of objects. Each object proceeds through a placement function that produces the set of storage nodes. These nodes have a finite capacity. Simulation stops at first storage failure: when there is no free space for the object in the storage node.

At the second stage, the topology is expanded with a few nodes, filled gray in Fig. 1. Each uploaded object from the first stage proceeds through placement function once again. If the locations at the second stage have changed, then the object migrated in the storage system.

Algorithm 1. Migration simulation

```
procedure MIGRATION(O, T1, T2)                ▷ Set of objects and two topologies
    Counter ← 0
    N ← length(O)
    for k ← 1 to N do
        P1 ← Placement(T1, Oi)                ▷ Set of storage nodes
        P2 ← Placement(T2, Oi)
        if P1 ≠ P2 then                       ▷ If the placement has changed,
            Counter ← Counter + 1             ▷ increase the counter
        end if
    end for
    return Counter ÷ N
end procedure
```

All experimental results presented in this paper provided with confidence intervals of Student's T distribution with $\alpha = 0.05$.

3 CRUSH-Based Storage with Dynamic Topology

3.1 Topology in Open Storage Systems

Paper [13] describes the CRUSH. This algorithm places objects in a distributed object storage system in a controlled and predictable way, achieving a uniform load distribution over nodes. Paper [3] proposes the use of the CRUSH in open object storage systems. Unlike proprietary closed systems, such systems work in untrusted dynamically changing environment without a single point of control over the network. Nodes could be provided by different unrelated authorities, therefore storage system becomes a platform for node owners to provide storage as a service.

Provide efficient data storage in such environment can be really tricky, but CRUSH defines weight coefficients for the storage nodes. In Eq. 1, CRUSH uses both hash distance from node i to the stored object x and the weight of node w_i itself to calculate a numeric characteristic of node c_i. Nodes with lowest c store the object.

$$c_i(x, r) = f(w_i)hash(x, r, i) \tag{1}$$

Storage system can determine weight function on its own. It could be represented as reputation value based on P2P interaction, node capacity parameter, or even storage pricing [1]. For the experiments we defined weight function as a multiplication of normalized node capacity c and storage price p, see Eq. 2.

$$f(w_i) = \bar{c_i}\bar{p_i} \tag{2}$$

With this weight function, we placed objects in the model and expanded the topology. Results presented in Fig. 2.

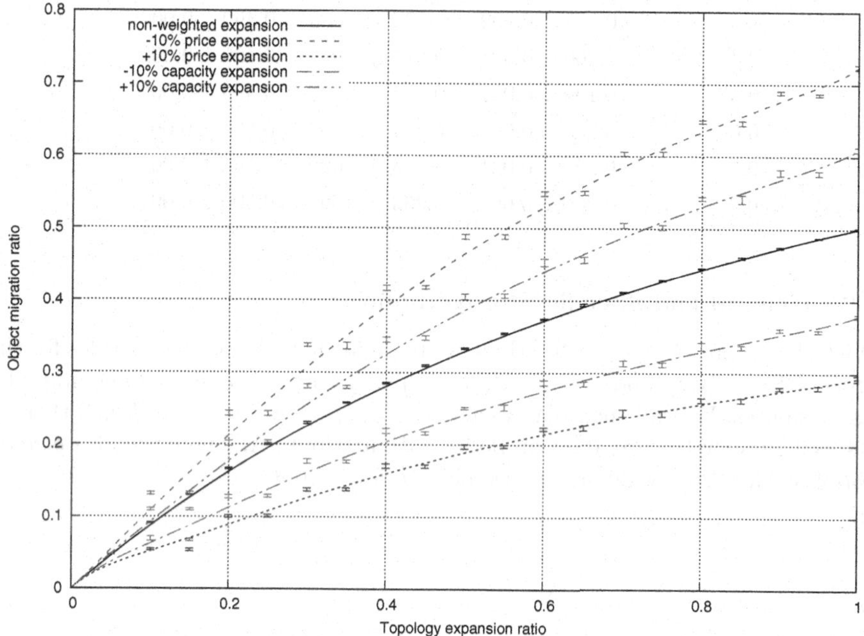

Fig. 2. Object migration ratio by nodes expansion with different weights

While all nodes are the same, 2x topology expansion leads to the 50% object migration. However, node weights influence migration ration: if there are more profitable nodes in the system, objects will try to migrate there. Deviation from the non-weighted expansion can be controlled by normalization function: capacity weight is less significant than price weight in this example.

3.2 Migration Ratio at Different System Load

In the first experiment, the model generated objects until storage failure: if the amount of stored objects is more than a capacity parameter. Overall system load was about 80% at the end of the first simulation stage. Model went from the transient state to the steady state. If we stop this stage at any specific system load limit, migration rate will be the same, see Table 1. This allows to exclude system load parameter from the model and make it more robust.

Table 1. Migration rate at different system load

System load	Migration ratio		
	25% expansion	50% expansion	100% expansion
0.1	0.201 ± 0.003	0.331 ± 0.002	0.499 ± 0.004
0.2	0.201 ± 0.002	0.332 ± 0.003	0.500 ± 0.003
0.3	0.201 ± 0.002	0.332 ± 0.001	0.499 ± 0.002
0.4	0.200 ± 0.001	0.334 ± 0.001	0.499 ± 0.002
0.5	0.199 ± 0.001	0.333 ± 0.002	0.499 ± 0.002
0.6	0.200 ± 0.001	0.334 ± 0.002	0.500 ± 0.002
0.7	0.200 ± 0.001	0.333 ± 0.002	0.500 ± 0.001
0.8	0.200 ± 0.001	0.333 ± 0.001	0.500 ± 0.001

3.3 General and Specific Storage Policies

With the graph topology CRUSH allows to define flexible placement rules for the objects. Since every node has attributes, the data owner can filter some nodes by its attributes. We call such placement rules as *specific policies.* In Fig. 3 there is a specific policy to store data on nodes with attribute A. There are also *general policies.* They do not define any specific node attributes.

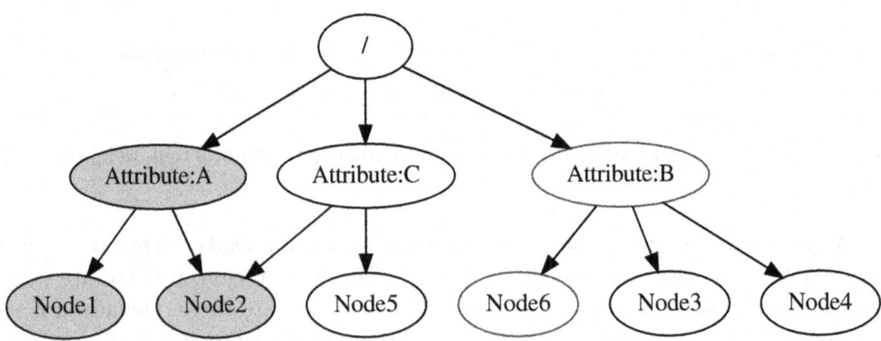

Fig. 3. CRUSH topology with specific policy (gray) and expanded node(red) (Color figure online)

Previously we expanded CRUSH topology in a symmetric way: at the 1.0 expansion ratio every node had a pair with the same attributes. But this is not very accurate in the real-world environment. We add new model parameter $\phi = \frac{E}{A}$ where E is a set of attributes in expanding nodes and A is a set of all attributes in the graph. In Fig. 3 there is a one expanded node with one out of three attributes, therefore $\phi = \frac{1}{3}$.

If the storage system store data by general policies, migration ratio does not affected by ϕ, see Fig. 4. However specific policies may significantly lower migration rate as soon as ϕ tends to zero. With $\phi = 1$ migration ratio will met worst case scenario with up to 50% migration if the topology graph size doubled.

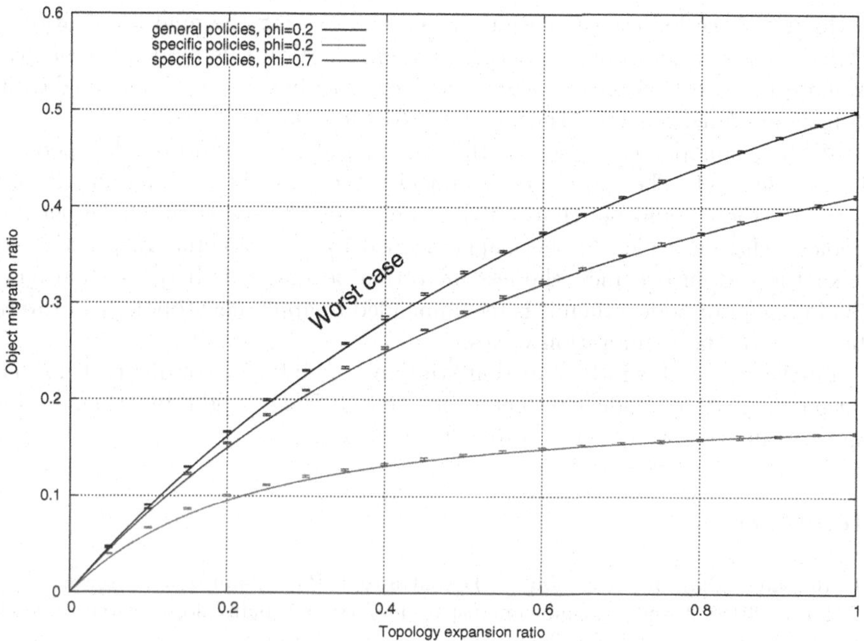

Fig. 4. Object migration ratio by nodes expansion with different ϕ and storage policies

3.4 DHT Approach

DHT and CRUSH are based on the hash function that used to decide the location of the object. Despite the fact that CRUSH is designed to predictably reduce randomness in the placement function, P2P networks still actively use DHT as a storage cache. To place the data, each node calculates a hash distance between hash of the data and the hash of the node (3). Nodes P with minimal hash distance will store the data (4).

$$distance(x, y) = |hash(x) - hash(y)| \quad (3)$$

$$N = \{N_1, N_2, \ldots, N_m\}, \min_{n \in N} distance(n, obj) = P \quad (4)$$

This called consistent hashing. CRUSH also uses consisting hashing after filtering nodes and attributes. Therefore DHT has the migration ratio results as CRUSH with general policies only. It is enough to build efficient distributed caches. They do not need active replication of stored data because cache miss is a regular situation. But flexible work with node attributes is mandatory for distributed object storage.

4 Conclusion

In this work, we built a simulation model for assessing the migration ration in a CRUSH-based distributed object data storage. With this model, the one can evaluate the effects of node weight coefficients to find the optimal weight and normalizing functions for a specific data storage system.

With this model we found out that the migration ratio does not depend on the system load. The more specific policies there are in the system, the less migration there could be. However, it depends on the structure of the network topology change, which in this paper defined by ϕ parameter. In practice the larger the system becomes, the less migration occurs in it. If the system has a small number of nodes, then it is recommended to apply new topology in several steps to avoid large migration waves.

Further research will be aimed at studying the influence of different normalization methods of node parameters for the CRUSH weight function. Performance evaluation will be done using this model.

References

1. Research Plan for Distributed Decentralized Blockchain-based Storage Platform (2018). https://github.com/nspcc-dev/research-plan/blob/master/research_plan.pdf. Accessed 5 Nov 2019
2. Yandex Cloud. How to use API .https://cloud.yandex.ru/docs/storage/s3/. Accessed 8 June 2020
3. Bogatyrev, A., Liubich, S., Wahle, F., Bogatyrev, S., Vanin, A.: The model of network map and data placement in the distributed decentralized storage platform. In: Proceedings of the 10th Majorov International Conference on Software Engineering and Computer Systems. CEUR-WS (2018)
4. Bogatyrev, V.: Fault tolerance of clusters configurations with direct connection of storage devices. Autom. Control Comput. Sci. **45**, 330–337 (2011). https://doi.org/10.3103/S0146411611060046
5. Kaashoek, M., Karger, D.: Koorde: a simple degree-optimal distributed hash table (2004). https://doi.org/10.1007/978-3-540-45172-3_9
6. Karger, D., Lehman, E., Leighton, T., Panigrahy, R., Levine, M., Lewin, D.: Consistent hashing and random trees: distributed caching protocols for relieving hot spots on the world wide web. In: Proceedings of the Twenty-Ninth Annual ACM Symposium on Theory of Computing, pp. 654–663. STOC 1997, Association for Computing Machinery, New York, NY, USA (1997). https://doi.org/10.1145/258533.258660, https://doi.org/10.1145/258533.258660

7. Muralidhar, S., et al.: f4: Facebook's warm blob storage system. In: 11th USENIX Symposium on Operating Systems Design and Implementation (OSDI 14), pp. 383–398 (2014)
8. Noghabi, S.A., et al.: Ambry: linkedin's scalable geo-distributed object store. In: Proceedings of the 2016 International Conference on Management of Data, pp. 253–265 (2016)
9. Paulsen, J.: Enormous Growth in Data is Coming - How to Prepare for It, and Prosper From It. https://blog.seagate.com/business/enormous-growth-in-data-is-coming-how-to-prepare-for-it-and-prosper-from-it. Accessed 1 June 2020
10. Ratnasamy, S., Francis, P., Handley, M., Karp, R., Shenker, S.: A scalable content-addressable network. In: Proceedings of the 2001 Conference on Applications, Technologies, Architectures, and Protocols for Computer Communications, pp. 161–172 (2001)
11. Spirovska, K., Didona, D., Zwaenepoel, W.: Optimistic causal consistency for geo-replicated key-value stores. 2017 IEEE 37th International Conference on Distributed Computing Systems (ICDCS), pp. 2626–2629 (2017)
12. Stoica, I., Morris, R., Karger, D., Kaashoek, M., Balakrishnan, H.: Chord: a scalable peer-to-peer lookup service for internet applications. ACM SIGCOMM Comput. Commun. Rev. **31** (2001). https://doi.org/10.1145/964723.383071
13. Weil, S.A., Brandt, S.A., Miller, E.L., Maltzahn, C.: Crush: controlled, scalable, decentralized placement of replicated data. In: SC 2006: Proceedings of the 2006 ACM/IEEE Conference on Supercomputing, pp. 31–31 (2006)
14. Weil, S., Brandt, S., Miller, E., Long, D., Maltzahn, C.: Ceph: a scalable, high-performance distributed file system, pp. 307–320 (2006)

Reduction of the Energy Consumption on Preliminary Signal Processing in Vibration Monitoring Networks

Vitaly Morozov and Konstantin Alikin(✉)

V.A. Trapeznikov Institute of Control Sciences of RAS, 65 Profsoyuznaya street, Moscow 117997, Russia
morbe36@mail.ru, ak-evmt@yandex.ru
https://www.ipu.ru

Abstract. Energy consumption reduction is of utmost importance for Internet of Things. This pertains especially to vibration monitoring networks usually consisting of plethora sensors and components for preliminary data processing. Such networks have to run uninterruptedly for long periods of time, for instance as a part of jerks and oscillation measurement systems used for earthquake registration or large building vibration monitoring. In these systems electronic data processing equipment is abundant. Lowering of the operating current and voltage in electronic equipment tends as a rule to increase the processing error. It is true particularly of the scaling amplifiers widely used in the course of data processing for preamplification of weak signals from analog sensors. In the paper, we propose an efficient method and corresponding circuit design for reproduction error suppression in micropower scaling amplifier by the additional feedback loop. The implementation of additive negative feedback by error signal supports substantial total scaling error reduction. Experiments and simulation carried out have verified manifold error reduction by the proposed structure in parallel with lowering of amplifiers energy consumption.

Keywords: Internet of things · Vibration sensors · Energy consumption · Scaling amplifier · Error suppression · Additive negative feedback

1 Introduction

Autonomous wireless sensors serve as the backbone of the Internet of Things (IoT) [1]. These sensors do not have an external power supply and their lifespan is typically limited by the lifespan of their battery nodes [2]. In [3] authors discuss a network of sensors that can extract (harvest) the energy from the surrounding environment (wind, solar, etc.) and highlight the importance of energy consumption optimization.

Wireless networks consisting of a great number of autonomous vibration sensors are widely used for various IoT applications (industrial arrangements

© Springer Nature Switzerland AG 2020
V. M. Vishnevskiy et al. (Eds.): DCCN 2020, CCIS 1337, pp. 472–483, 2020.
https://doi.org/10.1007/978-3-030-66242-4_37

monitoring, earthquake registration [4], structural health monitoring for critical infrastructure [2], etc.). While the energy consumption of the micro-controller unit and the wireless communication modules accompanying the sensors may be significantly reduced by employing triggered wake-up and data accumulation, the analog preprocessing equipment (preamplifiers, filters) has to operate uninterruptedly. Therefore, an urgent problem for vibration monitoring systems designer is lowering of the energy consumption of the analog electronic parts.

Most kinds of vibration sensors, for instance, electromagnetic produce relatively weak signals. Thus, preliminary amplification of the sensor signal by the factor of 50–100 is necessary before filtration and conversion into digital form. Consequently, the power consumption rises significantly, because amplifiers are a major part of the analog preprocessing hardware in the systems discussed.

Advances in semiconductor technology have led to a substantial decrease of microcircuit amplifiers power consumption. Unfortunately, it attended with a substantial rise of frequency dependent error occurred by the signal amplification [5].

In the paper, we suggest a circuit-configurational method for lowering processing errors caused by the signal amplification. Our method permits the use of micropower amplifiers in inverting scaling amplifiers instead of more power-consuming high-speed ones.

2 Structure of a Scaling Amplifier

Preamplifiers for sensor signals are based as a rule on operational amplifiers (OA) with negative feedback circuit. The commonly used structure of scaling amplifiers (SA) consisting of the operational amplifier OA with negative feedback resistors R_{in}, R_{f} is shown in Fig. 1. Such a configuration has a given amplification factor of $K_{\mathrm{g}} = R_{\mathrm{f}}/R_{\mathrm{in}}$.

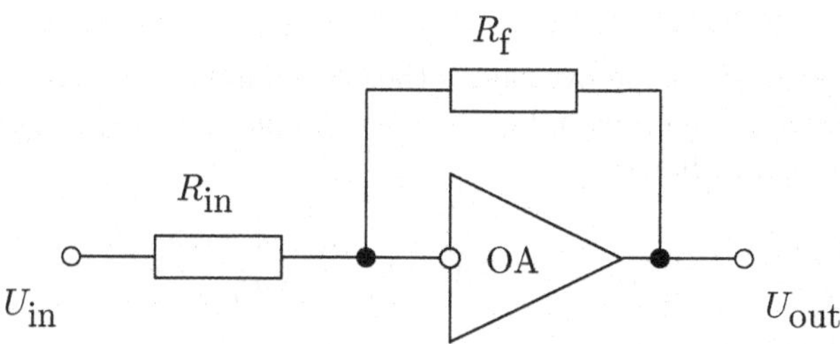

Fig. 1. Circuit of a commonly used scaling amplifier

In spite of a rather steady performance, variable error components occur in this structure during the signal processing, particularly:

1. An amplitude error caused by the inaccuracy of the amplification factor and nonlinearity of the transfer function.

2. A frequency dependent error leading to high frequency components weakening in the output signal spectrum.

The zero shift in the preamplifier output signal has a negative impact on system performance in some applications.

To mitigate the first of the aforementioned errors it is enough to hold the ratio R_f/R_{in} to the highest possible precision. The choice for precise resistive components on the market is sufficient. Additionally, the insertion of a regulated weak DC signal to the SA input eliminates the output voltage shift.

Next we go on considering the frequency dependent error. Let us define the amplification factor in the structure shown in Fig. 1 for the case when OA has a limited bandwidth in the high frequency region. The requirements of stability by deep degrees of negative feedback are imposed on the OA in this and similar circuits. Consequently, the transfer function for OA in Laplace operator form ought to be of first order: $K_0/(1+sT)$, where K_0 — direct current amplification factor of the OA, T — time constant for the OA as a first order system. Therefore, on the base of the well-known relations for control systems with negative feedback, the real amplification factor of the preamplifier (Fig. 1) would be expressible as a transfer ratio for a system with negative feedback coefficient equal to R_{in}/R_f:

$$K_r = \frac{K_0/(1+sT)}{1+K_0/(1+sT)(R_{in}/R_f)}, \tag{1}$$

differing from the rated meaning K_g by the value:

$$\Delta K = K_g - K_r. \tag{2}$$

After the substitution $K_g = R_f/R_{in}$ and some simplifications we can get from (1) and (2) the equation for the difference between given and real amplification factor values:

$$\Delta K = \frac{K_g{}^2(1+sT)}{K_0}. \tag{3}$$

The difference defined above causes the occurrence of the error component in the output signal of the preamplifier. This component grows with the amplified signal frequency as follows:

$$\Delta U_{out} = U_{in}\Delta K = U_{in}\frac{K_g{}^2(1+sT)}{K_0}. \tag{4}$$

The ratio of the retrieved error component to the full output voltage value U_{out} equals

$$\frac{\Delta U_{out}}{U_{out}} = \frac{K_g(1+sT)}{K_0}. \tag{5}$$

It is possible that the error voltage ΔU_{out} calculated from (4) is out of the allowable range defined by the system requirements, the type of amplifier chosen and the frequency needed. In the present state of electronic components, this is very possible, because of a tendency toward lowering amplifiers power consumption [5].

Let us consider as an example a micropower operational amplifier LTC2063 with supply current 2 µA, operating supply voltage 1.7 to 5.25 V, gain-bandwidth product $GBP = 20$ kHz, $K_0 = 140$ dB [6].

Before the error calculation in accordance with the formula (4) it is necessary to find the time constant T for the chosen amplifier. For this purpose a well-known relation: $T = K_0/(2\pi GBP)$ may be used, of which after the substitution of the data presented above follows that $T = 79.6$ sec. Further calculation shows that the frequency dependent relative error for scaling unit with $K_g = 100$ based on the chosen OA is equal to 0.05 or 5% by the frequency 10 Hz and further grows with the frequency roughly linearly according to (5). Taking into account that in many cases the frequencies of registered vibration signals are up to 100 Hz, the value of the frequency dependent error calculated above is completely unsuitable.

As an example, let us consider one more contemporary microcircuit, AD8502 consisting of two micropower operational amplifiers produced by complementary MOS technology and featuring a maximum supply current of 1 µA per amplifier. AD8502 can operate from a single-supply voltage of 1.8 V to 5.5 V or a dual-supply voltage of ±0.9 V to ±2.75 V. With its low power consumption the AD8502, according to the manufacturer recommendations, is ideally suited for a variety of battery-powered applications including smoke and fire detectors and vibration monitors. It is a perfect amplifier to use for the demonstration of the method proposed in the paper. It has following dynamic performance characteristics: $GBP = 7$ kHz, $K_0 = 120$ dB [7].

However, obtaining the time constant for this type of amplifier directly from the equation $T = K_0/(2\pi GBP)$ is rather cumbersome, because in the deduction of (4) we presumed the phase margin of 90°. But according to the datasheet for the amplifier the phase margin is 60°. It means that the approximation of the amplifier as a first order system is not valid for this amplifier. Rough estimation of the time constant for AD8502 gives the real value of $T \approx 15$ s. From the Eq. (4) the expected value of the reproduction error is approximately 1.3% 10 Hz. This value, similarly to the one calculated for LTC2063, is too high for the applications discussed.

The method for the extraction of the error component from the total output voltage of an inverting scaling amplifier was described in [8]. According to the method additive resistors equal to R_{in} and R_f are connected in series between the input and the output of the SA. The common point of the resistors is connected with the input of the additional non-inverting amplifier, that has an amplification factor of $K_g + 1$ (Fig. 2).

On the output of the structure shown in Fig. 2 occurs the error signal of the SA containing the error component ΔU_{out} defined in (4). It is clear that if the extraction of this component is possible, a design of an additive negative feedback loop for the error suppression will be realizable—see the Fig. 3a.

In this structure the EA is an amplifier of the error signal in terms as used in control systems theory [9]. Additive resistors chain R_{in}–R_f in Fig. 2 produces a

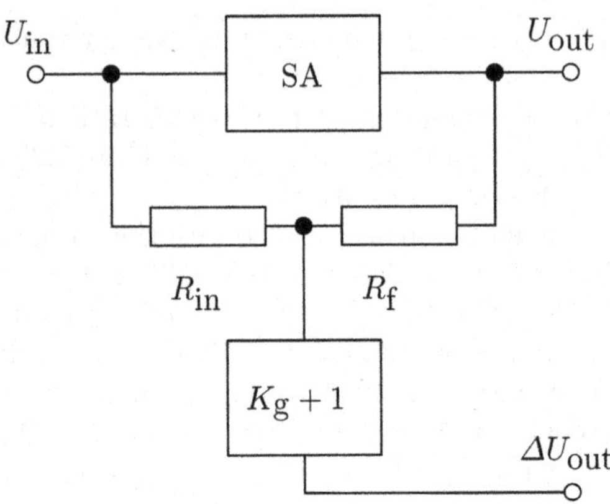

Fig. 2. Circuit for the scaling error extraction

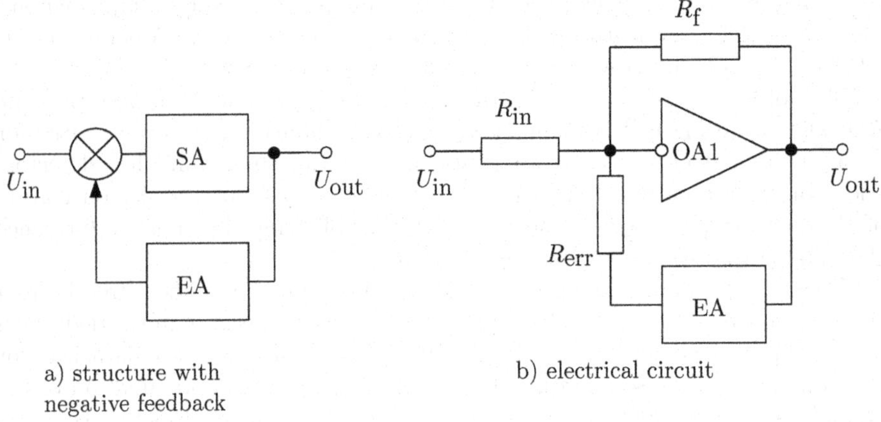

Fig. 3. Scaling amplifier with an error suppression loop

reference signal for error component extraction. Resistor chain deteriorates the component ΔU_{out} by $K_g + 1$ times additive amplifier regains it. Therefore, the amplification factor of the EA is equal to one and as the only amplifying element in the loop, we can use an SA with an additive input.

The summing input for error signal may be achieved by the connection of additive resistor R_{err} to the summing point of the main OA (Fig. 3b).

It is evident that in the structure presented in Fig. 3b the amplification factor of the additive negative feedback loop is equal to $K_{err} = K_{EA} \cdot R_f / R_{err}$. As mentioned above the EA is intended only for the extraction of the error signal ΔU_{out} without amplification so that $K_{EA} = 1$ and the equation $K_{err} = R_f / R_{err}$ is valid. Application of the well-known expression for the error signal suppression in systems with a negative feedback to the structure in Fig. 3b gives:

$$\frac{\varDelta U'_{\mathrm{out}}}{\varDelta U_{\mathrm{out}}} = \frac{1}{K_{\mathrm{err}}} = \frac{R_{\mathrm{err}}}{R_{\mathrm{f}}}, \tag{6}$$

where $\varDelta U'_{\mathrm{out}}$ is the error component value in the presence of the additive negative feedback.

Consequently, controlling the value of the error signal suppression is possible by the choice of a proper R_{err} value.

3 Simulation

The modeling was carried out using LTspice XVII software to determine the frequency dependent error in the circuit shown in Fig. 3b with and without a negative feedback. The manufacturer provided model of the micropower twin operational amplifier AD8502 [7] was used with the other components values as follows $R_{\mathrm{in}} = 100$ kΩ, $R_{\mathrm{f}} = 10$ MΩ, so that the rated amplification factor is $K_{\mathrm{g}} = 100$. The operating supply voltage was set to ± 2.75 V and the input signal amplitude to $U_{\mathrm{in}} = 20$ mV.

Scaling error—the relative difference between the given value of the output voltage equal to 2 V and the actual voltage for varying frequencies without additional error suppression we can see in Fig. 4, curve **a**.

The similarly defined difference in presence of an additional negative feedback loop ($R_{\mathrm{err}} = 500$ kΩ) is shown in Fig. 4, curve **b**. Hence it follows that a negative feedback by the error voltage with depth mentioned above reduces the scaling error 50 Hz frequency by the factor of 8.

4 Discussion and Recommendations

It follows from the previous results of simulation that negative feedback by error voltage with depth mentioned above does not reduce scaling error in the proposed circuit design by the factor calculated from (6). One can see comparing curves **a** and **b** in Fig. 4 that at 40–50 Hz frequencies real error reduction factor is approximately 10 instead of 20 determined by the ratio between component values used for the simulation $R_{\mathrm{err}}/R_{\mathrm{f}} = 1/20$.

The significant difference occurs because the amplification factor of the error signal extractor was assumed to be equal to one at all frequencies for the calculation of the error compensation ratio ($K_{\mathrm{EA}} = 1$, see formula (6) deduction).

In other words, assumption $K_{\mathrm{EA}} = 1$ means that the EA (Fig. 3b) precisely reproduces the scaling error signal value that is used thereafter as the negative feedback signal. That is not true in fact, because in the negative feedback loop there are two serially connected components: the SA with its additive input and the EA. Each of the components has a first order transfer function by s but in common, we have an oscillatory element in the negative feedback loop. This will disrupt the stability and the precision of the input signal reproduction in the SA as a whole.

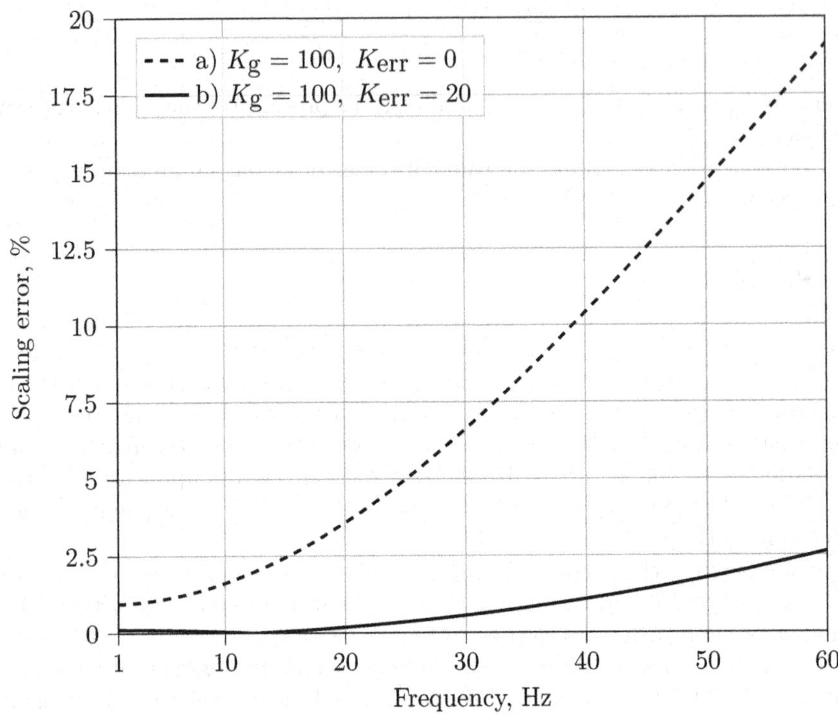

Fig. 4. Scaling error with and without an error suppression loop

Let us consider the situation more thoroughly. As it may be seen from Fig. 2 the error signal, attenuated in relation to ΔU_{out} value during extraction by additive chain R_{in}–R_{f} is subsequently amplified by a factor of $K_{\text{g}} + 1$. For discussed values of K_{g} EA amplification factor is $K_{\text{g}} + 1 \approx K_{\text{g}}$. Thus EA transfer function is practically identical with transfer function K_{EA} for the OA itself without an additive error suppression loop as shown in Fig. 1 (provided that the EA and OA1 are cells of a twin operational amplifier microcircuit). Therefore, the transfer function for error signal may be considered as of satisfying Eq. (1) with sufficient precision. On rearrangement and after simplification the result can be written as:

$$K_{\text{EA}} \approx \frac{K_{\text{g}}}{1 + sT \cdot K_{\text{g}}/K_0}. \tag{7}$$

Similar relation describes SA transfer function by additive input having amplification K_{err}:

$$K_{\text{SAerr}} \approx -\frac{K_{\text{err}}}{1 + sT \cdot K_{\text{err}}/K_0}. \tag{8}$$

Evidently, there are two serially connected elements in the feedback loop each of them having a first order transfer function. Let us denote time constants

$T_1 = T \cdot K_g/K_0$ for K_{EA} and $T_2 = T \cdot K_{err}/K_0$ for K_{SA} and write down the transfer function for closed negative feedback loop as a second order relation

$$K_{NF} = -\frac{K_g K_{err}}{(1+sT_1)(1+sT_2)} = -\frac{K_g K_{err}}{s^2 T_1 T_2 + s(T_1+T_2)+1}. \tag{9}$$

To analyze the denominator of the Eq. (9) let us substitute the values of $T_1 = 1.3 \cdot 10^{-3}$, $T_2 = 2.6 \cdot 10^{-4}$ obtained from the previously provided amplifier parameters. Further calculations show that the denominator of the Eq. (9) has complex-valued roots. Consequently the negative feedback loop presents an oscillatory unit. Therefore, SA frequency response will have a peak on a frequency corresponding with the pole of function (9).

Plot of the frequency-dependent amplification error for the SA with an error suppression loop obtained by simulation in extended frequency range is presented in Fig. 5. It is obvious that the frequency response peak discussed above has an adverse effect on the error at frequencies much lower than the corner frequency corresponding with the pole of the function (9). This is the reason why the scaling error plot in Fig. 5 differs significantly from results of calculation by Eq. (6).

Another peculiarity of the plot in Fig. 5 is the error value being limited at a level slightly exceeding 30%. This is caused by the saturation of the OA output stage. Because of the frequency response peak mentioned above, on some frequencies the OA becomes saturated and the output voltage reaches a value close to its power source voltage of 2.75 V. The difference with the rated undistorted value of 2 V ($U_{in} = 20$ mV, $K_g = 100$) is approximately 33%.

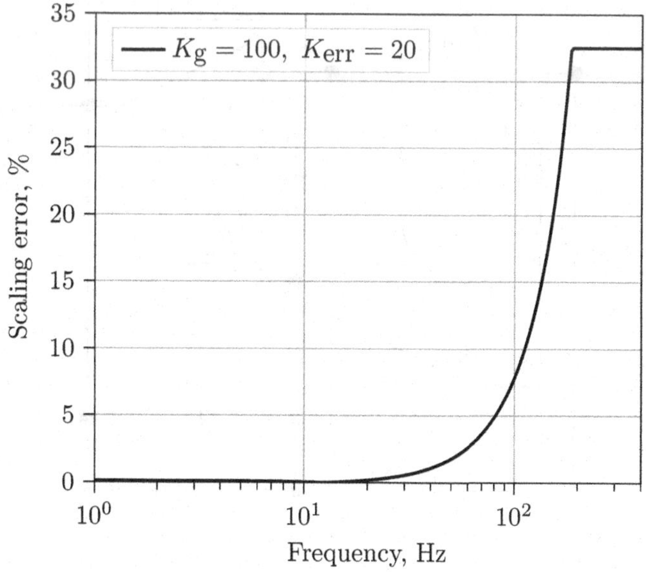

Fig. 5. Scaling error with an error suppression loop connected

Eliminating the peak mentioned above is possible with the help of the pole compensation method widely used in electronic design—insertion of a feed-forward element into the negative feedback circuit. For the pole compensation, it is practical to choose the negative feedback loop transfer function with the minimal time constant value. From the comparison between the values of T_1 and T_2 in (9) it is obvious that $T_2 \leq T_1$, because the value of K_g is usually higher than K_{err}.

Let us calculate time T_2 constant for AD8502 micropower operational amplifier with amplification factor $K_{err} = 20$ by additive input selected earlier as an example for the simulation of the error suppression loop performance (Fig. 4). The previously calculated value of the inherent time constant for the AD8502 amplifier is $T \approx 15$ s and for $K_{err} = 20$, $T_2 = T \cdot K_{err}/K_0 \approx 0.3$ ms. This is the time constant, which should be compensated to eliminate the undesired peak in the frequency response.

Pole compensation network can be implemented as a capacitor C connected in parallel across any serial resistor in the loop, particularly R_{err} or R_f. In the former case, the required capacity is $C = T/(2.2R_{err}) = 270$ pF, in the latter case $C \approx 14$ pF. It is obvious that the latter value is more convenient for the practical implementation.

In the complete electrical circuit of the SA with the additive negative feedback loop (Fig. 6), resistor R_f is shunted with a capacitor C by the non-inverting input of OA2.

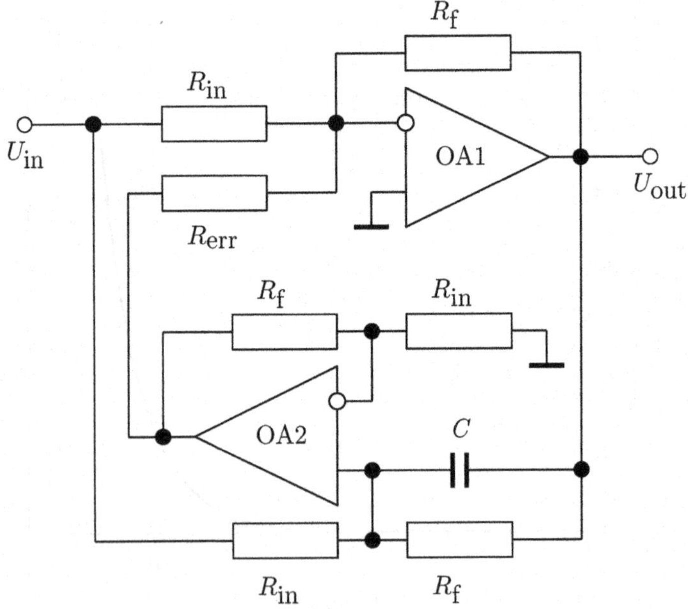

Fig. 6. Circuit of a scaling amplifier with a feedforward network R_fC

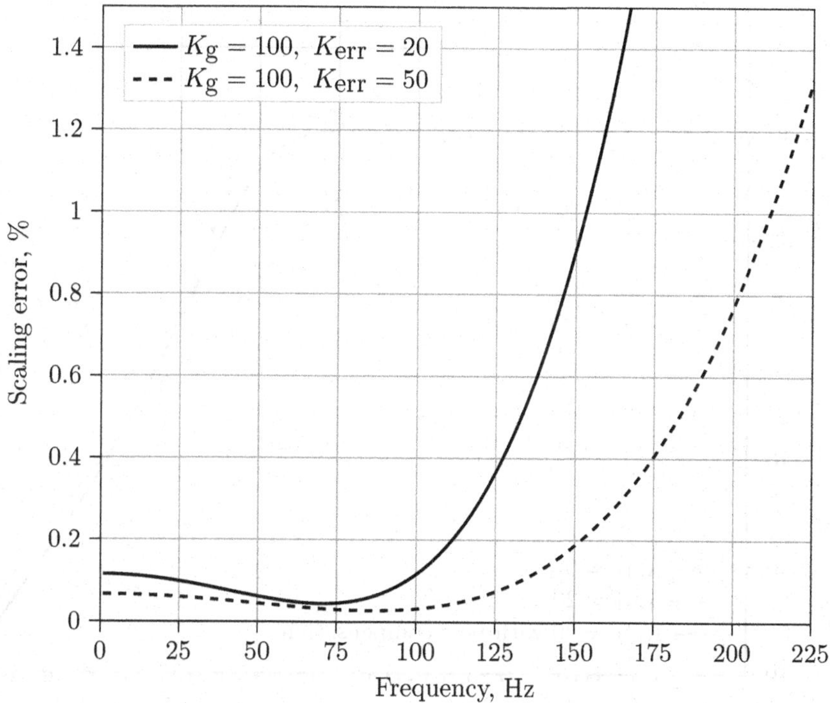

Fig. 7. Error suppression with different K_{err} values and pole compensation

The compensation of the pole having lower time constant T_2 in (9) by a 15 pF capacitor connected in parallel across a 10 MΩ resistor as shown in fig. 6 was verified by simulation results for $K_{err} = 20$ and 50 for AD8502 amplifier [7]. As shown in Fig. 7 the scaling error compared to a traditional scaling amplifier error (curve **a** in Fig. 4) goes down as Eq. (6) predicts.

Figure 8 demonstrates the changes in the frequency response of the preamplifier designed as an element for sensor signal processing. From this figure, one can see that the simple pole compensation circuit may effectively suppress the increase of the amplification above the given value at some frequencies. Consequently, the resulting frequency response characteristic is flat until 250 Hz. The exact frequency depends on the chosen values of K_{err} and the capacitor used for the pole compensation.

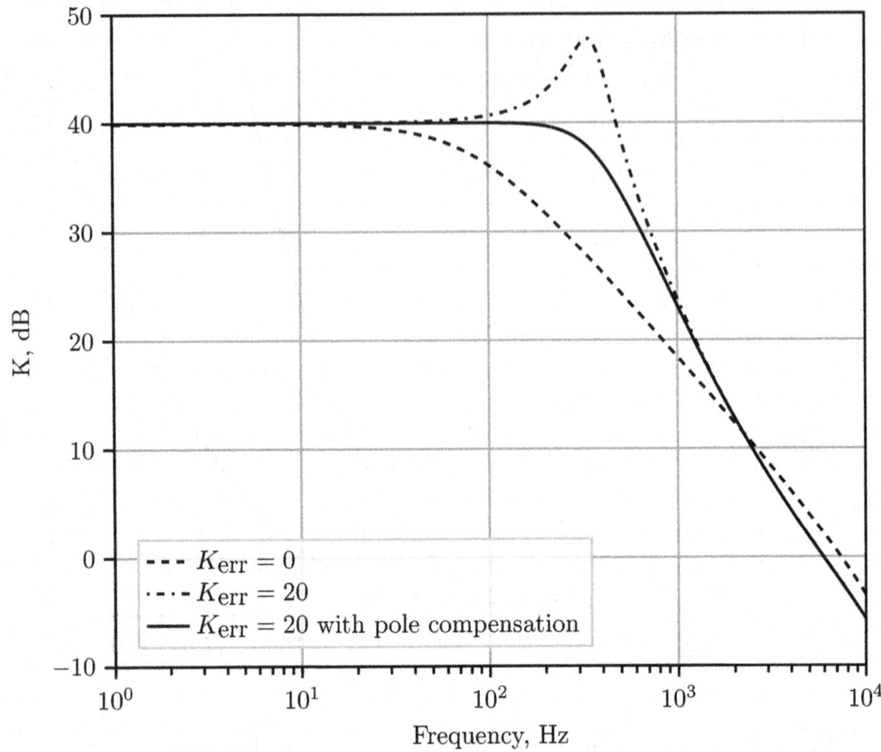

Fig. 8. Frequency responses of preliminary signal processor (preamplifier) as a whole with different error compensation variants

5 Conclusion

The method of the error suppression presented in the paper allows a substantial reduction of the signal processing error in inverting scaling amplifiers. It should be mentioned that not only the frequency dependent error component is suppressed by this method, but also the total difference of the output signal from the rated value.

The main additional element required in order to realize the loop of error suppression described above is another operational amplifier. An important point is that this additional component may be equal in power consumption to the main amplifier. Therefore, an effective application of readily available twin micropower operational amplifiers is possible.

Doubling of the supply current associated with the use of an additional operational amplifier does not substantially influence the effectiveness of the proposed method, because the saving in energy consumption from its application is significantly higher.

Acknowledgements. The reported study was funded by RFBR, project number 19-29-06043.

References

1. Atzori, L., Iera, A., Morabito, G.: The Internet of Things: a survey. Comput. Netw. **54**(15), 2787–2805 (2010). https://doi.org/10.1016/j.comnet.2010.05.010
2. Noel, A.B., Abdaoui, A., Elfouly, T., Ahmed, M.H., Badawy, A., Shehata, M.S.: Structural health monitoring using wireless sensor networks: a comprehensive survey. IEEE Commun. Surv. Tutorials **19**(3), 1403–1423 (2017). https://doi.org/10.1109/comst.2017.2691551
3. Dudin, A.N., Kim, C.S., Dudin, S.A.: Optimal control by a node of wireless sensor network with quality of transmission depending on the amount of harvested energy. In: Distributed Computer and Communication Networks: Control, Computation, Communications (DCCN-2019), pp. 29–36 (2019)
4. Kinoshita, S.: Kyoshin Net (K-Net), Japan. Int. Geophys. Ser. **81**(B), 1049–1056 (2003)
5. Kolombet, E.A.: Microelectronic Means of Analog Signal Processing. Radio i Svyaz', Moscow (1991). (in Russian)
6. LTC2063 datasheet and product info — analog devices. https://www.analog.com/en/products/ltc2063.html. Accessed 10 Sept 2020
7. AD8502 datasheet and product info — analog devices. https://www.analog.com/en/products/ad8502.html. Accessed 10 Sept 2020
8. Babayan, R.R., Morozov, V.P.: Bandwidth increase of an electromagnetic sensor signal amplifier (in Russian). Sensors Syst. **212**(3), 62–65 (2017)
9. Goodwin, G.C., Graebe, S.F., Salgado, M.E.: Control System Design. Prentice Hall, Upper Saddle River (2001)

Stability and Sustainability of Cryptotokens in the Digital Economy

N. V. Apatova, O. V. Boychenko, O. L. Korolyov, I. V. Gavrikov(✉), and T. K. Uzakov

Institute for Economics and Management, Crimean Federal University, Simferopol, Russia
painttool@gmail.com

Abstract. The paper examines the idea of cryptocurrencies and its sustainability in the digital economy. The specifics of mining are examined from several points of view: in terms of technical requirements, economic theory and current costs. An analysis of the role of cryptocurrencies in the economy is conducted. The impact of cryptocurrency mining on the economy and environment is described. Several solutions to the problem of cryptocurrency volatility and mining are presented, including smart electricity monitoring powered by internet of things (IoT) and smart energy grids, GIS-powered heat monitoring, and integration of cryptotoken mining into smart city infrastructure.

Keywords: Cryptocurrency · Digital economy · Sustainability · Smart cities · Internet of Things

1 Introduction

The concept of virtualization appeared in the 1990s. Virtualization involves replacing real objects and actions with their images and relationships, and it addresses two main aspects: 1) virtualization of society: traditional institutions prescribe to do real things and real actions, but people instead work with virtual objects – images, which makes social institutions a kind of virtual reality; 2) virtualization of social institutions, which includes economics, politics and culture.

"The digital economy is a socioeconomic activity mediated by software and enabled by telecoms infrastructure" [14]. It also includes the social sphere, which is reflected on the Internet as mobile health, mobile education, smart homes, and buildings powered by the Internet of Things.

Virtualization has a dual nature, as it is at the same time a system that consists of separate, interconnected elements, and a process that occurs both in each element and the system as a whole. In financial virtualization processes, three areas can distinguished. The first is the emergence of electronic payments, plastic payment cards and Internet banking. Second, financial markets, online

© Springer Nature Switzerland AG 2020
V. M. Vishnevskiy et al. (Eds.): DCCN 2020, CCIS 1337, pp. 484–496, 2020.
https://doi.org/10.1007/978-3-030-66242-4_38

trading in currency and securities. Third, the creation of cryptocurrency and the associated blockchain technology.

Bitcoin is the first fully virtual currency, simultaneously neither a regular commodity made to satisfy needs or desires, nor a virtual unit backed by existing currencies. An important metric for assessing the popularity of cryptocurrencies, of which there are now thousands, is their capitalization – the total market value of the tokens in circulation. Bitcoin's market capitalization in early January 2018 approached $300 billion, and resides at approximately $166 billion at the time of writing [3].

Blockchain technology, based on distributed data storage and the organization of direct connections between financial market agents, and the technology behind most modern cryptocurrencies, makes transactions safe and transparent, reducing or eliminating corrupt behavior in the financial sector and serving as a publicly auditable evidence base for currency flows. However, the emergence and spread of cryptocurrencies also means that issuing currency is no longer the sole prerogative of the state or a central banking authority, and threatens the very existence of traditional financial institutions, such as banks and exchanges, which are inherently speculative intermediaries.

2 Cryptocurrency Mining

Since the Bitcoin architecture, as formulated by Nakamoto, was by definition completely open (public) and had no concept of trust in individual nodes, it was necessary to formulate an algorithm for coordinating changes in such a system that would take into account possible malicious behavior of nodes. This algorithm was called the consensus algorithm [12].

In proof-of-work-based (PoW) cryptocurrencies, transactions are confirmed by means of "mining" – a decentralized process of solving cryptographic tasks. If the task is successfully solved by a "miner", a new block is created in the chain, which becomes its integral part and contains information about the last N transactions (N being a variable number with an average value of about 2000). Since all subsequent blocks are cryptographically linked to the current one, each of them is an additional confirmation of the transaction. In order to consider a transaction validated, at least one block containing it is required, but in a modern cryptocurrency context, a transaction is guaranteed to be confirmed when it has about 6 confirmations in a blockchain [9].

In cryptocurrencies that use public blockchains with a PoW consensus algorithm, miners are usually paid with units of that same cryptocurrency for creating a new block. However, along with the payment for creating a block, miners also receive a commission from the users of the cryptocurrency for each transaction, which serves as motivation for including that particular transaction in the block they are creating.

Other consensus algorithms also exist. One popular alternative to proof-of-work is proof-of-stake (PoS) algorithms, where the creator of each subsequent block in the chain is selected using a combination of random selection and selection based on their stake in the system, i.e., the amount of funds in the account.

The choice on the basis of only the account size will inevitably lead to centralization of rights to transaction verification, due to which there are several different subtypes of PoS algorithms, many of which select the stake based on the amount of funds and how long these funds were in the account for.

Other custom-built and proprietary consensus algorithms are out of scope for this paper. PoS algorithms do not use a mining process as such, and therefore are also out of scope. Proof-of-work algorithms, however, are very widespread and do use mining, which is why PoW-based cryptocurrencies, and Bitcoin as the archetypal PoW-based cryptocurrency, are the focus of this paper.

2.1 Quasi-legal Mining

Bitcoin mining operations are computationally expensive, and therefore necessitate the use of specialized equipment and the aggregation of computational power into pools. This aspect of the mining process will be discussed in more detail later on. However, there are cryptocurrencies other than Bitcoin, which, due to Bitcoin's hegemony in the market, are called "alternative" cryptocurrencies or "altcoins". These differentiate themselves from Bitcoin not only in name, but often also in the algorithms they use for achieving consensus. Often, this means that the mining process for these cryptocurrencies is significantly less computationally expensive. This presents an opportunity for extracting greater value using more conventional hardware, down to consumer devices like office PCs and smartphones.

For well-intentioned users and vendors, this means that a much wider audience can meaningfully participate in the cryptocurrency's lifecycle, working to mine it without needing to invest into expensive specialized hardware, and widens the cryptocurrency's community. Consensual, opt-in mining could transform revenue generation for content creators on the web.

Still, these opportunities are taken advantage of by malicious agents as well. "Cryptojacking" is a new kind of malware propagated primarily through advertisement networks and vulnerabilities like cross-site scripting (XSS). It involves uploading a cryptomining script to a website, preferably a high-traffic one, which mines cryptocurrency without the visitor's knowledge or consent. Monero is a cryptocurrency that is especially popular with cryptojacking malware, thanks to its heightened anonymity and privacy features. An in-depth systematic study of cryptojacking was performed by Hong et al. in [7].

3 Cryptocurrency and Value

Because cryptocurrency is a novel financial phenomenon, it can be difficult to estimate its true utility, price and value. Much of the research on the economics of cryptocurrencies focuses on Bitcoin as the archetypal cryptocurrency, but some studies consider other cryptocurrencies, or so-called "altcoins", as well.

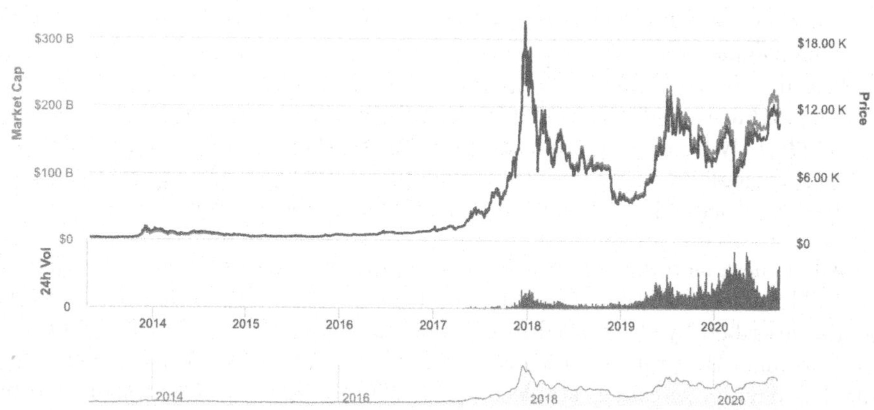

Fig. 1. Bitcoin market capitalization, price and 24-h trading volume, since mid-2013 [3]

3.1 Economic Role

The concept of cryptocurrency was initially born of a dream of decentralized money: a *currency* without a central bank or other controlling third party, issued algorithmically without even a strict need for trust between actors in the system. However, actual currencies fulfill the three basic functions of money:

1. Medium of exchange. Money is used as an intermediate medium of exchange for purchasing goods and services.
2. Unit of account. As money is used as a standard medium of exchange, it becomes the yardstick against which the value of goods and services is measured.
3. Store of value. Money as an asset retains value over time, subject to inflation and other adjustments.

Upon further analysis, however, modern cryptocurrencies often fail at fulfilling some or all of these functions. The volatility and instability of cryptocurrencies are the main causes of cryptocurrencies' low adoption as a means of payment.

1. Being a medium of exchange is arguably the easiest function to fulfill for any asset, as people can and do agree on using various assets and commodities as "money". However, cryptocurrency's ability to serve as a medium of exchange is often hampered by two factors: the relative technical difficulty of obtaining and exchanging them for fiat money, and the slow network speeds in cryptocurrency networks. Slow network speeds might not have been as significant of a problem were it not for the volatility of cryptocurrencies, which means that a transfer of cryptocurrency for a certain value may have an unpredictably different value upon arrival.

2. Volatility also influences cryptocurrencies' ability to act as a unit of account, as 1 BTC or 1 ETH at one point in time may imply a drastically different value compared to a different point in time.
3. Volatility and instability again prevent cryptocurrencies from serving as a meaningful store of value. In 2010, Bitcoin was used (as a means of exchange) to indirectly purchase pizza at a rate of 10,000 BTC for 25 USD. In December 2017, a single BTC was worth a little under $20,000. Already in December 2018, its price fell to $3,300 – a 76% drop from the previous year.

All of these factors paint traditional cryptocurrencies as a highly speculative asset. Movements in their price are often not tied to any market events, and poorly described by theoretical frameworks used for stocks and other assets.

To combat the problem of wildly unstable prices, the cryptocurrency community has devised so-called "stablecoins" – cryptocurrency assets designed to be pegged against a fiat currency (most commonly the US dollar). They attempt to fulfill the main functions of money, and usually succeed, albeit at the cost of other factors, such as transparency and long-term viability. Such stablecoins, exemplified by the Tether cryptotoken pegged to the US dollar, are stabilized against fiat currency using various methods – in the case of Tether, via off-chain collateral in the form of liquid currency reserves.

In a way, stablecoins serve as a medium of exchange, unit of account and store of value in the context of other cryptocurrencies, as opposed to the wider digital economy – they are used as the more "long-term" (on the scales typical of cryptocurrencies) store of value, and their being pegged to a fiat currency makes them a convenient medium of exchange and unit of account, as those functions are mirrored. The problems associated with stablecoins are often more regulatory in nature – in Tether's case, questionable accountability and the difficulty in verifying the actual volume of collateral reserves against the issuer's claims mar the otherwise viable cryptotoken concept.

3.2 Legal Framework

Today there is still no definite consensus in understanding the nature of cryptocurrency and its role in the digital economy. For example, a number of countries (Canada, Netherlands) consider cryptocurrency a currency, others (US, Germany) consider it a commodity. China and India have introduced a ban on the circulation of cryptocurrency, while the US and UK do not impose restrictions the development of cryptocurrency-related financial technologies.

Legislation of cryptocurrencies is further complicated by their popularity among hackers and other malicious actors. The decentralized nature of cryptocurrencies, i.e., their lack of central oversight and control, and their privacy-centric features make it difficult to establish a functional legal framework which requires personal accountability and central oversight.

In Russian legislation, cryptocurrencies were a grey zone since 2014. However, a new bill, "On digital financial assets", which will come into effect in 2021, defines and regulates cryptocurrencies and related digital technologies, including the distributed ledger and smart contracts, in Russian legislation. Cryptocurrencies are classified as commodities that may be purchased, exchanged, and sold for fiat currency, but may not be used as legal tender for payment to Russian residents. Cryptocurrencies are therefore classed as investments, and the Central Bank is tasked with limiting risks to unqualified investors purchasing cryptocurrency.

3.3 Theories of Value

There exists a multitude of different approaches to defining a cryptocurrency's value. In [10], an approach to evaluating the equilibrium price of Bitcoin using classical quantity theory of money is proposed. Li and Liao propose that the equilibrium between Bitcoin supply and demand, given B as the supply (amount awarded to a miner), equals

$$B = \frac{PY}{P_B V} + S \tag{1}$$

where P is the general price level of goods and services, Y is the quantity of goods and services traded using Bitcoin as a medium of exchange, P_B is the current dollar value per unit of Bitcoin, V is the velocity of Bitcoin (the frequency at which a single Bitcoin token is used for purchasing goods and services), and S is the amount of Bitcoin demanded for speculative purposes.

After several transformations, Li and Liao arrive at the equation for a socially optimal level of Bitcoin mining, balancing social benefit (non-speculative uses of Bitcoin) and cost of mining:

$$P_B(B - S) = \left(\sum_{i=1}^{M_s} N_i\right)\left(\frac{F}{D} + c\right) \tag{2}$$

where F is the mining equipment replacement cost, c is the average energy consumption per mining attempt, D is the total number of mining attempts per lifetime of mining equipment, N_i is the number of mining attempts per second for miner i, M_s is the socially optimal miner population. They contrast this equation with the equation for the *actual* level of bitcoin mining optimal for the mining business as a whole, as the mining business does not take into account whether the cryptocurrency is used speculatively or not:

$$P_B(B) = \left(\sum_{i=1}^{M} N_i\right)\left(\frac{F}{D} + c\right) \tag{3}$$

where M is the total miner population. Li and Liao then posit that $M_s < M$, and that as long as speculative demand for Bitcoin exists, the level of mining will necessarily be higher than socially desirable.

Hayes in [6] analyzes Bitcoin value formation from a cost of production point of view. First, he calculates the average amount of Bitcoin that can be mined by a single miner in a day (24 h):

$$BTC_{day} = \left(\frac{\beta\rho \times 3600}{\delta \times 2^{32}}\right) \times 24 \tag{4}$$

where β is the current block reward (in BTC/block), ρ is the hashing power of the miner in gigahashes, δ is the current difficulty of a block in gigahashes/block.

The dollar cost of mining per day is calculated by Hayes thus:

$$E_{day} = \frac{\rho}{1000} \times E \times C \times 24 \tag{5}$$

where ρ is the hashing power of the miner, E is the dollar price of a kilowatt-hour of energy, C is the energy consumption of the mining process in watts per gigahash.

Therefore, the price of production of Bitcoin, according to Hayes, equals

$$\frac{BTC_{day}}{E_{day}}$$

which serves as the lower bound of the Bitcoin market price, below which miners would be operating at a loss. Naturally, other factors are also at play when analyzing the value of Bitcoin for individual actors, such as personal speculative interest, ideological motivation, etc.

4 Mining Impact

Mining is potentially a very lucrative business, as depending on the price of a cryptocurrency, the block reward may be a significant sum. However, individual miners do not have sufficiently powerful equipment to make mining a reliably profitable business; additionally, it is more rational for individuals to share a smaller but guaranteed reward, as opposed to having a small chance at receiving a large reward.

Due to this reasoning, miners come together in so-called "mining pools", which are communities that combine individual miners' computing power to increase the chance of solving the task of creating a block. Over the last few years, such mining pools have grown rapidly in size: for example, a little under half of all computing power in the Bitcoin network belongs to just three major Chinese pools [2].

However, the mining process is not without cost. Although cryptocurrencies are fully virtual and do not require raw materials for their creation, they do require one resource – electricity. Still, estimating the total electricity cost of cryptocurrency mining is difficult, as most mining operations do not openly share the plans and specifications for their facilities. However, some approaches to estimating the electricity cost of mining have been formulated for Bitcoin.

Two such approaches were formulated in [17]. The first uses hashrates, power use and energy efficiency of specific mining equipment; this approach is useful for determining the lower bound of the total electricity cost, as it does not take into account the costs associated with cooling a large mining operation. Any computer produces waste heat, and specialised equipment designed for Bitcoin mining produces a significant amount of it at a constant rate. Because of this, de Vries formulated a second approach based on production cost and lifetime cost of a mining machine, taking into account electricity costs for cooling as well. Extrapolating from some publicly available information on the energy consumption of mining facilities, this approach yields an approximate value for the total electricity cost of Bitcoin, with a model for predicting future trends.

De Vries' study indicated a lower bound of 2.55 GW and an approximate actual value of 7.67 GW, which converts to approximately 67 TWh per year. According to Digiconomist's estimates [4], the total electricity cost of Bitcoin mining has remained relatively stable at 73 TWh per year since mid-2018, with a dip in early 2019 (shown on Fig. 2). This places Bitcoin squarely among the top 50 countries in terms of electricity consumption in the world, level with Israel and Greece.

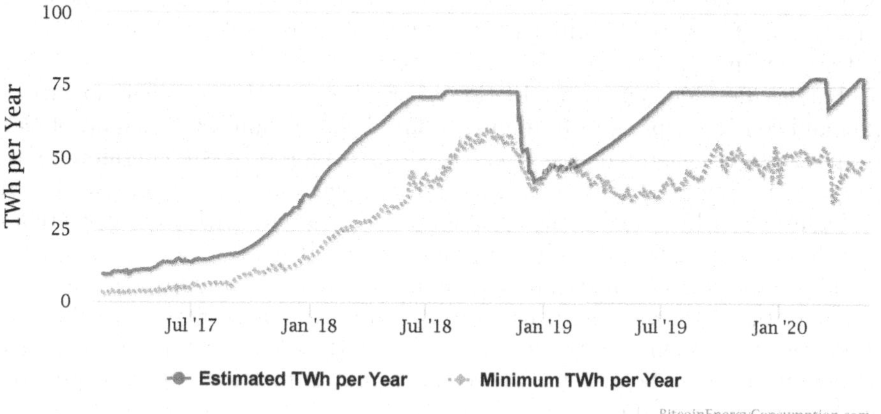

Fig. 2. Bitcoin energy consumption index chart [4]

The energy efficiency of a single transaction is similarly poor according to Digiconomist, a single Bitcoin transaction costing the equivalent of approximately 400,000 VISA transactions in terms of electricity.

The carbon footprint of cryptocurrency mining should also be taken into account, as many of the heaviest mining operations are located in China and other countries where coal and other non-renewables provide most of the available power. This negatively affects the environmental situation in the country, as well as other countries in similar situations – Digiconomist's current estimates place Bitcoin's carbon footprint level with that of the country of Syria. Because of this, the Chinese government has reportedly planned to ban cryptocurrency mining [5]. Alongside China, a ban on mining is being considered in Russia and India, while in Vietnam cryptocurrency mining activity is punishable by a fine. However, a ban on mining may lead to cryptocurrency's total transformation into a shadow business, the formation of underground pools, and increased rates of electricity theft.

An analysis by Australian entrepreneur Nick Gogerty shows that 90% of the cost of cryptocurrencies is attributable to the cost of electricity used to produce them [16]. For each megawatt of electricity used to produce Bitcoin, approximately 0.65 tons of CO2 are emitted into the atmosphere [8,13]. However, it is difficult to estimate the true ecological impact of cryptocurrencies, since mining may take place in data centers and on dedicated mining farms on one hand, and in residential spaces on the other. This fact makes assessing the carbon footprint very difficult, compounded by the fact that there are many different types of equipment in use for mining.

Cryptocurrency production equipment, while in constant operation, generates a large amount of waste heat. Without sufficient air conditioning of the mining facility, that equipment can overheat and fail. For the equipment to work properly, it is necessary to maintain a constant ambient temperature. Therefore, additional refrigeration equipment, air conditioners or fans are required, which in turn consume a significant amount of electricity.

Further development and adoption of cryptocurrencies requires modernization and constant upgrades to mining farms. Thus, a significant amount of electronic waste is generated. Electronic waste is a type of waste containing electronic and other electrical devices as well as their parts. Electronic waste can be highly hazardous due to the substances they contain, such as lead, mercury, polychlorinated biphenyls, polyvinyl chloride, etc. [15].

There are also so-called “environmentally responsible” cryptocurrencies. For example, the BitSeeds cryptocurrency allocates 10% of BitSeeds issued to the Rainforest Fund to finance rainforest restoration [11]. Another example of “green” cryptocurrencies is SolarCoin, whose main goal is to stimulate solar-powered energy producers [1,16].

5 Solutions

As mentioned earlier, cryptocurrencies were designed as a new form of fiat money. However, technical difficulties stand in the way of their widespread adoption. The algorithmic difficulty of mining means that transaction speed in the network is limited by how quickly new blocks are created – while this is the defining feature of cryptocurrency decentralization, it also makes cryptocurrencies significantly less competitive compared to traditional electronic financial networks. The solution to this would be designing a novel algorithm for proving work (or a different kind of consensus algorithm) in a cryptocurrency network, which would be similarly secure while being less computationally expensive.

Additionally, cryptocurrencies are extremely volatile assets due to a fixed supply coupled with a rapidly changing demand. This makes them questionable stores of value, and results in a (not unfounded) perception of cryptocurrencies as speculative assets. There is no single clear solution to this problem – specific kinds of cryptoassets require special approaches. For true cryptocurrencies, so-called "stablecoins" attempt to solve the problem of volatility by pegging their value to a fiat currency using some kind of collateral as a stabilization mechanism. The economics of this approach are sound, but transparency and accountability are required to make them truly stable and legitimate in the eyes of governments. A less drastic solution for cryptocurrencies would be devising an algorithmic issuing body, which would attempt to balance supply against rising or falling demand; however, further research would be required into the economics of such an algorithm – as replacing the algorithm with human actors would simply revert the cryptocurrency back to a regular currency with a central bank. For cryptotokens issued by companies and other entities, volatility is similar to that of stocks and other market assets, and the solutions are similar as well.

In addition to the above, cryptocurrencies are still in a legal grey zone, even if they are not banned in certain jurisdictions. A multitude of factors prevent their full legitimization – not the least of which are legislators' ignorance and inertia. However, reasons also include a paradoxical lack of transparency and accountability in cryptocurrencies designed to be transparent and auditable.

Because of the questionable legal status of cryptocurrencies, many mining operations operate clandestinely, or at least not fully openly. This has two manifestations:

1. Malicious actors perform illicit mining activities on the devices of unsuspecting users and companies.
2. Miners in general use powerful computing equipment for mining, which results in unexpected stress on power grids.

Because of the decentralized nature of cryptocurrencies, punitive measures against the former problem on an institutional level are unlikely to be effective. However, technical measures, such as blacklists and signature-based detection, are effective in deterring cryptojacking.

Automated control and monitoring systems for electricity – smart power grids – are promising candidates for alleviating the latter problem, especially coupled with internet-of-things-powered edge systems that feed into smart municipal information systems. This kind of data-driven infrastructure would be able to determine abnormal electricity consumption patterns and inform monitoring authorities almost immediately. The advent of 5G communications means it is possible for this infrastructure to communicate in real time, without appreciable delays between the information being gathered at the edge and being received for processing by a central information system, and even makes it possible for a smart power grid to regulate power supply to problematic areas itself, given a high enough level of integration.

Additionally, cryptocurrency mining hot-spots are often literal hot spots: mining operations can produce significant heat, which may also be detected as part of municipal smart infrastructure. Satellite-based monitoring and imaging systems as part of a GIS may be especially useful here.

One approach to managing the stress placed on the environment by cryptocurrency mining operations may be the introduction of mandatory registration of all mining farms and a special "green" tax in regions where electricity used for cryptocurrency mining is produced primarily by coal power plants. This tax can simultaneously make it possible to finance development of the renewable energy sector in a given region and minimize the difference in the cost of mining between regions that produce electricity from coal and regions that use alternative power sources.

To introduce this tax, it would be necessary to develop an appropriate regulatory framework, approve the tax base and its rate, and introduce punitive measures for those who do not pay this tax. Taxpayers would be monitored by state tax authorities, and the function of verifying the correctness of information reported by the miners would be delegated to energy companies.

The use of heat generated during the operation of mining farms in heating systems and for growing agricultural products in greenhouses may also help minimize environmental impact.

In the context of a smart city, cryptocurrency mining operations may be integrated into the infrastructure. For example, the waste heat generated by cryptocurrency mining equipment could be used as auxiliary source of heat at a municipal heating plant. The integration could be taken a step further if the tokens being mined were not simply commercial means of exchange like Bitcoin, but had a higher associated social benefit and carried some utility in the context of a smart city, e.g. were a means for consumers to pay for utilities (even the same electricity and heating) – in this case, the cryptocurrency becomes more than simply a currency, and "cryptotokens" becomes a more fitting name.

6 Conclusion

Cryptocurrencies are an interesting and potentially transformative development in financial virtualisation, brought about by the advent of digital economy. They have the potential to create a wholly new mode of trust between participants in the financial system, cutting out trusted third parties and intermediaries, like banks and exchanges, that society has come to rely on.

However, this new phenomenon is not without its drawbacks. Firstly, technical and theoretical challenges, and the widespread speculative applications of cryptocurrency tokens that are precipitated by them, mean that much of their use today lies outside of socially beneficial fields. Secondly, the imposed artificial scarcity of these tokens means that creation of new tokens necessitates "mining" operations – computationally expensive operations, often performed on specialised hardware; these mining operations have undesirable consequences, like extremely high energy consumption (with associated carbon emissions) and significant waste heat production.

The authors propose several solutions to these problems. To combat cryptocurrency volatility and inadequacy as money, further research into their technical and theoretical (economic) aspects is necessary, with the end goal of transforming existing algorithms and cryptocurrency economic theory to better serve the needs of a wider population of consumers.

From an ecological point of view, preventing excessive stress on municipal power grids is possible by means of IoT sensors and smart energy grids, reporting and potentially even correcting issues autonomously.

A special tax, given an adequate tax rate and a clear system of control over the activities of miners will minimize the negative consequences of cryptocurrency mining. Prohibitive measures, on the contrary, will not lead to positive results, and may also lead to the popularization of shadow mining and slow the development of the emerging cryptocurrency market.

The waste heat produced by cryptomining can also be repurposed in the context of a smart city: the waste heat may be used for municipal heating, and the cryptotokens being mined can be integrated into the smart city as a means of payment for utilities and other goods and services within the city.

However, in the long-term, the only solution that remains viable is one that combines transformation of economic theory with a dramatic increase in performance and power efficiency of cryptotoken systems, in order to make them competitive with existing electronic financial infrastructure.

References

1. Andoni, M., et al.: Blockchain technology in the energy sector: a systematic review of challenges and opportunities. Renew. Sustain. Energy Rev. **100**, 143–174 (2019)
2. Blockchain.info: Hashrate distribution: An estimation of hashrate distribution amongst the largest mining pools (2020). https://blockchain.info/pools
3. Coinmarketcap: Bitcoin price, charts, market cap, and other metrics (2020). https://coinmarketcap.com/currencies/bitcoin/
4. Digiconomist: Bitcoin energy consumption index (2020). https://digiconomist.net/bitcoin-energy-consumption
5. Goh, B., John, A.: China wants to ban bitcoin mining (2010). https://www.reuters.com/article/us-china-cryptocurrency/china-wants-to-ban-bitcoin-mining-idUSKCN1RL0C4
6. Hayes, A.S.: Cryptocurrency value formation: an empirical study leading to a cost of production model for valuing bitcoin. Telematics Inf. **34**(7), 1308–1321 (2017). https://doi.org/10.1016/j.tele.2016.05.005, http://www.sciencedirect.com/science/article/pii/S0736585315301118
7. Hong, G., et al.: How you get shot in the back: a systematical study about cryptojacking in the real world. In: Proceedings of the 2018 ACM SIGSAC Conference on Computer and Communications Security, pp. 1701–1713 (2018)
8. Karl, J.O., Malone, D.: Bitcoin mining and its energy footprint. In: Proceedings of the 25th Joint IET Irish Signals and Systems Conference 2014 and 2014 China-Ireland International Conference on Information and Communications Technologies, pp. 280–285 (2014)
9. Kawase, Y., Kasahara, S.: Transaction-confirmation time for bitcoin: a queueing analytical approach to blockchain mechanism. In: Yue, W., Li, Q.-L., Jin, S., Ma, Z. (eds.) QTNA 2017. LNCS, vol. 10591, pp. 75–88. Springer, Cham (2017). https://doi.org/10.1007/978-3-319-68520-5_5
10. Li, Z., Liao, Q.: Toward socially optimal bitcoin mining. In: 2018 5th International Conference on Information Science and Control Engineering (ICISCE), pp. 582–586 (2018)

11. Lia, J., Lia, N., Peng, J., Cuia, H., Wua, Z.: Energy consumption of cryptocurrency mining: a study of electricity consumption in mining cryptocurrencies. Energy **168**, 160–168 (2019)
12. Nakamoto, S.: Bitcoin: A peer-to-peer electronic cash system (2008). http://bitcoin.org/bitcoin.pdf
13. Palacio, N.S.: A technological tool for sustainable development or a massive energy consumption network. Bionatura, **3**(4) (2018). http://revistabionatura.com/2018.03.04.11.html
14. Ravindran, S., Ragoonanan, G.: Network virtualization is key to thriving in the digital economy (2015). https://www.telecomasia.net/content/network-virtualization-key-thriving-digital-economy
15. Scott, R.: The e-waste stream in the world-system. J. Globalization Stud. **3**(1), 79–94 (2012)
16. Solarcoin: Incentivizing a solar-powered planet (2020). https://solarcoin.org
17. de Vries, A.: Bitcoin's growing energy problem. Joule **2**(5), 801–805 (2018)

Identification System Model for Energy-Efficient Long Range Mesh Network Based on Digital Object Architecture

Dmitriy Sazonov(✉) and Ruslan Kirichek

The Bonch -Bruevich Saint-Petersburg State University of Telecommunications, 22 Prospekt Bolshevikov, 193232, St. Petersburg, Russian Federation
dim-saz@yandex.ru

Abstract. The analysis of the possibility of building a system for identifying Internet of Things devices based on the Digital Object Architecture in mesh networks is given. The basic principles of the LPWAN networks and mesh networks are described. A brief overview of Digital Object Architecture technology is given. Schemes for integration of the Digital Object Architecture platform into mesh network are proposed. Various configuration options for this system in order to increase productivity are considered. The model of the Handle resolution system for identifiers of digital objects in mesh networks as a queuing system is proposed. An analysis of the results is given.

Keywords: Internet of Things · Mesh networks · LPWAN · LoRaWAN · Digital object architecture · Handle system · Identification

1 Introduction

The rapid growth of the Internet of Things (IoT) technology has led to the proportional growth of the market for various applications that use this concept. The most popular areas are the following [1–4]:

- augmented reality applications;
- smart home applications;
- smart cities application.

For IoT applications, power consumption, low latency, edge device density, and communication security are important parameters [2,4].

One of the answers to these requirements for new generation networks was the creation of the concept of energy-efficient long-range networks (LPWAN).

The publication has been prepared with the support of the grant from the President of the Russian Federation for state support of leading scientific schools of the Russian Federation according to the research project NSh-2604.2020.9.

© Springer Nature Switzerland AG 2020
V. M. Vishnevskiy et al. (Eds.): DCCN 2020, CCIS 1337, pp. 497–509, 2020.
https://doi.org/10.1007/978-3-030-66242-4_39

LPWAN is a specification that allows Internet of Things devices to communicate and transmit data over significant distances (kilometers) with low power consumption [5,6]. However, LPWAN has a low data rate.

One of the implementation of the LPWAN concept is LoRaWAN technology. LoRaWAN based on LoRa modulation technology (Long Range) at the physical level [7–9]. LoRaWAN protocol provides a significant coverage area compared to analogs. Based on the use of LoRa technology at the physical level, the LoRaWAN protocol created and maintained by the LoRa Alliance [5,7].

LoRaWAN solutions based on star or star-of-stars topology, where edge devices interact through multiple gateways with the base network server [10–12]. Figure 1 shows a basic LoRaWAN network topology.

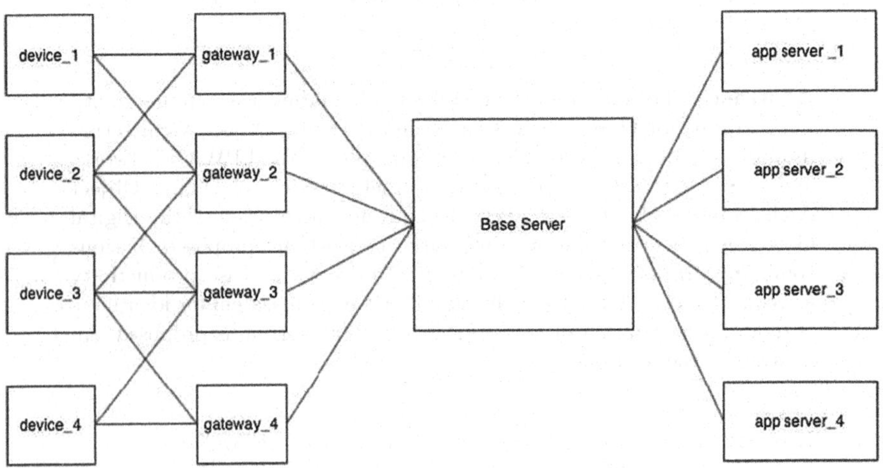

Fig. 1. LoRaWAN topology

Besides LPWAN concept and LoRaWAN technology there are many IoT solutions based on mesh topology [10,12,13]. Figure 2 shows basic mesh topology.

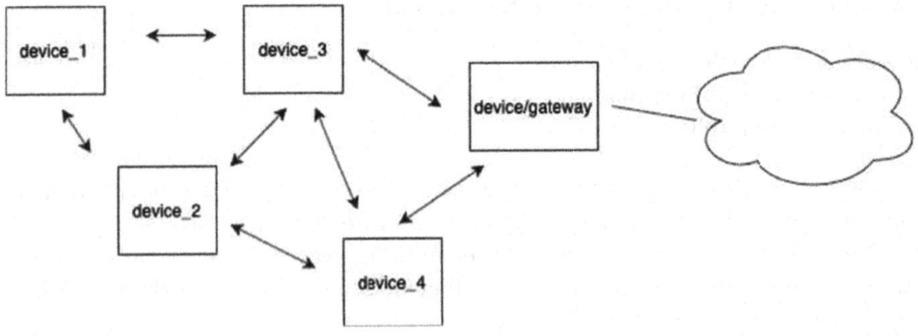

Fig. 2. Mesh topology

Features of mesh topology based networks:

- traffic control algorithms (for choosing the optimal route);
- route adaptation;
- self-organization and self-healing.

Popular technologies based on mesh topology are: ZigBee, Thread, 6LowPAN, Z-Wave and Bluetooth Low Energy 5.0 protocols. And LoRa technology allows you to increase the coverage of such networks [11,12].

A key feature of the mesh topology compared with the star topology or star-of-stars topology of LoRaWAN is the resilience of the network to the failure of individual components [12–14]. In this case, routing will adapt to the new structure and the transmission will not be interrupted. In the star topology, when the central hub (network server) fails, the network falls apart.

One of the possible solution to this problem is the switching from the star topology to mesh topology for LoRa-based applications. This gives us the possibility to achieve the fault tolerance of mesh topology and low power consumption and long distance transmitting of LoRaWAN. The idea of organization low power long range mesh networks was considered in [11,12]. This articles presented a model showing the possibility of organizing LoRaWAN devices in a mesh topology. In our work, we will focus on the problem of identifying edge devices in such hybrid network.

2 Identification Approaches in LoRa and Mesh Networks

In networks based on mesh topology each node can send and receive data. Nodes in mesh networks are divided into several types [12,13]:

- edge device;
- gateway (have a connection to an external network);
- network configurator (individual node, that manage network structure).

Configurators are responsible for assigning addresses in the mesh network and monitoring the overall status of the network.

Typically, device addresses are combined with their unique identifiers in mesh networks, i.e., those unique descriptors that are integrated into devices during the configuration phase of the entire network subsequently serve as addresses in the network when implementing hop forwarding messages between nodes.

LPWAN based solutions also have their own identification system. For example, in LoRaWAN networks, each edge device has the following set of identification information [7–9]:

- AppEUI;
- DevAddr;
- DevEUI.

There is no standard for implementing an identification system for IoT networks only recommendations and custom vendor implementations (e.g. LoRaWAN and its DevEUI addresses).

To implement identification system for a mesh networks for LoRa-based applications, it will be reasonable to use technology that would allow to use the existing approaches to identification in LoRaWAN and mesh networks and integrating them as part of own identification platform.

The aim of this work is to analyze the possibility of using the Digital Object Architecture as a platform for implementing an identification system for Internet of Things devices in such LoRa-based mesh networks.

3 DOA Based Identification

3.1 Digital Object Architecture

DOA technology was created by Corporation for National Research Initiatives - CNRI two decades ago [15–17].

Digital Object Architecture has proven itself worldwide as a system for identifying academic, professional and government information: it is the well-known DOI identifiers of articles and book publications [17–20].

The main structural elements of DOA are a digital object, digital object registry and repository and resolution system [16,18,19].

The digital object provides various information about the object it represents: information about the author, access requirements authentication, etc. [19,20]. DOA also provides a special subsystem for the management and administration of digital objects.

Digital object registries and repositories are metadata storages for searching and resolving unique identifiers into digital object data.

Resolution System is a subsystem that allows to resolve unique identifiers into specific digital information.

The concrete implementation of digital object architecture is the Handle System [16,19].

The Handle system is based on distributed network of locally administered databases (LHR) connected through global handle registries (GHR). Distributed system allows flexible configuration and management of the infrastructure, increases the fault tolerance [19–21]. The Handle system also include administration subsystem, handle resolutions subsystem and an user authentication and authorization subsystem.

The classical Handle system based on two-level architecture [16,19,21]. The first level is the global handle registries. The second level is a layer of local handle registers. To resolve the handle identifier the client first sends a request to the GHR, which responds with information about the LHR that manages the required digital object. Figure 3 shows Handle System architecture.

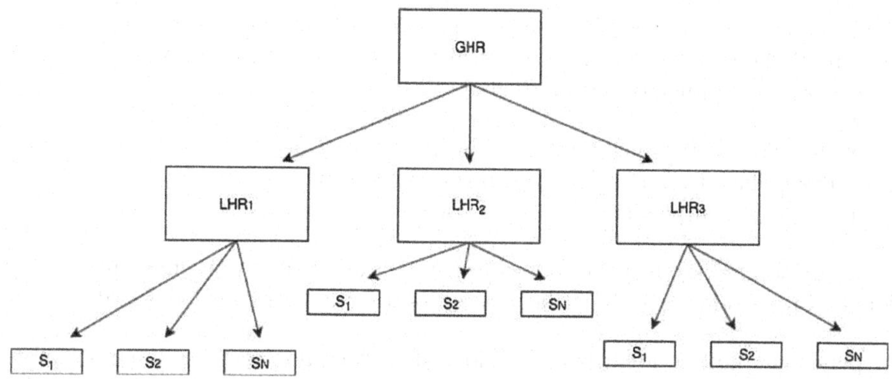

Fig. 3. Handle system architecture

3.2 DOA as Identification System

The International Telecommunication Union (ITU) has the following list of recommendations for identification systems for Internet of Things devices in narrowband wireless communication networks [22]:

- the uniformity and constancy of the identifier of the digital object x (i.e., only one identifier must correspond to each device);
- the system architecture should not limit the number of metadata objects that can be registered for a single device;
- object metadata should contain both links to registers containing information about the device, and a copy of the information itself;
- the possibility of interoperability between different resolution systems should be ensured;
- architecture of the system should be scalable;
- the system should be widely available and form the basis of an open resolution system architecture;
- structures describing meta information about a digital object should be extensible to provide flexible adaptation of the system to the changing requirements of end customers;
- the system should include administration mechanisms;
- authentication/authorization processes in the system should use certificates, electronic signatures or public key cryptographic systems;
- certificates, electronic signatures or keys should be certified by a higher authority with a high degree of trust;
- authorization data supporting the processes of management and interaction with digital objects are also stored as digital objects.

DOA technology provides a common approach to digital information management both on the Internet and other networks. It uses protocols that work on top of the classic IP networks [18,20].

DOA is based on the same basic principles that are embedded in the architecture of the modern Internet, which determined its success and the ability to evolve over a significant period of time [15,19,20]:

- open architecture (include protocols and interaction interfaces);
- independence from specific implementation technologies;
- minimizing complexity for the end user.

DOA technology has the following key characteristics that make it suitable for the implementing an identification system for IoT devices [16,18,20]:

- persistent unique identifier. The identifier does not contain dynamic information about the object. All information is stored in a digital object, the identifier only refers to it;
- universality - DOA system identifier allows you to identify not only physical objects, but also digital entities;
- DOA is based on distributed platform with different administration schemes at the local and global levels;
- network scalability;
- flexible digital object administration;
- high safety.

Identification systems based on DOA concept contains information about the class of the object, the subsystem of working with the object and the method of protection and access to data. Such systems allows you to build complex identification solutions, including interaction of various information systems within one system based on digital objects [19–21].

Handle system as a implementation of the DOA concept provides programming interface for clients (rest api) for managing records of digital objects and obtaining information (resolution) as well as a set of tools and libraries for writing custom applications [16,20]. It is possible to organize various schemes of client interaction with the Handle system to organize various identification scenarios for mesh networks.

3.3 DOA Based Identification for Mesh Networks

Consider the following architecture for identification system in a mesh network based on Handle System. The administrator (device owner) sends information about the IoT devices to the Handle System, providing the basic description and various metadata that will be required in the future to complete the identification process. In addition, it is possible to implement a role-based access model for devices, which can then be used at the stages of network configuration to split it into subnets. Next, the Handle System sends unique identifiers of the registered devices to the owner. Administrator flashes the identifiers to the device memory. When the devices starts to configure in a mesh network, identifiers are embedded in a message body from the edge devices and transmitted over the network.

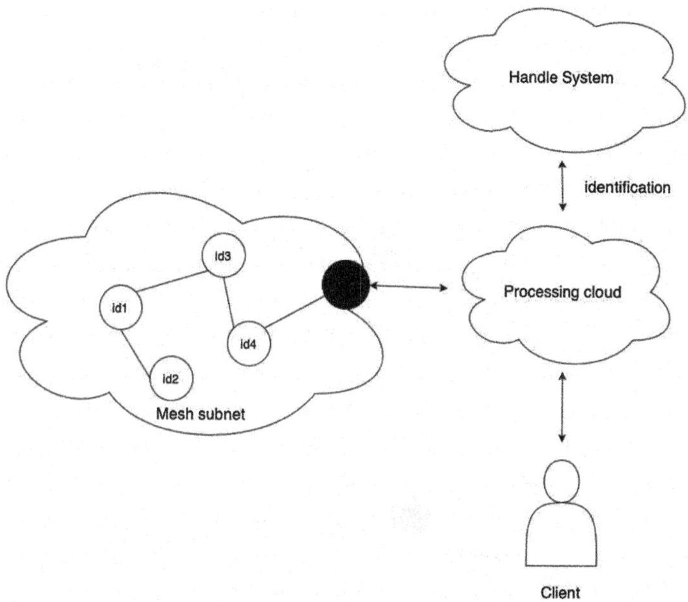

Fig. 4. Basic mesh network identification process

Through the gateway, such messages with identifiers are sent to the cloud for further processing and identification.

In the Fig. 4, the described process is shown.

Mesh subnet consisting of several edge devices and a gateway device (black). The authorization subsystem in the cloud processes the message flow, sends information to the Handle subsystem (or its local cache if it is already filled with previously processed identifiers) to obtain identification information. To speed up flow processing, authorization subsystem can be configured to periodically scan the message stream.

After checking the authorization information, the data from edge device is sent to further processing. If the request is not authorized, messages from the device are rejected. The authorization subsystem can send a notification request to the network coordinator to block an unauthorized device. The subsystem also can send notification to the network administrator about the presence of an unauthorized device. Figure 5 shows the diagram of the system interaction with an unauthorized device.

To analyze described process let us consider a queuing model of a identification system for a mesh network based on the DOA architecture.

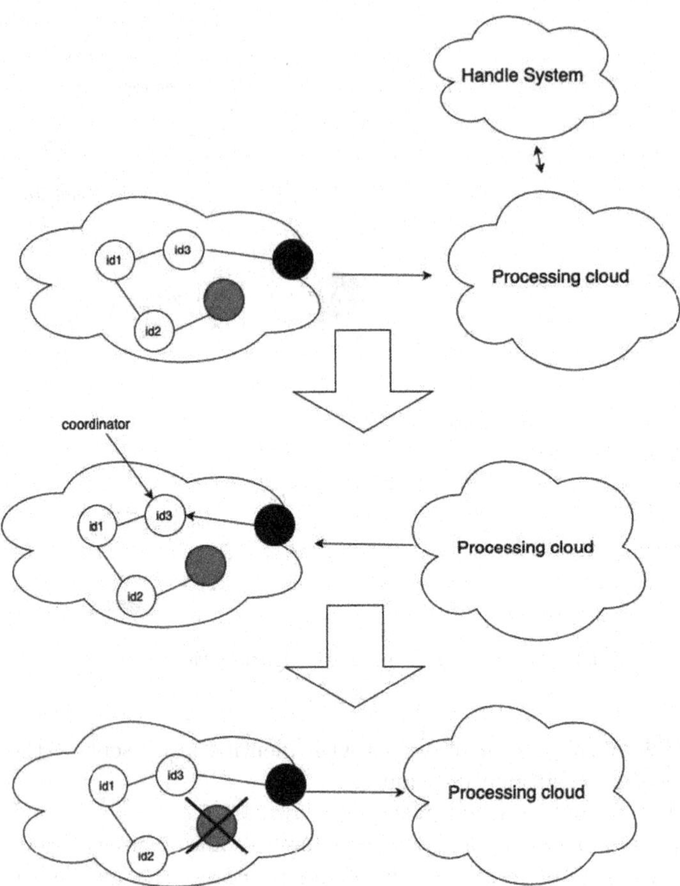

Fig. 5. Unauthorized device filtering

4 Queuing Model of Identification System for Mesh Network

The queuing model is based on the model of the DOA system considered earlier in [12,19,21]. We introduce a couple of assumptions into our model to be able to simulate it and further analyze it.

It is possible to represent a mesh network edge device in our model as a single entity with a buffer and a delay block. Figure 6 shows edge device model.

The total latency of the transmission from the edge device to the processing server is defined as the product of the delay in one device by the number of hops.

Within our model, we will consider the Handle System as a separate server consisting of a buffer and a delay block. Handle System block in queue system will be similar to edge device model on Fig. 6.

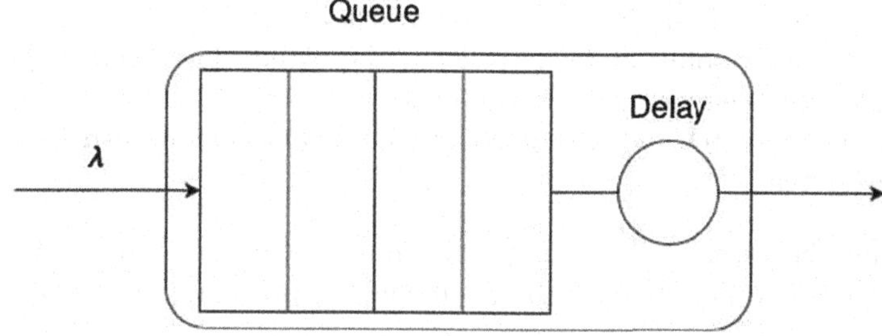

Fig. 6. Edge device model

Figure 7 shows a described queuing model of identification system built in the AnyLogic package.

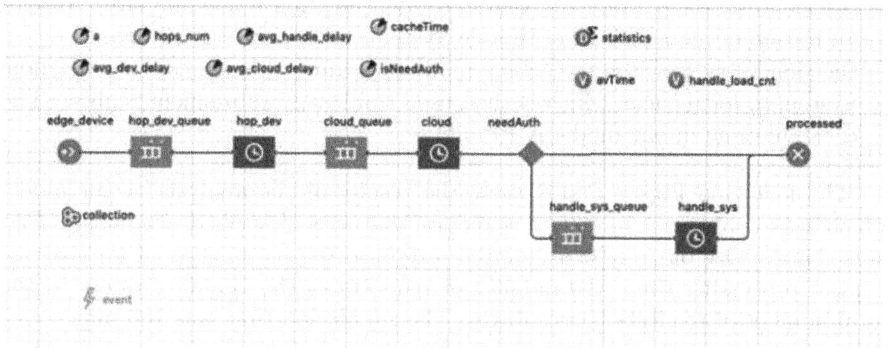

Fig. 7. AnyLogic model

AnyLogic model consists of the following components:

- edgeDevice - single edge device of the mesh network, transmitting data through intermediate devices to the gateway and further to the cloud;
- hopDevQueue and hopDev - model of intermediate edge devices (also include gateway). The number of devices in the chain is determined by the hopsNum parameter. In our model, it was 3 intermediate devices. HopDelay block in our model has the following parameters: capacity = hopsNum (each edge device in the mesh network usually can process one request at a time); processing time = hopsNum * avgDevDelay (where avgDevDelay is the average processing time of the request on the edge device, the value of 100 ms was selected in our model);
- cloudQueue + cloud - model of remote server that process all incoming request from mesh network. The average processing time of one request in our model is 200 ms. Server in our model can handle 15 requests in parallel (we assume that the service processes requests in 15 threads);

- handleSysQueue + handleSys - model of the Handle System remote service that processes resolution requests from the cloud. The average processing time (with the network delay) is 400 ms;
- needAuth - a switch block that control data flow that needed to resolve in Handle System.

The time between incoming requests from edge device is distributed exponentially with load parameter a. In our model, it varies discretely from 1 to 200 with step 0.5.

In this model we analyze several parameters: average processing time of the request from the edge device, the load of the Handle System and the total load of the identification system.

In the described model we considered several configurations of the identification system:

- every incoming request from the edge device will send to the Handle System;
- the cloud server periodically switch to identification mode and start to send all incoming requests to the Handle System. Scan period is 10 min;
- the cloud server switch to caching mode by adding all responses from Handle System to local cache which is then used to identify subsequent request. The cache lifetime in out model is 10 min.

The following Figs. 8 and 9 show the results of the analysis of our model which was configured in different modes that was described previously. Figures characterize our model system according to the parameters of the average request processing time, the number of requests to the Handle System to the total number of requests (percentage). The simulation time was 100 min.

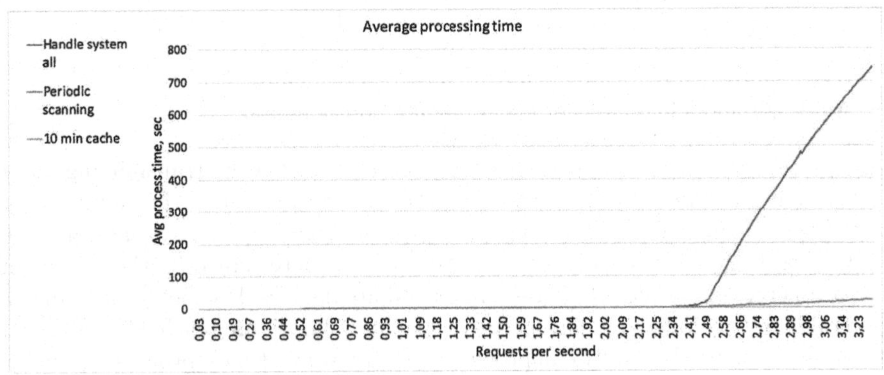

Fig. 8. Processing average time

From Fig. 8 it can be seen that when our model was process every incoming request with identification request to the Handle System, even at low load level

(3 requests per second), the average processing time for requests is around 700 s. When model switched to the periodic scan mode with a period of 10 min, it showed performance improve (about 25 s). When our model worked with the cache with 10 min lifetime, it showed the best results (0.5 s).

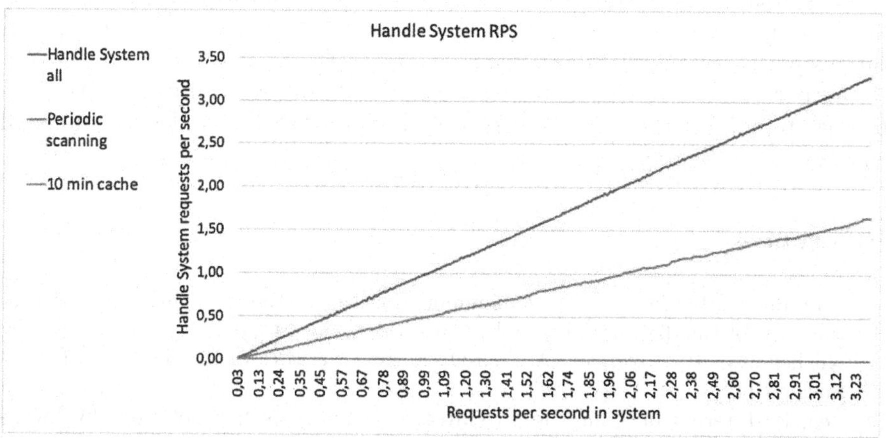

Fig. 9. Handle System loading

The Fig. 9 shows the dependence of the rps load (requests per second) received by the Handle System from the total number of rps that incomes to our system. It can be seen that when working with the cache, the load level on the Handle System stay constant (4 devices, cache 10 min). In other configurations, the dependence is linear.

5 Conclusion

In this paper, the possibility of implementing a DOA-based identification system in a hybrid network of LoRa devices combined in mesh topology was considered. An attempt to move from the classic star topology, which is typical for LoRaWAN to mesh, is caused by the desire to solve the problem with a single point of failure, typical for the star topology, and move to a distributed mesh network topology, which has a number of advantages listed in this work.

To build the identification system, the DOA platform was chosen, since it is quite promising, there are already solutions for identification based on DOA in other areas. In addition, the Digital Object Architecture as a platform meets the ITU recommendations for identification systems in narrow-band Internet of Things networks.

Possible schemes of the identification system in the described mesh network are considered. To analyze the system performance, a queuing model was built in the AnyLogic package.

Even an analysis of the simple model shows that the implementation of an identification system based on the Handle System for mesh networks is possible. Handle System is a client-server type system, and interaction and integration of third-party systems with it can occur via http requests (rest api).

Analysis of the described model shows that the operating mode of the identification system using the query cache shows the best performance. This operating mode of the system is possible, since device identifiers usually do not change throughout the life cycle of a system.

As part of the development of the described concept in future work, it is planned to build a test mesh network system with DOA-based identification and analyze the configuration of the system.

References

1. Borodin, A., Kucheryavy, A.: Communication Networks of the fifth generation as the basis of the digital economy. Electrosvyaz **5**, 45–49 (2017)
2. Kucheryavy, A., Prokopiev, A.: Self-Organizing Networks. Libavich, Saint Petersburg (2011)
3. Lea, P.: Internet of Things for Architects: Architecting IoT Solutions by Implementing Sensors, Communication Infrastructure, Edge Computing, Analytics, and Security. Packt Publishing, UK (2018)
4. Kucheryavy, A.: Internet of things. Elektrosvyaz **1**, 21–24 (2013)
5. Kumaritova, D., Kirichek, R.: Review and comparative analysis of LPWAN network technologies. Inf. Technol. Telecommun. **4**, 33–48 (2016)
6. Khutsoane, O., Isong, B., Abu-Mahfouz, A.: IoT devices applications based on lora/lorawan. In: IECON 2017–43rd Annual Conference of the IEEE Electronics Society, pp. 6107–6112 (2017)
7. History of LoRa technology. https://nekta.tech/en/technology/. Accessed 9 Sep 2020
8. Semtech Corporation. Sx1276/77/78/79 datasheet, rev. 6. Semtech Corporation (2019)
9. Semtech Corporation. SX1272/3/6/7/8 LoRa Modem Design Guide AN1200.13. Semtech Corporation (2013)
10. Nurilloev, I., Paramonov, A., Koucheryavy, A.: Connectivity estimation in wireless sensor networks. In: Galinina, O., Balandin, S., Koucheryavy, Y. (eds.) NEW2AN/ruSMART -2016. LNCS, vol. 9870, pp. 269–277. Springer, Cham (2016). https://doi.org/10.1007/978-3-319-46301-8_22
11. Pham, V., Dinh, T., Kirichek, R.: Method for organizing mesh topology based on LoRa technology. In: 10th International Congress on Ultra Modern Telecommunications and Control Systems and Workshops (ICUMT), pp. 1–6 (2018)
12. Pham, V., Gallyamov, D., Vorozheykina, O., Kitichek, R.: Model of energy-efficient long range mesh network. Elektrosvyaz **5**, 33–41 (2020)
13. Wireless Mesh Network - A well proven alternative to LPWAN. https://www.eot.dk/Files/Images/Elektronikmesse/2017/Konferencerne/Pre-fra-SPEAKERE/Wireless-Mesh-Network-a-well-proven-alternative-to-LPWAN.pdf. Accessed 9 Sep 2020
14. Difference Between Star and Mesh Topology. https://techdifferences.com/difference-between-star-and-mesh-topology.html. Accessed 9 Sep 2020

15. Kahn, R.E.: A framework for distributed digital object services. Int. J. Digit. Libr. **6**, 115–123 (2006)
16. The Handle System. https://www.dona.net/handle-system. Accessed 9 Sep 2020
17. Al-Bahri, M., Yankovsky, A., Borodin, A., Kirichek, R.: Testbed for identify IoT-devices based on digital object architecture. In: Galinina, O., Andreev, S., Balandin, S., Koucheryavy, Y. (eds.) NEW2AN/ruSMART -2018. LNCS, vol. 11118, pp. 129–137. Springer, Cham (2018). https://doi.org/10.1007/978-3-030-01168-0_12
18. Digital Object Protocol Specification, version 1.0, CNRI. https://www.dona.net/doipv1doc. Accessed 9 Sep 2020
19. Kirichek, R., Sazonov, D.: Digital object architecture as an approach to identifying internet of things devices. Distrib. Comput. Commun. Networks, **597–611** (2019)
20. Borodin, A., et al.: Identification of devices and systems of narrowband wireless networks of the Internet of things. Part I. lektrosvyaz **5**, 24–33 (2020)
21. Al-Bahri, M., Kirichek, R., Sazonov, D.: Modeling of the Internet of things device identification system based on the architecture of digital objects. Proc. Educ. Inst. Commun. **1**, 42–47 (2019)
22. Recommendation X.660 (07/11) X.660 : Information technology - Procedures for the operation of object identifier registration authorities: General procedures and top arcs of the international object identifier tree. https://www.itu.int/rec/T-REC-X.660-201107-I/en. Accessed 9 Sep 2020

Efficient Wireless Data Collection System Based on LoRaWAN Technology and Distributed Computation Approach

Yury Rassadin(✉) and Sergey Dushin

Trapeznikov Institute of Control Sciences, Profsoyuznaya 65, 117997 Moscow, Russia
rassadinj@gmail.com
https://energy.ipu.ru/

Abstract. In the paper a way to build a wireless sensor network based on LoRaWAN technology is considered. The proposed approach allows to increase the intensity of data collection compared to standard implementations of LoRaWAN network and save energy efficiency of wireless sensors. The main feature of the network is the edge computing algorithm of distributed data compression. The algorithm is based on controlling prediction error on the end device side. The prediction is completed by recursive linear prediction algorithm on the server side. Higher data collection rates are achieved along with acceptable level of energy efficiency for autonomous operation of sensors and low load of the physical communication channel. The effectiveness of the proposed solution can be illustrated by the results of simulation experiments and the network of temperature sensors for intensive temperature measurement in the problem of identifying the topology of the office building heating system. The network is based on open source software, ChirpStack server, LoRaWAN end-device stack from Semtech and additional modules developed by the authors.

Keywords: Wireless sensor network · High-intensity data collection · LoRaWAN · Edge computing · IoT

1 Introduction

Sometimes it seems that modern technologies have touched all areas of our life. The stronger is the contrast with the real level of technological equipment of the majority of buildings in operation today – introduction of modern automation and intelligent control technologies in old buildings is a very laborious and time-consuming problem. For buildings from XXth century automatization of energy systems, heating and climatic systems is related with various difficulties due to the fact that the necessary functionality was not provided at the design stage. In this paper we consider an approach to the design and deployment of a sensor network that allows to apply existing smart building techniques in poorly adapted areas. Our approach also provides a perspective for improving this class

© Springer Nature Switzerland AG 2020
V. M. Vishnevskiy et al. (Eds.): DCCN 2020, CCIS 1337, pp. 510–520, 2020.
https://doi.org/10.1007/978-3-030-66242-4_40

of methods by increasing the frequency of measurements. Our development was based on accessible hardware elements and open source software for maximum reproducible results.

Typical problems that affects the smart building efficiency are the power supply for sensors, availability of data cables, direct access to key building objects. The exact schemes of the facilities are often missing, and the maintenance is held under conditions of uncertainty and varying parameters. Wireless autonomous sensors are most likely non-alternative when connection to reliable power networks is lacking and cable infrastructure of the communication network is absent. Energy efficiency for this king of sensors is very important, so the use of high-speed wireless technologies, such as IEEE 802.11 (Wi-Fi) standards family, is most likely excluded. There are available solutions based on modern wireless LAN technologies, such as LoRaWAN, SigFox and NB-IoT [1–3]. They solve typical problems of data collection and equipment management quite effectively [4]. A common disadvantage of this kind of technologies for building wireless sensor network is the low bandwidth of the wireless channels used and the intensity of data exchange limitations for keeping the sensors battery life at an acceptable level [4,5]. The software methods of saving power for wireless sensors are the subject of recent works [6,7].

At the same time in real IoT applications it is often necessary to collect much more data than LPWAN technologies provide and currently there is scientific interest in such problems [8–10]. An interesting example of this problem is the high frequency temperature measurement of heat pipes that is necessary for identification of the building's heating system topology. To register temperature waves a measuring and registration system with a frequency of at 1 Hz is required. In the paper we propose a method for providing high-intensity data collection from wireless autonomous sensors, which solves the problem of intensive temperature measurement and can be used for identifying the topology of the building's heating subsystem, as well as in a variety of other applications that require intensive data collection by wireless sensors.

The paper is organized as follows. The second section describes the main idea of the work, the data compression algorithm for communication between peripherals and the central server, which allows to increase the polling frequency of LoRa channel. The third part describes the network architecture, server side, and software for end devices. The fourth section deals with network deployment issues, hardware lists and areas for further research. The fifth section is about simulation experiment that was held to estimate the effectiveness of proposed approach. The sixth section discusses some areas for further research.

2 The Intensive Data Collection Method

The authors take part in the development of the Center of Digital Solutions for Smart Grid [11]. For one of the sub-tasks of the Center, a system is required that is able to register the temperature wave spreading with a data collection frequency determined by the heat wave speed, in our case 1 Hz. The idea is to identify the topology of the heat network by creating a heat wave using the heat node control devices. Data on the each node temperature variations, synchronized in time, will be processed by profile specialists in the future.

2.1 Client-Server Interconnection Logic

The data compression algorithms are powerful and commonly used tools for increasing the amount of data transmitted through the communication channels and optimizing power consumption in communication networks [12–14]. However the use of traditional compression in LPWAN networks leads to unacceptable increases in data latency and power consumption on the sensor side as additional data processing is required. An interesting works toward to data compression in LPWAN can be found in [15–18]. This is why the information is transmitted in uncompressed form and the development of such algorithms for LPWAN must take into account the requirements and specific conditions of these networks.

The essence of the proposed approach is synchronous prediction of the input signal both on the server side and on the sensor side [7]. In this case, information from the sensor is sent when the prediction error exceeds the predefined threshold. This new measurement is used to correct the prediction on the server side. If the sensor is silent, the server decides that the prediction is accurate enough and uses it as real-time information. Prediction procedure uses linear algorithm [19] that is based on the Levinson-Durbin recursion. It is also important that the predictor filter coefficients [20,21] are calculated on the server side to save battery budget. They are transmitted to the sensor only if the error threshold value is exceeded. This algorithm is schematically shown in the Fig. 1.

Under these conditions the use of a physical transmitter is minimized (the most energy consuming functional part of an autonomous sensor). The server uses its input estimation as real-time data until corrective information from the sensor is received.

2.2 Data Prediction Method

To describe the math side of algorithm let consider linear prediction technique [20]. The idea of linear prediction is to represent current sample as a linear combination of past samples (1).

$$\hat{x}(i) = \sum_{n=1}^{N} a_n * x(i-n) \tag{1}$$

The formula (1) describes the convolution of N past samples of observed process and coefficients a_n. The coefficients can be considered as taps of finite impulse response filter, which is called filter-predictor.

The optimal coefficients of predictor in terms of minimizing MSE (means squire error) can be found to solve Yule-Walker Eq. (2).

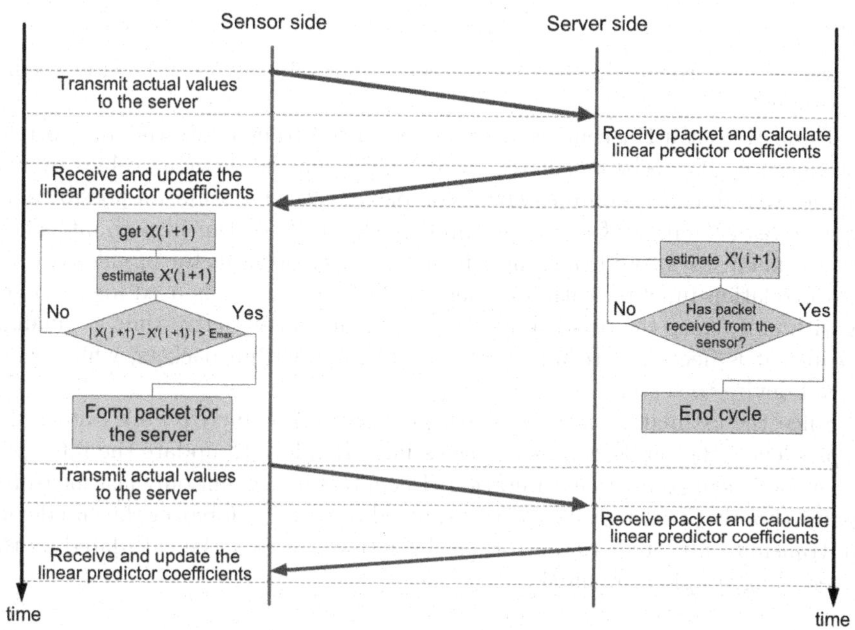

Fig. 1. Distributed data compression algorithm for LPWAN

$$\begin{pmatrix} R_0 & R_1 & \cdots & R_{N-1} \\ R_2 & R_0 & \cdots & R_{N-2} \\ \vdots & \vdots & \ddots & \vdots \\ R_{N-1} & R_{N-2} & \cdots & R_0 \end{pmatrix} * \begin{pmatrix} a_1 \\ a_2 \\ \vdots \\ a_N \end{pmatrix} = - \begin{pmatrix} R_1 \\ R_2 \\ \vdots \\ R_N \end{pmatrix} \tag{2}$$

where R_n - autocorrelation function of the estimated process, a_n - filter predictor coefficients.

Traditional methods of solving Yule-Walker equation have computational complexity $O(N^3)$. However, the autocorrelation matrix is Toeplitz matrix, so the fast computational algorithm called Levinson-Durbin recursion can be used [21]. This algorithm provides computational complexity $O(N^2)$. The main idea of Levinson-Durbin recursion is to solve Yule-Walker equation by blocks increasing of order equation from 1 to N. The generalized form of the recursion for arbitrary order $n+1$ [22] described by formulas (3)–(6).

$$k_{n+1} = \frac{\alpha_n}{e_n}, \tag{3}$$

$$A_{n+1}(z) = A_n(z) + k_{n+1} * z^{-1} * [z^{-n} * A_n(z)], \tag{4}$$

$$e_{n+1} = [1 - |k_{n+1}|^2] * e_n, \tag{5}$$

$$\alpha_n = R_{n+1} + a_{n,1} * R_n + a_{n,2} * R_{n-1} + ... + a_{n,n} * R_1, \tag{6}$$

where e_n - error prediction, k_{n+1} - reflection coefficient, $A_n(z)$ - filter predictor response.

To do Levinson-Dusrbin recursion we are need to estimate the autocorrelation function. It can be calculated recursively updating previous values by new samples to reduce the computational complexity [21]. Of course, a possible long term non-stationarity of estimated process should be taken in account. There are two main ways to do this: introducing the forgetting factor or calculate the autocorrelation function within a window. The second one is used for proposed algorithm. To keep the network payload and transmitter active time as small as possible, it is necessary to minimize the data size within packets, which sends by end device to server.

Since the estimated samples is accepted as well-predicted while end device keeps silence, the server can use these estimated values to update the filter predictor coefficients on the next iteration. Taken this in account only last measured sample (poorly predicted) can be transmitted to server. Moreover the end device can transmit only prediction error, as the actual value can be calculated by the server using simple formula (7).

$$x(i) = \hat{x}(i) + e_i \tag{7}$$

3 System Architecture

Same as the standard LoRaWAN network architecture [1], the modified system has server and client sides. The main functional blocks are shown in the Fig. 2. The server part consists of standard LoraWAN components (network bridge, network server, application server) and additional modules that provide information compression, improve sensor energy efficiency and channel utilization efficiency. At each time interval the sampling module implements the measurement prediction if the data from the sensor did not come or transfer the data directly if it was received from the sensor. If data is received from the sensor, which means that specified error threshold is exceeded, the module for recalculating the predictor filter coefficients generates new coefficient values according to the Levenson-Durbin algorithm. The Transmission Control Module sends predictor filter coefficients to the sensor and receives and processes packets from the sensor.

The physical and channel levels required for LoRaWAN as well as additional modules are implemented on the sensor side. The prediction module estimates input value using a formula identical to the server one. The predicted value is then compared to the result of real measurements. The measurement module has quite common design. Since NTC thermistors change their characteristic nonlinearly, the temperature values are stored as a tabular function. We use binary search in this table, intermediate values are calculated linearly, $t(r) = t_1 + (t_2 - t_1)(r - r_1)/(r_2 - r_1)$. The resistance corresponding to the temperature

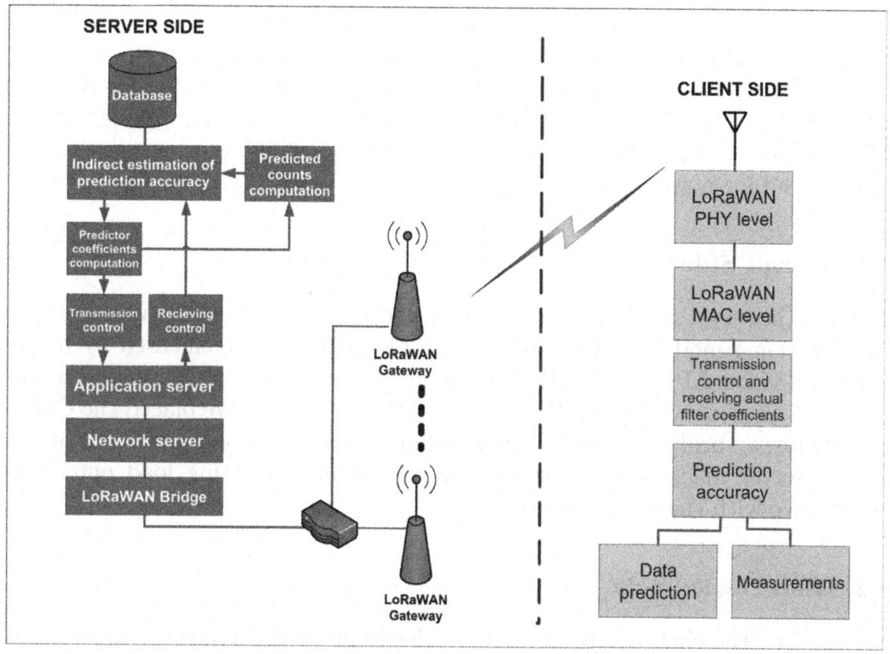

Fig. 2. Modified LoRaWAN network structure

of 25 °C is used as the zero iteration. The control module provides transmitting data if the prediction error [7] is greater than the predefined threshold, as well as receiving predictor filter coefficients from the server and updating them.

4 Intensive Temperature Measurement of the Building Heat Pipes

The process of deploying the described network can be divided into three stages. The first stage was the choice of equipment capable of solving the problems faced by the authors. The second stage was to create software for end devices and for the server part, testing and debugging their interaction. The third and final stage should include work on the deployment of the network in a particular building: scaling the results of the second stage on the entire building, providing reliable coverage by gateways, testing launches.

The required equipment list contains temperature sensors, server, gateways and network infrastructure for them. We use a virtual machine with OpenSUSE operating system as central node. The gateways were chosen to comply with Russian regional standards. An affordable solution is the MikroTik R11e-LR8 gateway based on the Semtech SX1301 processor. During the development process, we also used the Vega BS-1 gateway, which will be integrated into the working network later. The temperature sensor is controlled by the STM32L151 [23] microcontroller, which is declared by Semtech to be energy efficient. The choice of the LoRa transmitter manufacturer is obvious, we opted for Semtech SX1272 [24]. The temperature measuring module is a voltage divider with a NTC thermistor B57861. A battery with a voltage of 3.3–5 V (in our case, 3.6V) can be used for power supply.

4.1 Server Side

To build the server part, the open software package ChirpStack [25] was used, the computational module of the compression algorithm is implemented by means of the provided API. We use Python programming language to implement the predictor filter, as well as the WebSocket client and server. We placed the ChirpStack gateway bridge, network server and application server components on the central server machine in order to shift the main computing load onto it in accordance with the main idea of work.

4.2 End Device Side

For creating the client firmware we used the Semtech development environment and open-source libraries. The Semtech company gives full freedom to developers, including the commercial use of the results. Sensor firmware development was implemented in C language in the ac6 System Workbench for STM32 IDE environment on the basis of the LoRaWAN end device stack open project. Energy efficiency is achieved not only by reducing the number of data transmission sessions, but also by controlling power supply to peripheral devices. The SX1272 modem is capable of independently controlling the high-frequency transmitter module, but we decided to control it directly from the microcontroller maximize energy savings.

4.3 Equipment Layout

There should be two types of devices distributed throughout the building, sensors and gateways. The LoRaWAN protocol allows great freedom and convenience in the arrangement of end devices, so in the paper we consider the placement of sensors to depend on the specific problem. In the case of identification of the heating system, the sensors should be located in various rooms on the heating pipes mainly near the windows. The LoRaWAN Protocol has a mechanism for managing multiple gateways connected to a single network server, so the most difficult problem when deploying such a network is to find an optimal non-excess location of the gateways. The building where the experimental network was deployed is elongated, so we decided to place the base stations near the windows, covering the extended facades of the building separately. We have placed an additional base station in the basement of the building for coverage of the main nodes there.

5 Simulation

Matlab/Simulink models of server side and sensor side where built to estimate the proposed method efficiency [7]. The real data collected by sensor network deployed in Institute of control science of RAS is used during experiments. In particular, it is data from intrabuilding humidity and light sensors collected during two weeks. The actual collected data and its estimation on server side for humidity sensor are presented on Fig. 3. The same for light sensor is presented of Fig. 4. On the figures the original data is drawn by solid lines, the estimation one is drawn by dotted lines.

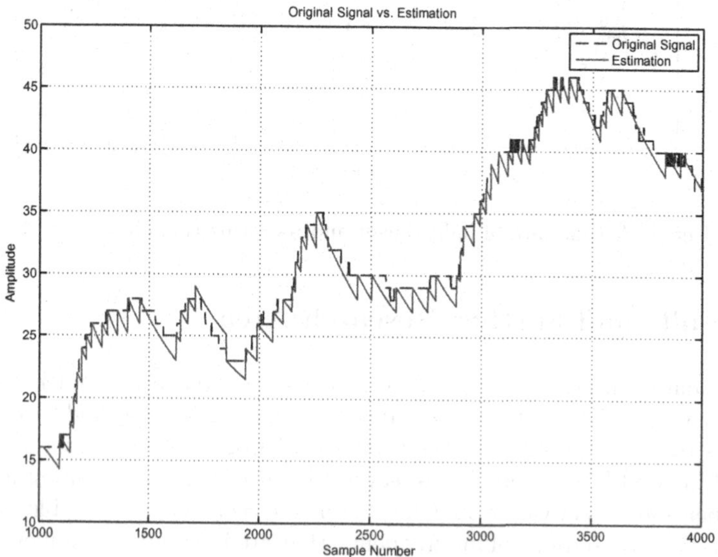

Fig. 3. Actual data of humidity sensor and its estimation on server side

In experiments, which results is shown on Figs. 3 and 4, the data were predicted by filter predictor with order of 5 and data window of 100 [7] and maximum prediction error was 5%. As can be seen from features, for both signals the original data can be well enough predicted during several time slots from resent original data arrivals. Although, the prediction error increases during prediction time without data updating from sensor side. The average data compression ration during experiments time is 8.4 for humidity sensor data and 6.5 for light sensor data.

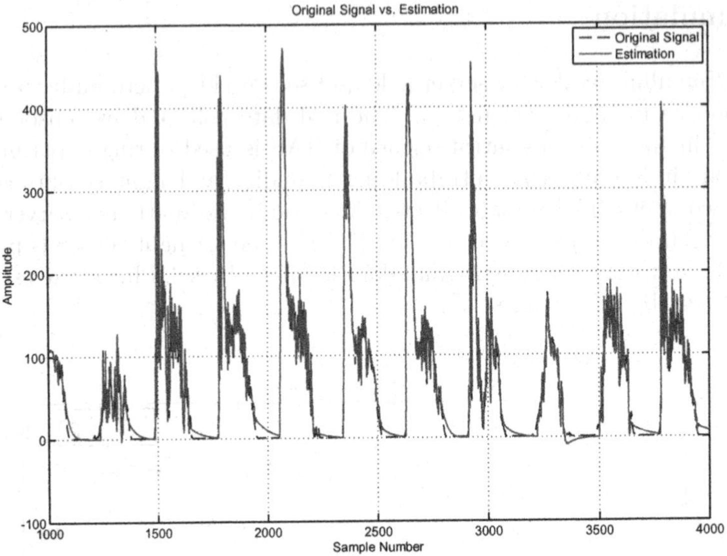

Fig. 4. Actual data of light sensor and its estimation on server side

6 Results and Further Research Area

Intensive data collection from small groups of sensors has been implemented and tested for today. The next step is to develop a guaranteed method to scale such segments to the required size. We want to orientate ourselves to the volumes declared in the LPWAN brochures, several hundreds of sensors simultaneously. Using open source software from ChirpStack and Semtech, we are able to bypass the limitations on sensor polling frequency that are inherent to commercial solutions, and the network is still evolving. One of the directions of development can be improvement of prediction algorithms using mathematical models of monitored objects. For some experiments the central server capacity may not be enough, so the server site can be transferred to a more powerful machine, but so far this has not been necessary. We hope that a deeper study of server software from ChirpStack will help us to optimize the proposed approaches.

7 Conclusion

A modified architecture of wireless sensor network is presented in the paper, which allows to increase the intensity of data collection in comparison with standard implementations of LoRaWAN networks. The main functional feature of the network is the use of the authors' algorithm for distributed data compression in LPWAN networks. The proposed approach achieves the main goal in terms of intensity while energy efficiency stays at an acceptable level for autonomous operation of sensors and load of the physical communication channel remains low. The algorithm involves prediction of input data on the server side, while the sensor only transmits corrective information in case of exceeding the predefined prediction error. This approach minimizes the use of standalone sensors and reduces the load on the physical communication channel. Lower power consumption allows to extend sensors lifetime until the batteries are replaced. The authors' contribution to network software development is at the level of server applications and sensor firmware of the devices. The effectiveness of the proposed approach will be investigated when solving the problem of intensive data collection of temperature sensors in the identification of the topology of the heating system of an office building.

References

1. Mekki, K., Bajic, E., Chaxel, F., Meyer, F.: Overview of cellular LPWAN technologies for IoT deployment: Sigfox, LoRaWAN, and NB-IoT. In: 2nd IEEE International Workshop on Mobile and Pervasive Internet of Things - Athens (2018)
2. Ayoub, W., Mroue, M., Nouvel, F., Samhat, A.E., Prevotet, J.: Towards IP over LPWANs technologies: LoRaWAN, DASH7, NB-IoT. In: 2018 Sixth International Conference on Digital Information, Networking, and Wireless Communications (DINWC) (2018). https://doi.org/10.1109/dinwc.2018.8356993
3. Jovanović, M.L., Koprivica, M., Neškovi, N.: Implementation of IoT system for securing telecommunications infrastructure based on LoRaWAN operator's network. In: IEEE EUROCON 2019–18th International Conference on Smart Technologies, Novi Sad, Serbia, pp. 1–6 (2019). https://doi.org/10.1109/EUROCON.2019.8861632
4. Bazenkov, N.I., et al.: Intensive data collection system for smart grid and smart building research. In: 2019 1st International Conference on Control Systems, Mathematical Modelling, Automation and Energy Efficiency (SUMMA), Lipetsk, Russia, pp. 411–415 (2019)
5. Pukrongta, N., Kumkhet, B.: The relation of LoRaWAN efficiency with energy consumption of sensor node. In: 2019 International Conference on Power, Energy and Innovations (ICPEI), Pattaya, Chonburi, Thailand, pp. 90–93 (2019)
6. Sacaleanu, D.I., Pericoara, L.A., Stoian, R., Lzarescu, V.: A new energy saving framework for long lasting wireless sensor nodes. In: 2015 7th International Conference on New Technologies, Mobility and Security (NTMS) (2015). https://doi.org/10.1109/ntms.2015.7266467
7. Dushin, S.V., Frolov, S.A.: Distributed data compression algorithm for low-power wide-area networks. In: Vishnevskiy, V.M., Samouylov, K.E., Kozyrev, D.V. (eds.) DCCN 2019. CCIS, vol. 1141, pp. 163–173. Springer, Cham (2019). https://doi.org/10.1007/978-3-030-36625-4_14
8. Benninger, S., Magno, M., Gomez, A., Benini, L: EdgeEye: a long-range energy-efficient vision node for long-term edge computing. In: Tenth International Green and Sustainable Computing Conference (IGSC), Alexandria, VA, USA, 2019, pp. 1–8 (2019)
9. Van den Abeele, F., Haxhibeqiri, J., Moerman, I., Hoebeke, J.: Scalability analysis of large-scale LoRaWAN networks in ns-3. IEEE Internet Things J. **4**(6), 2186–2198 (2017). https://doi.org/10.1109/JIOT.2017.2768498

10. Wu, Y., He, Y., Shi, L.: Energy-saving measurement in LoRaWAN based wireless sensor networks by using compressed sensing. IEEE Access **8**, 49477–49486 (2020). https://doi.org/10.1109/ACCESS.2020.2974879
11. Official cite of the center of digital solutions for smart grid. https://energy.ipu.ru/
12. Gibson, J.: Information Theory and Rate Distortion Theory for Communications and Compression. Morgan & Claypool, San Rafael (2013)
13. Helman, D.R., Langdon, G.G.: Data compression. IEEE Potentials **7**(1), 25–28 (1988). https://doi.org/10.1109/45.1889
14. Hanzo, L., Somerville, F.C.A., Woodward, J.P.: Speech signals and an introduction to speech coding. In: Hanzo, L., Somerville, F.C.A., Woodard, J., (eds.) Voice and Audio Compression for Wireless Communications. Wiley, Hoboken, pp. 1–28 (2007). https://doi.org/10.1002/9780470516034.ch1
15. Jang, Y.S., Usman, M.R., Usman, M.A., Shin, S.Y.: Swapped Huffman tree coding application for low-power wide-area network (LPWAN). In: 2016 International Conference on Smart Green Technology in Electrical and Information Systems (ICSGTEIS), Bali, pp. 53–58 (2016). https://doi.org/10.1109/ICSGTEIS.2016.7885766
16. Abdelfadeel, K.Q., Cionca, V., Pesch, D.: Dynamic context for static context header compression in LPWANs. In: 2018 14th International Conference on Distributed Computing in Sensor Systems (DCOSS), New York, NY, pp. 35–42 (2018). https://doi.org/10.1109/DCOSS.2018.00013
17. Pongpunpurt, P., Khawsuk, W., Sutthisangiam, N.: Development of Huffman code for Lora technology. In: IEEE SmartWorld, Ubiquitous Intelligence & Computing, Advanced & Trusted Computing, Scalable Computing & Communications, Cloud & Big Data Computing, Internet of People and Smart City Innovation (SmartWorld/SCALCOM/UIC/ATC/CBDCom/IOP/SCI). Leicester, United Kingdom, 2019, pp. 1882–1887 (2019). https://doi.org/10.1109/SmartWorld-UIC-ATC-SCALCOM-IOP-SCI.2019.00331
18. Szalapski, T., Madria, S.: Real-time data compression in wireless sensor networks. In: 2010 Eleventh International Conference on Mobile Data Management, Kansas City, MO, pp. 307–308 (2010). https://doi.org/10.1109/MDM.2010.32
19. Jackson, L.B.: Digital Filters and Signal Processing, 2nd edn., pp. 255–257. Kluwer Academic Publishers, Dordrecht (1989)
20. Vaidyanathan, P.P.: The Theory of Linear Prediction. Morgan & Claypool Publishers, San Rafael (2008)
21. Ramirez, M.A.: A Levinson algorithm based on an isometric transformation of Durbin's. IEEE Signal Process. Lett. **15**, 99–102 (2008)
22. Hannan, E.J., Quinn, B.G.: The determination of the order an autoregression. J. Roy. Stat. Soc. Ser. B. **41**, 190–195 (1979)
23. Official site of STMicroelectronics. https://www.st.com/en/microcontrollers-microprocessors/stm32l151-152.html
24. Official site of Semtech Corporation. https://www.semtech.com/products/wireless-rf/lora-transceivers/sx1272
25. Official site of Chirpstack project. https://www.chirpstack.io/

Computing Load Distribution by Using Peer-to-Peer Network

Anton Mamonov, Ruslan Varlamov, and Soltan Salpagarov(✉)

Peoples' Friendship University of Russia (RUDN University),
6 Miklukho-Maklaya St., Moscow 117198, Russian Federation
anton.mamonov.golohvastogo@mail.ru, ravarlamov@mail.ru,
salpagarov-si@rudn.ru

Abstract. Modern computation problems arise that cannot be solved by increasing the number and quality of computers alone. In this work, we develop distributed computing methods.

By applying knowledge from different areas of science, such as queueing theory and theory of computation, we create our own distributed computing system. The main idea behind our approach is the equality of hierarchy and decentralization, and generalization of calculations, covering more high-performance computation areas.

Usage of peer-to-peer architecture requires a thoughtful design but offers several advantages, which are discussed in this work.

To compose our system, we set a mathematical model of the target situation, draw attention to its weaknesses, and develop software principles to resolve them. After solving several design issues, we create simulation models to analyze the effectiveness of the future system.

After analyzing the models, we select the next steps to improve and finalize our system.

Keywords: p2p · Distributed computing · Queueing theory · Theory of computation · Program design

1 Introduction

The last decade has seen such major changes as computation. The only thing that can outpace the performance of the equipment is the increased software requirements. With each new generation of computing or networking devices, a dozen more high-performance technologies are emerging. In other words, there is a race between supply and consumption in the IT sphere [1].

As a result of this protracted situation, we have a myriad of computing devices of various capacities, most of which have long since ceased to meet the requirements. Providers constantly update and increase the quantity of their

The publication has been prepared with the support of the "RUDN University Program 5-100" (A. Mamonov, original draft preparation; R. Varlamov, visualization; S. Salpagarov, project administration).

© Springer Nature Switzerland AG 2020
V. M. Vishnevskiy et al. (Eds.): DCCN 2020, CCIS 1337, pp. 521–532, 2020.
https://doi.org/10.1007/978-3-030-66242-4_41

products. Thus, there is a large pool of computing devices that are used only on a part-time basis and are not suitable for modern high-performance tasks. These trends have been observed and studied for some time [2]. This is a great example of using distributed computing, for example [3].

Despite this, many devices are not connected to any of them, and much of the computing power is wasted. The number of these computing capacities can reach the performance of several supercomputers, while the device itself can be used to some extent.

There are several reasons for this. It is most difficult to configure and connect to distributed systems. Even with the growing familiarity with high technology among ordinary users, it is still difficult for the average person to work with the program outside of the clicking scenario.

So we have a situation where, despite the huge amount of idle computers, the cost of deploying distributed computing systems is too high for any normal use. In our view, this is due to the most commonly used centralized architecture, which is not an optimal method of distributing computation. Another possible analogue for this is the peer-to-peer architecture, already used to some extent for distributed computing, for example [4]. So we decided to develop a new non-centralized method of dissemination.

This paper is structured as follows. Section 2 elaborates currently existing methods of high-performance computing. In Sect. 3 we present a conceptual model of PDCS. In Sect. 4 we outline the theory of algorithms in complexity estimation. In Sect. 5 we outline some design problems and decisions in network establishment. Section 8 describes the distribution process of PDCS. In Sect. 6 we evaluate the resulting system. Finally, in Sect. 7 we summarize our contributions.

2 High-Performance Computing Overview

The problem of high-performance computing is not new. It has a long history of finding solutions. It finds them in various fields of science. Before proceeding to develop our solution, we considered the most significant trends in this direction.

The use of graphics processors for general purpose computing has recently become popular, as the graphics processor has good computing power, which is usually idle [5,6]. But this is not suitable for all problems, the architecture of the system becomes more complex and not as energy efficient as CPU and FPGAs.

You can also use special high-level, numerically oriented programming languages such as Matlab, Scilab etc. A lot of scientific and developing efforts are focused on their constant upgrading, which helps computer algebra systems stay up-to-date [7]. Although this approach can save a lot of time, it is not used for any task, and even though tasks will be solved faster, this time gain is not satisfactory for some of them.

There are also some fundamental researches in the computing area. The most promising technology right now is quantum computers [8]. However, even though promising, quantum computing is far from the point where it will be accessible to everyone. Much more common, but still inaccessible for the majority of scientific society tools for high-performance computing are supercomputers [9].

This article specifically delves into distributed computing technologies. Work even though on distributed computing started with a very applied purpose - for military needs, and namely, the automation of secret communication processes and intelligence processing information has been conducted intensively in the United States since the 1960s. This technology is currently used in many volunteer and commercial projects [10].

For example, BOINC - The Berkeley Open Infrastructure for Network Computing, is an open-source middleware system for volunteers and grid computing [11]. Currently, in mid-2020, it has an average daily performance of about thirty thousand petaflops, from eight hundred thousand of computers. But to use even a fraction of this, newcomers will need to learn how to build up their project, and then draw attention to it. On the other end, while participants, donating their performance to the system, are not burdened with such difficulties, but also has no benefits from it.

In case you want to set up a more equal system, you will need either learn how to modify one of the existing solutions or, more likely, build it up yourself. Which, even with the usage of specific libraries and utilities, is no easy task. JPPF [12] – an open-source Grid Computing platform is written in Java, written with sole propose of easing said process, still requires well-honed programmer's skills.

Although all those solutions are valid in their areas, there is a lack of user-friendly technologies, ones that would not require additional studies or financial investment. So, we present PDCS - Peer-to-peer distributed computing system, a network which use knowledge from teletraffic engineering and computational complexity analysis to manage beneficial calculation distribution for all participants.

3 Concept

3.1 Mathematical Basis

To design an efficient distributed computing system, we raise the issue of load balancing. To do this, we're using models from the queueing theory [13]. It is a science field that studies queueing systems, and a computing system can be easily interpreted as such.

Consider the work of the office with v employees, each of whom uses a personal computer for periodic calculations. In the terminology of the queuing theory, each employee is a stream of applications and personal computers are the service devices. With the flow rate of applications λ, the processing speed μ, and with the maximum number of applications in the queue of the device r, it is possible to compose an Erlang model and calculate the time characteristics [14].

However, for the standard situation, when each employee uses only his personal computer, not one common system, but a set of v separate systems is formed. In a situation like this, when λ is small and μ is large, there is an imbalance between the individual devices - while half of them are free, the other half are in the queue and vice versa.

But if you use middleware for communication between devices, it is possible to combine the threads into a single thread with flow rate λ and the total queue capacity $R = r * v$ (Fig. 1).

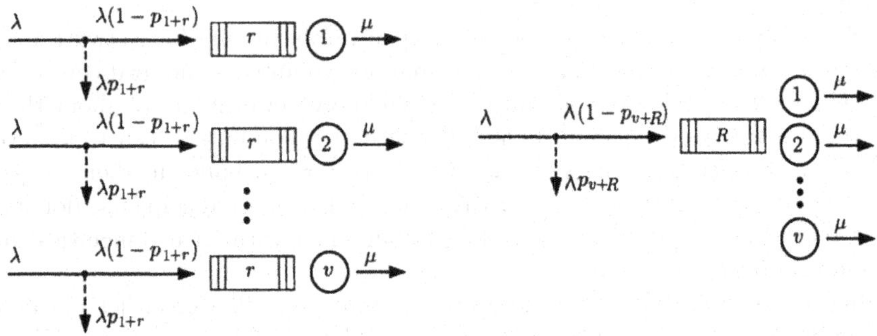

Fig. 1. Model on divided (left) and distributed (right) system

Even with the additional time spent on transferring information between devices, the benefit of using a distributed system is obvious. The shorter total time spent in the queue reflects more efficient use of the available computing resource.

To further increase the efficiency of calculations, it is necessary to take into account the heterogeneous nature of devices and applications in a real situation. Employee requests can vary in complexity, just as the computers used can vary in power.

To take this into account, you can turn to multiservice models. Assuming the presence of K types of services, it is necessary to convert the scalar quantities λ and μ into vectors of dimension K and compose a vector of comparative complexity b. The traditional model of a multiservice system involves assigning a standard unit to the least complex request and displaying the complexity of the rest through comparison with the standard unit. After that, in the mathematical model, the processing of an application of the i-type requires the work of the b_i number of devices.

With the addition of a difference in the power of computers in this model, a vector q of dimension v was formed - the number of applications of minimal complexity performed by one of v devices per unit time.

To implement such a model in practice, it is necessary to organize a distributed computing system. To do this, you need to write software that will read user requests, generate applications and distribute them over the network. The main principles that should be met by the software are the generalization of calculations, the equality between network nodes and the transparency of working with the system.

3.2 Generalization

Generalization of computation stands for the reduction of requests received by the system to a set of generalized operations.

In the terminology of Post, which was set in one of the first articles concerning the theory of algorithms [15], the concept of a general problem, which is consisting of a class of specific problems, is introduced. Thus, giving an algorithm for solving one general problem, we obtain a solution for a whole set of specific problems. This is fairly obvious for simple cases such as arithmetic operations, but the boundaries of generalization are easily extended if recursion is allowed.

So, for example, calculating the factorial is essentially a special case of multiplying a natural number by the factorial of the previous one. The calculation of the number of combinations is several divisions of certain factorials. And this means that the calculation of one or another probability can be reduced to a set of arithmetic basic operations.

The developed system is based on the distributed computing system considered in [4]. This system was used to calculate the probability of universal transmission in a streaming television system. These calculations include multiple computations of combinatorial formulas and their time complexity grows exponentially depending on M - the number of viewers and N - the number of channels. The distribution of calculations in this paper was proposed to be carried out following one of two methods - the first involves the separation of each combinatorial formula into factorial components and their distribution over the network. The second proposes the separation of the formula by the external sum operator.

Both of these methods solve one specific problem, each time for specific input. And both of them can be generalized to sequential applications of arithmetic primitives, some of which are performed centrally on one node, and some are distributed throughout the system.

By giving the system the ability to compute primitives nested into each other, all that remains is to provide a generalization process – the formation of these primitives from the user's request.

3.3 Equalization

Equality between network nodes stands for the implementation of Peer-to-peer architecture during system creation.

Peer-to-peer technology which has been developing over the past decade, continues to find new uses. For example, peer-to-peer architecture is not new for the field of time-consuming computing. Although the most common form of distributed computing today is the grid, the traditional one is implemented with a centralized architecture, you can already find examples of peer-to-peer computing [10].

Nowadays many initially centralized platforms and distributed computing tools are gradually integrating the ability to implement partial or full peer-to-peer. This is because peer-to-peer networks have many advantages over classic client-server counterparts. Among them, easy scalability, high fault tolerance, simple moderation, etc. In several scenarios, these qualities can greatly outweigh the natural flaws of decentralized systems - poor connectivity, complexity in development, etc.

In the situation under consideration, the most valuable will be the fact that the organization of a network using a peer-to-peer architecture will allow us to avoid the difficulties of installing and separately maintaining a server. This should provide a way to popularize peer-to-peer computing, which, in turn, will provide peer-to-peer systems with the necessary capacities.

3.4 Transparency

Transparency of the system means that its internal workflow is invisible to users.

The main source of resources for any peer-to-peer system is its users. A large number of nodes means more memory, higher speeds for file sharing and more computing power for distributed computing. To provide our system with users' resources, we make system transparency one of the main development goals.

It's not enough to provide significant time gains in time-consuming calculations. The system also should not force users to learn local syntax or make complicated queries. Any advantage in the speed of actual calculations might be lost if a comparable amount of time is needed to organize them. In such a situation, the audience of the system will significantly narrow, which will reduce the total number of nodes and, as a result, the output of the system.

To provide a user-friendly interface and low entry threshold, it was decided to change the focus of development from creating a domain-specific language in favour of using the well-known general-purpose language of mathematical formulas. There is no need to have special education to know that "=" means "equal to", that the dependent parameters are listed in brackets after the name of the formula. Just having an education is enough.

4 The Algorithmic Analysis

During the construction of the system, the question arose about the efficient allocation of resources, but for this, it was first necessary to decide how the complexity of a given task would be assessed. Fortunately, we are not the first to encounter such a problem and there is a very large range of works devoted to the algorithmic analysis [16].

Our system should be able to calculate the complexity of the task, regardless of the class of algorithms used in it. When analyzing the behaviour of the complex function of an algorithm, the asymptotic notation [17] accepted in mathematics is often used to show the growth rate of a function, masking specific coefficients. Such an estimate of the complexity of the algorithm is called the complexity of the algorithm and will determine the preferences in using one or another algorithm for large values of the dimension of the source data in our system.

For our system, the best-known estimate of the complexity of the Big O notation algorithm is best suited, since the system must allocate enough resources so that one node does not delay all the others.

There are several methods for determining the operating time of a software implementation of an algorithm. Since the system implies solving problems of any type, the most suitable method is that can determine the implementation time of any algorithm. The most straightforward and at the same time effective way of estimating the complexity of the algorithm is the operational analysis, the meaning of which is to obtain the operational function of the complexity for each of the elementary operations used by the algorithm taking into account data types. And then experimentally determine the average execution time of this elementary operation on a particular computer. The expected lead time is calculated as the sum of the products of the operational complexity for the average time of operations:

Since the system can be used in various fields of scientific research, calculations can use a huge range of elementary operations. Therefore, at the beginning of its work, the computer will calculate the approximate time for all these operations. Also, the computer that sends a request for solving the problem will count the number of simple operations involved in it. Based on these data, the task will be most effectively divided using the distribution mechanism.

5 The Network

While peer-to-peer architecture offers some important advantages, it also poses several challenges to network maintenance. There is no centralized server to receive and sort out all system messages. Without a universal proxy, numbers of connections between nodes are increasing in arithmetic progression. That imposes a huge restriction on the amount of information exchanged by nodes.

To exchange such messages without overloading network, it is convenient to use UDP - user datagram protocol. It is a common communications protocol, mainly used for connections with low latency and loss tolerance between applications. By using UDP it's easy to maintain receiving and sending of messages - datagrams, with only one port for any possible addresses.

But, while UDP doesn't create any logical binds on connected nodes, it also doesn't support transmission reliability. To send a message, UDP needs to cut it in parts of appropriated size, and it doesn't guarantee that all of those parts will arrive on the recipient in the right order or arrive at all. Fortunately, it possible to avoid fragmentation altogether, if the original message is small enough to fit completely in one datagram. To find out the exact maximum size of the message, the respective specifications can be used.

The specification for Internet Protocol [18] tells us that a minimum IP packet length that all IPv4 members should support is of 576 bytes. Therefore, to avoid fragmentation and possible loss of UDP packets, and also that they will be received by any host, the size of the data in datagrams should not exceed minimum IP packet length minus maximum IP header length, minus UDP header size [19]. In numbers, 576 − 60 − 8 = 508 bytes for each message. In most programs the encoding of characters holds 2 bytes for each char, thus up to 254 characters can be transmitted within a single datagram. While formulas and values of variables do not exceed this restriction, it's safe from any incidental losses by fragmentation.

As was mentioned earlier, PDCS is working with natural mathematical distribution. For example, a sum operation, placed over the expression F with dependency on a variable i, in range from the lower limit i_0 to the upper limit N.

$$Sum(N, F(i)) = \sum_{i=i_0}^{N} F(i) \tag{1}$$

It is abstract enough to cover anything from integral sums to probability calculations. To move from abstraction to practical implementation, we will need only one line that represents the exact formula of F and values of lower and upper limits of summarizing. Even some advanced queries with multiple variables can easily fit into 254 characters, and the answer even larger.

The implementation of these design decisions present itself in simple and effective way. The network part of PDCS, consisting a virtual room and participating nodes, is mapped by two appropriate classes: Room and Node. So the Node class is a virtual client, and the Room class is a virtual server. To operate properly, each PDCS client needs only to contain a list of Nodes as a Room class, and some vital information about Nodes, such as their IP addresses, measured performance, and current state.

When the software starts, a new empty room is automatically created, where the first node immediately turns on, under the computer's own IP address. Broadcast network initialization information about this new PDCS client via the dedicated port. Other clients already on the network send an invitation to join in response. After completing the virtual handshake, both newly connected clients create a new record in their virtual rooms.

During the operation, reception and sending of messages, the virtual room updates the information about its status, including the set of nodes therein, their current occupation, the queue of requests for calculations. If one of the nodes does not reply to the system messages sent to it, including the special periodic signal "ping", it is excluded from the room. Upon receipt of a set of requests for calculations, the room sends them to free nodes, waiting for confirmation of receipt of the request and, if necessary, repeating the sending until all requests are satisfied.

6 The Mechanism of Distribution

The overall efficiency of any system depends to a large extent on the distribution of the workload. Especially in targeted systems, and their request-report process on one node inevitably involves other nodes. Distributed computing is a striking example of these systems.

But simply installing the messaging described in the previous section will not solve all distribution problems in our system. To work properly and efficiently, equal nodes will have to share the same information about each other's condition. Each time one node assigns itself to a task, it needs to inform the entire system about that, same for the moment the task is completed. In addition, it is necessary to exchange a standard ping message to confirm the online status of each node. If one of the nodes does not respond or does not execute the echo request in time, the workload distribution should be changed immediately.

In practice, total knot failure is unlikely. But as long as the PDCS client operates as a background process, the user can and will use his computer in other actions, which will cause a dynamic change in the performance of the host. For information to remain relevant, nodes must check their performance during each calculation, as described in Sect. 4. In case a node does not have queries for some time or the client performance is changed directly by the user, an automatic query must be formed.

Now, with the installed network and updated system status information, the PDCS nodes can execute the actual distribution for the actual calculations. For a practical example, set expression F to the formula for the probability that the number of successes for a set of independent experiments N equals i. With each independent success or failure respectively equal to p and q, probability of at least i_0 successes will be:

$$Sum(N) = \sum_{i=i_0}^{N} p^i q^{N-i} C_N^i \tag{2}$$

Such request can be entered by user in string form and will look like :

input : sum(i = i0, N)(p ∧ i * q ∧ (N-i) * comb(N, i)

Which will be broken down by a parser and then used to form a secondary network query, one for each member of the sum. These requests form a queue at the node and are sent to free nodes. If the number of requests exceeds the number of free nodes, some of them remain unaccounted for until the first requests are fulfilled.

solve : (i = i0)(p ∧ i * q ∧ (N-i) * comb(N, i)
solve : (i = i0+1)(p ∧ i * q ∧ (N-i) * comb(N, i)
e.t.c.
solve : (i = N)(p ∧ i * q ∧ (N-i) * comb(N, i)

As it is common for calculation of combinations, the bigger difference between N and i values is, the bigger result will be. So, it makes sense to send the first half of secondary requests to more powerful nodes, and the second half to the others. Thus, the responses will be more synchronized.

Once the free node receives one of the secondary requests, it sends back a confirmation, then informs the whole system that it is currently occupied, and starts the calculations. If case origin node doesn't receive a confirmation for a specific request, it withdraws this request back to the queue. The same happens if confirmation is received, but the response does not arrive in twice the time approximated.

When all secondary requests are executed and sent back to the source node, it combines them in a final sum and notifies the user with a numerical output.

7 System Estimation

Before developing the prototype of the system, it was necessary to check superiority of our distribution method over the others, as well as to measure its approximate effectiveness: the time saved by the system, stability of the system. For this, it was decided to develop a system model using a suitable simulation program. The first model was to be based on individual computers that solve problems on their own, and the second was to represent our approach (Fig. 2).

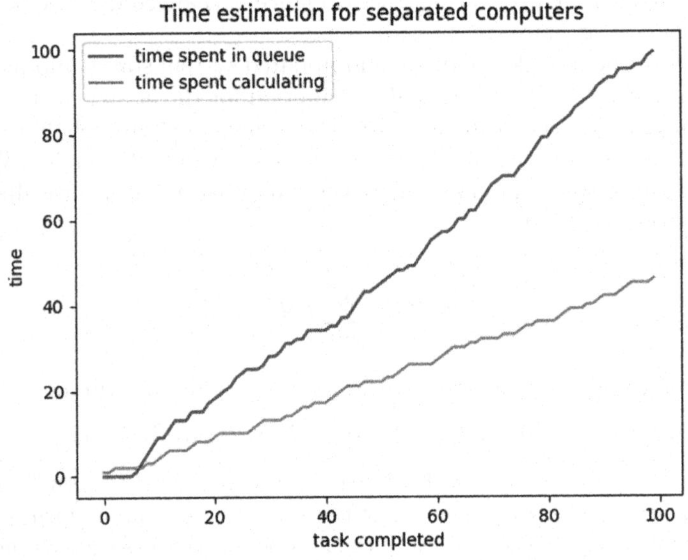

Fig. 2. Model of separate computing

The choice fell on AnyLogic. It is a multimethod simulation modelling tool. It supports agent-based, discrete event, and system dynamics simulation methodologies. It is suitable for our task, as it has fairly extensive functionality that allows you to create dynamic models with different amounts of input and output data. At the same time, it allows you to track each task individually and has built-in statistics tools. Also, the program itself is written in Java, just like our system, which saved time on building models (Fig. 3).

The P2P model showed overall high level of reliability and validity while the comparison of models was used to measure the potential time gains.

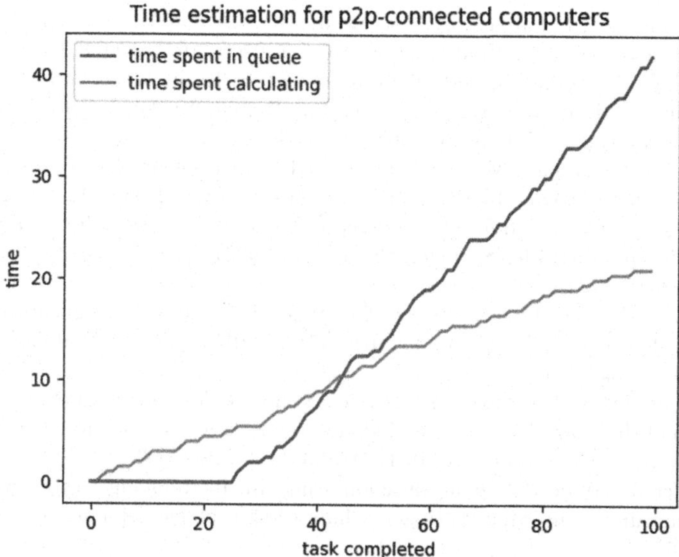

Fig. 3. Model of p2p computing

8 Conclusion

In this article, we have studied a problem with high-performance computing, and various ways of its solution. Concluding that most solutions are not user-friendly or easily accessible, we have developed our paradigm of the computing system and created a mathematical model for it. With the usage of different scientific fields, we have improved our design and created simulating models to analyze the resulting system.

In further research, we aim to use obtained models to create a software template of PDCS - peer-to-peer distributed computer system and conduct a comparative analysis between our system and other solutions for high-performance computing problem.

References

1. Trattner, A., Hvam, L., Forza, C., Herbert-Hansen, Z.N.L.: Product complexity and operational performance: a systematic literature review. CIRP J. Manufact. Sci. Technol. **25**, 69–83 (2019). https://doi.org/10.1016/j.cirpj.2019.02.001
2. Post, D.: The future of computing performance. Comput. Sci. Eng. **13**(4), 4–5 (2011). https://doi.org/10.1109/MCSE.2011.69
3. Liu, G., Xiao, Z., Tan, G., Li, K., Chronopoulos, A.T.: Game theory-based optimization of distributed idle computing resources in cloud environments. Theoret. Comput. Sci. **806**, 468–488 (2020). https://doi.org/10.1016/j.tcs.2019.08.019

4. Khachumov, K.M., Salpagarov, S.I., Mamonov, A.A., Varlamov, V.A.: Combinatorial problem solving method by allocating resources. In: Selected Papers of the IX Conference "Information and Telecommunication Technologies and Mathematical Modeling of High-Tech Systems", Moscow, Russia, 19 Apr 2019 (2019). http://ceur-ws.org/Vol-2407/paper-07-165.pdf
5. Dubey, S.P., Kumar, M.S., Balaji, S.: GPU computing for compute-intensive scientific calculation. In: Das, K.N., Bansal, J.C., Deep, K., Nagar, A.K., Pathipooranam, P., Naidu, R.C. (eds.) Soft Computing for Problem Solving. AISC, vol. 1057, pp. 131–140. Springer, Singapore (2020). https://doi.org/10.1007/978-981-15-0184-5_12
6. Mittal, S., Vetter, J.S.: A survey of methods for analyzing and improving GPU energy efficiency. ACM Comput. Surv. **47**(2) (2015). 23 https://doi.org/10.1145/2636342. Article 19
7. Kulyabov, D.S., Korol'kova, A.V., Sevast'yanov, L.A.: New features in the second version of the Cadabra computer algebra system. Program. Comput. Softw. **45**(2), 58–64 (2019). https://doi.org/10.1134/S0361768819020063
8. Zulehner A., Wille R.: Quantum computing. In: Introducing Design Automation for Quantum Computing. Springer, Cham (2020). https://doi.org/10.1007/978-3-030-41753-6_2
9. Gertsenberger, K., Rogachevsky, O.: Mock data challenge for the MPD/NICA experiment on the HybriLIT cluster. In: EPJ Web Conference, vol. 173, p. 04006 (2018). https://doi.org/10.1051/epjconf/201817304006
10. Ivashko, E., Chernov, I., Nikitina, N.: A survey of desktop grid scheduling. IEEE Trans. Parallel Distrib. Syst. 2882–2895 (2018). https://doi.org/10.1109/TPDS.2018.2850004
11. Anderson, D.P.: BOINC: a platform for volunteer computing. J. Grid Comput. **18**(1), 99–122 (2019). https://doi.org/10.1007/s10723-019-09497-9
12. Pengembangan Prototipe Sistem Network Rendering Alternatif Berbasis JPPF (2018). https://doi.org/10.22146/jnteti.v7i1.395
13. Naumov, V.A., Gaidamaka, Y.V., Samouylov, K.E.: Computing the stationary distribution of queueing systems with random resource requirements via fast fourier transform. Mathematics **8**, 772 (2020). https://doi.org/10.3390/math8050772
14. Basharin, G.P.: Lectures on the mathematical theory of teletraffic
15. Post, E.: Finite combinatory processes-formulation 1. J. Symbol. Logic **1**(3 Sep), 103–105 (1936). https://doi.org/10.2307/2269031
16. Cormen, T.H., Leiserson, C.E., Rivest, R.L.: Introduction to Algorithms, 3rd edn., p. 47. MIT Press, Cambridge (2009). ISBN 978-0-262-53305-8
17. Hildebrand, A.J.: Asymptotic Notations. Asymptotic Methods in Analysis. Math 595 (2009)
18. Postel, J.: INTERNET STANDARD RFC 791 Internet Protocol (1981)
19. Postel, J.: INTERNET STANDARD RFC 768 User Datagram Protocol (1980)

Firewall Simulation Model with Filtering Rules Ranking

Anatoly Botvinko[1] and Konstantin Samouylov[1,2(✉)]

[1] Peoples' Friendship University of Russia (RUDN University), 6 Miklukho-Maklaya St., Moscow 117198, Russian Federation
{bottvinko_ab,samuylov_ke}@rudn.ru

[2] Federal Research Center "Computer Science and Control" of the Russian Academy of Sciences (FRC CSC RAS), 44-2 Vavilov St., Moscow 119333, Russian Federation

Abstract. The article has been written in continuation of a series of works on the evaluation of the probabilistic and time characteristics of firewalls while ranking a set of filtering rules. The problem under consideration is the efficiency reduction of filtering the information flows caused by: a) using a sequential circuit for verifying packet compliance with rules; b) heterogeneous character and variability of network traffic. By using the developed model, the main firewall performance indicators for various traffic behavior scenarios were evaluated. The model proposed allows to evaluate the effectiveness of filtering rules ranking methods in order to improve the firewall performance.

Keywords: Firewall · Filtering rules ranking · Network traffic · Simulation model · Queuing system

1 Introduction

A firewall (FW) is a local or functionally-distributed tool implementing the control over the information that inputs to the automatic system (AS) and/or outputs from the AS and provides protection of the AS by filtering the information, i.e. its analysis on the set of the criteria and making a decision on its distribution [1].

FW is one of the principal means used to ensure information security (IS) of the AS. That includes the fact that FW allows to solve the tasks of preventing unauthorized access to protected information, its destruction, modification, blocking, copying, providing it to the third parties and its spreading [2].

The permanent maintenance of high performance while filtering network traffic is required to ensure the stable functioning of the critical information infrastructure in conditions of: a) a sharp increase in the information flows volume of common use networks, b) high heterogeneity and variability of the network traffic characteristics; c) wide use of various multimedia protocols and d) an increase in the number of computer attacks. Therefore, the actual task is to increase the

© Springer Nature Switzerland AG 2020
V. M. Vishnevskiy et al. (Eds.): DCCN 2020, CCIS 1337, pp. 533–545, 2020.
https://doi.org/10.1007/978-3-030-66242-4_42

efficiency of filtering network traffic flows, as well as to maintain it at the level sufficient to ensure the stable operation of the AS.

The fundamental idea of the FW's rules optimization methods is to provide the increase in the efficiency of traffic filtering by reducing the time for searching the rules that match network packets. The most of the existing rule optimization methods for organizing a rule set use filtering rules ranking and base on a) the current characteristics of information flows [3–6] or b) on the use of various data structures different from a linear list [5,7–10]. In a number of works, there are the methods used for searching and correcting errors in creating the sets of filtering rules [11–13]. These methods don't take the dynamics of changes in information flows into account. Because of this, they are less adaptive to changes in the nature of the flows filtered. Therefore, when frequent changes in the flow characteristics happen, it is rational to apply ranking methods that: a) are based on the use of statistical estimation and b) take the dynamics of changing in the information flow characteristics into account.

The objective of this work is to create a mathematical model for evaluating the firewall performance with an altering set of rules. Unlike earlier works [14–16], the designed model allows to obtain estimates of FW performance while filtering various types and scenarios of network traffic behavior. To achieve the objective, the methods of imitational modelling and representing the FW in the form of a complex system and queuing systems at separate time intervals are used in the work.

The results obtained in the work reflect the possibility of increasing the FW performance and the rationality of using the ranking method developed in [16,17] for various traffic scenarios, including the FW functioning with significant overload that is close to the packet loss limit.

2 Method for Ranking Firewall Filtering Rules

Filtering rules are: a) a list of conditions according to which and by using the given filtering criteria the permission and/or prohibition of further packets transmission is carried out; b) a list of actions performed for registration and/or for additional protective functions. The following parameters, attributes, and characteristics of network traffic flows are usually used as filtering criteria: a) service fields of packets that contain network addresses, identifiers, interface addresses, port addresses, and other significant data; b) external characteristics, e.g. time and frequency parameters, and data volumes. Each rule prohibits or permits the transmission of information of a certain kind between subjects and objects. As a result, the subjects from one AS get the access only to permitted information objects from another AS. The interpretation of a rule set is carried out by a sequence of filters that permit or prohibit the transfer of data (packets) to the next filter or to the protocol level [1]. The number of used filters depends on the FW type.

So, after receiving the network packet, the verification of the compliance of the packet parameters with the conditions of the first rule takes place;

if the packet doesn't meet the conditions in the rule fields, the verification of compliance with the fields of the next rule starts. The first rule that the packet meets will determine the action applied to the package. In case if the packet doesn't meet any rule, the actions determined by the default policy given by the rule of the last set will be performed.

3 Firewall Model with Filtering Rules Ranking

Modern FWs are complex and multifunctional systems using various hardware platforms, operating systems and types of filters. Also, for all FWs of various manufacturers, such as Cisco Systems, Juniper Networks, the process of information flow filtering includes traffic filtering on the network and transport levels of the OSI model. Thus, in this work a FW model related to the type of packet filters was designed [18–20].

The filtering process is a sequential check of parameters of input packets compliance with the conditions of the ruleset. The packet check is sequentially performed for all of the rules from the set until the first match is found. Only one packet can be checked simultaneously; the rest of the received packets wait for the service start. The packets that don't meet the permission rules of filtering are rejected (the default prohibition policy is considered).

By rule set ranking we will mean the organization of the rules in decreasing order of their weights obtained in accordance with the estimation of the characteristics of information flows filtered. While setting the weights for each rule, a time series is composed and it reflects the number of packets that have been filtered in accordance with the conditions of the rule within several time intervals. With the accumulation of a sufficient number of data, the estimation of the average values of the number of packets for subsequent time intervals takes place. While predicting the values, the nonparametric method of local approximation (MLA) is used [21]. Examples of time series and an algorithm for rule ranking are contained in the work [14].

A model with filtering rules ranking is a complex stochastic system, thus, to describe the model, the approaches to the systems description applied in the theory of queuing systems were used [22–24].

Let's represent the FW model as a system $\mathrm{M} = \{Z, \mathbf{A}, T, F, G, X\}$, the moment of transition of which from one state to another one is depicted in Fig. 1.

Let's describe the system functioning by introducing the following notations:

1. $Z = \{\mathbf{z}_0, ..., \mathbf{z}_k, ...\}$ – an array of the system states; $\mathbf{A} = (\mu_0, \mu, N, C)$ – an array of the system states; T – the functioning time interval of the system; the change of system states occurs at discrete time moments $t_k^- = t_k - 0, t_k \in T, k \geq 1$; F – the system states transitions operator; G – the system output operator; $X = \{\mathbf{x}_0, ..., \mathbf{x}_k, ...\}$ – an array of input signals entering the system; $Y = \{\mathbf{y}_1, ..., \mathbf{y}_k, ...\}$ – an array of output signals of the system.

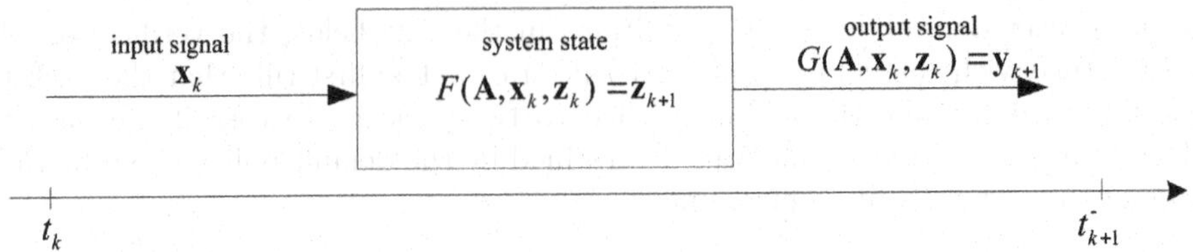

Fig. 1. Scheme of the FW model with filtering rules ranking.

2. μ_0 – the service intensity during the initial processing of the packet by the FW network card; μ – the service intensity during the checking of the compliance of the packet with one of the rules from filtering set; C – the storage capacity in the system; N – the number of rules in the filtering set.
3. $\mathbf{z}_k = (\mathbf{r}_k, \mathbf{d}_k)$ – the system state on the time interval $[t_k, t_{k+1})$, where $\mathbf{r}_k = (r_1^k, ..., r_N^k)$ is the set of rules in which the component $r_i^k, i = 1, ..., N$ is a rule located in i-place of the set; $\mathbf{d}_k = (d_1^k, ..., d_N^k)$ is the vector of packet service times, in which $d_i^k, i \in 1, ..., N$ corresponds to the processing time of i-type packets on the interval $[t_k, t_{k+1})$.
4. $\mathbf{x}_k = (x_1^k, ..., x_N^k) \in X$ the input signal; the component $x_i^k, i = 1, ..., N$ – a random variable characterizing the number of i-type packets, received in the system on the time interval $[t_k, t_{k+1})$.
5. $\mathbf{p}_k = (p_1^k, ..., p_N^k)$ is a vector of the rule weights, set in accordance with the MLA estimation of the information flows characteristics; the component $p_i^k, i = 1, ..., N$ corresponds to the weight of the rule standing in i-place on the interval $[t_k, t_{k+1})$.
6. $\mathbf{y}_k = (q_k, v_k, w_k, u_k) \in Y$ – the output signal, the components of which correspond to the estimation of performance indicators on the time interval $[t_k, t_{k+1})$: q_k – the average number of packets in the drive, v_k – the average service time of a packet in the system, w_k – the average time of waiting for the start of packet servicing in the system and u_k – the average time the packet stays in the system.
7. $F(\mathbf{A}, \mathbf{z}_k, \mathbf{x}_k) = \mathbf{z}_{k+1}$ – the operator of system states transitions at the moment of time t_{k+1}^-: calculates $\widehat{\mathbf{x}}_k$ that evaluates the information flows characteristics and determines the vector of the rule weights $\mathbf{p}_k$ according to $\widehat{\mathbf{x}}_k$.
8. $\mathbf{z}_{k+1} = (\mathbf{r}_{k+1}, \mathbf{d}_{k+1})$ determines the system state, where the vector $\mathbf{r}_{k+1}$ is calculated by ranking the rule set $\mathbf{r}_k$ according to the weights $\mathbf{p}_k$, and the vector $\mathbf{d}_{k+1}$ is calculated in accordance with the obtained set $\mathbf{r}_{k+1}$ and intensities μ_0, μ of the packet servicing.
9. $G(\mathbf{A}, \mathbf{z}_k, \mathbf{x}_k) = y_{k+1}$ at the moment of time t_{k+1}^- determines the system performance indicators on the time interval $[t_k, t_{k+1})$.

Considering the functioning of the system M on each interval $[t_k, t_{k+1})$ separately, let's represent the system M as a queuing system (QS) with a distribution function (DF) $B_k(t)$ of the requests service duration, that depends on the filtering rules order on the time interval $[t_k, t_{k+1})$. The request flow according to the packet flow entering the FW is received by the QS. Given QS at the request Poisson flow according to the Basharin-Kendall classification will have the $M_N/PH/1/C$ type.

Receiving the characteristics of given QS is performed by imitational modeling, and the results of it are given in the fourth section of this article.

Further, there is a terminology used in this work and it is accepted by queuing theory: the packets entering the model and being the entities in the imitational model are considered as QS requests.

4 Designing of an Imitational FW Model with Rules Ranking

The imitational model is intended to evaluate the FW performance indicators with rules ranking. Imitational methods using allows to remove restrictions on the DF of the incoming request flow and the request service of model M. So, the imitational model makes it possible to receive the estimation of FW performance while filtering various types and scenarios of network traffic behavior. Also, the imitational model reproduces the process of the system M functioning:

- Packets filtering by FW corresponds to request servicing in the QS of $M_N/PH/1/C$ type;
- the calculation of weights, and rules ranking corresponds to the operator F application;
- the calculation of the estimations of FW performance indicators corresponds to the operator G application.

To design the imitational model, the discrete-event modeling approach is used. The choice of this approach was made because of: a) strictly defined list of events arising during the functioning process; b) the necessity of receiving statistical information that are collected in discrete-event modeling systems (average service time and waiting time in the queue of requests, average number of requests in the queue, etc.); c) its wide application in queuing theory [25].

For the software implementation, the Simulink graphical environment of imitational modeling was used alongside with the application of the SimEvents discrete state library using the queuing theory apparatus of and queuing systems. The Simulink environment is integrated into the MATLAB matrix computing system, and that allowed to use built-in mathematical algorithms and data processing means to implement the features of the system M functioning, i.e. rules ranking, calculating MLA estimations and changing system states at different time intervals. The usage of MATLAB provided a high system ability for scaling and expanding the functionality without significant changes in the program architecture.

The input data in the model are the requests serviced in accordance with the FCFS principle (First-Come, First-Served). Only one request can be served simultaneously. The DF of request service time is a phase-type function.

The requests waiting for the service are in the final queue with the FIFO service discipline (first in, first out). If the queue is full, the incoming request leaves the system without servicing.

The output data after the modeling are the estimations of the QS performance indicators presented in graphical and tabular form, as well as comparative indicators of the effectiveness of rules ranking, that are calculated on their basis for various scenarios of traffic entrance.

All of the time-depending values in the imitational model have microsecond measuring. Thus, there is a possibility for the user to change the dimension.

The software implementation of the imitational model consists of the Simulink model file and MATLAB command sequence files. The model file contains visual-oriented programming blocks that apply QS elements, display the information about the progress of modeling and record the input signal into the MATLAB workspace (see Fig. 2). The command sequence files contain program code applying the model functionality: a) setting the initial system parameters; b) calculating the MLA estimations of information flow characteristics; c) setting the weights; d) rules ranking; e) processing data received while modeling; f) graphical and tabular presentation of modeling results; g) model launch commands in the Simulink environment.

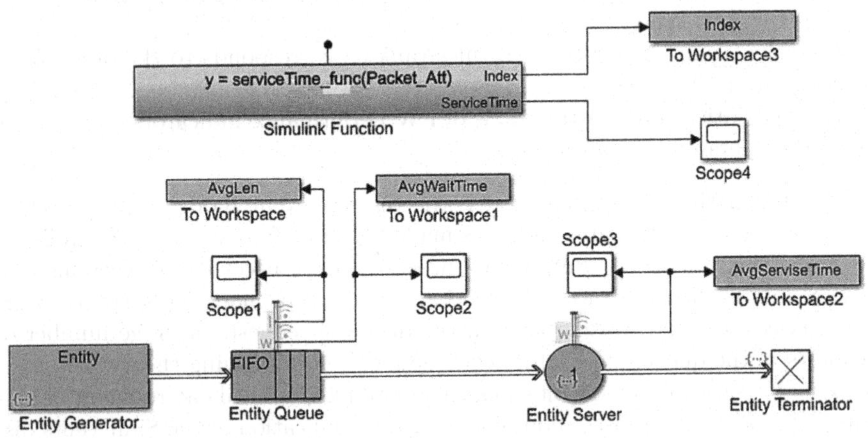

Fig. 2. Flowchart of Simulink FW model with filtering rules ranking.

While starting the modelling, the initial parameters of the Simulink model are entered into the M-file – the number of the rules, the initial rule set, the number of request types, the storage capacity of the request store, DF, the intensity of requests' receiving and servicing, etc. Then, the modeling of the packet package filtering by FWs is performed in accordance with the set parameters. Generation of various request types is performed by using the "Entity Generator" block. The calculation of the service time is performed by using the "serviceTime_func" function implemented in the "Simulink Function" block (see Fig. 3).

The "serviceTime_func" function is a custom Simulink function calculating the service time in accordance with the set model parameters. The input arguments of the function are: a) "Packet_Att" – is the type of request filtered; b) "TimeDMA" – is the service intensity during initial packet processing by the FW network card; c) "TimeOnRule" – is the service intensity while checking of the compliance of the packet with one rule of the filtering set; d) "RuleSet" – is a current rule set; e) "CountRule" – is the number of filtering rules. The output arguments are: a) the packet filtering time by FW and b) the "Index" signal for the calculation of the number of packets that match for each rule type.

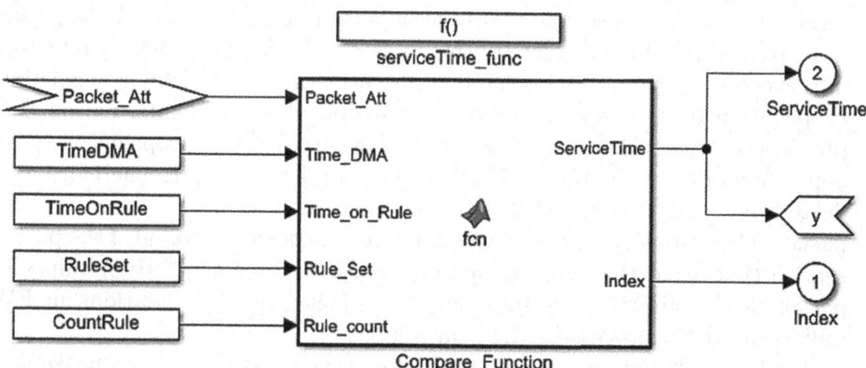

Fig. 3. Custom "serviceTime_func" function for calculating the service time.

The algorithm for calculating the service time is implemented in the M-function task block "Compare_Function" and is shown in Listing 1.

Listing 1. "Compare_Function" function

```
function [ServiceTime, Index] = ...
fcn(Packet_Att,Time_DMA, Time_on_Rule, Rule_Set, Rule_count)
Index       = -1;
ServiceTime =  0;
 for i = 1:1:Rule_count-1
   if Packet_Att == Rule_Set(i) && Packet_Att ~= 0
   Index = i;
   ServiceTime=gamrnd(Index,Time_on_Rule)+gamrnd(1,Time_DMA);
   break
   elseif Packet_Att ~= Rule_Set(i) && i == Rule_count-1 ...
                      && Packet_Att ~= 0
   Index = 0;
   ServiceTime=gamrnd(Rule_count,Time_on_Rule)+gamrnd(1,Time_DMA)
   end
 end
end
```

5 Evaluation of the Firewall Performance Indicators

Usually, FWs are located at entry points at the borders of protected networks or network segments. Being a network node that processes large volumes of incoming and outgoing traffic, FW significantly influences on network performance, and FW performance indicators largely correspond to key characteristics of border network devices - routers, switches, etc. [26].

These key features include:

- packet service delay – the time FW that has received the packet takes to start its transmission to the next node in the network;
- serialization delay – the time that takes a network card to transmit the packet at a given width of bandwidth. In this work, the serialization delay isn't taken into account, since for a wide bandwidth the serialization delay is negligible;
- propagation delay – the time for transmission of a bit of information in a physical communication line. The bit is transmitted with a propagation speed depending only on the physical environment of data transfer. Therefore, this delay type is not considered in this work;
- packet delay variation (Jitter) – characterizes the deviation value of the packet service time from the average value of packet service time. High values of packet delay variation may be caused by queue lengths variations in FW buffers or packet processing time variations;
- packet loss – the number of lost packets or the proportion of lost packets to the total number of transmitted packets. With sharp bursts in packet income intensity, large packet queues can be formed and they need to be processed. Since the buffer memory in the FW network card is limited, a part of the packets may be lost. Also, the packet losses are possible when the information distortion is detected. In this work, it is assumed that packets are transmitted without any information distortion;
- packet transmission delay (or packet delay) – is the time that the FW spends on processing and transmitting the packet. Packet transfer delay is the sum of the serialization, propagation and service delays. Since the serialization delay and the propagation delay are not considered in the work, and the packet service delay is a separate efficiency indicator, the packet transmission delay is not considered as well.

So, the packet service delay is used as the primary indicator of the FW performance. It consists of: a) the time required for the packet initial processing; b) the time the packet stayed in the buffer memory and c) the time for searching the rule corresponding to the packet filtered.

To evaluate the effectiveness of rules ranking, comparative indicators are used. They are calculated between the initial time interval (where the ranking wasn't performed) and the time intervals where the rule set was ranked: $\Delta Q_S = q_s - q_1$, $\Delta V_S = v_s - v_1$, $\Delta W_S = w_s - w_1$, where s is a number of the time interval where the evaluation was calculated.

The network traffic in the model is generated by a source summarizing stationary request Poisson flows with different intensities. In this case m packet types enter with high intensity $\lambda_i = \lambda^k_{high}$, and the rest of the requests – with low intensity $\lambda_i = \lambda^k_{low}$, where $i = 1, ..., m$ is the request type; k is the number of time interval. Let's call the request with high intensity as priority requests, and requests with low intensity as non-priority requests.

Three scenarios of traffic behavior are considered under the imitational modeling:

- the intensities of priority and non-priority requests are constant at each time interval, and the QS functions without any overloads;
- the request intensities increase on each time interval $\lambda^k_{low} = \lambda^{k-1}_{low} + 0.1\lambda^1_{low}$, $\lambda^k_{high} = \lambda^{k-1}_{high} + 0.1\lambda^1_{high}$, the QS starts functioning with a slight overload on the interval $[t_k, t_{k+1}), k = 1, ..., 11$ (the queue size doesn't exceed 2.5% of the storage drive capacity) and a significant overload on the interval $[t_k, t_{k+1}), k = 12, ..., 15$ (the queue size is 45%, ..., 80% of the storage drive capacity);
- the request intensities increase on each time interval $\lambda^k_{low} = \lambda^{k-1}_{low} + 0.1\lambda^1_{low}$, $\lambda^k_{high} = \lambda^{k-1}_{high} + 0.1\lambda^1_{high}$, the QS starts functioning with a significant overload approaching the limit of request losses because of the overflow of the storage drive at all intervals $[t_k, t_{k+1}), k = 1, ..., 15$ (the queue size is 95%, ..., 99% of the storage drive capacity).

Raw data used to conduct the imitational modeling of the network traffic filtering process are presented in Table 1.

The rules ranking while modeling starts on the interval $[t_5, t_6)$ and further, on each time interval. The evaluation of the performance indicators received during the modeling is presented in Fig. 4.

As it can be seen from the figures, the rules ranking leads to reducing the average packet servicing time for all traffic scenarios. The largest difference between request servicing is $\Delta V_{12} = 0.078$ [ms].

For the 1st scenario (the functioning mode without packet losses caused by buffer memory overflows) the highest-ranking effect is reached, i.e. all of the comparative indicators show the increase in the FW performance. In the 2nd scenario than can be conditionally considered as the start of a DDoS attack, the ranking has allowed to decrease the growth rate of the system load coefficient. The average waiting time for the service start and the packet residence time in the system approaches the values of the indicators of the 3rd scenario on the interval $[t_{13}, t_{14})$. In the 3rd scenario (DDoS attack with high intensity of incoming packets), the rules ranking provided a decrease of the average packet servicing time and, as a result, the average packet residence time decreased. In this case the comparative indicators have larger values than in the 1st and 2nd scenarios because of the high initial values of performance indicators.

Table 1. Raw data

N	m	μ_0^{-1}	μ^{-1}	C	$t_{k+1} - t_k$	$\lambda_{low}^1, \lambda_{high}^1$	$\lambda_{low}^1, \lambda_{high}^1$	$\lambda_{low}^1, \lambda_{high}^1$
5000 [Rules]	750 [Rules]	$2.7 \cdot 10^{-3}$ [RPS]	$5 \cdot 10^{-5}$ [RPS]	1000 [Requests]	1000 [ms]	0.68, 1.37 [RPS]	3.16, 6.32 [RPS]	14.7, 28.9 [RPS]

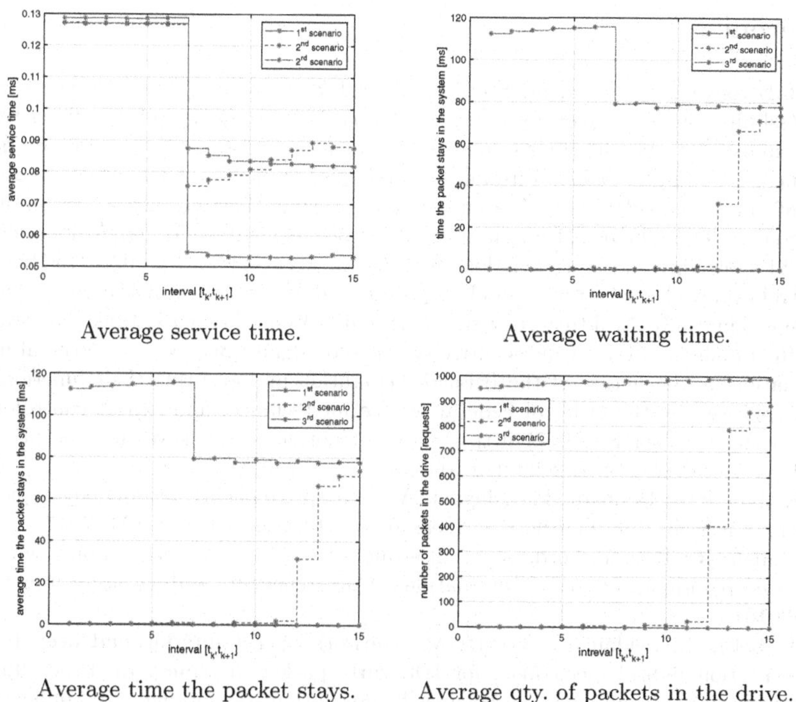

Average service time. Average waiting time.

Average time the packet stays. Average qty. of packets in the drive.

Fig. 4. Firewall performance indicators

6 Conclusion

So, in the work, an imitational model was designed, and that allows to evaluate the main firewall performance indicators. The estimates obtained demonstrate that, for various traffic behavior scenarios, the ranking has increased the FW performance by 10%, ..., 17% compared to the traffic filtering without ranking. The best of all effect of ranking is achieved when simulating the functioning of the queening system in conditions of overload, storage overflow and subsequent packet loss. That is typical for a high-volume and long-term DDoS attack. Therefore, the results of the work confirm the assumptions about the usefulness of applying the rules ranking to increase the traffic filtering efficiency, including the cases while DDoS attacks take place.

References

1. Rukovodyashhij dokument. Sredstva vychislitelnoj texniki. Mezhsetevye ekrany. Zashhita ot nesankcionirovannogo dostupa k informacii. Pokazateli zashhishhennosti ot nesankcionirovannogo dostupa k informacii [Governing document. Computer aids. Firewall. Protection against unauthorized access to information. Indicators of security against unauthorized access to information] (1997). https://fstec.ru/tekhnicheskaya-zashchita-informatsii/dokumenty/114-spetsialnye-normativnye-dokumenty/383. Accessed 25 Sep 2020. (In Russian)
2. Trebovaniya k sozdaniyu sistem bezopasnosti znachimyx obektov kriticheskoj informacionnoj infrastruktury rossijskoj federacii i obespecheniyu ix funkcionirovaniya [Requirements for creating security systems for significant objects of critical information infrastructure of the Russian Federation and ensuring their functioning]. (Approved by the order of Federal service for technical and export controlof, V. No 235, 21 December 2017). https://fstec.ru/normotvorcheskaya/akty/53-prikazy/1589. Accessed 25 Sep 2020. (In Russian)
3. Kadam, P.S., Tambade, P.S., Jayant, A.J.: Adaptive packet filtering techniques for Linux firewall. Int. J. Adv. Res. Ideas Innov. Technol. **31**, 171–174 (2017)
4. Mohan, R.: Dynamic ordering of firewall rules using a novel swapping window-based paradigm. In: ACM International Conference Proceeding Series, pp. 11–20 (2016)
5. Trabelsi, Z., Zeidan, S., Masud, M.: Network packet filtering and deep packet inspection hybrid mechanism for IDS early packet matching. In: Proceedings - International Conference on Advanced Information Networking and Applications, AINA, pp. 808–815 (2016)
6. Zhu, Y.C.: Optimization design and implementation of gateway based on firewall for access control. In: ICIST 2016 (2016)
7. Saleous, H., Trabelsi, Z.: Enhancing firewall filter performance using neural networks. In: 15th International Wireless Communications and Mobile Computing Conference implementation of Gateway Based on Firewall for Access Control. 6th International Conference on Information Science and TechnWCMC 2019 (2019)
8. Trabelsi, Z., et al.: Improved session table architecture for denial of stateful firewall attacks. IEEE Access **6**, 35528–35543 (2018)
9. Trabelsi, Z., Zeidan, S., Masud, M.M.: Hybrid mechanism towards network packet early acceptance and rejection for unified threat management. IET Inf. Secur. **11**, 104–113 (2017)
10. Hung, N.M., Nhat, V.D.: B-tree based two-dimensional early packet rejection technique against DoS traffic targeting firewall default security rule. In: Proceedings of the 2014 7th IEEE Symposium on Computational Intelligence for Security and Defense Applications, CISDA 2014 (2015)
11. Zhang, L., Huang, M.: A firewall rules optimized model based on service-grouping. In: Proceedings of 12th Web Information System and Application Conference, WISA 2015, 2016 (2015)
12. Abbes, T., Bouhoula, A., Rusinowitch, M.: Detection of firewall configuration errors with updatable tree. Int. J. Inf. Secur. (2016)
13. Saadaoui, A., Ben, Y.B.S.N., Bouhoula, A.: FARE: FDD-based firewall anomalies resolution too. J. Comput. Sci. **23**, 181–191 (2017)

14. Botvinko, A.Yu., Samujlov, K.E.: Razrabotka i issledovanie metodov dinamicheskoj optimizacii pravil filtracii mezhsetevyx ekranov [Development and research of methods for dynamic optimization of firewall filtering rules]. In: 3nd Young Researchers International Conference on the INternet of THings and ITs ENablers: "IoT and 5G", pp. 28–33 (2017). (In Russian)
15. Botvinko, A.Yu., Samujlov, K.E.: Matematicheskaya model raboty mezhsetevogo ekrana dlya multimedijnogo trafika [Mathematical model of firewall service for multimedia traffic]. T-Comm Telecommun. Transp. **9**(12) (2015). (In Russian)
16. Botvinko, A.Yu., Samujlov, K.E.: Adaptivnoe ranzhirovanie nabora pravil mezhsetevogo ekrana metodom lokalnoj approksimacii [Adaptive ranking of the firewall rule set using local approximation]. In: Distributed Computer and Communication Networks: Control, Computation, Communications, DCCN 2018, pp. 334–341 (2018). (In Russian)
17. Botvinko, A.Yu.: Optimizaciya nabora pravil filtracii v mezhsetevyx ekranax [Optimizing the set of filtering rules in firewalls]. In: Vserossijskoj konferencii s mezhdunarodnym uchastiem Moskva [All-Russian conference with international participation Peoples' Friendship University of Russia], Moscow, 24–28 April 2017. (In Russian)
18. Lebed, S.V.: Mezhsetevoe ekranirovanie. Teoriya i praktika zashhity vneshnego perimetra. [Firewall protection. Theory and practice of external perimeter protection]. Bauman Moscow State Technical University Publication (2002). (In Russian)
19. Laponina, O.R.: Osnovy setevoj bezopasnosti [The foundation of network security]. Moscow. Publishing house of the national Open University INTUIT. Publ. (2014). 377 p. (In Russian)
20. Ivanov, K.V.: Tutubalin P.I.: Markovskie modeli zashhity avtomatizirovannyx sistem upravleniya specialnogo naznacheniya [Markov models of protection of automated control systems for special purposes]. Kazan. Publishing house of GBU Republican center for monitoring the quality of education (2012). 216 p. (In Russian)
21. Katkovnik, V.Ya.: Neparametricheskaya identifikaciya i sglazhivanie dannyx: metod lokalnoj approksimacii [Non-parametric data identification and smoothing: local approximation method]. The science. Main editorial office of physical and mathematical literature Publ. (1985). (In Russian)
22. Buslenko, N.P.: Modelirovanie slozhnyx sistem [Modeling complex systems]. The science. Main editorial office of physical and mathematical literature Publ. (1978). 400 p. (In Russian)
23. Apanasovich, V.V., Tixonenko, O.M.: Cifrovoe modelirovanie stoxasticheskix sistem [Digital modeling of stochastic systems]. Minsk. University Publ. (1986). 126 p. (In Russian)
24. Kotkova, E.A., Antyuxov, V.I.: Cistemnyj podxod k raschetu parametrov slozhnyx sistem [Systematic approach to calculating the parameters of complex systems], vol. 3. Credo new. Publ. (2019). (In Russian)
25. Akopov, A.S.: Imitacionnoe modelirovanie: uchebnik i praktikum dlya akademicheskogo bakalavriata [Simulation modeling: textbook and practicum for academic baccalaureate students]. Moscow. Urait Publ. (2016). 389 p. (In Russian)
26. Saadaoui, A., Ben, Y.B.S.N., Bouhoula, A.: FARE: FDD-based firewall anomalies resolution tool. J. Comput. Sci. **23**, 181–191 (2017)

Performance of Forward Error Correction in Transport Protocol at Intrasegment Level

P. V. Pristupa(✉), P. A. Mikheev, and S. P. Suschenko

Tomsk State University, Tomsk, Russia
pristupa@gmail.com, doka.patrick@gmail.com, ssp.inf.tsu@gmail.com

Abstract. In this paper we propose a transport connection model managed by reliable transport protocol with forward error correction (FEC) in the modes of selective and group reject in the form of a discrete-time Markov chain. It takes into account the influence on the transport connection throughput by protocol parameters (window size and acknowledge time-out duration), error level in the communication channels of the data transmission path, round-trip delay, and forward error correction process parameters. We assessed the benefits of the transport protocol with FEC over the classic transport protocol in the multidimensional feature space of influencing variable values.

Keywords: Transport protocol · Forward error correction · Data transmission path · Markov chain · Transport connection throughput · Window size · Time-out duration · Round-trip delay · Loss value

1 Introduction

Transport connection throughput is the most important quality indicator of interactions between network applications and respective software and hardware. This performance indicator is mostly defined by transport protocol and its parameters – window size and time-out duration [1]. Subscriber connection modeling and its potential analysis was conducted in [2–11] et al. Mathematical modeling of a classic transport protocol with a decision feedback was conducted in [2–8], simulation modeling of protocol performance was conducted in [9,10]. But these results were obtained under significant limitations on protocol parameters and factors, that define the performance of transport connection. In [11] the model of subscriber competition over the transport connection throughput is proposed, but only for the case of selective retry. In the modern transport protocols the FEC technologies [12–19] are becoming commonplace. An example of the like technology is the QUIC protocol by Google [12]. In [13,14] the researchers performed a simulation modeling of a series of transport protocol scenarios with the capabilities of the simplest FEC mechanisms for wired and wireless communication environments. FEC technology appropriateness in the

© Springer Nature Switzerland AG 2020
V. M. Vishnevskiy et al. (Eds.): DCCN 2020, CCIS 1337, pp. 546–556, 2020.
https://doi.org/10.1007/978-3-030-66242-4_43

real-time applications for multimedia traffic is illustrated in the full-scale studies [15–18]. Complexity of corrupted data recovery, costs and possible technological restrictions in various communication media are also estimated there. A mathematical model of transport protocol with FEC in the intersegment space is discussed in [19]. We found the areas of preferable usage for this technology over the reduced feature space of parameters, characteristics and factors that define the FEC efficiency. But the research on the efficiency of FEC methods [12–19] was conducted primarily on a qualitative level, using numeric and test-bench experiments with significant limitations over communication channel characteristics and parameters of data transfer protocols and error correction protocols. In the paper we address the mathematical model of transport connection in the phase of information transfer managed by reliable transport protocol with a FEC mechanism in the form of a discrete-time Markov chain. On the basis of this model we suggest the analysis of conditions that grant an advantage over the classic transport protocol in transport protocol performance with FEC technology in the space of each segment.

2 Transport Connection Model

Let's examine the data transfer process between the subscribers of the transport protocol based on the algorithm with a decision feedback [1] in the modes of selective or group reject. TCP protocol [1], dominant in the contemporary computer networks, is considered as an example of a family of the like reliable protocols. In the selective reject mode the source has to retransmit only the protocol data blocks (segments) not received, but in the group reject mode the source has to retransmit every block starting from a not received one [1]. We assume the interacting subscribers have unlimited data stream for transfer using the protocol data blocks (segments) of the same length. The receiver's acknowledgments about the correct receiving of data are transmitted in the segments of counterstream.

We assume the managing transport protocol has additional logic for intrasegment FEC. This way, before transmitting each segment, the source divides each segment into $A \geq 1$ fragments of the same size, adds $B - A$, $B \geq A$ redundant fragments of the same length and transmits an extended segment consisting of B fragments into transport connection. Each fragment is accompanied by service data allowing to reveal possible errors and restore the original segment from B fragments in the reception point. If up to $B - A$ arbitrary fragments of extended segment get corrupted, the receiver is still able to recover the original segment and avoid initiating retransmissions (see Fig. 1).

Let f_f and f_r be confidence of fragment transmission along the transport connection from source to receiver and back, respectively. This way the probability of original segment delivery to the receiver, provided the receiver can recover it using the FEC mechanism, is defined as $\Psi_f = \sum_{i=A}^{B} C_i^B {f_f}^i (1 - f_f)^{B-i}$, confidence of sender receiving the acknowledgment about the safe receipt is

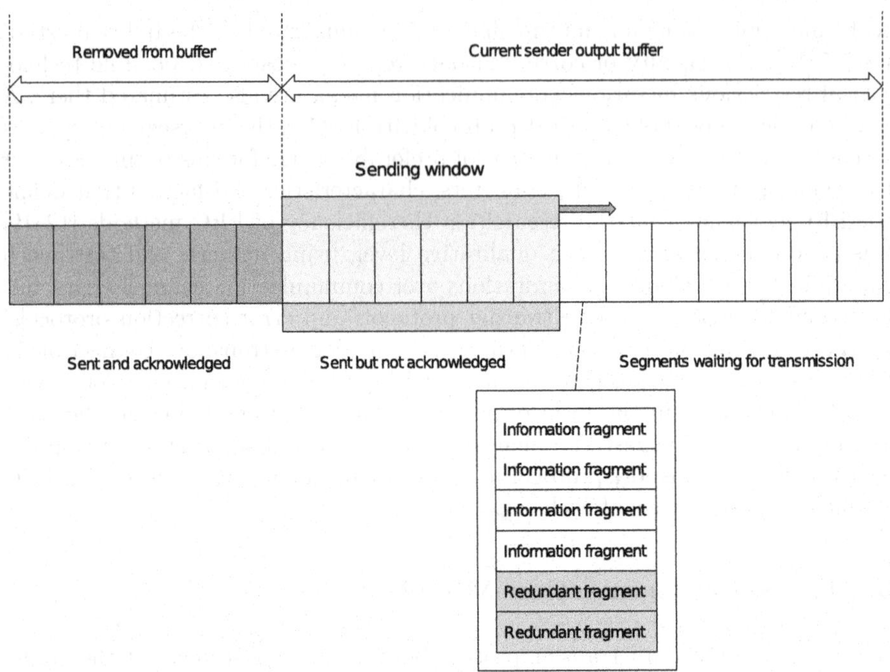

Fig. 1. Transport protocol with intrasegment FEC.

$\Psi_r = \sum_{i=A}^{B} C_i^B {f_r}^i (1 - f_r)^{B-i}$. These parametric dependencies from A and B define the mathematical model of different FEC methods. We consider that the links along the data transmission path have the same speed of operation in both directions, and duration of extended segment transmission cycle in a individual link equals t. Authors believe that assuming homogeneous links with the same transfer duration t is safe enough as in case of different duration for different links (t_1, t_2, ..., t_n) we always able to consider links as a number of smaller links connected in series that have a smaller duration t dividing each of t_i.

In a general case the length of path from source to receiver, used for transmitting the data stream, could be different from the reverse path, used for transmitting the acknowledgments about receiving the segments. Let's assume the length of data transmission path, expressed in the number of hops in the forward direction equals $D_f \geq 1$. The reverse path, used for transmitting acknowledgments about the correct receiving of segment sequences to the sender, is of length $D_r \geq 1$. Let's assume there is no segment loss due to buffer memory blockage in the path nodes. Data stream management is enabled by the mechanism of sliding window [1] of size $W \geq 1$.

The data transfer process in the transport connection could be expressed as a Markov process with discrete time expressed in multiples of cycle duration t, because the time between receiving acknowledgments has a geometric distribution with parameter Ψ_r [2]. The Markov chain defines the sender activity and the changes in the size of a queue of extended segments awaiting confirmation in the source. Number of Markov chain states is defined by acknowledge time-out duration S, expressed in number of cycles of duration t. Time-out duration is related to the data transmission path length and window size by inequalities $S > W$, $S \geq D_f + D_r$. The sum of forward and reverse data transmission paths could be interpreted as round-trip delay $D = D_f + D_r$, measured in durations t (not including the loss of protocol blocks during transmission along the path). Once a segment is sent, the sender stores the segment ID in a queue. The more segments are sent, the bigger the queue becomes. After getting an acknowledgment the sender removes the corresponding segment IDs from the queue. The states of Markov chain $i = \overline{0, W}$ represent the number of segments that were transmitted by the sender, but not acknowledged, the states $i = \overline{W+1, S-1}$ represent the time periods, when the sender is not active and waits to receive the acknowledgments about the correct receiving of segment sequence consisting of W segments. The source moves from the zero state into $D-1$th state along each cycle t with a probability of determined event. In the states $i \geq D-1$ after the next discrete cycle t expires, the sender starts to receive the acknowledgments and, depending on the state of receipt, the sender either transmits the new segments (if receipt is confirmed), or the sender resends the corrupted segments. Finishing the cycle in the state $D-1$ corresponds to the time of transmitting the first segment to the receiver and receiving the acknowledgment of receipt. Further increase in the state index occurs with the probability of acknowledgment corruption $1 - \Psi_r$ in the reverse path. In the states $i \geq D-1$ in the selective reject mode the confirmation triggers moving to the state $D-1$, provided $W \geq D$, or to the state $D+W-2-i$, provided $W \leq D$. In the group reject mode the initial states $i \geq D-1$ return in the states $D-1$ (provided $W \geq D$) or $D+W-2-i$ (provided $W \leq D$) upon receiving the acknowledgment only in case of successful transmission of $i-D+1$ extended segments to the receiver, otherwise the sender returns to the zero state, because the queue of transmitted, but not acknowledged, segments resets to zero. As the source stops transmitting the segments in the states $i \geq W$, receiving the acknowledgments in the states $i = \overline{W, D+W-3}$ results in moving into the states $D+W-2-i$, and from the states $i = \overline{D+W-2, S-2}$ – into the zero state. It is true for selective reject mode, but in a group reject mode these changes in the states are implemented after receiving the positive acknowledgments. In the state $S-1$ the time-out of waiting for acknowledgment from the receiver about receiving the segments in a correct way expires, and the sender moves into the zero state in any reject mode.

3 Probabilistic and Temporal Characteristics of the Transport Connection

Markov chain transition probabilities define the description of an information stream transmission process from the initial state i into the resulting state j

with a FEC technology in the group reject mode:

$$\pi_{ij} = \begin{cases} 1, j = i+1, i = \overline{0, D-2}; \\ 1-\Psi_r, j = i+1, i = \overline{D-1, S-2}; \\ \Psi_r{\Psi_f}^{i-D+2}, j = D-1, i = \overline{D-1, W-1}, W \geq D; \\ \Psi_r(1-{\Psi_f}^{i-D+2}), j = 0, i = \overline{D-1, W-1}, W \geq D; \\ \Psi_r{\Psi_f}^{i-D+2}, j = D+W-2-i, i = \overline{D-1, D+W-3}, W \leq D; \\ \Psi_r(1-{\Psi_f}^{i-D+2}), j = 0, i = \overline{D-1, D+W-3}, W \leq D; \\ \Psi_r{\Psi_f}^{i-D+2}, j = D+W-2-i, i = \overline{W, D+W-3}, W \geq D; \\ \Psi_r, j = 0, i = \overline{D+W-2, S-2}; \\ 1, j = 0, i = S-1. \end{cases} \quad (1)$$

For the selective reject mode transition probabilities could be obtained from this equation provided $\Psi_f = 1$, as the transitions from the confidence state of segment delivery to the receiver are independent. Probability distribution of states for the Markov chain with the described transition probability structure is defined by relations between the protocol parameters W, S, and round-trip delay duration D. It has a functional view, obtained in [2].

The most important operational characteristic of a protocol is its throughput, defined by the data transmission path parameters, overhead and features of the protocol procedures of transmission management [1,2]. Normalized performance of transport connection is defined by the average number of uncorrupted segments delivered to the receiver (taking into account the reject mode [1]) over the average time between two consequent incoming acknowledgments [2]. As the time between two incoming acknowledgments is distributed geometrically with the parameter Ψ_r, the average time between two incoming acknowledgments in the cycle durations t will be defined as $T = 1/\Psi_r$. Thus the throughput for the selective reject procedure is defined as [2]:

$$Z_s(W, S, D, A, B) = \Psi_r \sum_{i=D-1}^{D+W-2} (i - D + 2)\Psi_f P_i + W\Psi_f \sum_{i=D+W-1}^{S-1} P_i. \quad (2)$$

Transport connection throughput for group reject mode taking into account the retransmission of every segment starting from the first not received one [2] is defined as:

$$Z_g(W, S, D, A, B) = \Psi_f\Psi_r \sum_{i=D-1}^{D+W-2} \frac{1-{\Psi_f}^{i-D+2}}{1-\Psi_f} P_i + \frac{1-{\Psi_f}^W}{1-\Psi_f} \sum_{i=D+W-1}^{S-1} P_i. \quad (3)$$

4 Areas of Appropriateness for Applying the Forward Error Correction

Let's examine the efficiency of FEC technology on the protocol data blocks on the level of reliable transport protocol in the space of each segment. According to the FEC technology procedure the sender divides every data segment into A fragments of the same length, adds $B - A \geq 0$ redundant fragments of the same size to these fragments, allowing to restore the segment on the receiving side even if up to $B - A$ arbitrary fragments of extended segment get corrupted. The receiver, upon receiving the extended segment with corrupted fragments, tries to recover it and, if successful, sends an acknowledgment to the sender, packed in the similar extended data segment of counterstream.

Efficient use of FEC technology assumes finding the values for segment fragmentation coefficient A and number of additional redundant fragments $B - A$ for error correction, that grant maximum performance of transport connection with predetermined characteristics and protocol parameters. Redundant fragments in the transmitted sequence increase the probability of segment delivery to the receiver, but this increase is achieved by increase in overhead in the form of time needed for transmitting the additional data. This rises the need to find in the multidimensional feature space an area of transport connection characteristic values (D, Ψ_r, Ψ_f), transport protocol parameters (W, S) and FEC mechanism (A, B), that grant the superiority of FEC management procedure over the classic protocol procedure with a decision feedback without employing the error correction. Let's compare the management procedures provided the subscribers' data streams have the same transmission rate. Let's define the gain in performance due to employing the FEC mechanism in comparison with the classic protocol procedure with a decision feedback in the form of: $\Delta(A, B) = Z(W, S, D, A, B) - Z(W, S, D, 1, 1)$. Let's analyze the gain values not including the overhead related to adding extra headers and trailers to each fragment of the source segment for error diagnostics and correct segment assembling from a set of fragments. In the general case the comparative analysis could be performed only computationally. In several cases the area of positive values of gain could be found in a simple analytic form thanks for lowering the number of dimensions of feature space.

In the selective reject mode with the completely reliable reverse data transmission path ($f_r = 1$), gain takes a simple analytic form, invariant to the window size $\Delta_s(A, B) = \frac{A\Psi_f}{B} - {f_f}^A$, assuming $W \geq D$, and dependent on the window size $\Delta_s(A, B) = \frac{1}{D-W+1}(\frac{A\Psi_f}{B} - {f_f}^A)$, assuming $W < D$. Thus, for the FEC parameters satisfying the condition $B = A + 1$, $A \geq 1$, it is evident that an area of positive values of gain $\Delta_s(A, A+1) = {f_f}^A(A-1-\frac{A^2 f_f}{A+1})$ exists for $A \geq 2$ over the interval $f_f \in (0, 1 - \frac{1}{A^2})$. The maximum value of gain $\Delta_s(A, A+1) = \frac{(A-1)^{A+1}}{A^A(A+1)}$ assuming $W \geq D$ and $\Delta_s(A, A+1) = \frac{(A-1)^{A+1}}{A^A(A+1)(D-W+1)}$ assuming $W < D$ is achieved for $f_f = 1 - \frac{1}{A}$. As the FEC parameter A increases, the maximum of gain moves to the right on the coordinate f_f, and the maximum gain value grows from $\Delta_s(2, 3) = 1/12$ and $\Delta_s(2, 3) = 1/12(D - W + 1)$ assuming $W \geq D$ and $W < D$ respectively (see Fig. 2).

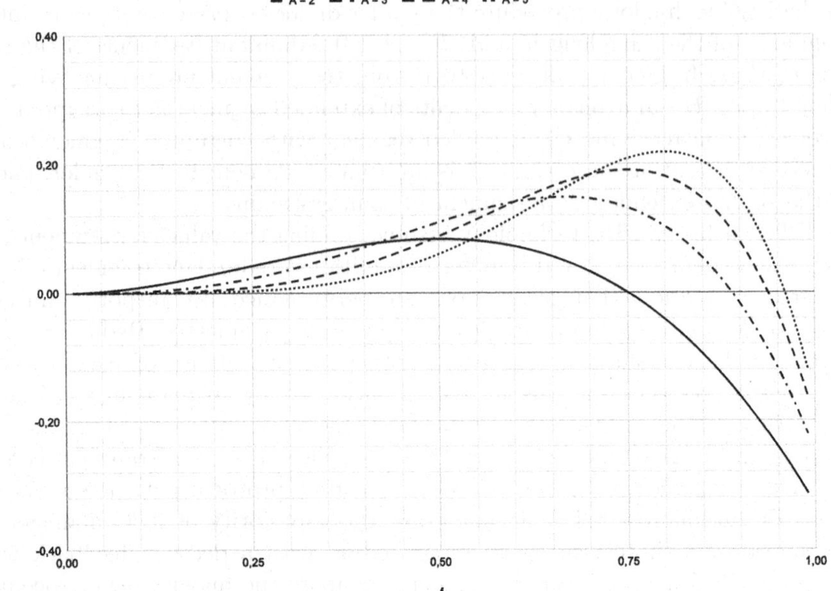

Fig. 2. Benefit function $\Delta_s(A, B, f_f)$ where $f_r = 1$, $W \geq D$, $B = A + 1$ for different values of A. The positive values area of the benefit function determines the applicability of using FEC.

For the FEC process parameters $B = A + 2$, $A \geq 1$ the positive values of gain are achieved on the interval $f_f \in (0, \frac{(A-1)(A+2)}{A(A+1)})$, $A \geq 2$, and the maximum of gain is achieved for $f_f = 1 - \sqrt{\frac{2}{A(A+1)}}$ (see Fig. 3).

In the group repeat mode for a completely reliable reverse data transmission path ($f_r = 1$) the gain becomes $\Delta_g(A, B) = \frac{A\Psi_f}{B[D-(J-1)\Psi_f]} - \frac{f_f{}^A}{D-(J-1)f_f{}^A}$, where $J = W$ provided $W < D$ and $J = D$ provided $W \geq D$. Then the positive values of gain, provided $A = 1$, $B = 2$, are achieved for confident fragment transmission in the forward data path of transport connection on the intervals $f_f \in (0, \frac{D-2}{D-1})$, $W \geq D$, $D > 2$ and $f_f \in (0, 2 - \frac{D}{W-1})$, $1 + \frac{D}{2} < W < D$ (see Fig. 4, Fig. 5 and Fig. 6).

On the Fig. 5 we can see that for $W = 3$ the benefit function is negative on the whole range of f_f. In such cases it practically means that FEC mechanism should not be used.

For loaded transport connection ($W \geq D$) as the round-trip delay D increases the area of gain positive values expands, maximum of gain increases over the curve with saturation and shifts to the area of higher confidence of fragment delivery f_f. The dependency of gain on parameter A is an extreme one provided $B = A + 1$, $A \geq 1$. For underloaded transport connection ($1 + \frac{D}{2} < W < D$) with the window extension from $W = 1 + \frac{D}{2}$ to $W = D - 1$ the area of positive

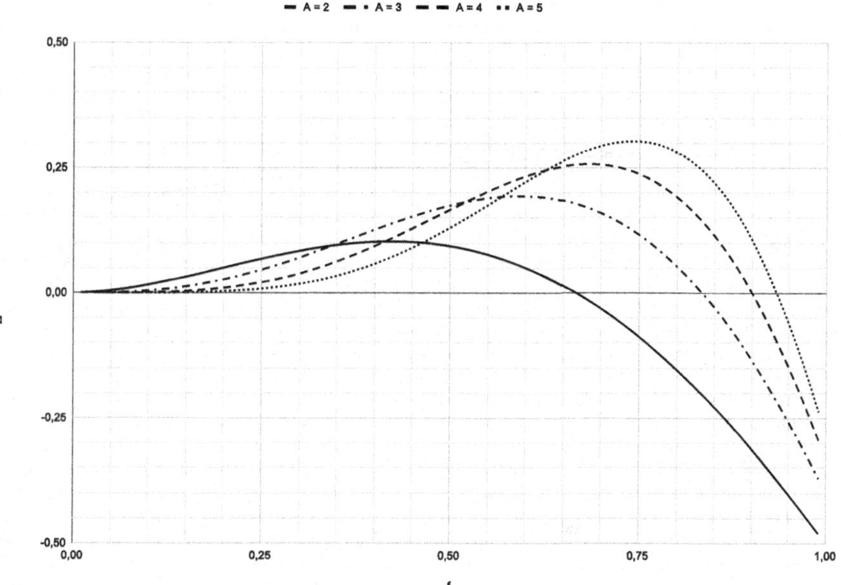

Fig. 3. Benefit function $\Delta_s(A, B, f_f)$ where $f_r = 1$, $W \geq D$, $B = A + 2$ for different values of A.

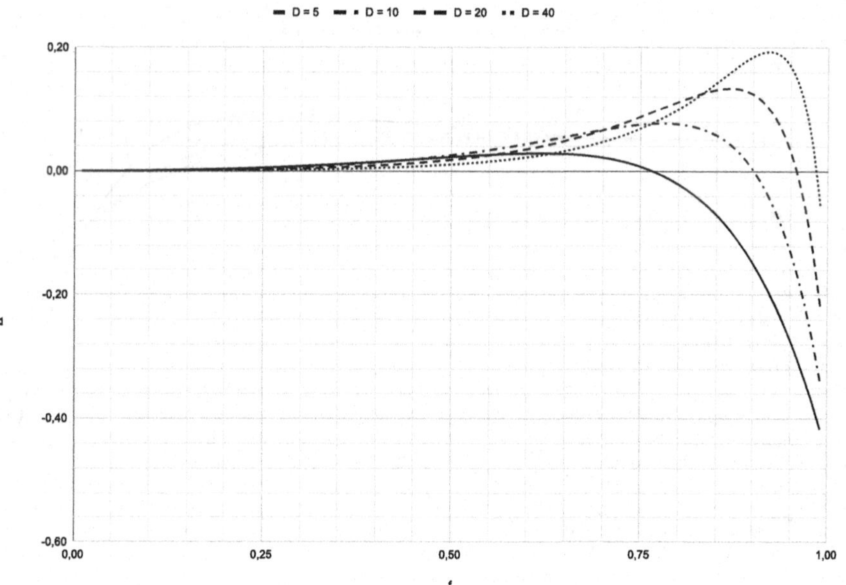

Fig. 4. Benefit function $\Delta_g(A, B, f_f)$ where $f_r = 1$, $W \geq D$, $B = A + 1$ for different values of D.

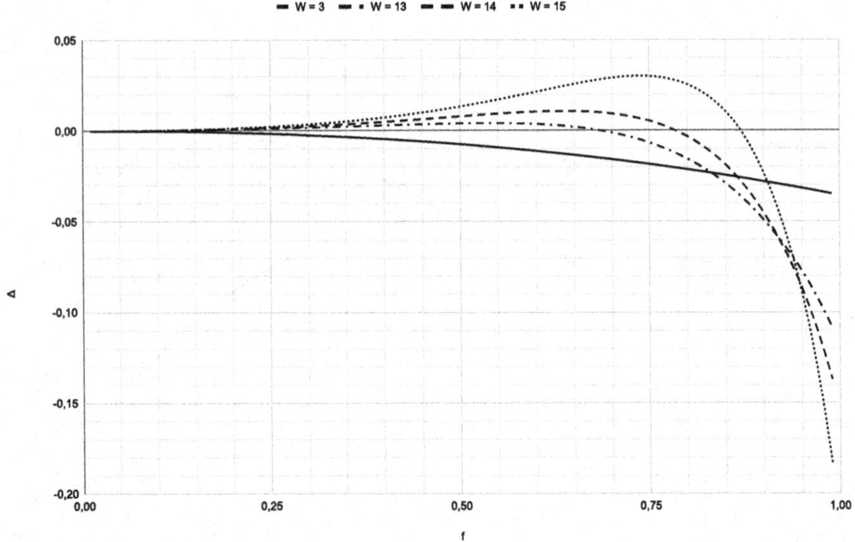

Fig. 5. Benefit function $\Delta_g(A, B, f_f)$ where $f_r = 1$, $W < D$, $D = 16$, $B = A + 1$ for different values of W.

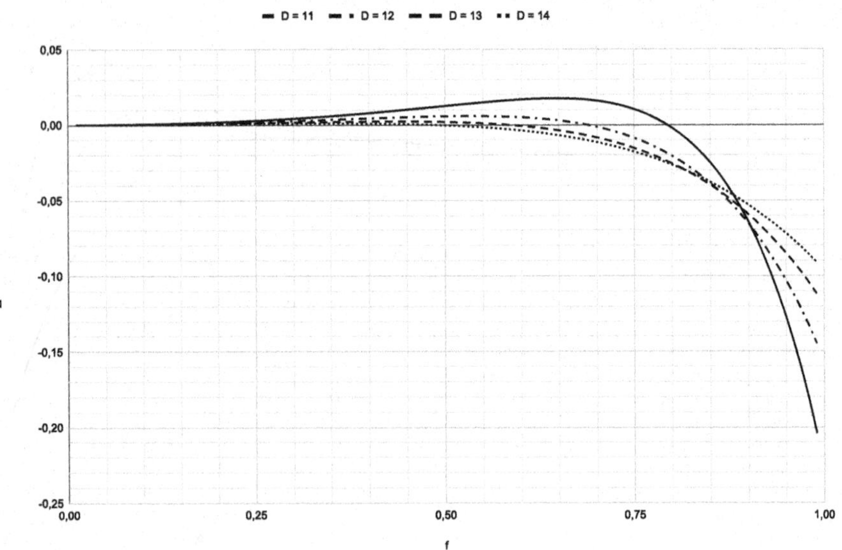

Fig. 6. Benefit function $\Delta_g(A, B, f_f)$ where $f_r = 1$, $W < D$, $W = 10$, $B = A + 1$ for different values of D.

values of gain extends on the coordinate f_f, maximum of gain increases and shifts to the right. The same effect has an increase in a FEC parameter A for $B = A + 1$. If the condition $W = 1 + \frac{D}{2}$ is satisfied, increase in round-trip delay duration D makes the area of positive values of gain narrower on the coordinate f_f, lowering the benefits of FEC technology. It should be noted that, other things being equal, size of the fragments decreases with the increase of coefficient of segment division into fragments (parameter A), so the confidence of delivery to the recipient f_f increases as well. Thus, in order to perform a correct benchmark of gain on different values of the parameter A, we need to employ the indicator of delivery reliability for an individual bit, byte or word.

5 Conclusion

In this paper we propose a model for a process of data segment transmission in a transport connection, managed by reliable transport protocol with FEC technology, implemented in the space of every segment, and acknowledgment of data, received after applying the FEC procedure. It is shown, that for every mode of repeating, FEC mechanism usage is appropriate and especially beneficial in the fully loaded transport connections ($W \geq D$) with a significant round-trip delay (D). In general, the FEC benefit in any reject mode is mostly defined by confidence of fragment delivery (f_f and f_r), source segment fragmentation parameters (A) and redundant data (B), as well as ratio between the window size and round-trip delay duration. Further analysis may include the problem of researching the operational characteristics of transport connections, under contention between subscribers for available bandwidth of shared links in the network path. Furthermore, an important task is generalizing the results obtained in this paper over the case of competitive use of connection path throughput over various subscriber connections with shared channels of data transmission path. Also, it seems reasonable to consider buffer memory blockage in the path nodes as an additional aspect of the model.

References

1. Fall, K., Stevens, R.: TCP/IP Illustrated, Volume 1: The Protocols, 2nd edn., p. 1017. Addison-Wesley Professional Computing Series (2012)
2. Kokshenev, V.V., Mikheev, P.A., Sushchnenko, S.P.: Comparative analysis of the performance of selective and group repeat transmission models in a transport protocol. Autom. Remote Control **78**(2), 247–261 (2017)
3. Kokshenev, V., Mikheev, P., Suschenko, S., Tkachyov, R.: Analysis of the throughput in selective mode of transport protocol. In: Vishnevskiy, V.M., Samouylov, K.E., Kozyrev, D.V. (eds.) DCCN 2016. CCIS, vol. 678, pp. 168–181. Springer, Cham (2016). https://doi.org/10.1007/978-3-319-51917-3_16
4. Bogushevsky, D., Mikheev, P., Pristupa, P., Suschenko, S.: The time-out length influence on the available bandwidth of the selective failure mode of transport protocol in the load data transmission path. In: Vishnevskiy, V.M., Kozyrev, D.V. (eds.) DCCN 2018. CCIS, vol. 919, pp. 120–131. Springer, Cham (2018). https://doi.org/10.1007/978-3-319-99447-5_11

5. Kassa, D.F.: Analytic Models of TCP Performance, p. 199. PhD Thesis, University of Stellenbosch (2005)
6. Giordano, S., Pagano, M., Russo, F., Secchi, R.: Modeling TCP startup performance. J. Math. Sci. **200**(4), 424–431 (2014)
7. Kravets, O.Y.: Mathematical modeling of parameterized TCP protocol. Autom. Remote Control **74**(7), 1218–1224 (2013)
8. Arvidsson, A., Krzesinski, A.: A model of a TCP link. In: Proceedings of the 15th International Teletraffic Congress Specialist Seminar (2002)
9. Olsen, Y.: Stochastic modeling and simulation of the TCP protocol. Uppsala Dissertations in mathematics, vol. 28, p. 94 (2003)
10. Mikheev, P., Pristupa, P., Suschenko, S.: Performance of transport connection with selective failure mode when competing for throughput of data transmission path. In: Vishnevskiy, V.M., Samouylov, K.E., Kozyrev, D.V. (eds.) DCCN 2019. CCIS, vol. 1141, pp. 89–103. Springer, Cham (2019). https://doi.org/10.1007/978-3-030-36625-4_8
11. Nikitinskiy, M.A., Chalyy, D.J.: Performance analysis of trickles and TCP transport protocols under high-load network conditions. Autom. Control Comput. Sci. **47**(7), 359–365 (2013)
12. Langley, A., et al.: Protocol: design and internet-scale deployment. In: SIGCOMM 2017, Los Angeles. CA, USA 2017, pp. 183–196 (2017)
13. Lundqvist, H., Karlsson, G.: TCP with end-to-end FEC, pp. 152–156. International Zurich Seminar on Communications (2004)
14. Barakat, C., Altman, E.: Bandwidth tradeoff between TCP and link-level FEC. Comput. Netw. (39), pp. 133–150 (2002)
15. Shalin, R., Kesavaraja, D.: Multimedia data transmission through TCP/IP using hash based FEC with AUTO-XOR scheme. ICTACT J. Commun. Technol. **03**(03), 604–609 (2012)
16. Ribadeneir, A.F.: An Analysis of the MOS under condition of delay, jitter and packet loss and an analysis of the impact of introducing piggybacking and Reed Solomon FEC for VOIP. Master's thesis, Georgia State University (2007)
17. Matsuzono, K., Detchart, J., Cunche, M., Roca, V., Asaeda, H.: Performance analysis of a high-performance real-time application with several AL-FEC schemes. In: Proceedings of the IEEE 35th Conference on Local Computer Networks, LCN 2010, pp. 1–7 (2010)
18. Herrero, R.: Modeling and comparative analysis of Forward Error Correction in the context of multipath redundancy. Telecommun. Syst. Modell. Anal. Des. Manag. **65**(4), 783–794 (2017)
19. Mikheev, P., Suschenko, S., Tkachyov, R.: Estimation of high-speed performance of the transport protocol with the mechanism of forward error correction. In: Vishnevskiy, V.M., Samouylov, K.E., Kozyrev, D.V. (eds.) DCCN 2017. CCIS, vol. 700, pp. 259–268. Springer, Cham (2017). https://doi.org/10.1007/978-3-319-66836-9_22

Author Index

Printed in the United States
By Bookmasters